Units and Conversion Factors

Quantity	Units (and Name)	
Acceleration	m/s²	1 in./sec² = 0.0254 m/s²
Area	m²	1 in.² = 645.2(10)⁻⁶ m²
Density	kg/m³	1 slug/in.³ = 890,600 kg/m³
Force	N (newton)	1 lb = 4.448 N
Frequency	1/s (Hz, hertz)	
Length	m (meter)	1 in. = 0.0254 m
Mass	kg (kilogram)	1 slug = 14.59 kg
Moment	N·m	1 in.·lb = 0.1130 N·m
Moment of inertia (area)	m⁴	1 in.⁴ = 416.2(10)⁻⁹ m⁴
Moment of inertia (mass)	kg·m²	1 slug·in.² = 0.009415 kg·m²
Power	N·m/s = J/s (W, watt)	1 in.·lb/sec = 0.1130 W
Pressure, stress	N/m² (Pa, pascal)	1 psi = 6895 Pa
Spring constant	N/m	1 lb/in. = 175.1 N/m
Time	s (second)	
Velocity	m/s	1 in./sec = 0.0254 m/s
Volume	m³	1 in.³ = 16.39(10)⁻⁶ m³
Work, energy	N·m (J, joule)	1 in.·lb = 0.1130 N·m

SI Unit Prefixes

Prefix	Symbol	Factor	Example
giga	G	10^9	1 GPa = 10^9 Pa
mega	M	10^6	1000 MPa = 1 GPa
kilo	k	10^3	1 km = 1000 m
centi	c	10^{-2}	100 cm = 1 m
milli	m	10^{-3}	10 mm = 1 cm
micro	μ	10^{-6}	1 μm = 10^{-3} mm

Some Useful Data

Atmospheric pressure:	$p \approx 100$ kPa
Acceleration of gravity:	$g \approx 9.8$ m/s/s
Mass density of water:	$\rho \approx 1000$ kg/m³
Weight density of water:	$\gamma \approx 9800$ kN/m³
Mass to weight conversion:	$w = mg$

ADVANCED MECHANICS OF MATERIALS

SECOND EDITION

ADVANCED MECHANICS OF MATERIALS
SECOND EDITION

ROBERT D. COOK
University of Wisconsin, Madison

WARREN C. YOUNG

PRENTICE HALL
UPPER SADDLE RIVER, NEW JERSEY 07458

Library of Congress Cataloging-in-Publication Data

Cook, Robert Davis.
  Advanced mechanics of materials / Robert D. Cook, Warren C. Young. — 2nd ed.
    p. cm.
  ISBN 0-13-396961-4
  1. Strength of materials. I. Young, Warren C. (Warren Clarence), 1923– . II. Title.
TA405.C843 1998
620.1'12—dc21   98-24374
                                                                                        CIP

Acquisitions editor: WILLIAM STENQUIST
Editorial production: BOOKMASTERS, INC.
Editor-in-Chief: MARCIA HORTON
Assistant VP of production and manufacturing: DAVID W. RICCARDI
Managing editor: EILEEN CLARK
Full service/manufacturing coordinator: DONNA M. SULLIVAN
Creative director: JAYNE CONTI
Cover design: KAREN SALZBACH
Copy editor: CHERYL WILMS
Editorial assistant: MEG WEIST

Reprinted with corrections November, 1999.

The author and publisher of this book have used their best efforts in preparing this book. These efforts include the development, research, and testing of the theories and programs to determine their effectiveness. The author and publisher make no warranty of any kind, expressed or implied, with regard to these programs or the documentation contained in this book. The author and publisher shall not be liable in any event for incidental or consequential damages in connection with, or arising out of, the furnishing, performance, or use of these programs.

© 1999 by Prentice-Hall, Inc.
Upper Saddle River, New Jersey 07458

All rights reserved. No part of this book may be
reproduced, in any form or by any means,
without permission in writing from the publisher.

Printed in the United States of America

10 9 8 7 6 5 4

ISBN 0-13-396961-4

Prentice-Hall International (UK) Limited, *London*
Prentice-Hall of Australia Pty. Limited, *Sydney*
Prentice-Hall Canada Inc., *Toronto*
Prentice-Hall Hispanoamericana, S.A., *Mexico*
Prentice-Hall of India Private Limited, *New Delhi*
Prentice-Hall of Japan, Inc., *Tokyo*
Prentice-Hall Asia Pte. Ltd., *Singapore*
Editora Prentice-Hall do Brasil, Ltda., *Rio de Janeiro*

Contents

PREFACE   *xi*

1   ORIENTATION. REVIEW OF ELEMENTARY MECHANICS OF MATERIALS   **1**

     1.1   Methods of Stress Analysis   1
     1.2   Terminology   3
     1.3   Properties of a Plane Area   5
     1.4   Axial Loading. Pressure Vessels   9
     1.5   Torsion   11
     1.6   Beam Stresses   12
     1.7   Beam Deflections   14
     1.8   Symmetry Considerations. Static Indeterminacy   15
     1.9   Plastic Deformation. Residual Stress   18
     1.10   Other Remarks   19

2   STRESS, PRINCIPAL STRESSES, STRAIN ENERGY   **28**

     2.1   Preliminary Remarks   28
     2.2   Principal Stresses: Theory   30
     2.3   Principal Stresses: Calculations   33
     2.4   Octahedral and Maximum Shear Stress   35
     2.5   Stress-Strain-Temperature Relations   37
     2.6   Strain Energy Density   39
     2.7   Stress Concentration   41
     2.8   Contact Stress   44

3   FAILURE AND FAILURE CRITERIA   **51**

     3.1   Introduction   51

3.2	Isotropic Material, Brittle Failure	52
3.3	Isotropic Material, Ductile Failure	56
3.4	Anisotropic Materials	58
3.5	Introduction to Fracture Mechanics	59
3.6	Introduction to Fatigue	63

4 APPLICATIONS OF ENERGY METHODS   77

4.1	Introduction	77
4.2	Reciprocal Theorems	82
4.3	Strain Energy in Terms of Loads	84
4.4	Unit Load Method	87
4.5	Example Solutions with Straight Members	90
4.6	Curved Members	93
4.7	Statically Indeterminate Problems	97
4.8	Formulas from Geometric Considerations	103
4.9	Strain Energy in Terms of Displacements	105
4.10	Principle of Stationary Potential Energy	107
4.11	Rayleigh-Ritz Method	110
4.12	Trigonometric Series	114
4.13	Finite Element Method	116

5 BEAMS ON AN ELASTIC FOUNDATION   133

5.1	Introduction	133
5.2	Basic Equations, Winkler Foundation	135
5.3	Semi-Infinite Beams with Concentrated End Loads	138
5.4	Infinite Beams with Concentrated Loads	140
5.5	Example Problems	143
5.6	Uniformly Distributed Load	147
5.7	Beams of Finite Length	150

6 CURVED BEAMS   161

6.1	Introduction. Strain Distribution	161
6.2	Circumferential Stress	164
6.3	Numerical Examples	167
6.4	Radial Stress and Shear Stress	170
6.5	Sections Having Thin Flanges	174
6.6	Closed Sections with Thin Walls	179
6.7	Deflections of Sharply Curved Beams	180

7 ELEMENTS OF THEORY OF ELASTICITY   190

7.1	Introduction	190
7.2	Displacements, Strains, and Compatibility	193

7.3	Equilibrium Equations and Boundary Conditions 195	
7.4	Stress Field Solution for Plane Stress Problems 198	
7.5	Polynomial Solutions in Cartesian Coordinates 201	
7.6	Displacements Calculated from Stresses 203	
7.7	Plane Stress Problems in Polar Coordinates 205	
7.8	Circular Hole 207	
7.9	Concentrated Force Loading 210	
7.10	Thermal Stress 213	
7.11	Torsion. Warping Function 216	
7.12	Torsion. Prandtl Stress Function 219	
7.13	Inelastic Material Behavior 221	

8 PRESSURIZED CYLINDERS AND SPINNING DISKS 231

8.1	Governing Equations 231	
8.2	Pressurized Cylinders 233	
8.3	Shrink Fits. Compound Cylinders 237	
8.4	Spinning Disks, Part I 242	
8.5	Spinning Disks, Part II 245	
8.6	Plastic Action in Thick-Walled Cylinders 247	
8.7	Plastic Action in Spinning Disks 251	

9 TORSION 261

9.1	Introduction. Saint-Venant Torsion Theory 261	
9.2	Membrane Analogy 263	
9.3	Membrane Analogy for Thin-Walled Open Sections 265	
9.4	Remarks and Formulas 268	
9.5	Pure Twist of Thin-Walled Tubes of One Cell 271	
9.6	Pure Twist of Thin-Walled Multicell Tubes 275	
9.7	Torsion with Restraint of Warping: I Sections 278	
9.8	Sectorial Area 283	
9.9	Sectorial Properties 286	
9.10	Warping of Thin-Walled Cross Sections in Pure Torsion 289	
9.11	Stresses in Open Sections When β is Not Constant 292	
9.12	Torsion with Restraint of Warping: General Sections 293	
9.13	Bimoment 295	
9.14	Large Rotation, Axial Load, and Pretwisting 298	
9.15	Torsion with Plastic Action 302	

10 UNSYMMETRIC BENDING AND SHEAR CENTER 314

10.1	Unsymmetric Bending of Straight Beams 314	
10.2	Numerical Examples 317	
10.3	Beam Deflections in Unsymmetric Bending 319	
10.4	Transverse Shear Stress and Shear Flow 323	

10.5	Shear Center: Introduction	327
10.6	Shear Center: Special Open Sections	330
10.7	Shear Center: General Open Sections	333
10.8	Shear Center: Remarks	336

11 PLASTICITY IN STRUCTURAL MEMBERS AND COLLAPSE ANALYSIS   346

11.1	Introduction	346
11.2	Stress and Deflection in Loading and Unloading	347
11.3	Plastic Action in Bending	350
11.4	Plastic Hinges in Beams	353
11.5	Collapse Analysis of Beams	356
11.6	Collapse Analysis of Plane Frames and Arches	361
11.7	General Theorems	365
11.8	Interaction: Bending and Axial Force	367
11.9	Interaction: Some Additional Relations	370

12 PLATE BENDING   379

12.1	Introduction. Assumptions and Limitations	379
12.2	Governing Equations in Cartesian Coordinates	381
12.3	Uniform Bending. Boundary Conditions	383
12.4	Series Solution for Rectangular Plates. Selected Results	387
12.5	Governing Equations for Axisymmetric Plates	389
12.6	Solution of Simple Axisymmetric Problems	392
12.7	Selected Solutions for Axisymmetric Plates	395
12.8	Superposition Solutions for Axisymmetric Plates	397
12.9	Plastic Collapse Analysis of Plates	399

13 SHELLS OF REVOLUTION WITH AXISYMMETRIC LOADS   411

13.1	Introduction	411
13.2	Geometry and Terminology	412
13.3	Membrane Forces and Stresses	413
13.4	Applications of Membrane Theory	416
13.5	Membrane Shells of Uniform Strength	419
13.6	Bending Theory of Cylindrical Shells	421
13.7	Bending Formulas for Edge Loads	423
13.8	Applications of Bending Theory	425
13.9	The Reinforcing Ring	430

14 BUCKLING AND INSTABILITY   443

14.1	Introduction. Elementary Theory	443
14.2	An Energy Method	446
14.3	Beam-Columns: Classical Theory	449

14.4	Beam-Columns: Approximations 451	
14.5	Inelastic Buckling 453	
14.6	Plates: Buckling, Effective Width, and Failure 455	
14.7	Additional Buckling Problems 460	
14.8	Remarks about Stability, Buckling, and Collapse 462	

REFERENCES   *471*

INDEX   *477*

Preface

This book is intended as a text for a second course in mechanics of materials and assumes the reader has satisfactorily completed a first course in the subject. For the most part, topics in the book are treated by going a step or two beyond elementary mechanics of materials when possible without complicated analytical or numerical methods.

Among instructors, there is little agreement as to what topics should comprise a second course. Accordingly, the book discusses a variety of topics and contains enough material for a two-semester course sequence. Topics included have practical value, either because the physical situation is often encountered or because the concepts and procedures of analysis have value beyond the particular problem to which they are applied. Each chapter uses notation commonly found in handbooks and in advanced texts and papers about the subject addressed. Unavoidably, this means that there are some inconsistencies of notation from one chapter to another. Historical notes come largely from S. Timoshenko, *History of Strength of Materials,* McGraw-Hill Book Co., New York, 1953, and from J. M. Gere and S. Timoshenko, *Mechanics of Materials,* 4th ed., PWS-Kent Publishing Co., Boston, 1997.

Computer solutions play an increasing role in stress analysis. Widely available software includes handbook formulas and finite element programs for analysis. The computational approach is well suited to many problems. However, pervasive use of computers in engineering education may have unfortunate consequences. Students may devote their efforts to writing or operating software rather than to carefully and correctly judging the computed results. Physical understanding may be neglected, and an ability to arrive at a ball-park answer may never develop. In engineering design, this narrow way of using computers can be very costly.

Regardless of the solution method, a problem must be understood before it can be solved. A principal goal in the study of mechanics of materials, perhaps more important than learning specific solutions or procedures, is *to learn what the problems are—* that is, to develop a physical understanding of how bodies of various shapes respond to loads of various kinds. Only then can a problem be properly analyzed and the results checked for correctness.

As in the first edition, topics are treated by extending concepts and procedures of elementary mechanics of materials, assisted when necessary by advanced methods such as theory of elasticity. In this second edition, the first three chapters include material from Chapter 1 of the first edition, with enlarged treatment of fatigue. Energy methods and curved beams are now treated earlier in the book. Theory of elasticity, thick cylinders and spinning disks, plates, and shells are all treated later. Rectangular plates, plastic action in plates, and buckling have been given more attention. Readers of the first edition will notice many smaller changes. Less obvious are innumerable details: All of the first edition was examined and revised where clarity could be improved. Most homework exercises in the second edition are new or revised. Answers to homework exercises and worked-out solutions appear in the Solutions Manual, which will be provided by the publisher on request to instructors who adopt the book as a course text.

Good advice to the student is given by C. E. Smith, in the preface to his text *Applied Mechanics: Statics and Dynamics,* John Wiley & Sons, Inc., New York, 1982:

"Very few people can learn mechanics solely by observing the analysis carried out by someone else. At some point (the sooner the better) the student must discard the observer's role and, making and correcting the inevitable mistakes, attempt to work problems and carry out derivations with increasing independence. For this reason, these books will be relatively ineffective in the lap of someone sitting in an easy chair. They must be studied at a desk with an ample supply of scratch paper at hand. The paper will be needed for attempting solutions of example problems before given solutions are read, for carrying out steps of analysis that are omitted from the books, and for making supplemental sketches. Students who work in this way, always attempting to relate solution procedures to basic ideas, will achieve an understanding that is satisfying as well as professionally useful."

ADVANCED MECHANICS OF MATERIALS

SECOND EDITION

CHAPTER 1

Orientation. Review of Elementary Mechanics of Materials

This chapter includes a selective and *brief* review of important assumptions, procedures, and results from a first course in mechanics of materials. Some items of importance are incorporated in subsequent chapters rather than appearing here. The reader is encouraged to consult a textbook of elementary mechanics of materials for detailed treatment of material reviewed in this chapter.

1.1 METHODS OF STRESS ANALYSIS

Typical questions posed in stress analysis are: Given the geometry of a body or structure, as well as its material properties, support conditions, and time-independent loads applied to it, what are the stresses and what are the displacements? A solution may be obtained by analytical, numerical, or experimental methods. Analytical methods include *mechanics of materials* and *theory of elasticity*. This book considers both, and places emphasis on the first.

Mechanics of materials is the engineer's way of doing stress analysis. The method involves the following steps.

1. Consider deformations produced by load, and establish (or approximate) how they are distributed over the body. This may be done by experiment, intuition, symmetry arguments, and/or prior knowledge of similar situations.
2. Analyze the geometry of deformation to determine how strains are distributed over a cross section.
3. Determine how stresses are distributed over a cross section by applying the stress-strain relation of the material to the strain distribution.
4. Relate stress to load. This step involves drawing a free-body diagram and writing equations of static equilibrium. The result is a formula for stress, typically in terms of applied loading and geometric parameters of the body.
5. Similarly, relate load to displacement, either by integration of the strain distribution determined in step 2 or by using energy arguments that relate work done by applied loads to elastic strain energy stored.

Results of a mechanics of materials analysis may be exact, or good approximations, or rough estimates, depending mainly on the accuracy of assumptions made in the first step. Examples of the foregoing analysis are reviewed in subsequent sections, which point out that a substantial list of restrictions is needed if the resulting formulas are to be valid.

Theory of elasticity is the mathematician's way of doing stress analysis. In this method, one seeks stresses and displacements that simultaneously satisfy the requirements of equilibrium at every point, compatibility of all displacements, and boundary conditions on stress and displacement. In contrast to the mechanics of materials method, this method does not operate under any initial assumption or approximation about the geometry of deformation. Therefore theory of elasticity can solve a problem for which deformations cannot be reliably anticipated, such as the problem of determining stresses around a hole in a plate. However, the technique is more difficult than the mechanics of materials method and cannot be successfully applied to as great a variety of practical problems. Often, a practical problem is treated by a mixture of elasticity and mechanics of materials techniques.

Many problems of stress analysis are best solved numerically, on computers that range from PCs to supercomputers. Numerical analysis software is powerful and versatile; it has become comparatively easy to use and presents results graphically with great polish. None of this analytical power assures that results are even approximately correct. An analyst might easily blunder in deciding what simplifications are appropriate, in choosing the specific computational procedures to use, or in preparing input data. Computed results may contain large errors and, in any case, must be checked against results obtained in some other way. Mechanics of materials analysis serves well for checking, even in cases where it provides only a rough approximation. Regardless of the analysis method, success in solving a problem depends mainly upon the analyst's having clear insight into the phenomenon under study.

An analysis, by any method other than experiment, is applied to a model of reality rather than to reality itself. One cannot possibly take full account of the numerous details of the actual problem. Accordingly, the model is an idealization, in which geometry, loads, and/or support conditions are simplified, based on the analyst's understanding of which aspects of the actual problem are unimportant for the purpose at hand. Thus, a stress raiser may be temporarily neglected, weight of the body may be ignored, or a distributed load may be regarded as acting at a point. (As a practical matter, even the magnitude of loading is not usually known with much precision.) After devising a model, one must do all appropriate analyses. For example, one must not stop with stresses if buckling is also a possible mode of failure. Accordingly, a goal of studying stress analysis is to learn what idealizations and analysis goals are appropriate, which implies that one must learn how bodies of various shapes and support conditions respond to various loads.

Finally, some words about derivations. Why study the derivation of a formula? First, it makes the formula plausible. A more important reason is that a derivation makes clear the assumptions and restrictions needed in order to obtain the formula. Thus, by knowing the derivation, one can recognize situations in which a formula should *not* be applied.

1.2 TERMINOLOGY

The following list is far from exhaustive. Terms listed are used throughout this book.

Beam: An elongated member, usually slender, intended to resist lateral loads by bending.

Body force: A loading that acts throughout a body rather than only on its surface. Self-weight and the inertia force of spinning about an axis are instances of body force.

Boundary conditions: Prescribed displacements at certain locations; for example, the stipulation that the supported end of a cantilever beam neither translates nor rotates. These boundary conditions may also be called *support conditions*. The term "boundary conditions" may also indicate prescribed stresses, forces, or moments. For example, at the unsupported end of a cantilever beam loaded only by its own weight, transverse shear force and bending moment must both vanish.

Brittle behavior: A material failure in which fracture surfaces show little or no evidence that failure has produced permanent deformation.

Cold working: Deformation that results in residual stresses. (In contrast, *hot working* is deformation at high enough temperature that stresses quickly dissipate by annealing.) Cold working by *shot peening* is the bombarding of an object by metal shot (roughly 0.2 mm to 4 mm in diameter) thrown at substantial velocity (roughly 70 m/s), the purpose being to produce residual compressive stresses in the surface layer.

Curvature: The reciprocal of the radius of curvature ρ, that is, $\kappa = 1/\rho$; used in beam theory.

Ductile behavior: Material behavior in which appreciable permanent deformation is possible without fracture.

Elastic: Material behavior in which deformations produced by load disappear when load is removed.

Elastic limit: The largest uniaxial normal stress for which material behavior is elastic. (Compare *yield strength.*)

Elastic modulus: The ratio of axial stress σ_a to axial strain ϵ_a in uniaxial loading; $E = \sigma_a/\epsilon_a$. Restricted to a linear relation between σ_a and ϵ_a. Also called *modulus of elasticity* or *Young's modulus*.

Fixed: A boundary condition in which all motion is prevented. Also called *built-in, clamped,* or *encastre*.

Flexure: Bending.

Frame: A structure built of bars, in which relative rotation between bars is prevented at joints, as by welding bars together where they meet. Bending of the bars is usually important in the calculation of stresses. (Compare *truss.*)

Homogeneous: Having the same material properties at all locations.

Isotropic: Having the same properties (stiffness, strength, conductivity, etc.) in every direction. As examples, glass is isotropic, wood is not isotropic. (Compare *orthotropic.*)

Lateral: Directed to the side; thus, directed normal to the axis of a beam or normal to the surface of a plate or a shell.

Nonlinear problem: A problem in which deflections or stresses are not directly proportional to the load that produces them. An example is the contact stress where a train wheel meets the rail. The area of contact grows as load increases. Another example is an initially flat membrane, like a trampoline. Lateral load is resisted by forces in the membrane that are functions of both the amount of deflection and the deflected shape.

Orthotropic: Having different stiffness (or other properties) in different directions, with the directions of maximum and minimum stiffness being mutually perpendicular. (Compare *isotropic*.)

Permanent set: Deformation that remains after removal of the load that produced it.

Plastic: A state of stress or deformation that results in permanent set if the load is removed.

Poisson's ratio: Designated by ν, where $\nu = -\epsilon_t/\epsilon_a$, and ϵ_t and ϵ_a are respectively the transverse and axial strains produced by a uniaxial stress σ_a below the proportional limit.

Principal stress: A normal stress σ, acting on an area A (or dA) when A (or dA) is free of shear stress. In this book, numerical subscripts on principal stresses indicate algebraic ordering, maximum to minimum; that is, $\sigma_1 \geq \sigma_2 \geq \sigma_3$.

Prismatic member: A straight bar with identical cross sections. In other words, a uniform straight member; the solid generated by translating a plane shape along a straight axis normal to its plane.

Proportional limit: The largest uniaxial normal stress for which stress is directly proportional to strain. (Compare *yield strength*.)

Safety factor (*SF*): The number by which the working load (the maximum load anticipated in normal service) must be multiplied to produce the design load (the load that causes failure). If the loading has more than one component force or moment, all components must change proportionally if this definition is to apply. If stress is the quantity indicative of failure, and if stress is directly proportional to applied load, then *SF* can also be regarded as the number by which the stress that causes the material to fail must be divided in order to obtain the allowable stress, which is the maximum stress to be allowed in service. Typically, design codes prescribe allowable stresses. The number chosen for *SF* is influenced by uncertainties about loads, material properties, quality of fabrication, and accuracy of design procedures; by the cost of failure; and by the cost of adopting a large *SF*.

Saint-Venant's principle: The proposition that two statically equivalent loadings, applied (separately) to the same region of a body, each produces essentially the same state of stress and deformation in the body at distances from the loaded region greater than the larger dimension of the loaded region. (*Caution:* This principle is not reliable for thin-walled construction or for some orthotropic materials.)

Shaft: An elongated member, usually slender and straight, intended to resist torsional loads.

Shear modulus: The ratio of shear stress τ to shear strain γ; $G = \tau/\gamma$. Restricted to a linear relation between τ and γ. Also called *modulus of rigidity*.

Simply supported: A boundary condition in which lateral displacements are prevented but rotations are allowed. A simple support applies no moment to a structure. A simple support may also be called *pinned* or *hinged*.

Static indeterminacy: A condition in which one is unable to calculate all support reactions, or all internal forces or stresses, by use of only the conditions of static equilibrium. (Deformations must also be considered in order to obtain a complete solution.)

Static load: A load that does not vary with time. A more precise term would be "quasi-static load," because a truly static load could be neither applied nor removed.

Superposition: The principle that two or more static loads, applied sequentially in any order, produce the same final result as obtained by applying all loads simultaneously. The principle is not applicable in instances of nonlinearity of response, under either an individual load or combinations of loads.

Transverse: Across. Thus, for load or deflection, the same as *lateral*.

Truss: A structure built of bars in which each bar is idealized as a two-force member, as if ends of bars were connected together by frictionless pins. (Compare *frame*.)

Yield strength: The maximum uniaxial tensile stress that can be applied without exceeding a specified permanent set upon release of load. It may also be called *yield stress*. The specified permanent set is often taken as an axial strain of 0.002. In a metal, numerical values of the elastic limit, proportional limit, and yield strength are usually quite similar.

1.3 PROPERTIES OF A PLANE AREA

Properties of a plane area are often needed, particularly for beam problems. The more essential properties and manipulations are reviewed here.

Definitions. Consider a plane area A, with rectangular Cartesian coordinates st in the same plane, Fig. 1.3-1a. By definition,

$$I_s = \int_A t^2 \, dA \qquad I_t = \int_A s^2 \, dA \qquad I_{st} = \int_A st \, dA \qquad (1.3\text{-}1)$$

I_s and I_t are *moments of inertia*, about s and t axes respectively. I_{st} is the *product of inertia*. I_s and I_t are always positive, but I_{st} can be positive, negative, or zero. Contributions $st \, dA$ are positive for areas dA in the first and third quadrants and negative for areas dA in the second and fourth quadrants (Fig. 1.3-1b). If s or t is a symmetry axis of A, then $I_{st} = 0$. The argument is shown in Fig. 1.3-1b. Each contribution $+st \, dA$ is matched by a contribution $-st \, dA$. Summing over A, we obtain $I_{st} = 0$.

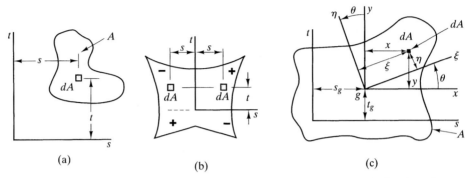

FIGURE 1.3-1 (a) Arbitrary plane area A. (b) Plane area symmetric about the t axis. Quadrants bear signs corresponding to their contribution to I_{st}. (c) Plane area with centroidal axes xy and $\xi\eta$.

Parallel Axis Theorems. These theorems relate quantities in Eq. 1.3-1 to corresponding quantities referred to *parallel* axes in the plane of A whose origin is at the *centroid of area A*. In Fig. 1.3-1c, let x and y be rectangular centroidal axes of A, respectively parallel to axes s and t, and in the plane of A. The parallel axis theorems are

$$I_s = I_x + At_g^2 \qquad I_t = I_y + As_g^2 \qquad I_{st} = I_{xy} + As_g t_g \tag{1.3-2}$$

where $I_x = \int y^2\, dA$, $I_y = \int x^2\, dA$, and $I_{xy} = \int xy\, dA$. Distances s_g and t_g are the coordinates of centroid g in the st system. These distances carry algebraic signs (both are positive in Fig. 1.3-1c). The argument for the last of Eqs. 1.3-2 is as follows. Substitute $s = s_g + x$ and $t = t_g + y$ into Eq. 1.3-1, and note that $\int x\, dA$ and $\int y\, dA$ both vanish because the xy system is centroidal. Thus

$$\begin{aligned} I_{st} &= \int_A (x + s_g)(y + t_g)\, dA = \int_A xy\, dA + 0 + 0 + s_g t_g \int_A dA \\ &= I_{xy} + As_g t_g \end{aligned} \tag{1.3-3}$$

The remaining two theorems in Eqs. 1.3-2 are proved in similar fashion.

Centroidal Principal Axes. In general, equations for principal axes do not require that axes be centroidal. However, in what follows, the origin of coordinates is placed at the centroid of area A because centroidal coordinates are the most useful.

Consider Fig. 1.3-1c. Systems xy and $\xi\eta$ are both rectangular, centroidal, and coplanar with A. The orientation of system xy can be chosen for convenience; for example, parallel to straight sides if area A happens to have them. System $\xi\eta$ is oriented at arbitrary angle θ with respect to system xy. Coordinates of a point in the rotated system $\xi\eta$ are $\xi = y \sin\theta + x \cos\theta$ and $\eta = y \cos\theta - x \sin\theta$. Thus we can obtain the following expressions by integration and substitution of trigonometric identities for $\sin^2\theta$, $\cos^2\theta$, and $\sin\theta\cos\theta$ (see Eqs. 1.10-1).

$$I_\xi = \int_A \eta^2\, dA \quad \text{yields} \quad I_\xi = \frac{1}{2}(I_x + I_y) + \frac{1}{2}(I_x - I_y)\cos 2\theta - I_{xy}\sin 2\theta \tag{1.3-4a}$$

$$I_{\xi\eta} = \int_A \xi\eta\, dA \quad \text{yields} \quad I_{\xi\eta} = \frac{1}{2}(I_x - I_y)\sin 2\theta + I_{xy}\cos 2\theta \quad (1.3\text{-}4\text{b})$$

One can select θ so that the moment of inertia of A becomes a maximum about either the ξ axis or the η axis. If I_ξ is the maximum I, it happens that I_η is the minimum I, and vice versa. The maximum I and the minimum I are called *principal moments of inertia* and their corresponding axes are called *principal axes*. The value of θ that maximizes (or minimizes) I_ξ is called θ_p. It is determined from the equation $dI_\xi/d\theta = 0$, which yields

$$\tan 2\theta_p = \frac{2I_{xy}}{I_y - I_x} \quad (1.3\text{-}5)$$

Angle θ_p has two values, $\pi/2$ apart, one for I_{max}, the other for I_{min}. By using Eq. 1.3-5 in Eq. 1.3-4a, we obtain the *principal moments of inertia:*

$$I_{max,min} = \frac{I_x + I_y}{2} \pm \sqrt{\left(\frac{I_x - I_y}{2}\right)^2 + I_{xy}^2} \quad (1.3\text{-}6)$$

Substitution of Eq. 1.3-5 into Eq. 1.3-4b yields $I_{\xi\eta} = 0$. That is, *the product of inertia is zero for principal axes*. The converse is also true: if $I_{\xi\eta} = 0$, then axes ξ and η are principal. Therefore, *if ξ or η is an axis of symmetry, then ξ and η are principal axes.*

From Eq. 1.3-6, we see that $I_{max} + I_{min} = I_x + I_y$. This relation can be useful in calculation, for example to determine I_{min} when I_{max}, I_x, and I_y have already been calculated. It may be physically obvious which of the two angles in Eq. 1.3-5 refers to the I_{max} axis, as in Fig. 1.3-2b. Otherwise the candidate angle can be substituted into Eq. 1.3-4a to see if I_ξ turns out to be I_{max} or I_{min}. Or, adapting a formula developed for stress transformation (see below Eq. 2.2-5), the counterclockwise angle θ_p from the x axis to the axis about which I is maximum is given by $\tan \theta_p = (I_x - I_{max})/I_{xy}$.

If $I_{max} = I_{min}$, angle θ does not matter. Then all centroidal axes yield the same I, and the product of inertia is zero for all these axes (Fig. 1.3-2c).

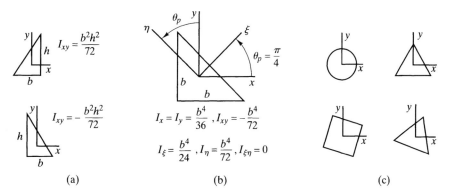

FIGURE 1.3-2 Various plane areas with centroidal axes xy. (a) Right triangles. (b) Isosceles right triangle. (c) Circle, square, and equilateral triangles. For each, $I_x = I_y$ and $I_{xy} = 0$.

Chapter 1 Orientation. Review of Elementary Mechanics of Materials

Polar Moment of Inertia J. Let r be the distance from the origin of xy coordinates to an element of area dA. Then $r^2 = x^2 + y^2$, and with respect to the "pole" at $x = y = 0$,

$$J = \int_A r^2 \, dA \quad \text{yields} \quad J = \int_A (x^2 + y^2) \, dA \quad \text{or} \quad J = I_y + I_x \quad (1.3\text{-}7)$$

The latter formula may be useful as a calculation device. Also, J is used in the torsional analysis of bars of circular cross section.

EXAMPLE

For the plane area in Fig. 1.3-3 we will determine I_x, I_y, I_{xy}, locate the principal centroidal axes, and determine the principal moments of inertia.

The centroid of A, at $x = y = 0$, has already been located, by means of calculations explained in textbooks about statics. For convenience in the following calculations, the cross section is arbitrarily divided into parts 1 and 2, as shown. Centroids of these parts are at $x = y = -15$ mm for part 1, and $x = y = 25$ mm for part 2. Equation 1.3-2 yields

$$I_x = \left[\frac{20(100)^3}{12} + 2000(-15)^2 \right] + \left[\frac{60(20)^3}{12} + 1200(25)^2 \right] \quad (1.3\text{-}8)$$

where the two bracketed expressions come from parts 1 and 2, respectively. I_y is obtained from a similar calculation and I_{xy} is

$$I_{xy} = [0 + 2000(-15)(-15)] + [0 + 1200(25)(25)] \quad (1.3\text{-}9)$$

Collecting results, we have

$$I_x = 2.907(10^6) \text{ mm}^4 \quad I_y = 1.627(10^6) \text{ mm}^4 \quad I_{xy} = 1.200(10^6) \text{ mm}^4 \quad (1.3\text{-}10)$$

From Eq. 1.3-5, we calculate the orientation of a principal axis.

$$\tan 2\theta_p = \frac{2(1.200)}{1.627 - 2.907} \quad \text{which yields} \quad \theta_p = -31.0° \quad (1.3\text{-}11)$$

which is the clockwise angle shown in Fig. 1.3-3. The other possible angle, $\theta_p + 90°$, is a 59° counterclockwise angle from the x axis to the η axis. From Eq. 1.3-6,

$$I_{max} = 3.627(10^6) \text{ mm}^4 \quad I_{min} = 0.907(10^6) \text{ mm}^4 \quad (1.3\text{-}12)$$

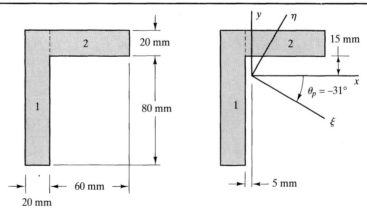

FIGURE 1.3-3 A plane area. Axes xy are centroidal. Axes $\xi\eta$ are centroidal and principal.

In this example it is clear by inspection of Fig. 1.3-3 that $I_\xi = I_{max}$ rather than $I_\eta = I_{max}$. Angle $\theta_p = -31°$ to the I_{max} axis, shown in Fig. 1.3-3, is verified by the formula $\tan \theta_p = (I_x - I_{max})/I_{xy}$.

1.4 AXIAL LOADING. PRESSURE VESSELS

Straight Bars. Consider a prismatic bar loaded by centroidal axial force P, Fig. 1.4-1a. The basic assumption about deformation is that plane cross sections remain plane when load P is applied. Thus any two cross sections a distance dx apart increase their separation an amount du (Fig. 1.4-1b), and axial strain is $\epsilon = du/dx$ at all points in a cross section. If the same stress-strain relation prevails throughout a cross section (that is, if the material is homogeneous), then axial stress σ is also the same at all points in a cross section. Equilibrium of axial forces requires that $\sigma A = P$. Thus the stress formula becomes $\sigma = P/A$. This result is not valid close to points of load application, where it is obvious that plane cross sections do not remain plane. According to Saint-Venant's principle, $\sigma = P/A$ should be an accurate formula at distances greater than ℓ from the loaded points, where ℓ is shown in Fig. 1.4-1c. The resultant force provided by a uniform stress distribution acts at the centroid of a cross section. For any cross section, load P must be collinear with this resultant. Therefore, if σ is to be uniformly distributed over a cross section, load P must be directed through centroids of cross sections. Accordingly, the bar cannot be curved. Taper, if not pronounced, causes little departure from the basic assumption; then σ is almost uniform over a cross section and is a function of axial coordinate x.

In uniaxial stress, a linearly elastic material has the stress-strain-temperature relation

$$\epsilon = \frac{\sigma}{E} + \alpha \Delta T \qquad (1.4\text{-}1)$$

where α is the coefficient of thermal expansion and ΔT is the temperature change. From the strain expression $\epsilon = du/dx$, an increment of axial displacement is $du = \epsilon \, dx$. Combining this expression with Eq. 1.4-1 and integrating, we obtain

$$u = \int_0^L \left(\frac{\sigma}{E} + \alpha \Delta T \right) dx \qquad (1.4\text{-}2)$$

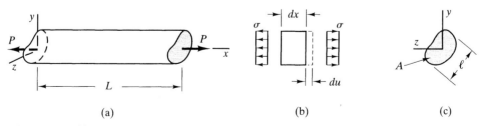

FIGURE 1.4-1 (a) Prismatic bar under centroidal axial load P. (b) Axial deformation, and axial stress σ. (c) Typical cross section.

as the axial deformation over a length L. (The symbol u is used in preference to δ or Δ in order to agree with notation in subsequent chapters, where u, v, and w denote displacement components in x, y, and z directions respectively.) Any of the quantities in parentheses in Eq. 1.4-2 may be a function of x. For the uniform bar in Fig. 1.4-1, with $\Delta T = 0$, Eq. 1.4-2 reduces to the familiar expression $u = PL/AE$. The presence of E in this formula—or in any other formula—makes it obvious that the formula is restricted to linearly elastic conditions.

Pressure Vessels. Let the cylindrical tank in Fig. 1.4-2 be thin walled, which customarily means that $r_i \approx 10t$ or more. Internal pressure causes points to displace radially but not circumferentially. Radial displacement u, greatly exaggerated, is shown in Fig. 1.4-2b. The initial length of arc CD is $r_i\, d\theta$. Its final length, after radial displacement u, is $(r_i + u)d\theta$. Its change in length is therefore $u\, d\theta$, and its circumferential strain is $\epsilon = (u\, d\theta)/(r_i\, d\theta) = u/r_i$. It is reasonable to assume that all points through the thickness have almost the same radial displacement u. Therefore, because all points also have almost the same radius, circumferential strain is almost uniform through the vessel wall. If the material is homogeneous, uniform strain implies uniform stress. Hence, summing forces in the direction of pressure p in Fig. 1.4-2c, we obtain

$$p(2r_i\, dx) = 2(\sigma t\, dx) \quad \text{from which} \quad \sigma = \frac{pr_i}{t} \qquad (1.4\text{-}3)$$

In similar fashion one can obtain axial stress $pr_i/2t$ in the cylindrical tank and stress $pr_i/2t$ in any surface-tangent direction in a spherical tank. These formulas are not reliable, even for thin-walled pressure vessels, near changes in geometry such as AA and BB in Fig. 1.4-2a, which are circles where end caps are connected to the cylindrical vessel.

If the vessel were thick walled, we could not conclude that circumferential strains are almost uniform through the vessel wall. Imagine, for example, that $t = r_i$. Then, for circumferential strain to be the same both inside and outside, radial displacement of the outer surface would have to be twice that of the inner surface. This conclusion is unreasonable. In fact, the inside displaces somewhat more than the outside. Thus, if the wall is thick, the inner surface carries higher strain and therefore higher stress than the outer surface. Considerations from theory of elasticity are needed to obtain expressions for stresses in a thick-walled cylinder under internal pressure.

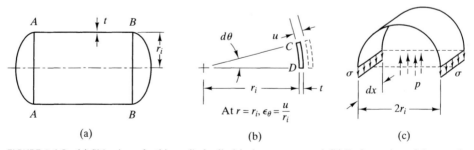

FIGURE 1.4-2 (a) Side view of a thin-walled cylindrical pressure vessel. (b) Deformation of the vessel wall due to internal pressure, viewed axially. (c) Circumferential stress, exposed by a cutting plane that contains the axis of the cylinder.

1.5 TORSION

Consider a prismatic bar of circular cross section, whose material is homogeneous and isotropic. The geometry of deformation, which may be established by experiment or by symmetry arguments, is that initially plane cross sections remain plane when the bar is twisted. Also, radial straight lines remain straight, and rotate about the axis. The diameter and length of the bar do not change. From all this one deduces that radial, circumferential, and axial normal strains are absent, and that shear strain γ varies linearly with distance r from the axis but is independent of the circumferential and axial coordinates. If a rectangular grid is drawn on the surface of the bar, one finds that twisting produces the deformed grid shown in Fig. 1.5-1a. All right angles of the grid change by the same amount. This amount is the value of shear strain γ at radius $r = c$.

Let the shear stress versus shear strain relation be linear, $\tau = G\gamma$. Then, since shear strain γ varies linearly with distance from the axis, so does shear stress τ: symbolically, $\tau = kr$, where k is a constant. To relate τ to the torque T that produces it, we consider equilibrium of moments about the axis of the bar. Thus, from Fig. 1.5-1b.

$$T = \int_A r(\tau \, dA) \quad \text{or} \quad T = k \int_A r^2 \, dA = kJ \tag{1.5-1}$$

Hence $k = T/J$, and the expression $\tau = kr$ becomes $\tau = Tr/J$, which is the standard torsion formula. Note that τ acts on longitudinal planes as well as on transverse planes, as shown in Fig. 1.5-1c.

Figure 1.5-1c leads to a formula for θ, the angle of twist of one end of the bar relative to the other. Angles γ and $d\theta$ are small, so

$$ds = \gamma \, dx = r \, d\theta \quad \text{hence} \quad \theta = \int_0^L \frac{\gamma}{r} \, dx \tag{1.5-2}$$

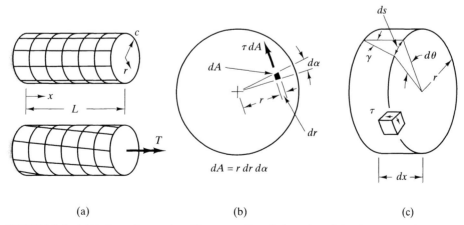

FIGURE 1.5-1 (a) Deformation produced by torque T applied to a bar of circular cross section. (b) Force increment $\tau \, dA$ produces torque increment $r(\tau dA)$. (c) Geometry of deformation that leads to a formula for angle of twist θ.

This result does not require that the material be linearly elastic. But if it is, we can substitute $\gamma = \tau/G = Tr/GJ$, whereupon the integrand in Eq. 1.5-2 becomes $T\,dx/GJ$. If T, G, and J are independent of x, we obtain the familiar expression $\theta = TL/GJ$. The presence of G in this formula makes it obvious that the formula is limited to linearly elastic conditions.

The manner of support or torsional load application, or the presence of stress raisers such as circumferential grooves, causes only local disturbances of stress, in accord with Saint-Venant's principle. These disturbances have little effect on the angle of twist. If the bar is tapered, then $J = J(x)$. The formula $\tau = Tr/J$ has little error provided the taper is slight. Changes that would invalidate our simple formulas, and the reasons why, are as follows. Orthotropy, unless it is polar about the axis of the bar, would make γ and τ depend on the circumferential coordinate as well as on r. The same effect would be produced by material properties that vary circumferentially, and by a noncircular cross section (see Section 7.11). A sharply curved geometry, as for the coil of a massive helical spring, would make γ larger toward the inside of the coil (see Section 6.1).

1.6 BEAM STRESSES

Bending. Consider a prismatic beam, whose material is homogeneous and isotropic. We require that the beam have a plane of symmetry, and that the beam be bent to an arc in this plane (Fig. 1.6-1). The geometry of deformation can be established by experiment or by symmetry arguments: Initially plane cross sections remain plane when bending moment is applied. Arbitrary cross sections AB and CD have the relative rotation $d\theta$. At coordinate y, axial strain is $-\epsilon$ and axial displacement is $-\epsilon\,dx$, negative because ϵ is compressive when y is positive. With ρ the radius of curvature, the small angle $d\theta$ can be expressed in two ways.

$$d\theta = \frac{-\epsilon\,dx}{y} \quad \text{and} \quad d\theta = \frac{dx}{\rho} \quad \text{hence} \quad \epsilon = -\frac{y}{\rho} \quad (1.6\text{-}1)$$

Thus we see that ϵ varies linearly with y. It is reasonable to assume a uniaxial state of stress. If the stress-strain relation is linear, then axial stress σ is $\sigma = ky$, where k is a constant. Two equilibrium conditions are applicable: The stress distribution provides zero axial force and bending moment M. That is,

$$0 = \int_A \sigma\,dA \quad \text{hence} \quad 0 = k \int_A y\,dA \quad (1.6\text{-}2a)$$

$$M = -\int_A y(\sigma\,dA) \quad \text{hence} \quad M = -k \int_A y^2\,dA = -kI \quad (1.6\text{-}2b)$$

Equation 1.6-2a demands that $\int y\,dA = 0$, which means that the z axis, at $y = 0$, passes through the centroid of the cross section. From Eq. 1.6-2b we obtain $k = -M/I$, where I is the moment of inertia of cross-sectional area A about its centroidal axis z. Hence the expression $\sigma = ky$ becomes $\sigma = -My/I$, which is the standard flexure formula. Typically we write simply $\sigma = My/I$, because the algebraic sign of σ at a given y is obvious from the direction of the bending moment.

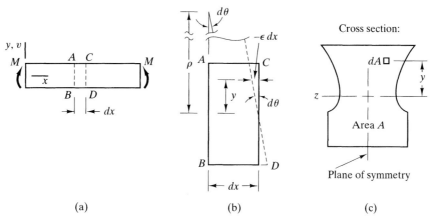

FIGURE 1.6-1 (a) Beam bent in the *xy* plane. (b) Deformations in the *xy* plane. (c) Arbitrary (but symmetric) cross section of area *A*. The symmetry plane of the cross section is normal to the paper.

Common situations to which the flexure formula is not applicable, or applicable only after modification, are as follow. If there is no symmetry plane, we cannot presume that axial strain ϵ is independent of z (axis z is shown in Fig. 1.6-1c). That is, Eq. 1.6-1 is no longer correct. This situation is called unsymmetric bending and is discussed in Chapter 10. If the beam has pronounced initial curvature before load is applied, plane cross sections still remain plane, but we cannot conclude that ϵ varies linearly with y (see Chapter 6). If the material is not linearly elastic, then $\sigma \neq ky$, and the latter forms of Eqs. 1.6-2 no longer apply. Similarly, if the material is not homogeneous, then $\sigma \neq ky$. Therefore we cannot use $\sigma = My/I$ to analyze a reinforced concrete beam. Finally, if the cross section is wide we must consider that the body is a plate rather than a beam (Chapter 12).

Transverse Shear. If bending moment M is not constant, a transverse shear force V exists in a straight beam. Force V produces transverse shear stress, which acts on transverse planes and on longitudinal planes. Formulas for shear flow and shear stress are derived from the flexure formula and are therefore subject to the same restrictions. From the shear flow formula, usually written as $q = VQ/I$, we obtain the *average* transverse shear stress $\tau = q/t = VQ/It$, where t is a thickness measured in the plane of the cross section. This average shear stress may be quite accurate or quite inaccurate, depending on circumstances. For example, in Fig. 1.6-2b, shear stress on plane AB is small because t in $\tau = VQ/It$ is large (here t is the width of the flange). Moreover, if plane AB is moved very close to the inner flange surface, τ must approach zero on that portion of the inner flange where the adjacent surface is free of stress. On the portion of plane AB immediately adjacent to the web, τ approaches the transverse shear stress on plane CD, where t is the web thickness and $\tau = VQ/It$ is accurate. The largest transverse shear stress in the flange is exposed by a vertical cutting plane such as EF, where t in VQ/It is the flange thickness. Details of these matters, and of how to use the formula VQ/It, appear in textbooks of elementary mechanics of materials.

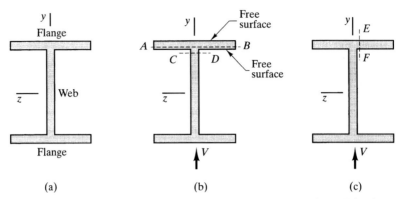

FIGURE 1.6-2 Cross section of a beam that carries transverse shear force V. Cutting planes AB, CD, and EF are normal to the yz plane.

1.7 BEAM DEFLECTIONS

Briefly, the formula that relates lateral deflection v to bending moment M is developed as follows. We use notation in Fig. 1.6-1. If $|dv/dx| \ll 1$, as is usual in practical beams, then the curvature of the deformed beam can be written as $1/\rho = d^2v/dx^2$. Also, for a linearly elastic material, Eq. 1.6-1 and the flexure formula $\sigma = -My/I$ yield another expression for curvature: $1/\rho = -\epsilon/y = -(\sigma/E)/y = -(-My/EI)/y = M/EI$. Equating the two expressions for curvature, we obtain

$$\frac{d^2v}{dx^2} = \frac{M}{EI} \qquad (1.7\text{-}1)$$

Restrictions on this formula include those on the flexure formula. Also, deflections must be sufficiently small that slope $\theta = dv/dx$ of the deformed beam is everywhere much less than unity in magnitude. Transverse shear deformation has been neglected. Equation 1.7-1 actually says that M/EI is equal to the *change* in curvature. This viewpoint may become important for a beam having initial curvature before load is applied. For a beam initially straight and then bent to radius ρ, the initial curvature is zero, and the change in curvature is $(1/\rho - 0) = 1/\rho$.

An alternative form of Eq. 1.7-1 can be written, as follows. Equations of static equilibrium, applied to Fig. 1.7-1a, yield $dM/dx = V$ and $dV/dx = q$, where q is the intensity per unit length of distributed lateral load. Hence $d^2M/dx^2 = q$. For M we can substitute $EI(d^2v/dx^2)$ from Eq. 1.7-1. Thus

$$\frac{d^2}{dx^2}\left(EI\frac{d^2v}{dx^2}\right) = q \quad \text{or} \quad EI\frac{d^4v}{dx^4} = q \quad \text{if} \quad EI \text{ is independent of } x \qquad (1.7\text{-}2)$$

The latter form will be useful in subsequent chapters.

One can determine beam deflections (or solve statically indeterminate beam problems) by integrating Eq. 1.7-1 and making use of support conditions to evaluate constants of integration (and redundant reactions). Usually it is easier to solve these problems by use of tabulated beam formulas and the superposition principle. Indeed,

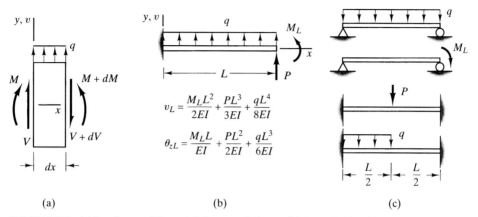

FIGURE 1.7-1 (a) Loads on a differential element of a beam. (b) Formulas for tip deflection and tip rotation of a uniform cantilever beam. (c) Problems of deflection, rotation, or static indeterminacy solvable by use of formulas in (b).

the few formulas in Fig. 1.7-1b are sufficient to solve most common problems of straight beams, including all those shown in Fig. 1.7-1c. An example problem is solved in Section 1.8. Like Eq. 1.7-1, formulas in Fig. 1.7-1b require that $|\theta| \ll 1$ throughout the beam.

1.8 SYMMETRY CONSIDERATIONS. STATIC INDETERMINACY

Symmetry Considerations. Sometimes one can exploit symmetry to obtain internal forces, determine support conditions, or reduce the effort required for analysis. For example, consider the simply supported beams in Fig. 1.8-1. Both have symmetry of geometry, elastic properties, and support conditions with respect to a plane normal to the beam axis at its center. The beams differ only in loading. In Fig. 1.8-1a, a mirror reflection of either half in the symmetry plane yields the other half in geometry, elastic properties, loading, support reactions, deformations, and internal forces at the symmetry plane. For antisymmetric loading, Fig. 1.8-1b, one half yields the other half after reflection and *reversal* of loading, support reactions, deformations, and internal forces at the symmetry plane. These considerations, in combination with the action-reaction nature of internal forces exposed by cutting open the beam, preclude the existence of shear forces V_C for symmetric loading and bending moments M_C for antisymmetric loading. Thus in either case the number of unknowns is immediately reduced by half.

The same considerations can be used in three dimensions. The semicircular beams in Fig. 1.8-2a lie in the xy plane. For each, there is symmetry of geometry, elastic properties, and support conditions about the yz plane. For symmetric loading, Fig. 1.8-2a, symmetry considerations dictate that at midpoint C there is no x direction displacement, no rotation about the y axis or the z axis, no transverse shear force in the y direction or the z direction, and no torque about the x axis. These conditions are listed in Fig. 1.8-2a. Unknowns at C are displacements v and w, rotation θ_x about the x axis, axial force F_x, and bending moments M_y and M_z about the y and z axes. These unknowns could be

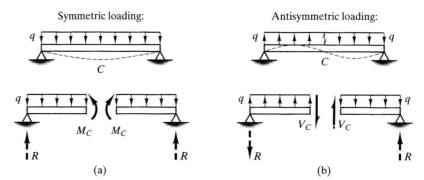

FIGURE 1.8-1 Uniform simply supported beams, showing internal moments and forces at center C. Supports apply negligible horizontal force if deflections are small.

determined by analysis of either half of the semicircular beam. In Fig. 1.8-2b two of the forces P and Q are reversed, so the load is antisymmetric. Symmetry considerations dictate the zero quantities listed in Fig. 1.8-2b. Again, analysis of either half of the beam is sufficient to determine the unknown quantities at C, which are $u, \theta_y, \theta_z, V_y, V_z$, and T_x.

The foregoing arguments are not immediately obvious. The reader is urged to consider these examples patiently, and to make supplementary sketches that show internal forces and moments.

Static Indeterminacy. The term is defined in Section 1.2. Calculations are illustrated by the following examples.

The stepped bar in Fig. 1.8-3a is all of the same material. It is to be uniformly heated from its stress-free temperature while confined between rigid walls. Statics tells us only that the walls apply forces P of equal magnitude. To determine them we must use a compatibility condition, which here is that the bar has no net change in length from end to end. Thus, taking P positive in tension and presuming that conditions are linearly elastic, we write

$$\alpha L \, \Delta T + \alpha L \, \Delta T + \frac{PL}{AE} + \frac{PL}{(2A)E} = 0 \tag{1.8-1}$$

where α is the coefficient of thermal expansion and ΔT is the temperature change. Solving for P and then for stresses $\sigma_1 = P/A$ and $\sigma_2 = P/2A$, we obtain

$$P = -\frac{4EA\alpha \, \Delta T}{3} \qquad \sigma_1 = -\frac{4E\alpha \, \Delta T}{3} \qquad \sigma_2 = -\frac{2E\alpha \, \Delta T}{3} \tag{1.8-2}$$

Note that axial strains are not zero, even though the overall change in length is zero. For example, in part 1, $\epsilon_1 = (\sigma/E) + \alpha \, \Delta T = -\alpha \, \Delta T/3$. Note also that modest temperature change can produce large stress. In the present example, if the bar is steel and $\Delta T = 100°C$, then σ_1 is about 320 MPa in magnitude.

As a second example, consider the beam in Fig. 1.8-3b. It is statically indeterminate to the second degree. Symmetry considerations can be used to reduce the degree of indeterminacy. Imagine that M_C is applied as two couples $M_C/2$, an infinitesimal dis-

Section 1.8 Symmetry Considerations. Static Indeterminacy

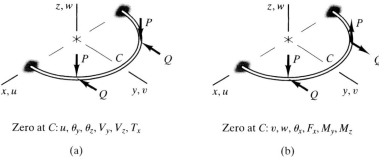

FIGURE 1.8-2 Uniform semicircular beams in the xy plane.

tance apart, and straddling point C. The loading is antisymmetric, so at C there is a transverse shear force V_C but zero bending moment and zero vertical displacement (Fig. 1.8-3c). Using formulas in Fig. 1.7-1b to state that the transverse displacement is zero at C, we solve for V_C and then for moment M_B at the wall.

$$-\frac{(M_C/2)a^2}{2EI} + \frac{V_C a^3}{3EI} = 0 \quad \begin{cases} V_C = \dfrac{3M_C}{4a} \\ M_B = V_C a - \dfrac{M_C}{2} = \dfrac{M_C}{4} \end{cases} \qquad (1.8\text{-}3)$$

Finally, having resolved the indeterminacy, we can use Fig. 1.7-1b again to determine the rotation at C.

$$\theta_C = \frac{(M_C/2)a}{EI} - \frac{V_C a^2}{2EI} \quad \text{hence} \quad \theta_C = \frac{M_C a}{8EI} \qquad (1.8\text{-}4)$$

Problems such as those in Fig. 1.8-3 are probably called to mind by the term "statically indeterminate analysis." However, the term is also appropriate for the derivation of conventional stress formulas such as $\sigma = My/I$: an equilibrium equation, such as the first of Eqs. 1.6-2b, yields the second only when it is known how stress varies over the cross section. The variation is obtained by consideration of displacements.

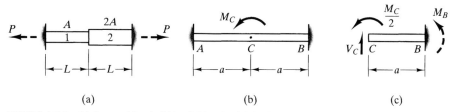

FIGURE 1.8-3 (a) Stepped bar held by rigid walls. (b) Statically indeterminate beam. (c) Right half of the beam, with symmetry considerations exploited.

1.9 PLASTIC DEFORMATION. RESIDUAL STRESS

Here we review the problem of plastic torsion for a shaft of solid circular cross section. Other instances of plastic action are considered in subsequent chapters.

The stress-strain relation in Fig. 1.9-1a is linearly elastic up to stress τ_Y and flat-topped thereafter. This idealized behavior is called "elastic–perfectly plastic" and is appropriate for low-carbon steel. Because strain hardening is ignored, calculations provide a maximum or "fully plastic" torque T_{fp} that is less than the actual maximum torque. We now ask for T_{fp} and the pattern of residual stress upon unloading.

As twist increases, yielding eventually begins. It spreads from the outer surface toward the axis of the shaft. To calculate T_{fp}, we assume that twist is sufficiently great that practically all the material has yielded. Thus, shear stress is the constant value τ_Y throughout, and the first of Eqs. 1.5-1 provides

$$T_{fp} = \tau_Y \int_0^{2\pi} \int_0^c r(r\, dr\, d\alpha) = \tau_Y \frac{2\pi c^3}{3} \quad \text{hence} \quad T_{fp} = \frac{4}{3}\left(\tau_Y \frac{\pi c^3}{2}\right) \quad (1.9\text{-}1)$$

where the latter expression in parentheses is the torque that initiates yielding, obtained from the torsion formula for linearly elastic conditions; that is, $\tau = Tr/J$ with $\tau = \tau_Y$ at $r = c$. This result shows that torque can be increased 33% after yielding begins.

Unloading can be accomplished by superposing on T_{fp} a torque of equal magnitude but reversed in direction. Anticipating that unloading will be elastic, we obtain the stress distribution in Fig. 1.9-1c from the reversed torque $T = T_{fp}$ and the elastic stress formula $\tau = Tr/J$. At first glance this calculation may appear wrong because the largest stress exceeds τ_Y. However, stresses in Fig. 1.9-1c always appear in *combination* with stresses in Fig. 1.9-1b. In combination, τ never exceeds τ_Y in magnitude, so unloading does not produce further yielding. If torque T_{fp} is again applied, residual stresses combine with the reverse of stresses in Fig. 1.9-1c to produce again the fully plastic stress pattern of Fig. 1.9-1b, but without renewed yielding.

The residual angle of twist after unloading cannot be calculated because we have not specified how much the shaft was twisted in producing T_{fp}. An *infinite* angle of twist would be required to bring inelastic strains all the way to $r = 0$.

What is the range of torque for which conditions are linearly elastic? If there are no residual stresses, a torque $T = \tau_Y J/c = \tau_Y \pi c^3/2$ could be applied in either direction without yielding, for an elastic range of $\tau_Y \pi c^3$. If the residual stresses in Fig. 1.9-1d pre-

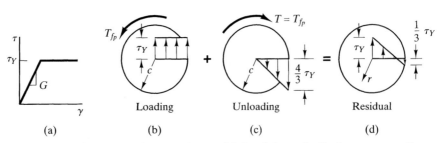

FIGURE 1.9-1 (a) Elastic–perfectly plastic material. (b,c,d) Stress distributions corresponding to fully plastic torque, unloading (reversed elastic) torque, and resultant (zero) torque.

vail, we could apply a torque T_{fp} in the original direction or $(2/3)(\tau_Y J/c)$ in the reversed direction without renewed yielding, for an elastic range of $\tau_Y \pi c^3$. Thus the magnitude of the elastic range has not changed.

1.10 OTHER REMARKS

Stress Transformation. For reference, and for use in chapters that follow, two-dimensional stress transformation equations are shown in Fig. 1.10-1. These equations may be restated in other forms, for which the following trigonometric identities are useful.

$$\sin 2\theta = 2 \sin \theta \cos \theta \qquad \cos^2\theta = \frac{1}{2}(1 + \cos 2\theta)$$

$$\cos 2\theta = \cos^2\theta - \sin^2\theta \qquad \sin^2\theta = \frac{1}{2}(1 - \cos 2\theta)$$

(1.10-1)

Dimensional Homogeneity. In the calculation of stresses and deflections, it is often best to obtain a numerical result as the final step of solution, by substitution of data into a symbolic result. Thus we avoid manipulating numbers for some quantities that may cancel if manipulated as symbols. A more important reason is that a symbolic result permits a partial check on the correctness of the solution. A valid result is dimensionally homogeneous. For example, in Fig. 1.7-1b let $[v_L]$ and $[\theta_L]$ denote the respective dimensions of deflection and rotation. With F and L used here to denote dimensions of force and length respectively, dimensions of terms that contain M_L in the formulas of Fig. 1.7-1b are

$$[v_L] = \frac{(FL)L^2}{(F/L^2)L^4} = L \qquad \text{and} \qquad [\theta_L] = \frac{(FL)L}{(F/L^2)L^4} = 1 \qquad (1.10\text{-}2)$$

These dimensions are correct: length units for v_L and dimensionless (radians in this case) for θ_L. This result does not prove the formulas to be correct, but had we obtained any other dimensions we would know for sure that the result is wrong.

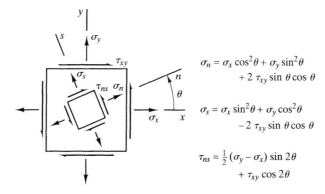

FIGURE 1.10-1 Transformation of stresses in a plane.

Units. Example problems and homework problems serve as vehicles to convey concepts, principles, and procedures. Accordingly, the system of units used for numerical problems is of little importance. SI units are used in this book. Note that the average stress due to a 1 MN force on a square meter can be written in the following forms:

$$\sigma = \frac{10^6 \text{N}}{1 \text{ m}^2} = 10^6 \text{ Pa} = 1 \text{ MPa} \quad \text{or} \quad \sigma = \frac{10^6 \text{N}}{(1000 \text{ mm})^2} = 1 \text{ MPa} \quad (1.10\text{-}3)$$

The latter form, which is used in subsequent chapters, avoids the conversion factor of 10^6. That is, forces in newtons, dimensions in millimeters, and stresses and moduli in megapascals form a consistent set of units, without need for conversion factors. However, if mass must be considered, as for inertia force loading, it will be easier to use meters rather than millimeters.

Classification by Problem Geometry. A slender member is usually called a *bar*, *beam*, or *shaft*, depending on whether the load is axial, lateral, or torsional. These problems are called one-dimensional, even though stress varies over a cross section as well as axially under bending or twisting load. A flat body whose thickness is much less than its other dimensions provides a two-dimensional problem. It is usually called a *plane* problem if loads have no lateral (thickness-direction) component, and a *plate* or *plate bending* problem if they do. In general, stresses in plane and plate problems vary with both of the in-plane directions. Stresses also vary in the thickness direction of a plate under lateral load. A floor slab is a familiar example. A *shell* is like a plate, but curved; familiar examples include an egg shell and a water tank. A shell can carry both surface-tangent and surface-normal loads. Many shells, and many solids too thick to be called shells, are symmetric about an axis and have loading that is also axisymmetric. Then nothing varies in the circumferential direction and analysis is simplified. Such a body is called a *shell of revolution* if it is thin-walled or a *solid of revolution* if it is not. An example of the latter is a turbine disk of strongly varying thickness that rotates at constant speed.

Connections. In this book, as in most other books about stress analysis, we may simply state that members are connected together, without saying how, and perhaps even disregarding stress concentrations associated with the connection. Thus we limit the scope of the book. Unfortunately the reader may then infer that connections are unimportant, which is far from the case. The behavior of a real structure may depend as much on its connections as on its individual members. *Connections are often the weakest parts of a stucture.*

Bolted or riveted connections are sometimes analyzed in a first course in mechanics of materials. One learns that several modes of failure are possible and that analysis can be tedious, despite simplifying assumptions that neglect stress concentrations, friction and possible slipping, making and breaking of contacts, misalignment, initial stresses, and damage to the material from cutting, bending, and punching holes. Other complexities arise if we consider gluing, welding, and shrink fits. Practical analysis and design of connections may be done using accepted codes and procedures that differ according to type of joint, and which vary considerably with type of industry. The study of connections is an important specialty in stress analysis. References include [1.1–1.5].

Handbooks. Many useful formulas for stress analysis do not appear in textbooks but may be found in handbooks or their computer software equivalents. The existence of this information does not erase the need for ability in stress analysis. Formulas can be used successfully only if the engineer understands the physical problem well enough to know what sort of formula to seek, understands the assumptions that underlie a formula, and is able to judge whether an answer produced by the formula is reasonable. Useful handbooks include [1.6, 1.7] for widespread coverage of stress and deflection, [1.8] for pressure vessels and the ASME code for them, [1.9] for buckling of bars, frames, plates, and shells, and [1.10] for modes and frequencies of vibration.

Codes. Engineering societies have produced codes that mandate allowable stresses, design procedure, and methods for testing, construction, operation, and maintainance of plants and equipment. Much of this information has grown out of experience with costly failures [1.11, 1.12]. Codes and specifications may receive little mention in engineering education, but it would be shortsighted to ignore them. Indeed, the engineer is often legally bound to follow one or more codes. Also, in situations where a code is applicable, it is likely to be the easiest route to an acceptable design. Students of structural engineering are probably familiar with design specifications of the American Institute of Steel Construction. There are a great many other codes and specifications, so many that space does not permit us to list them all.

The value of codes is illustrated by the history of boiler accidents. About the year 1900, on average, one boiler explosion occurred every day in the United States. Subsequently, codes for the design, construction, and operation of boilers were written and widely adopted. Today boiler explosions are rare despite a fifteen-fold increase in operating pressure since 1900 [1.13].

PROBLEMS

The following problems can be solved using the review material presented in this chapter, although many of the problems are less familiar or more challenging than those usually seen in an elementary textbook. Assume that materials are linearly elastic unless a nonlinear stress-strain relation is provided. State results symbolically in terms of loads, dimensions, properties of cross sections, and material constants, unless a numerical answer is required or other instructions are given.

1.4-1. A prismatic bar is loaded by an axial force P. Show that P must be directed through centroids of cross sections if axial stress σ is not to vary over a cross section.

1.4-2. Springs in the structure shown are linear and are unstressed when displacement v is zero. Determine an expression for v without assuming that $v \ll L$. With $L = 100$ mm and $k = 20$ N/mm, obtain numerical values of P for displacements v of 10 mm, 40 mm, and 50 mm. Show that superposition using the first two results does not yield the third. Plot P versus v.

1.4-3. Two slender rings, one aluminum and the other steel, just fit together at temperature $T = 0°C$, as shown. What is the contact pressure between them when $T > 0°C$?

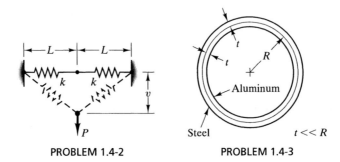

PROBLEM 1.4-2

PROBLEM 1.4-3

1.5-1. A shaft of solid circular cross section is loaded by torque T. Consider a half-cylinder cut from the shaft by three cutting planes (see sketch). Show that stresses exposed by the cutting planes keep the half-cylinder in static equilibrium.

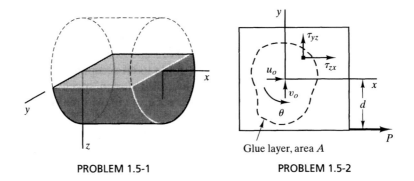

PROBLEM 1.5-1

PROBLEM 1.5-2

1.5-2. A flat plate is attached to a flat surface by a thin layer of glue of arbitrary shape and comparatively low modulus. An x-parallel load P is applied to the plate (see sketch). Axes xy are centroidal axes of the glue layer. What are shear stresses τ_{yz} and τ_{zx} in the glue layer? (Suggestion: Assume that these stresses are proportional to displacement components of the plate, and that the plate has rotation θ and translation components u_o and v_o at $x=y=0$. Area A and its properties will appear in the solution.)

1.6-1. The sketch shows the post-buckling shape of a slender bar that was initially straight. Load F is known, and the shape $y=f(x)$ of the buckled bar is accurately known. What is the easiest way to determine support reactions at ends of the bar?

PROBLEM 1.6-1

1.6-2. In the beam of Fig. 1.6-1, it is proposed that flexural stress has the form $\sigma=ky$. Show that the cross-sectional area must have a zero product of inertia if this equation is to be correct.

1.6-3. (a) Consider a prismatic beam, and conjecture that plane cross sections do *not* remain plane when bending moment is applied. Without equations, devise arguments that refute the conjecture.

(b) Similarly, consider a prismatic bar of circular cross section. Refute the conjectures that cross sections warp and radial lines become curved when torque is applied.

(c) The flexure formula $\sigma = My/I$ follows from the condition that plane cross sections remain plane in pure bending. A cantilever beam under transverse tip load experiences transverse shear deformation, and plane cross sections do *not* remain plane. Yet the flexure formula loses no accuracy. How can this be?

1.6-4. Let a prismatic beam have a rectangular cross section, b units wide and h units deep. The material has elastic moduli E_t in tension and E_c in compression. Derive expressions that relate stress to bending moment. The expressions should reduce to the conventional flexure formula if $E_t = E_c$.

1.6-5. The uniform beam shown has weight q per unit length. It rests on a rigid horizontal surface. If one end is lifted by a force $F < qL/2$, what is the maximum bending moment in the beam in terms of F and q?

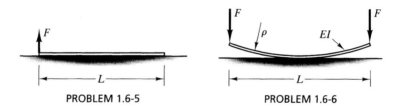

PROBLEM 1.6-5   PROBLEM 1.6-6

1.6-6. When not loaded, the uniform beam shown has constant radius of curvature ρ, where $\rho \gg L$. Downward forces F are then applied to the ends.

(a) What value of F reduces curvature at the center of the beam to zero?

(b) For larger F, a central portion of length s becomes flat. Obtain an expression for s.

1.6-7. The beam shown has a slight taper. For what value of h_L/h_o does the largest flexural stress appear at $x = L/2$? What then is the ratio of flexural stress at $x = L/2$ to flexural stress at $x = L$?

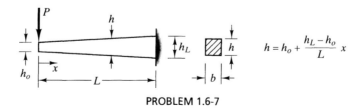

PROBLEM 1.6-7

1.7-1. A cantilever beam is loaded by uniform shear stress τ applied to its upper surface only, as shown. Obtain expressions for x-direction normal stress at A and at B. Neglect stress concentration effects. Also determine the deflection components of point C.

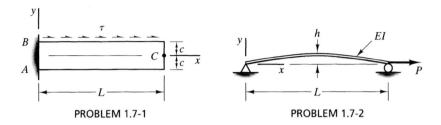

PROBLEM 1.7-1   PROBLEM 1.7-2

1.7-2. The slender bar shown is initially curved, so that with no load its axis has the equation $y = (4h/L^2)(Lx - x^2)$. What center deflection v_c is produced by force P? Assume that $v_c \ll h \ll L$.

1.7-3. Let the cantilever beam of Problem 1.7-1 be thermally loaded, such that the temperature varies linearly from ΔT on the lower surface to $-\Delta T$ on the upper surface. Obtain an expression for the deflection of point C due to ΔT.

1.7-4. It is proposed that a beam be constructed with a joint consisting of two horizontal links, as shown, so that the joint will transmit bending moment but no transverse shear force. Will this construction work as intended when load P is applied? Explain.

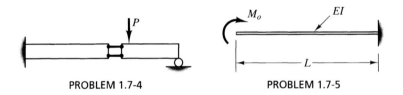

PROBLEM 1.7-4   PROBLEM 1.7-5

1.7-5. The cantilever beam shown is so slender that its material remains linearly elastic even when displacements are large. Obtain expressions for the horizontal and vertical displacement components of the tip. Show that these expressions reduce to the expected small-deflection results when $M_o L/EI$ is small.

1.7-6. For what value of a/b will the two parts of the beam shown have the same slope at hinge B when moment M_o is applied at A?

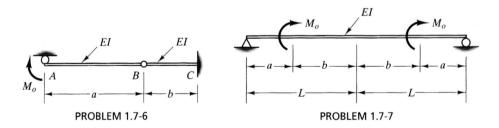

PROBLEM 1.7-6   PROBLEM 1.7-7

1.7-7. It is desired that both ends of a uniform beam remain horizontal when moments M_o are applied as shown. For what value of a/b will this be so?

1.7-8. Each of the beams shown is to be made with a small initial curvature, such that a load F moving across the beam will have no vertical displacement. What should be the initial shape $y = f(x)$ of each beam?

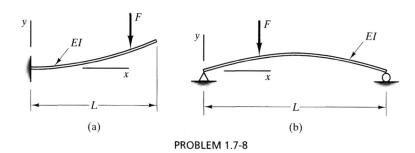

PROBLEM 1.7-8

1.7-9. Two identical rollers of average radius R are to be pushed together by end forces P, as shown. It is desired that the contact force between them be uniformly distributed along length L. Thus the rollers should not be quite cylindrical. How should R vary with x? Assume that the rollers are compact rather than quite slender, but that transverse shear deformation can be neglected.

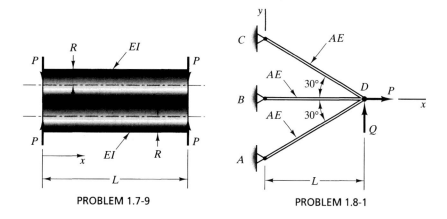

PROBLEM 1.7-9   PROBLEM 1.8-1

1.8-1. Members of the three-bar truss shown are identical except for length. Determine the displacement of joint D due to each of the following loadings. (a) $P = 0$, $Q > 0$. (b) $P > 0$, $Q = 0$. (c) $P = Q = 0$; all bars uniformly heated an amount ΔT.

1.8-2. Let gears of different sizes be fastened to either end of a prismatic shaft of circular cross section. Let there be two such shafts, set parallel so that gears of radius R and $2R$ engage in the manner shown. Frictionless bearings, not shown, ensure that the shafts twist without bending. What torsional stiffness T/θ is seen by torque T?

1.8-3. A square frame is made by welding together four identical slender bars of circular cross section. The frame is placed horizontally atop corner supports that can exert vertical

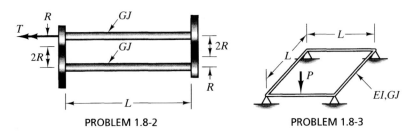

PROBLEM 1.8-2 PROBLEM 1.8-3

force but no moment. A vertical load P is applied to the middle of one side, as shown. What is the displacement of the loaded point and of the point opposite it? Let $G = E/2$; thus $EI = GJ$.

1.8-4. Two slender beams are built-in to a rigid disk and to rigid walls, as shown. Through what angle does the disk rotate if a small torque T is applied?

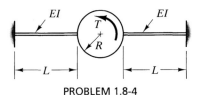

PROBLEM 1.8-4

1.8-5. Use formulas in Fig. 1.7-1b to determine the center deflection of each beam in Fig. 1.7-1c. In the second case, determine also the rotation at the right end.

1.8-6. A bimetal beam is constructed by bonding together two slender beams of rectangular cross section. Material properties of the component beams differ, including thermal expansion coefficients α_1 and α_2. With $\alpha_1 < \alpha_2$, uniform heating an amount ΔT causes the deformation shown. Write, but do not solve, sufficient equations to determine radius of curvature ρ, internal forces P_1 and P_2, and internal bending moments M_1 and M_2. Also, write expressions for axial stresses at upper and lower surfaces of the composite beam in terms of the internal forces and moments.

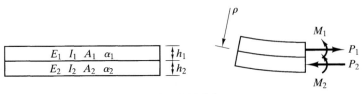

PROBLEM 1.8-6

1.8-7. Let several vertical posts of diameter D be arrayed in a straight line with distance L between them. A long slender beam is woven between the posts, as shown. Determine the maximum flexural stress in the beam.

1.8-8. A long straight beam has weight q per unit length. The beam is laid atop a small cylinder, as shown. Over what span $2L$ is the beam not in contact with the horizontal rigid floor?

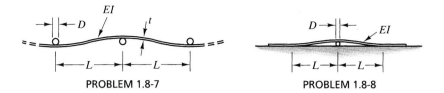

PROBLEM 1.8-7   PROBLEM 1.8-8

1.8-9. Two flat rigid walls include a small angle θ between them. A beam of rectangular cross section is just in contact with the walls, as shown. What uniform temperature increase ΔT is sufficient to place both ends of the beam in full contact with the walls? Express ΔT in terms of $\theta, h, \alpha,$ and L.

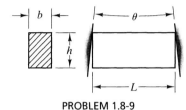

PROBLEM 1.8-9

1.9-1. The bar shown has uniform cross-sectional area A and is fixed at both ends. An idealized stress-strain relation is also shown. Assume that the relation is valid in compression as well as in tension.
  (a) Determine the value of load P that initiates yielding.
  (b) Determine the fully plastic load P_{fp}.
  (c) Determine the state of residual stress after load P_{fp} is removed.

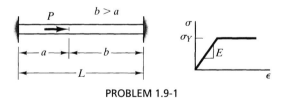

PROBLEM 1.9-1

1.9-2. Let the bar in Fig. 1.8-3a have the stress-strain relation used in Problem 1.9-1. Ends are fixed to the walls. Starting from the stress-free state, lower the temperature of the entire bar 1.5 times the amount ΔT that initiates plastic action.
  (a) What then are the axial stresses? Express answers in terms of σ_Y.
  (b) What are the residual stresses, and the residual displacement at the step, if the temperature is restored to its original value? Express answers in terms of $\sigma_Y, L,$ and E.

1.9-3. For the three-bar truss of Problem 1.8-1, let $Q=0$ and let the stress-strain relation be as depicted in Problem 1.9-1. Determine the fully plastic load P_{fp}. Also construct a dimensionless plot of P versus the horizontal displacement u_D of point D, using $P/A\sigma_Y$ as ordinate and $Eu_D/L\sigma_Y$ as abscissa.

CHAPTER 2

Stress, Principal Stresses, Strain Energy

2.1 PRELIMINARY REMARKS

The main topics of this chapter are the calculation of significant normal and shear stresses at a point and the calculation of strain energy per unit volume. These quantities are important in *failure criteria* (Chapter 3), which predict whether a given state of stress is safe or will cause the material to yield or to fracture. In the present section we review stress definitions, notation, and sign conventions.

Consider a cutting plane A in a stressed body, Fig. 2.1-1a. Force increment ΔF acts on a typical area increment ΔA. In general, ΔF differs in both magnitude and direction for other locations of ΔA in plane A. In Fig. 2.1-1b, Cartesian coordinates xyz are established, with plane A in the xy plane. Force ΔF is resolved into components ΔF_x, ΔF_y, and ΔF_z in the coordinate directions. Normal stress on the xy plane is defined as

$$\sigma_z = \lim_{\Delta A \to 0} \frac{\Delta F_z}{\Delta A} = \frac{dF_z}{dA} \tag{2.1-1}$$

Similarly, shear stresses are $\tau_{zx} = dF_x/dA$ and $\tau_{zy} = dF_y/dA$. Additional cutting planes isolate an infinitesimal volume element, Fig. 2.1-1c, which displays additional stresses.

Stresses on hidden faces in Fig. 2.1-1c are assumed to have the same magnitudes as the stresses shown. For example, the hidden face normal to the x axis carries normal stress σ_x and shear stresses τ_{xz} and τ_{xy}. Thus, for the present study of "state of stress at a point" we neglect small changes in stress over the infinitesimal span of the volume element.

Dual subscripts on shear stresses τ have the following meanings: the first subscript indicates the axis that is normal to the plane on which τ acts; the second indicates the axis to which the τ arrow is parallel. Equilibrium of moments about the x axis in Fig. 2.1-1c requires that

$$(\tau_{yz} dx\, dz) dy - (\tau_{zy} dx\, dy) dz = 0 \tag{2.1-2}$$

Similarly we can write equations of moment equilibrium about y and z axes. The three moment equilibrium equations yield

Section 2.1 Preliminary Remarks

$$\tau_{yz} = \tau_{zy} \qquad \tau_{zx} = \tau_{xz} \qquad \tau_{xy} = \tau_{yx} \qquad (2.1\text{-}3)$$

Thus, at the intersection of two planes that meet at a right angle, shear stresses directed perpendicular to the line of intersection are equal, and shear arrows are either both directed away or both directed toward the line of intersection. Because of Eqs. 2.1-3, the order of subscripts on a shear stress does not matter. Accordingly, a general state of stress consists of six stresses, three normal and three shear, not the nine stresses seen in Fig. 2.1-1c.

Stresses are considered positive when they have the directions shown in Fig. 2.1-1c. That is, positive stresses have arrows pointing in positive directions on positive faces. Positive faces are those visible in Fig. 2.1-1c.

Stress transformation is the process of determining stresses that act on a plane of arbitrary orientation. At any point in a general state of stress, one can determine an orientation such that the plane is free of shear stress. There are three such planes, mutually perpendicular. These are *principal planes*, and normal stresses that act on them are *principal stresses*. One principal stress is the algebraically largest, another the algebraically smallest, of all normal stresses on planes of all orientations at the point in question. The *principal stress problem* asks that we determine these stresses and their orientations. This result is an alternative description of the state of stress rather than a different state of stress.

Equilibrium is the only physical consideration needed to solve the principal stress problem. Material properties are not used; therefore the material need not be linearly elastic or even solid. The only assumption of importance is that the body is a continuum or can be idealized as one by "smearing over" small-scale inhomogeneities. As examples, we ignore the crystalline structure of metal and the cemented-particle structure of concrete. Because the derivation of principal stress equations requires no other restrictive assumptions that must be remembered, derivations that follow are presented compactly. Elementary textbooks present detailed derivations for the two-dimensional principal stress problem.

Strain transformation is analogous to stress transformation. Its two-dimensional form is useful in experimental mechanics, to obtain the state of strain from strain gage readings. It is discussed in textbooks on elementary mechanics of materials.

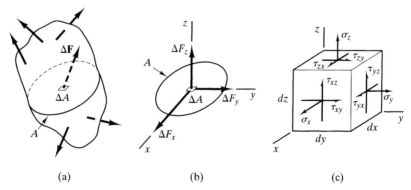

FIGURE 2.1-1 (a) Stressed body of arbitrary shape. (b) Components of a force $\Delta \mathbf{F}$ that acts on an increment of area ΔA. (c) Three-dimensional state of stress on an element of volume $dx\,dy\,dz$.

2.2 PRINCIPAL STRESSES: THEORY

The two-dimensional problem is reviewed first. The vector notation used here is not used in elementary textbooks, but the analytical process is the usual one of writing equations of static equilibrium, and we obtain the usual results. Vector notation is adopted because it permits a direct and easy extension from two dimensions to three.

Two Dimensions. In a two-dimensional principal stress problem, one principal plane and its principal stress are already known. In Fig. 2.2-1a, let the plane parallel to the paper carry no shear stress. It is therefore a principal plane. The principal (normal) stress on this plane may have any value. Often this normal stress is zero, because points of interest often lie on the surface of a body at locations where there are no externally applied normal forces.

In Fig. 2.2-1b the orientation of a face normal to the xy plane, but otherwise arbitrarily oriented, is defined by angle θ between its normal and the x axis. Lengths of perpendicular faces of the triangular element are related to the length of the hypotenuse by the sine and cosine of angle θ. Hence, taking the element thickness as constant, we conclude that if the face normal to the paper along the hypotenuse has area dA, then the other two faces normal to the paper have areas $\ell\, dA$ and $m\, dA$, as shown. Forces acting on the x-normal face are $\sigma_x \ell\, dA$ leftward and $\tau_{xy} \ell\, dA$ downward. Forces acting on the y-normal face are $\tau_{xy} m\, dA$ leftward and $\sigma_y m\, dA$ downward. Force $d\mathbf{R}$, on the inclined face, has components dR_x rightward and dR_y upward. Thus, force vectors on the three faces are

$$d\mathbf{R}_a = -\sigma_x \ell\, dA\, \mathbf{i} - \tau_{xy} \ell\, dA\, \mathbf{j}$$
$$d\mathbf{R}_b = -\tau_{xy} m\, dA\, \mathbf{i} - \sigma_y m\, dA\, \mathbf{j} \qquad (2.2\text{-}1)$$
$$d\mathbf{R} = dR_x \mathbf{i} + dR_y \mathbf{j}$$

where $\mathbf{i}$ and $\mathbf{j}$ are unit vectors in x and y directions respectively. (Following convention, we use boldface type to denote a vector quantity.) The element is in equilibrium if *forces* (not stresses!) sum to zero. That is, $d\mathbf{R} + d\mathbf{R}_a + d\mathbf{R}_b = \mathbf{0}$, which yields

$$dR_x = (\sigma_x \ell + \tau_{xy} m)dA \qquad \text{and} \qquad dR_y = (\tau_{xy}\ell + \sigma_y m)dA \qquad (2.2\text{-}2)$$

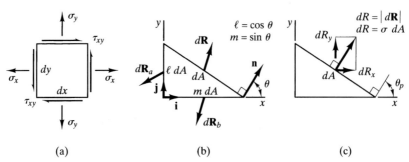

(a) (b) (c)

FIGURE 2.2-1 (a) Plane state of stress. (b) Force vectors that act on an arbitrary differential element. (c) Components of a force $d\mathbf{R}$ on a principal face.

We now ask that the inclined face be a principal face; that is, with normal stress σ but without shear stress. Accordingly, as shown in Fig. 2.2-1c,

$$dR_x = \ell(\sigma\, dA) \qquad dR_y = m(\sigma\, dA) \tag{2.2-3}$$

Equating dR_x and dR_y expressions in Eqs. 2.2-2 and 2.2-3, we obtain

$$\begin{aligned}\ell(\sigma_x - \sigma) + m\tau_{xy} &= 0 \\ \ell\tau_{xy} + m(\sigma_y - \sigma) &= 0\end{aligned} \quad \text{or} \quad \begin{bmatrix} \sigma_x - \sigma & \tau_{xy} \\ \tau_{xy} & \sigma_y - \sigma \end{bmatrix}\begin{Bmatrix} \ell \\ m \end{Bmatrix} = \begin{Bmatrix} 0 \\ 0 \end{Bmatrix} \tag{2.2-4}$$

Equations 2.2-4 have the trivial solution $\ell = m = 0$. But this result is not possible; there is no angle in the xy plane for which $\ell = m = 0$. Therefore, according to the theory of algebraic equations, a solution of Eq. 2.2-4 is possible only if the determinant of the coefficient matrix vanishes:

$$\begin{vmatrix} \sigma_x - \sigma & \tau_{xy} \\ \tau_{xy} & \sigma_y - \sigma \end{vmatrix} = 0 \quad \text{hence} \quad \sigma = \frac{\sigma_x + \sigma_y}{2} \pm \sqrt{\left(\frac{\sigma_x - \sigma_y}{2}\right)^2 + \tau_{xy}^2} \tag{2.2-5}$$

The latter result, which the reader should verify, is obtained by expansion of the determinant. The same result is obtained in an alternative way in an elementary course in mechanics of materials. The alternative way shows that the two values of σ in Eq. 2.2-5 are indeed a maximum and a minimum, and also yields a formula for the orientation of principal planes, namely $\theta_p = 0.5\,\text{arc tan}\,[2\tau_{xy}/(\sigma_x - \sigma_y)]$. A formula for the counterclockwise angle θ_p from the x axis to the direction of the maximum principal stress in the xy plane, here called $\sigma_{xy\text{max}}$, is $\tan\theta_p = (\sigma_{xy\text{max}} - \sigma_x)/\tau_{xy}$ [2.1]. This formula comes from Eq. 2.2-4 with $\sigma = \sigma_{xy\text{max}}$, $\ell = \cos\theta_p$, and $m = \sin\theta_p$.

Three Dimensions. The foregoing analysis is easily extended to three dimensions by adding terms to equations and accounting for force components parallel to the z axis. The appropriate differential element is the tetrahedron shown in Fig. 2.2-2. It is a corner sliced off the cube in Fig. 2.1-1c by a cutting plane. The orientation of the cutting plane is arbitrary and is described by its unit normal vector **n**.

$$\mathbf{n} = \ell\mathbf{i} + m\mathbf{j} + n\mathbf{k} \tag{2.2-6}$$

where **i**, **j**, and **k** are unit vectors shown in Fig. 2.2-2, and ℓ, m, and n are direction cosines of **n** with respect to x, y, and z directions respectively. If the inclined face has area dA, then faces normal to x, y, and z axes have the respective areas $\ell\, dA$, $m\, dA$, and $n\, dA$. In Eqs. 2.2-1, the change from two dimensions to three requires that additional forces appear on the right hand sides: $-\tau_{zx}\,\ell\, dA\,\mathbf{k}$ in the first equation, $-\tau_{yz}\,m\, dA\,\mathbf{k}$ in the second equation, and $+dR_z\mathbf{k}$ in the third equation. Also, a force $d\mathbf{R}_c = -\tau_{zx}\, n\, dA\,\mathbf{i} - \tau_{yz}\, n\, dA\,\mathbf{j} - \sigma_z\, n\, dA\,\mathbf{k}$ acts on the face normal to the z axis. The equilibrium equation is $d\mathbf{R} + d\mathbf{R}_a + d\mathbf{R}_b + d\mathbf{R}_c = \mathbf{0}$. Equations 2.2-2 become, for the three-dimensional case,

$$\begin{aligned} dR_x &= (\sigma_x\ell + \tau_{xy}m + \tau_{zx}n)dA \\ dR_y &= (\tau_{xy}\ell + \sigma_y m + \tau_{yz}n)dA \\ dR_z &= (\tau_{zx}\ell + \tau_{yz}m + \sigma_z n)dA \end{aligned} \tag{2.2-7}$$

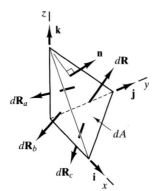

FIGURE 2.2-2 Arbitrary differential element in a three-dimensional state of stress, with unit normal vector **n** on oblique face.

If the inclined face is principal, force $d\mathbf{R}$ on the inclined face is parallel to unit normal **n** because the face carries normal stress σ but no shear stress. Components of $d\mathbf{R}$, analogous to components in Eqs. 2.2-3, are then

$$dR_x = \ell(\sigma\, dA) \qquad dR_y = m(\sigma\, dA) \qquad dR_z = n(\sigma\, dA) \qquad (2.2\text{-}8)$$

From Eqs. 2.2-7 and 2.2-8, we obtain

$$\begin{bmatrix} \sigma_x - \sigma & \tau_{xy} & \tau_{zx} \\ \tau_{xy} & \sigma_y - \sigma & \tau_{yz} \\ \tau_{zx} & \tau_{yz} & \sigma_z - \sigma \end{bmatrix} \begin{Bmatrix} \ell \\ m \\ n \end{Bmatrix} = \begin{Bmatrix} 0 \\ 0 \\ 0 \end{Bmatrix} \qquad (2.2\text{-}9)$$

which is analogous to Eqs. 2.2-4. The trivial solution $\ell = m = n = 0$ is not possible because direction cosines satisfy the relation $\ell^2 + m^2 + n^2 = 1$. Therefore the determinant of the 3×3 matrix in Eq. 2.2-9 must vanish. This condition yields three roots, σ_1, σ_2, and σ_3, which are the three principal stresses. We adopt the following convention for the algebraic ordering of principal stresses:

$$\left.\begin{array}{ll} \text{Maximum:} & \sigma_1 \\ \text{Intermediate:} & \sigma_2 \\ \text{Minimum:} & \sigma_3 \end{array}\right\} \text{ thus } \quad \sigma_1 \geq \sigma_2 \geq \sigma_3 \qquad (2.2\text{-}10)$$

For example, we might have $\sigma_1 = 30$ MPa, $\sigma_2 = -20$ MPa, and $\sigma_3 = -40$ MPa. If any two of the three principal stresses are equal, whether they are positive or negative, the state of stress is called hydrostatic in a plane. Principal directions are not unique in this plane. Nor are they unique if $\sigma_1 = \sigma_2 = \sigma_3$, which is an entirely hydrostatic state in which *all* directions are principal; that is, in an entirely hydrostatic state there is no shear stress on any plane, however oriented.

Even when a problem is regarded as two dimensional, as in Fig. 2.2-1, there are always three principal stresses. The principal stress normal to the analysis plane may be σ_1, σ_2, or σ_3. The ordering cannot be assigned until all three principal stresses are known.

2.3 PRINCIPAL STRESSES: CALCULATIONS

Solution Methods. For the three-dimensional principal stress problem, all solution methods are analytical. There is no graphical solution procedure analogous to the Mohr circle method used for plane problems.

One analytical solution method treats the principal stress problem as an eigenvalue problem. Readers familiar with matrices and linear algebra will recognize Eq. 2.2-9 as a standard eigenvalue problem. In this case, there are three eigenvalues, namely principal stresses σ_i for $i = 1,2,3$. To each σ_i there corresponds an eigenvector, namely direction cosines $\{\ell_i\, m_i\, n_i\}$, which define the orientation of the normal to the ith principal plane. This normal and the σ_i arrow have the same direction cosines. Standard mathematical software can solve an eigenvalue problem. The software user need only follow instructions and enter data (with care!).

Other analytical solution methods address the cubic equation obtained by setting to zero the determinant of the 3×3 matrix in Eq. 2.2-9.

$$\sigma^3 - I_1\sigma^2 + I_2\sigma - I_3 = 0 \qquad (2.3\text{-}1)$$

where

$$\begin{aligned}
I_1 &= \sigma_x + \sigma_y + \sigma_z \\
I_2 &= \sigma_x\sigma_y + \sigma_y\sigma_z + \sigma_z\sigma_x - \tau_{xy}^2 - \tau_{yz}^2 - \tau_{zx}^2 \\
I_3 &= \sigma_x\sigma_y\sigma_z + 2\tau_{xy}\tau_{yz}\tau_{zx} - \sigma_x\tau_{yz}^2 - \sigma_y\tau_{zx}^2 - \sigma_z\tau_{xy}^2
\end{aligned} \qquad (2.3\text{-}2)$$

Quantities I_1, I_2, and I_3 are known as *stress invariants* because the numerical value of each is independent of how coordinates xyz are oriented relative to the physical body. If another set of mutually orthogonal axes is established, with axes coincident with principal stress directions, then

$$\begin{aligned}
I_1 &= \sigma_1 + \sigma_2 + \sigma_3 \\
I_2 &= \sigma_1\sigma_2 + \sigma_2\sigma_3 + \sigma_3\sigma_1 \\
I_3 &= \sigma_1\sigma_2\sigma_3
\end{aligned} \qquad (2.3\text{-}3)$$

At a given point, regardless of what state of stress prevails at that point, Eqs. 2.3-2 and 2.3-3 yield the same numerical value for each of the three invariants.

Equation 2.3-1 can be solved for its roots (the principal stresses) by closed form expressions and by special Fortran programs [2.2, 2.3]. Alternatively, one can use general mathematical software or a versatile calculator. Or, one can use a *simple* calculator, with trial and/or iteration, as follows.

Equation 2.3-1 can be written in the iterative form

$$\sigma_{i+1} = [I_1\sigma_i^2 - I_2\sigma_i + I_3]^{1/3} \qquad (2.3\text{-}4)$$

Here $i = 1, 2, 3, \ldots, n$ refers to the trial or iteration number, *NOT* to the ranking of principal stresses in Eq. 2.2-10. For the first iteration, $i = 1$, we guess an approximate numerical value σ_1 and solve for σ_2, which is a better approximation of the root. Next we substitute σ_2 and calculate σ_3, which is still better. The process repeats until convergence. The number of iterations can be reduced by "jumping ahead" when the trend of iterates is known. When we obtain $\sigma_{i+1} = \sigma_i$, by iteration or by good guessing, we have obtained a root, which is one of the principal stresses. These calculations, as well as the calculation of principal stress directions, are illustrated as follows.

EXAMPLE 1

Determine principal stresses associated with the following state of stress. (Negative stresses are represented by arrows reversed from directions shown in Fig. 2.1-1c.)

$$\sigma_x = 20 \text{ MPa} \qquad \sigma_y = 30 \text{ MPa} \qquad \sigma_z = -10 \text{ MPa}$$
$$\tau_{xy} = 40 \text{ MPa} \qquad \tau_{yz} = 25 \text{ MPa} \qquad \tau_{zx} = -30 \text{ MPa} \tag{2.3-5}$$

Equations 2.3-2 provide $I_1, I_2,$ and I_3. Equation 2.3-4 becomes

$$\sigma_{i+1} = [40\,\sigma_i^2 + 3025\,\sigma_i - 89{,}500]^{1/3} \tag{2.3-6}$$

As the initial guess for one of the principal stresses, we arbitrarily adopt the largest number in Eqs. 2.3-5, even though it is associated with a shear stress. Iterates are

i	1	2	3	[restart]	1	2	
σ_i	40.0	45.7	51.0		70.0	68.3	(2.3-7)
σ_{i+1}	45.7	51.0	55.2		68.3	67.2	

where the restart value of 70.0 is a guess based on the previous trend of results. The converged value is 65.3 MPa. A similar iterative sequence for another principal stress, based on -30.0 as the initial guess, converges to -51.8 MPa. Finally, using the first of Eqs. 2.3-3 with $I_1 = 40.0$ MPa, we obtain the third principal stress $40.0 - 65.3 - (-51.8) = 26.5$ MPa. Ranked according to Eq. 2.2-10, the principal stresses are therefore

$$\sigma_1 = 65.3 \text{ MPa} \qquad \sigma_2 = 26.5 \text{ MPa} \qquad \sigma_3 = -51.8 \text{ MPa} \tag{2.3-8}$$

Principal stress directions may not be of interest. When they are, they can be calculated as illustrated by the following example.

EXAMPLE 2

We seek principal stress directions for the state of stress shown in Fig. 2.3-1a. The problem is two-dimensional because one of the principal planes and its principal stress are already known. Therefore Eq. 2.2-5 is applicable; it yields 80 MPa and -20 MPa, which happen to be σ_1 and σ_2 respectively. The third principal stress is $\sigma_3 = -120$ MPa. When σ_1 and the given values of $\sigma_x, \sigma_y,$ and τ_{xy} are substituted into Eq. 2.2-9, there results

$$\begin{bmatrix} -90 & -30 & 0 \\ -30 & -10 & 0 \\ 0 & 0 & -200 \end{bmatrix} \begin{Bmatrix} \ell_1 \\ m_1 \\ n_1 \end{Bmatrix} = \begin{Bmatrix} 0 \\ 0 \\ 0 \end{Bmatrix} \tag{2.3-9}$$

The last of these three equations yields $n_1 = 0$, which means that σ_1 has no component parallel to the z axis. Either of the first two equations yields $m_1 = -3\ell_1$. Also, we know that $\ell_1^2 + m_1^2 + n_1^2 = 1$. Hence, $\ell_1 = 0.316$ and $m_1 = -0.949$ (or $\ell_1 = -0.316$ and $m_1 = 0.949$). The two sets of val-

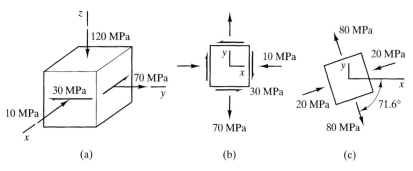

FIGURE 2.3-1 (a) A two-dimensional principal stress problem. (b) The same problem, represented conventionally. (c) Principal stresses in the xy plane.

ues for ℓ_1 and m_1 indicate the directions of the two 80 MPa arrows in Fig. 2.3-1c. The angle 71.6° in Fig. 2.3-1c is associated with the first of these two sets.

In general, all three direction cosines of a principal stress σ_i may have values other than 0 and 1. Then, when Eq. 2.2-9 is written numerically using a principal stress, one can solve for direction cosines of that principal stress by assigning a numerical value to any direction cosine (provided its correct value is not zero), solving for the other two from any two of the three equations, and finally scaling ℓ_i, m_i, and n_i so that the sum of their squares is unity. This process must be done for two of the three principal stress directions. Since σ_1, σ_2, and σ_3 are mutually perpendicular, the third direction can be established from the other two by taking the cross product of their unit vectors.

2.4 OCTAHEDRAL AND MAXIMUM SHEAR STRESS

Octahedral Stresses. By definition, an *octahedral plane* is a plane that makes equal angles with principal stress directions. *Octahedral shear stress* is shear stress on an octahedral plane. It is used in failure criteria (Chapter 3).

Let axes x, y, and z be parallel to the directions of principal stresses σ_1, σ_2, and σ_3 respectively (Fig. 2.4-1a). This choice of axes is made for convenience in deriving an expression for octahedral shear stress, as follows. Consider force $d\mathbf{R}$ on an arbitrarily oriented plane, as in Fig. 2.2-2.

$$d\mathbf{R} = \sigma_1(\ell\, dA)\mathbf{i} + \sigma_2(m\, dA)\mathbf{j} + \sigma_3(n\, dA)\mathbf{k} \tag{2.4-1}$$

in which $\ell\, dA$ is the area of the face that carries σ_1, and so on. Components of $d\mathbf{R}$ normal and tangent to the inclined plane are shown in Fig. 2.4-1b. These components provide normal and shear stresses on the inclined plane as follows.

$$\sigma_n = \frac{d\mathbf{R}\cdot\mathbf{n}}{dA} \quad \text{and} \quad \tau_s = \frac{dR_s}{dA} = \frac{1}{dA}\sqrt{dR^2 - dR_n^2} \tag{2.4-2}$$

Unit normal $\mathbf{n}$ is defined by Eq. 2.2-6. If the inclined plane is the octahedral plane, Fig. 2.4-1a, then direction cosines ℓ, m, and n of $\mathbf{n}$ are equal ($\ell^2 = m^2 = n^2 = 1/3$),

36 Chapter 2 Stress, Principal Stresses, Strain Energy

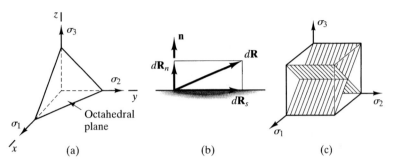

FIGURE 2.4-1 (a) The octahedral plane. (b) Normal and shear components of force $d\mathbf{R}$ on an arbitrary plane, viewed normal to the plane of $\mathbf{n}$ and $d\mathbf{R}$. (c) Planes that carry the absolute maximum shear stress τ_{max}.

and stresses in Eq. 2.4-2 become σ_{oct} and τ_{oct}. Octahedral normal stress is $\sigma_{oct} = (\sigma_1 + \sigma_2 + \sigma_3)/3$. Octahedral shear stress is

$$\tau_{oct} = \frac{1}{3}[(\sigma_1 - \sigma_2)^2 + (\sigma_2 - \sigma_3)^2 + (\sigma_3 - \sigma_1)^2]^{1/2} \tag{2.4-3}$$

The direction of τ_{oct} on the octahedral plane is not of interest.

By using the invariants in Eq. 2.3-3, one can show that

$$\tau_{oct} = \frac{1}{3}[2I_1^2 - 6I_2]^{1/2} \tag{2.4-4}$$

But I_1 and I_2 have the same numerical values in any other set of mutually orthogonal coordinates. Therefore we may substitute I_1 and I_2 from Eqs. 2.3-2 into Eq. 2.4-4, with the result

$$\tau_{oct} = \frac{1}{3}[(\sigma_x - \sigma_y)^2 + (\sigma_y - \sigma_z)^2 + (\sigma_z - \sigma_x)^2 + 6(\tau_{xy}^2 + \tau_{yz}^2 + \tau_{zx}^2)]^{1/2} \tag{2.4-5}$$

for which x, y, and z need not be principal stress directions. Equations 2.4-3 and 2.4-5 yield the same value of τ_{oct}, but the latter form does not require that principal stresses be determined in advance. As an exercise, the reader should verify that $\tau_{oct} = 48.7$ MPa for the first numerical example in Section 2.3, whether one uses Eqs. 2.3-5 and 2.4-5 or Eqs. 2.3-8 and 2.4-3.

Maximum Shear Stress. The largest shear stress to be found on any plane containing the point in question is

$$\tau_{max} = \frac{\sigma_1 - \sigma_3}{2} \tag{2.4-6}$$

Stress τ_{max} acts on planes inclined at 45° to the directions of both σ_1 and σ_3, as shown in Fig 2.4-1c. Note that σ_1 and σ_3 are the *extreme* principal stresses; intermediate principal stress σ_2 is not used in the calculation of τ_{max}. Algebraic signs must be observed in the calculation. For example, using Eqs. 2.3-8, $\tau_{max} = [65.3 - (-51.8)]/2 = 58.6$ MPa.

Invariably, $\tau_{max} > \tau_{oct}$ (unless both are zero). The radical in Eq. 2.2-5 may be called τ_{max} in elementary textbooks, but it is only the "in plane" τ_{max}. If "in plane" principal stresses calculated by Eq. 2.2-5 are not the extreme principal stresses σ_1 and σ_3, then the radical is not the absolute maximum shear stress.

2.5 STRESS–STRAIN–TEMPERATURE RELATIONS

Actual structural materials are usually composed of crystals, fibers, or cemented particles, so they are not homogeneous on a small scale. Directional bias may result from rolling in the manufacture of metal, or from alignment of fibers in wood and composite plastics. Thus a material is often anisotropic to some degree. An elastic material is not necessarily *linearly* elastic: Cast iron and concrete display stress-strain plots that are slightly curved thoughout. Indeed both display permanent set after removing the first loading, and have truly elastic behavior up to a certain stress only after several cycles of loading up to that stress. Despite these complications, for most analyses of machine parts and structures it is satisfactory to assume that the material is homogeneous and linearly elastic, and employ stress-strain-temperature relations as follows.

Isotropy. In three dimensions, for a material that is isotropic and linearly elastic,

$$\epsilon_x = \frac{1}{E}(+\sigma_x - \nu\sigma_y - \nu\sigma_z) + \alpha\,\Delta T \qquad \gamma_{xy} = \frac{\tau_{xy}}{G}$$

$$\epsilon_y = \frac{1}{E}(-\nu\sigma_x + \sigma_y - \nu\sigma_z) + \alpha\,\Delta T \qquad \gamma_{yz} = \frac{\tau_{yz}}{G} \qquad (2.5\text{-}1)$$

$$\epsilon_z = \frac{1}{E}(-\nu\sigma_x - \nu\sigma_y + \sigma_z) + \alpha\,\Delta T \qquad \gamma_{zx} = \frac{\tau_{zx}}{G}$$

where E is the elastic modulus, G is the shear modulus, ν is Poisson's ratio, α is the coefficient of thermal expansion, and ΔT is temperature change relative to a reference temperature at which stresses are zero. Subscripts indicate coordinates that are orthogonal but not necessarily rectilinear. For example, x, y, and z might be replaced by cylindrical coordinates r, θ, and z. Note that if material properties are temperature-dependent and ΔT is a function of the coordinates, then the material is inhomogeneous in the sense that E, G, ν, and α become functions of the coordinates. Also, $\alpha\,\Delta T$ must then be replaced by the integral of $\alpha\,dT$ from 0 to ΔT.

For a *plane* state of stress in (say) the xy plane, $\sigma_z = \tau_{yz} = \tau_{zx} = 0$, and Eqs. 2.5-1 simplify. The results, in forms explicit in strains and explicit in stresses, are

$$\epsilon_x = \frac{1}{E}(\sigma_x - \nu\sigma_y) + \alpha\,\Delta T \qquad \sigma_x = \frac{E}{1-\nu^2}(\epsilon_x + \nu\epsilon_y) - \frac{E\alpha\,\Delta T}{1-\nu}$$

$$\epsilon_y = \frac{1}{E}(\sigma_y - \nu\sigma_x) + \alpha\,\Delta T \qquad \sigma_y = \frac{E}{1-\nu^2}(\epsilon_y + \nu\epsilon_x) - \frac{E\alpha\,\Delta T}{1-\nu} \qquad (2.5\text{-}2)$$

$$\gamma_{xy} = \frac{1}{G}\tau_{xy} \qquad\qquad\qquad\qquad \tau_{xy} = G\gamma_{xy}$$

The latter set of normal stress equations cannot be extrapolated to three dimensions simply by adding terms. For example, the correct expression for σ_x in three dimensions is

$$\sigma_x = \frac{E}{(1+\nu)(1-2\nu)}[(1-\nu)\epsilon_x + \nu\epsilon_y + \nu\epsilon_z] - \frac{E\alpha \Delta T}{1-2\nu} \qquad (2.5\text{-}3)$$

In elementary mechanics of materials, it is shown that E, G, and ν are not independent. They satisfy the relation

$$G = \frac{E}{2(1+\nu)} \qquad (2.5\text{-}4)$$

Poisson's ratio ν is little affected by temperature. Modulus E is affected more: For stainless steel, E decreases about 20% if the temperature rises from 0°C to 450°C. Similarly, α may vary appreciably with temperature. E, G, ν, and α are almost independent of stress. For example, a 350 MPa hydrostatic compression increases the moduli of steel and aluminum about 0.8% and 2.6%, respectively. Moduli are increased by high strain rates, sometimes appreciably for rubberlike materials but negligibly for metals unless strain rates are extreme, as in an explosive forming process. Wood is somewhat sensitive to strain rate and quite sensitive to humidity.

Anisotropy. Anisotropy must be considered for various materials, such as fiber-reinforced composites. The stress-strain-temperature relation for any linearly elastic material, whether isotropic or anisotropic, can be written in matrix form as

$$\{\epsilon\} = [C]\{\sigma\} + \{\epsilon_o\} \qquad (2.5\text{-}5)$$

where $\{\epsilon\}$ contains the six strains, $\{\sigma\}$ contains the corresponding six stresses, $\{\epsilon_o\}$ contains the "initial" or thermal strains, and $[C]$ is a 6×6 matrix of material constants. For an *isotropic* material, normal strains and stresses are not coupled to shear strains and stresses; $[C]$ contains 24 zero terms, and 12 nonzero terms that depend on E, G, and ν. Thus, for an isotropic material, $[C]$ contains only two independent material constants (see Eq. 2.5-4). For a general *anisotropic* material, $[C]$ is a full matrix, which means that all the strains are coupled to all the stresses. Because of symmetry, $[C]$ contains 21 independent material constants (not 36) for a general anisotropic material. For an *orthotropic* material, defined in Section 1.2, there are only nine independent constants. If these nine constants are written in the form of elastic moduli and Poisson ratios, it is possible for some of the Poisson ratios to exceed 1/2, which is not possible for an *isotropic* material.

Extensions. Sometimes elastic constants can be contrived so as to simplify the analysis of a structure that has a regular pattern of geometric complexities, such as a perforated plate. A perforated plate contains a regular pattern of circular holes, arrayed in a square or a triangular pattern. By using "effective" elastic constants, whose values depend on the size of the holes relative to their spacing, the actual plate can be replaced by a substitute plate without holes. Stresses in the actual plate can be determined from stresses in the substitute plate by means of data provided [2.4].

2.6 STRAIN ENERGY DENSITY

"Strain energy density," denoted by U_o, is strain energy per unit volume. It plays a role in determining whether the associated state of stress will cause a material to fail (Chapter 3). U_o can be integrated over the volume of a structure to produce the total strain energy U in a structure. The use of U in structural and stress analyses is explained in Chapter 4.

Consider first a linear spring of stiffness k. Let the spring be stretched by a force f that increases from zero to a final value F. Simultaneously, displacement u of the force increases from zero to a final value D. Because the spring is linear, $f = ku$ and $F = kD$. Work done by the force in stretching the spring to displacement D is

$$U = \int_0^D f\, du = \int_0^D ku\, du = \frac{kD^2}{2} \quad \text{or} \quad U = \frac{F^2}{2k} \qquad (2.6\text{-}1)$$

where the latter form comes from the substitution $D = F/k$. Work done is stored as strain energy U.

Any linearly elastic structure loaded by a single force (or moment) can be represented as a linear spring of stiffness k. For example, consider a uniform cantilever beam of length L, tip-loaded by a transverse force P. Transverse tip displacement is $v_L = PL^3/3EI$. By definition, stiffness is force divided by displacement of the loaded point. Hence, $k = P/v_L = 3EI/L^3$. If instead the load were tip moment M_L, the beam could be represented as a linear spring of rotational stiffness $k = M_L/\theta_L = EI/L$. In cases where the loaded point does not displace parallel to the load, the component of displacement parallel to the load is used to calculate stiffness.

Simple States of Stress. Consider a cube of isotropic and linearly elastic material under uniaxial stress σ_x. Such a cube is viewed normal to an unloaded face in Fig. 2.6-1a. The cube can be represented as a linear spring of stiffness k. If each side of the cube has unit length, then $F = \sigma_x$, $D = \epsilon_x$, $k = \sigma_x/\epsilon_x = E$, and Eq. 2.6-1 yields the strain energy density

$$U_o = \frac{E\epsilon_x^2}{2} \quad \text{or} \quad U_o = \frac{\sigma_x^2}{2E} \qquad (2.6\text{-}2)$$

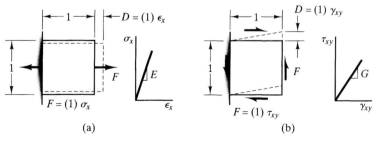

FIGURE 2.6-1 Unit cubes of linearly elastic material, with deformations shown by dashed lines. (a) Uniaxial stress σ_x. (b) Shear stress τ_{xy}.

40 Chapter 2 Stress, Principal Stresses, Strain Energy

Here $U = U_o$ because the structure has unit volume and is uniformly stressed. Similarly, for a state of pure shear, Fig. 2.6-1b,

$$U_o = \frac{G\gamma_{xy}^2}{2} \quad \text{or} \quad U_o = \frac{\tau_{xy}^2}{2G} \tag{2.6-3}$$

Multiaxial States of Stress. Consider first an isotropic and linearly elastic body in a state of plane stress. Nonzero stresses are σ_x, σ_y, and τ_{xy}. Let these stresses be applied one after another. If σ_x is applied first, with σ_y and τ_{xy} both zero, U_o is given by Eq. 2.6-2. If σ_y is now added, it produces the strains

$$\epsilon_y = \frac{\sigma_y}{E} \quad \text{and} \quad \epsilon_x = -\nu\epsilon_y = -\nu\frac{\sigma_y}{E} \tag{2.6-4}$$

and the following contribution to U_o.

$$\int_0^{\epsilon_y} \sigma_y \, d\epsilon_y + \sigma_x \epsilon_x = \frac{\sigma_y^2}{2E} - \nu\frac{\sigma_x \sigma_y}{E} \tag{2.6-5}$$

No integration is needed to obtain $\sigma_x \epsilon_x$ in Eq. 2.6-5 because σ_x remains constant as x-direction strain is produced by σ_y. Another contribution to U_o comes from Eq. 2.6-3. The final result for a state of plane stress, from Eqs. 2.6-2, 2.6-3, and 2.6-5, is

$$U_o = \frac{1}{2E}[\sigma_x^2 + \sigma_y^2 - 2\nu\sigma_x\sigma_y] + \frac{\tau_{xy}^2}{2G} \tag{2.6-6}$$

The foregoing argument can be extended to the fully three-dimensional case by adding the stresses σ_z, τ_{yz}, and τ_{zx}. The result is

$$U_o = \frac{1}{2E}[\sigma_x^2 + \sigma_y^2 + \sigma_z^2 - 2\nu(\sigma_x\sigma_y + \sigma_y\sigma_z + \sigma_z\sigma_x)]$$
$$+ \frac{1}{2G}[\tau_{xy}^2 + \tau_{yz}^2 + \tau_{zx}^2] \tag{2.6-7}$$

Strain Energy of Distortion. In an arbitrary state of stress, the average normal stress is

$$\sigma_a = \frac{1}{3}(\sigma_x + \sigma_y + \sigma_z) = \frac{1}{3}(\sigma_1 + \sigma_2 + \sigma_3) \tag{2.6-8}$$

which, incidentally, is the normal stress on an octahedral plane. *Deviatoric stresses* are given the symbol s and are defined as follows.

$$s_x = \sigma_x - \sigma_a \quad s_y = \sigma_y - \sigma_a \quad s_z = \sigma_z - \sigma_a$$
$$s_{xy} = \tau_{xy} \quad s_{yz} = \tau_{yz} \quad s_{zx} = \tau_{zx} \tag{2.6-9}$$

An arbitrary state of stress can be represented as the sum of two states: (1) a hydrostatic state in which principal stresses are $\sigma_1 = \sigma_2 = \sigma_3 = \sigma_a$, and (2) a state in which all stresses are deviatoric. No change of shape is produced by the hydrostatic state. No change of volume is produced by the deviatoric state.

Strain energy of distortion, associated with the deviatoric state, can be used to predict the onset of yielding. An expression for the strain energy of distortion per unit volume, U_{od}, is obtained from Eq. 2.6-7 by replacing the written stresses by the deviatoric stresses. Also, we eliminate E by using Eq. 2.5-4. Thus

$$U_{od} = \frac{1}{12G}[(\sigma_x - \sigma_y)^2 + (\sigma_y - \sigma_z)^2 + (\sigma_z - \sigma_x)^2 + 6(\tau_{xy}^2 + \tau_{yz}^2 + \tau_{zx}^2)] \quad (2.6\text{-}10)$$

Alternative expressions for U_{od} are

$$U_{od} = \frac{3}{4G}\tau_{oct}^2 \quad \text{and} \quad U_{od} = \frac{1}{6G}\sigma_e^2 \quad (2.6\text{-}11)$$

where τ_{oct} is given by Eq. 2.4-3, 2.4-4, or 2.4-5, and σ_e is an "effective" stress defined as

$$\sigma_e = \frac{1}{\sqrt{2}}[(\sigma_x - \sigma_y)^2 + (\sigma_y - \sigma_z)^2 + (\sigma_z - \sigma_x)^2 + 6(\tau_{xy}^2 + \tau_{yz}^2 + \tau_{zx}^2)]^{1/2} \quad (2.6\text{-}12)$$

From Eqs. 2.4-5 and 2.6-12, $\tau_{oct} = \sqrt{2}\,\sigma_e/3$. Conveniently, σ_e reduces to $\sigma_e = \sigma_x$ if σ_x is a uniaxial state of stress. If the state of stress is hydrostatic, then $\sigma_e = 0$. It is possible that $\sigma_e > \sigma_1$: for example, if principal stresses are $\sigma_1 = -\sigma_3$ and $\sigma_2 = 0$, then $\sigma_e = \sqrt{3}\sigma_1$. The von Mises failure criterion is often stated in terms of σ_e.

2.7 STRESS CONCENTRATION

Stress in a solid is rarely uniform. It rises to local peaks because of material inhomogeneity or abrupt changes in geometry. Material inhomogeneities include crystal boundaries in metal, small inclusions of foreign material, small voids, the various constituents of concrete, and the cell structure of wood. Unintentional and random changes in geometry include tool marks and surface scratches. *Intentional* changes in geometry are common, such as threads on a bolt, teeth on a gear, an oil hole, and a keyway in a shaft. Peak stress associated with a change in geometry is easy to calculate if the geometry is accurately known and the associated stress concentration factor has been tabulated.

Stress Concentration Factors. Consider a central circular hole in a plate under tension, Fig. 2.7-1a. It is obvious that stress must be greater than σ_o somewhere on a cross section containing the hole, because there the axial force $P = \sigma_o Dt$ must be carried by a reduced area. However, the mechanics of materials method cannot provide a formula for σ_{max} because we have no reliable way of predicting the geometry of deformation. The problem can be solved (with difficulty) by the theory of elasticity method. Most stress concentration problems are too complicated for either method. Many have been solved experimentally. Results have been tabulated for an isotropic and linearly elastic material, in the form of *stress concentration factors* K_t. Using them is simple. For the problem in Fig. 2.7-1a,

$$\sigma_{max} = K_t \sigma_{nom} \quad \text{where} \quad \sigma_{nom} = \frac{P}{A_{net}} = \frac{P}{(D - 2r)t} \quad (2.7\text{-}1)$$

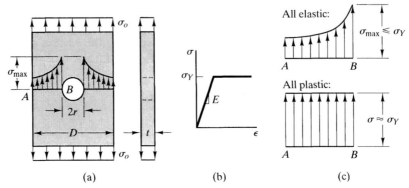

FIGURE 2.7-1   (a) Stress distribution in a plate with a circular hole. (b) Idealized stress-strain relation. (c) Approximate stress distributions for linearly elastic and fully plastic conditions.

K_t depends on D and r. (It would also depend on material properties if the material were anisotropic.) Note that nominal stress σ_{nom} is based on the "net" section. In general, for a body of finite dimensions, σ_{nom} is based on the smallest cross section adjacent to the point of highest stress. For other geometries, and for other loads such as bending and twisting, tabulated data provide numerical values for K_t [1.6, 1.7, 2.5]. For the particular case in Fig. 2.7-1a [1.6],

$$K_t = 3.00 - 3.13\left(\frac{2r}{D}\right) + 3.66\left(\frac{2r}{D}\right)^2 - 1.53\left(\frac{2r}{D}\right)^3 \qquad (2.7\text{-}2)$$

This result was obtained by fitting a curve to many experimental data.

A stress concentration may almost disappear if large load is applied and the material behavior is ductile. If the stress-strain relation is idealized as shown in Fig. 2.7-1b, increasing static load on the plate eventually causes all material on the net section to yield, so that the stress distribution is almost uniform (Fig. 2.7-1c). Thus there is almost no concentration of stress, although axial *strain* at B is approximately K_t times its value at A if $r \ll D$. The axial load P_Y that causes yielding to impend, and the "fully plastic" axial load P_{fp} that causes complete yielding, are respectively

$$P_Y = \sigma_{nom}(D - 2r)t = \frac{\sigma_Y}{K_t}(D - 2r)t \quad \text{and} \quad P_{fp} = \sigma_Y(D - 2r)t \qquad (2.7\text{-}3)$$

(The ratio of these two loads is $P_{fp}/P_Y = K_t$.) Stress concentration may not be of great concern with static loading and ductile behavior because yielding redistributes stresses and the part does not fail. Cyclic loading is of more concern because fatigue cracks may originate at stress concentrations and propagate a bit with each cycle until a member fails.

Figure 2.7-2b shows peak stresses at an elliptical hole in an infinitely wide plate under uniaxial stress σ_o away from the hole. This solution was obtained by theory of elasticity and is therefore considered exact. A crack in a plate can be regarded as an ellipse for which a/b approaches infinity; the formula then predicts infinite stress. Another special case of interest is the circular hole, $a = b$, for which $K_t = 3$ and the

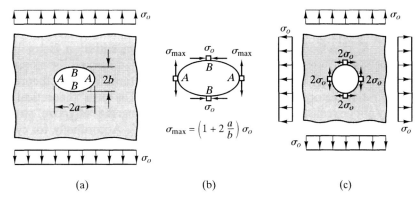

FIGURE 2.7-2 (a) Elliptical hole in an infinite plate under uniaxial tension. (b) Significant stresses at the elliptical hole under uniaxial tension. (c) Circular hole in an infinite plate under biaxial tension.

peak stress is $3\sigma_o$ at points A. (Equation 2.7-2 also yields $K_t = 3$ when $r \ll D$.) By superposing two solutions for the circular hole, with σ_o stresses at right angles, we obtain Fig. 2.7-2c, where stress $2\sigma_o$ appears at all points around the hole. Thus, $K_t = 2$ for a circular hole in a plane field of hydrostatic tension.

Remarks. What can be done to reduce K_t? Mainly, *round internal and re-entrant corners*. In Fig. 2.7-3a, for example, if a hole of width w must be present, an elliptical hole with long axis parallel to the load axis is better than a circular hole. But an elliptical hole is expensive to cut. The array of circular holes in Fig. 2.7-3b is better than a single circular hole. In the shrink-fit constructions of Fig. 2.7-3c, there are high stresses at the sharp corners where disk 1 meets the shaft. Peak stresses are less with the geometries used for disks 2 and 3, even though these geometries are not free of stress concentration. The effects of stress concentration can sometimes be reduced by favorable residual stresses that are introduced by cold working before loading. For example, a

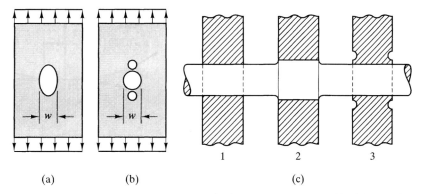

FIGURE 2.7-3 Ways to reduce a peak stress. (a,b) Flat plates in tension. (c) Disks, seen in cross section, are attached to a shaft by shrink-fit.

residual compressive stress near B in Fig. 2.7-1 would combine algebraically with tensile stress due to tensile load, thus reducing the net tensile stress.

K_t may be called a "geometric" stress concentration factor because its value depends only on the geometry and manner of loading (and the assumptions of homogeneity, isotropy, and linear elasticity). Another term for K_t is "theoretical" stress concentration factor, which explains the subscript t. The term "theoretical" suggests that K_t may not be realistic, which is true for cyclic loading: the "effective" stress concentration factor used in fatigue calculations is less than K_t. Accordingly, for cyclic loading K_t is replaced by a factor K_f (which, however, is derived from K_t—see Section 3.6).

2.8 CONTACT STRESS

Consider first a *rigid* circular disk of radius a, one face of which is pressed against the flat surface of a linearly elastic medium by a force P [6.1]. The resulting contact pressure is shown in Fig. 2.8-1a. The area of contact remains constant, and the contact stress is directly proportional to P. However, when two *elastic* bodies are pressed together they contact over a small area whose size depends on the load. Therefore stresses are not directly proportional to load. Stresses in the neighborhood of the contact between two elastic bodies are known as "Hertz contact stresses," or simply as "Hertz stresses." (Hertz, best known for work with radio waves, published his pioneer contact stress analysis in 1881, when he was 24.) Practical applications include ball and roller bearings, wheels on rails, cams, and gears. None of these applications are static: the contact area moves relative to the body; therefore stresses are cyclic, and fatigue damage is possible.

Hertz considered two linearly elastic, homogeneous, and isotropic bodies, each described by two principal radii of curvature. He assumed that there is no friction in the contact area and that dimensions of the contact area are small in comparison with the size of the bodies and their radii of curvature. He obtained expressions for the size of the contact area and the contact pressure, *neither* of which is directly proportional to

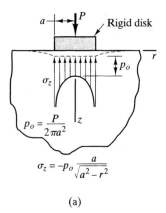
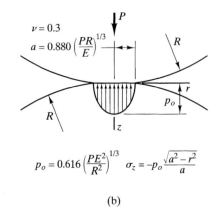

(a)   (b)

FIGURE 2.8-1 Two axisymmetric contact problems. Contact stresses between (a) a *rigid* circular disk and an elastic solid, and (b) two *elastic* spheres.

the load. Stresses other than contact pressure, and subsurface stresses, were studied by subsequent investigators.

Lengthy discussion and many formulas for Hertz stresses can be found in the literature [1.6, 1.7, 2.2, 2.6, 2.7]. What follows here are simple examples and a brief summary of their practical relevance. Poisson's ratio is taken as $\nu = 0.3$ in all formulas.

Spheres in Contact. Results for the special case of two identical spheres appear in Fig. 2.8-1b. The contact area is circular and the maximum contact pressure p_o appears at its center. Stress is proportional to $P^{1/3}$, which means that eight times the load will only double the stress.

Not all normal stresses at or near the contact are compressive. At the edge of the circular area of contact, the radial, circumferential, and z-direction stresses are principal stresses, where

$$\sigma_r = +0.13 p_o \qquad \sigma_\theta = -0.13 p_o \qquad \sigma_z = 0 \qquad (2.8\text{-}1)$$

Thus the edge of the contact circle is in a state of pure shear, with $\tau_{max} = 0.13 p_o$.

A larger shear stress appears on the z axis below the contact surface. Principal stresses along the z axis are plotted in Fig. 2.8-2a. They provide the maximum shear stress

$$\tau_{max} = \left(\frac{\sigma_r - \sigma_z}{2}\right)_{max} = 0.31 p_o \qquad \text{at} \qquad z \approx 0.50 a \qquad (2.8\text{-}2)$$

In cyclic loading, shear stress may be associated with the development of cracks. The foregoing results show that cracks may begin below the surface, where they cannot be seen. Similar conclusions apply to other geometries, such as contact regions between ball bearings and their races.

Parallel Cylinders in Contact. For cylinders of length L the contact area is a rectangle of dimensions L by b. For two cylinders of modulus E, Poisson's ratio $\nu = 0.3$, and radii R_1 and R_2 respectively, dimension b and the maximum contact pressure p_o are

$$b = 3.04 \sqrt{\frac{P}{LE}\left(\frac{R_1 R_2}{R_1 + R_2}\right)} \qquad p_o = 0.418 \sqrt{\frac{PE}{L}\left(\frac{R_1 + R_2}{R_1 R_2}\right)} \qquad (2.8\text{-}3)$$

where L should be about $5b$ or more if the formulas are to be reliable. The formulas can also be used for a cylinder resting on a flat surface or in a cylindrical trough. The radius of a trough is taken as negative. From Eq. 2.8-3 we see that four times the load will only double the stress.

In Fig. 2.8-2b, contact pressure on one of the two cylinders is viewed looking parallel to the cylinder axis. At locations indicated by black squares, the shear stress has magnitude $0.26 p_o$. With rolling contact, this shear stress cycles between $-0.26 p_o$ and $+0.26 p_o$ and can produce fatigue damage.

Hertz contact analysis shows no tensile stress anywhere in the contact region between parallel cylinders. Tensile stress may occur, however, if a component of load is directed tangent to the contact area.

Remarks. Static overload can produce plastic action that permanently dents bearing surfaces. Hertz contact formulas do not account for plastic action or for the altered

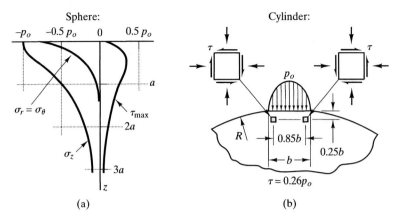

FIGURE 2.8-2 (a) Hertz contact between two spheres, with stresses below the contact surface along the load axis. (b) Hertz contact between two cylinders, with significant shear stresses below the contact surface.

geometry of a dent. Overload during operation can produce a surface layer with residual compressive stress, which is accompanied by a subsurface layer with residual tensile stress. Fatigue cracks can propagate along the subsurface layer until metal spalls off. Usually, bearings are replaced because their surfaces have deteriorated, signaled by operation that is noisy, rough, or loose.

With gears and cams, Hertz contact stresses are complicated by the presence of loads tangent to the surface. Such complication is also present with ball bearings, because the trough of a race has a radius little greater than the ball itself. With comparable radii, the contact area is increased and stresses are reduced, but more area in contact means more area over which some slipping occurs. Slipping generates heat, which causes thermal stresses and necessitates a lubricant. A lubricant film alters the distribution of contact pressure. Because of all these complications, the Hertz contact pressure p_o is useful mainly as a parameter in empirical design formulas. Recommendations as to capacity and proper use of bearings are provided by their manufacturers.

PROBLEMS

2.3-1. A principal element can be viewed parallel to each of its three principal stress directions. From each view we obtain a Mohr circle. All three circles can be plotted on one set of axes. Describe the state of stress for the following arrangements of Mohr circles.
   (a) Two of the circles have equal diameter and are mutually tangent at the origin of coordinates.
   (b) One circle has zero diameter.
   (c) All circles have the same diameter.

2.3-2. Determine the principal stresses associated with the following states of stress (in MPa). Also, for each state, verify that Eqs. 2.3-2 and 2.3-3 do indeed provide the same numerical values of I_2 and I_3.
   (a) $\sigma_x = 0, \sigma_y = 75, \sigma_z = -45, \tau_{xy} = 0, \tau_{yz} = 30, \tau_{zx} = 0$
   (b) $\sigma_x = -80, \sigma_y = 40, \sigma_z = -40, \tau_{xy} = -40, \tau_{yz} = 120, \tau_{zx} = 80$

(c) $\sigma_x = 55, \sigma_y = 85, \sigma_z = -120, \tau_{xy} = -33, \tau_{yz} = 75, \tau_{zx} = 55$
(d) $\sigma_x = 180, \sigma_y = 120, \sigma_z = -80, \tau_{xy} = -140, \tau_{yz} = 80, \tau_{zx} = 110$
(e) $\sigma_x = \sigma_y = \sigma_z = 0, \tau_{xy} = 20, \tau_{yz} = 10, \tau_{zx} = 0$
(f) $\sigma_x = \sigma_y = \sigma_z = 0, \tau_{xy} = 40, \tau_{yz} = 20, \tau_{zx} = 50$
(g) $\sigma_x = \sigma_y = \sigma_z = 0, \tau_{xy} = \tau_{yz} = \tau_{zx} = 100$

2.3-3. For the states of stress in Problem 2.3-2, determine the direction cosines of the maximum principal stress.

2.3-4. For the states of stress in Problem 2.3-2, determine the direction cosines of all three principal stresses.

2.3-5. In a certain state of stress, invariants are $I_1 = -80$ MPa and $I_2 = I_3 = 0$. Determine the principal stresses and describe the state of stress by name.

2.4-1. (a) Use Eqs. 2.4-1 and 2.4-2, and the known orientation of a τ_{max} plane, to show that τ_{max} planes also carry the normal stress $(\sigma_1 + \sigma_3)/2$.

(b) Show that Eq. 2.4-4 yields Eq. 2.4-3.

2.5-1. (a) Derive the second form of Eqs. 2.5-2 (explicit in stress) from the first form (explicit in strain).

(b) Derive (or otherwise verify) Eq. 2.5-3.

(c) Bulk modulus B of an isotropic material is defined as $B = -p/(\Delta V/V)$, where ΔV is the change in a volume V produced by hydrostatic pressure p. Derive an expression for B in terms of E and ν. Why does the result demand that ν be less than 0.5?

(d) Derive Eq. 2.5-4.

2.5-2. If $\sigma_x = 60$ MPa, $\sigma_y = -20$ MPa, $\tau_{xy} = 35$ MPa, and $\sigma_z = \tau_{yz} = \tau_{zx} = 0$, what are the principal strains? Let $G = 70$ GPa and $\nu = 0.30$.

2.5-3. A circular hole is to be cut in a flat plate of isotropic and linearly elastic material. Away from the hole, the plate is in a state of hydrostatic tension $\sigma_x = \sigma_y = \sigma_o$ (see Fig. 2.7-2c). If the edge of the hole is reinforced by a circular ring of the same material and a certain cross-sectional area A, as shown in cross section, the state of stress in the plate will remain $\sigma_x = \sigma_y = \sigma_o$ up to the edge of the reinforcing ring. In terms of σ_o, r_i, t, and ν, what should A be, and what circumferential stress is carried by the ring? What assumptions are contained in your analysis?

2.5-4. Thin sheets of aluminum are bonded to the sides of a thick steel plate, as shown. Initially, all stresses are zero. Obtain an expression for the uniform temperature increase ΔT that will produce a stress σ_Y in the aluminum. State the assumptions made. Using typical data, obtain an approximate numerical value of ΔT if σ_Y is the yield strength of aluminum.

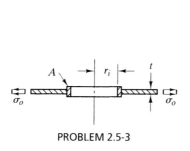

PROBLEM 2.5-3

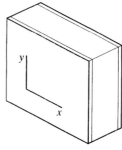

PROBLEM 2.5-4

2.5-5. A thin-walled tube is bonded to a solid cylinder of different material, as shown. Axial force P is applied to a rigid end cap. Stresses developed in the cylinder are radial (σ_{rc}), circumferential ($\sigma_{\theta c}$), and axial (σ_{zc}). Analogous stresses σ_{rt}, $\sigma_{\theta t}$, and σ_{zt} appear in the tube. Assume that friction at the base of the end cap and at the lower support can be neglected. Write enough equations to determine the aforementioned six stresses in terms of P, r_i, t, and material properties. Do not actually solve. Suggestions: In the cylinder, $\sigma_{rc} = \sigma_{\theta c}$ for all r. In the tube, $\sigma_{\theta t}$ depends on σ_{rc}, which acts as an internal pressure on the tube.

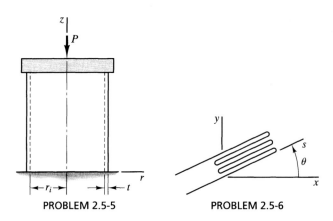

PROBLEM 2.5-5   PROBLEM 2.5-6

2.5-6. The strain gage shown is bonded to the surface of a linearly elastic body. The gage measures strain ϵ_s in direction s. Axes x and y are principal stress directions ($\tau_{xy} = 0$). Show that the gage responds to ϵ_x but not to ϵ_y if $\tan^2\theta = \nu$. Also determine σ_x in terms of ϵ_s, E, and ν. Suggestion: Use the stress transformation equation $\sigma_s = \sigma_x \cos^2\theta + \sigma_y \sin^2\theta$.

2.5-7. When unpressurized, a spherical balloon has radius a and wall thickness t. Under internal pressure p, a increases to r. Assume that Eqs. 2.5-2 are valid (which is not actually the case for rubber). Also, use the engineering definition of stress: force divided by *original* cross-sectional area. Obtain an expression for p in terms of a, r, t, E, and ν. For what r/a is p maximum? Plot $p/p_{\max}$ versus r/a for $0 < r/a < 4$.

2.6-1. For each state of stress shown, determine the following.
 (a) Principal stresses.
 (b) Principal stress directions.
 (c) Octahedral and absolute maximum shear stresses.
 (d) "Effective" stress σ_e.
 (e) Strain energy of distortion per unit volume U_{od}, in terms of G.

2.6-2. For each state of stress in Problem 2.3-2, determine the following.
 (a) Absolute maximum shear stress.
 (b) Octahedral shear stress.
 (c) "Effective" stress σ_e.
 (d) Deviatoric stresses in the given Cartesian system.
 (e) Deviatoric stresses in a Cartesian system whose axes coincide with principal stress directions.
 (f) U_{od} from deviatoric stresses in part (d). [Results of parts (b), (c), and (f) should agree with Eqs. 2.6-11.]

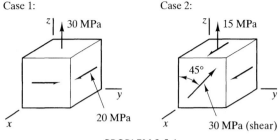

PROBLEM 2.6-1

2.7-1. In Fig. 2.7-1a, what is K_t if the hole diameter is almost as large as D? Might K_t depend on the magnitude of the load when span AB is much less than total width D? Explain. If "yes," what value would you expect K_t to have if the material is stressed to just below yield?

2.7-2. In Fig. 2.7-1a, remove the axial stretching load. Instead apply moments M to top and bottom edges so that the bar becomes a beam of rectangular cross section D by t bent in the plane of the paper. Stress concentration factor K_t for point B is 2.0, independent of r/D [1.6, 1.7]. For nominal stress use My/I, where $I = t[D^3 - (2r)^3]/12$. What is the approximate value of r/D for which nominal stress at A and maximum stress at B are equal? What then is the ratio of σ_{max} to the largest flexural stress away from the hole?

2.7-3. Consider a flat sheet of material in a state of plane stress. In-plane principal stresses are $\sigma_1 = 2\sigma_2$. An elliptical hole is to be cut in the sheet, with its axes in principal stress directions. What should be the aspect ratio a/b of the ellipse if peak stresses are to be as low as possible? What then is the peak stress, in terms of σ_1?

2.7-4. The sketch shows a flat plate of uniform thickness, confined between frictionless rigid walls, with a small circular hole in the middle. Compressive stress σ_o is applied as shown.

(a) What are the principal stresses at A and B on the small circular hole? For what value of ν is point A free of stress?

(b) Let the small circular hole be replaced by a small elliptical hole, with its axes in principal stress directions. What should be the orientation and aspect ratio a/b of the ellipse if principal compressive stresses at A and B are to be equal? What are these stresses, in terms of σ_o and ν?

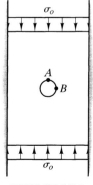

PROBLEM 2.7-4

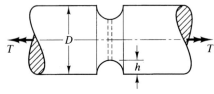

PROBLEM 2.7-5

2.7-5. (a) Use data in Fig. 2.7-2 to show that $4\tau_o$ is the largest normal stress at the edge of a small circular hole in a field of pure shear stress τ_o.

(b) For the semicircular groove in the circular shaft shown, $h/D = 0.2$, and $K_t = 1.69$ for shear stress. A small circular hole is drilled through the center of the shaft at the location shown. What is the largest normal stress at the edge of the hole, in terms of T and D?

(c) If the material has a yield strength τ_Y in shear and does not strain harden, what is the fully plastic torque, in terms of τ_Y and D?

2.7-6. In Fig. 2.7-3a,b, is there an alternative drilling or cutting procedure that could reduce K_t to unity? Explain.

2.7-7. Let the bar in Fig. 2.7-1 be loaded until all material across AB is plastic. If all load is then removed, and unloading is elastic, what is the residual stress at B, in terms of K_t and σ_Y?

2.8-1. An elastic cylinder of radius R is pressed against a parallel cylindrical surface. Calculate the maximum contact pressure in terms of P, R, E, and cylinder length L if the cylindrical surface is also elastic and (a) has radius R, (b) is flat (infinite radius), and (c) is a trough of radius $1.05R$.

2.8-2. Two identical wires of circular cross section are twisted together, as shown. Helix angle ϕ is small. Assume that stresses are zero before axial load P is applied.

(a) What torque T is required to keep the loaded end from rotating?

(b) What is the Hertz contact stress between the wires? Suggestion: Draw a small arc of the wire that shows the contact load, note that a wire axis has radius of curvature $\rho \approx 2R/\phi^2$ and recall how the pressure vessel formula $\sigma = pr_i/t$ is derived.

(c) For $\phi = 0.10$ radian and $E = 200$ GPa, what is the axial tensile stress if the maximum shear stress in Hertz contact is equal to the shear stress at the axis of a wire on a plane inclined at about 45° to the axis of the helix?

PROBLEM 2.8-2

2.8-3. Imagine that, for a certain load, the contact pressure between two elastic spheres of radius R is to be made uniform rather than varying with coordinate r as in Fig. 2.8-1b. For uniform contact pressure to exist, the shape cannot be spherical; R must be perturbed by an amount ΔR in the contact zone, where ΔR is a function of r. Analysis software exists that can calculate surface displacements of a sphere in response to any surface load. How could you use the software to determine the required ΔR?

CHAPTER 3

Failure and Failure Criteria

3.1 INTRODUCTION

Failure. A structure is said to fail if it does not perform as intended. Failure may be unrelated to the load-carrying ability of the structure: A structure no longer useful because of excessive wear or unsuitability to new conditions might be said to have failed. In mechanics of materials we are more restrictive. We are concerned with modes of failure that can be related to stress, strain, and stability. A sampling of such modes is as follows.

- *Fracture.* Fracture is a process in which new cracks develop or existing cracks are extended. Included are development and gradual extension of cracks under cyclic load (fatigue), and sudden breaking of a member under gradually increasing load (brittle fracture). Little or no permanent deformation is evident near fracture surfaces. If the member contains cracks large enough to be seen, brittle fracture may occur even if the material is ductile in a standard tension test. Brittle fracture surfaces are perpendicular to the load axis in uniaxial tension and are inclined to the load axis in uniaxial compression.
- *Yielding.* It is convenient to regard the yield strength σ_Y determined from a tension test as the uniaxial stress at which yielding begins, although a small amount of permanent deformation has already taken place when σ_Y is reached. Widespread yielding may exhaust the load-carrying capacity of a structure without breaking it. Or, yielding may cause too large a change in shape; for example, a bent door hinge may still carry load, but it has failed if the door will no longer close.
- *Low Stiffness.* Moving parts, such as vehicle drive shafts, may have undesirable resonant frequencies if they are too flexible. If too flexible, machine tools may be unable to hold tolerances and cutting tools may chatter.
- *Instability.* Buckling is an instability characterized by an abrupt decrease in stiffness as load increases. Buckling deformations may be elastic or plastic and may be widespread or quite localized. Stresses need not be high for buckling to occur.

- *Creep.* Creep is an increase of deformation under constant stress. It may be of concern only when temperatures are high, as for example with turbine blades. However, some materials creep at ordinary temperatures. Another aspect of creep is *relaxation,* which is a decrease of stress at constant deformation. Relaxation may loosen a bolted connection that operates at high temperature.

Failure Criteria. A failure criterion, also known as a theory of failure, addresses failure of a *material,* not failure of a structure. Commonly used failure criteria serve to predict whether a given state of stress will (a) cause the material to yield, or (b) cause the material to fracture. Usually a particular failure criterion is used for category (a) or category (b) but not for both. Thus, despite the generality implied by the name "failure criteria," the term has a narrow meaning.

Typically one knows the yield or fracture strength of a material from a laboratory test in uniaxial tension or uniaxial compression. A failure criterion then acts as an "interpreter" between known behavior in a state of uniaxial stress and behavior to be predicted in a more general state of stress. Without a failure criterion, the more general state of stress would have to be tested in the laboratory to determine whether it is safe. One cannot call for testing of all states of stress that may be encountered because the cost would be prohibitive.

In order to choose an appropriate failure criterion, one must know whether the failure mode will be ductile or brittle. The choice is not always obvious (if the material exhibits slight ductility, is it ductile or brittle?). An arbitrary rule is to classify a material as normally ductile if it exhibits more than 5% permanent elongation in a standard tension test. If we use a failure criterion based on tension-test data, we presume that the mechanism causing failure in static tension will also cause failure in any other static state of stress. However, it is conceivable that a material behaves in (say) a ductile manner in both uniaxial tension and uniaxial compression but in a brittle manner in the state of stress found in service. In this unfortunate circumstance, a failure criterion cannot relate tension and compression tests to the service condition; one must test the material under the state of stress found in service.

Failure criterion are not derivable laws. Rather, they are phenomenological rules, each contrived to fit experimentally observed behavior, and usually restricted to linearly elastic conditions. No single criterion is best for all materials or under all circumstances. For example, higher temperature or superposition of a large hydrostatic compression can change the behavior of some materials from brittle to ductile.

In the following sections we consider the more commonly used failure criterion. It is helpful to remember that each criterion operates by making a comparison between the test state and the given state. As a hypothetical example, imagine an "F criterion" for ductile failure, in which F is a certain function of the stresses. We can obtain a numerical value of F by using σ_Y, the yield strength determined by a tension test. We obtain a second value of F by using stresses that prevail in the given state. If the two values of F are equal, this "F criterion" predicts that yielding impends in the given state.

3.2 ISOTROPIC MATERIAL, BRITTLE FAILURE

We consider a material without macroscopic cracks. (A macroscopic crack is large enough to be seen by the unaided eye.) Failure criteria discussed in the present section allow the presence of *micro*scopic cracks and flaws. They may be inherent in the nature of the material, as are graphite flakes in gray cast iron. In brittle failure, a fracture surface develops. In tension, a fracture surface is normal to the axis of the tensile stress. In compression, a fracture surface is inclined to the direction of largest compressive stress, although usually not at 45° to it. Two failure criteria commonly applied to brittle behavior are the following.

Maximum Normal Stress Criterion. This criterion postulates that regardless of the state of stress, failure of an isotropic material occurs when the numerically largest principal stress reaches a limiting value. Consider first the case in which this principal stress is σ_1, a tensile stress. The limiting tensile stress σ_{tf} for a given material can be determined by a test to failure in uniaxial tension. In any other state of stress, failure is predicted if σ_1 of that state reaches or exceeds σ_{tf}. That is,

$$\text{Failure predicted when} \quad \frac{\sigma_1}{\sigma_{tf}} \geq 1 \qquad (3.2\text{-}1)$$

Thus, the lesser principal stresses σ_2 and σ_3 are assumed to play no role.

If the criterion is applied to a compressive state of stress, Eq. 3.2-1 is replaced by $|\sigma_3|/\sigma_{cf} \geq 1$, where $|\sigma_3|$ is the magnitude of the minimum principal stress ($\sigma_3 < 0$ here) and σ_{cf} is the strength determined by testing a specimen in uniaxial compression (conventionally, σ_{cf} is recorded as a positive number). Thus, in the name "maximum normal stress criterion," the word "maximum" refers only to magnitude. Uniaxial states of stress, tensile or compressive, are special cases of the Mohr criterion discussed next.

The maximum normal stress criterion may be quite inaccurate if all three principal stresses are compressive. Experiments show that materials can sustain very large hydrostatic compression without yielding or fracturing.

Mohr Criterion. Imagine that we test a material to failure in each of three states of stress: uniaxial compression, uniaxial tension, and pure shear, so as to obtain the limiting stresses σ_{cf}, σ_{tf}, and τ_f. Each state is described by a Mohr circle (Fig. 3.2-1a). An "envelope" can be drawn tangent to the circles. Mohr postulated that an arbitrary state

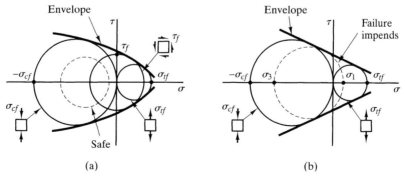

FIGURE 3.2-1 (a) Mohr failure envelope for tests to failure in compression, torsion, and tension. (b) Mohr failure envelope for tests to failure in compression and tension.

of stress is safe from failure if the Mohr circle representing that state of stress lies within the envelope. Thus the intermediate principal stress σ_2 is assumed to play no role. The Mohr criterion can be applied to ductile failure as well as to brittle failure [3.1]. However, it is usually applied only to brittle failure, and in the following simplified form.

If the shear test is omitted, the envelope consists of two straight lines (Fig. 3.2-1b). Let σ_1 and σ_3 be the extreme principal stresses of an arbitrary state of stress, and let σ_{tf} and σ_{cf} be failure strengths of the material in tension and compression tests respectively. Then

$$\text{Failure predicted when} \quad \frac{\sigma_1}{\sigma_{tf}} - \frac{\sigma_3}{\sigma_{cf}} \geq 1 \qquad (3.2\text{-}2)$$

Here σ_1 and σ_3 are signed quantities, as usual, but σ_{tf} and σ_{cf} are positive *magnitudes* of failure strengths. When equality prevails in Eq. 3.2-2, the dashed circle in Fig. 3.2-1b becomes tangent to the failure envelope. Proof of this result is left as an exercise.

Equation 3.2-2 is best suited to failure analysis of isotropic materials that fail in a brittle manner and are much stronger in compression than in tension. Accuracy is greatest for a state of stress whose Mohr circle lies between the tension and compression circles in Fig. 3.2-1b. Accuracy may be low for a state of large and almost hydrostatic compression. If the state of stress is uniaxial, the Mohr criterion reduces to the maximum normal stress criterion.

For a cohesionless material such as sand, the Mohr failure envelope can be approximated from results of a single test in which all principal stresses are compressive (Fig. 3.2-2). Angle ϕ is associated with the coefficient of dry friction, $\mu = \tan \phi$, and is the maximum possible slope of a sand surface [3.1].

Safety Factor. If stress is directly proportional to load, safety factor SF can be regarded as the number by which a given state of stress must be multiplied to produce failure. Thus, for Eqs. 3.2-1 and 3.2-2 respectively,

$$\frac{(SF)\sigma_1}{\sigma_{tf}} = 1 \quad \text{and} \quad \frac{(SF)\sigma_1}{\sigma_{tf}} - \frac{(SF)\sigma_3}{\sigma_{cf}} = 1 \qquad (3.2\text{-}3)$$

The latter equation incorporates the assumption that σ_1 and σ_3 change by the same factor when the load is changed in intensity. Failure of the material is predicted when stresses are such that SF is less than unity.

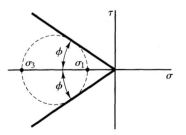

FIGURE 3.2-2 Mohr failure envelope for a loose granular material.

EXAMPLE

For a certain isotropic material, tests to failure in uniaxial tension and uniaxial compression provide the respective failure strengths $\sigma_{tf} = 14$ MPa and $\sigma_{cf} = 120$ MPa. At a point in a part made of this material, there exists the state of plane stress $\sigma_x = 0$, $\sigma_y = -18$ MPa, and $\tau_{xy} = 20$ MPa. Determine the safety factor of this state of stress and whether failure is predicted. Consider the maximum normal stress criterion and the Mohr criterion.

Equation 2.2-5 yields principal stresses $\sigma_1 = 12.9$ MPa and $\sigma_3 = -30.9$ MPa. The intermediate principal stress is $\sigma_2 = 0$. According to the maximum normal stress criterion,

$$\frac{(SF)12.9}{14} = 1 \quad \text{hence} \quad SF = 1.09 \tag{3.2-4}$$

Because $SF > 1$, we predict that there will be no failure. According to the Mohr criterion,

$$\frac{(SF)12.9}{14} - \frac{(SF)(-30.9)}{120} = 1 \quad \text{hence} \quad SF = 0.85 \tag{3.2-5}$$

Because $SF < 1$, we predict that there will be failure. The two criteria disagree. Without additional information we cannot say which is the more reliable.

Remarks. Experiment has shown that when material behavior is brittle, a reduction in size of a member does not produce as much of a reduction in strength as predicted by application of scale factors. A reason for this "size effect" is that a material tends to have a certain number of flaws per unit volume, so a small member is less likely to include a flaw of critical size.

Also, if we compare large and small members of similar shape, loading, and peak stress, we see that the large member has a smaller stress gradient and hence a greater volume of material subjected to near-peak stresses (Fig. 3.2-3). Accordingly, the large member is more likely to have a flaw of critical size in a highly stressed region. A terse summary of this argument is that the strength of a brittle material increases as stress gradients increase [3.2].

Flaws tend to be random in location, orientation, and size. It is therefore not surprising that results of experiments to determine the strength of a brittle material show considerable scatter.

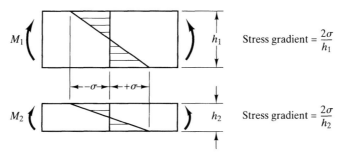

FIGURE 3.2-3 Flexural stress distributions in beams of different size but carrying the same maximum stress.

3.3 ISOTROPIC MATERIAL, DUCTILE FAILURE

In the present context, ductile failure means the onset of yielding. Yielding occurs when there is slipping within the material structure but without fracture. In this context a failure criterion may also be called a *yield criterion*.

Maximum Shear Stress Criterion. This criterion postulates that yielding of an isotropic material begins when the maximum shear stress reaches a limiting value τ_Y. If τ_{max} is the maximum shear stress in an arbitrary state of stress,

$$\text{Yielding predicted when} \quad \frac{\tau_{max}}{\tau_Y} \geq 1 \quad (3.3\text{-}1)$$

A simple way to determine τ_Y is from a tension test. In terms of principal stresses, the maximum shear stress is $\tau_{max} = (\sigma_1 - \sigma_3)/2$. Accordingly, if yielding in uniaxial tension is regarded as beginning at the conventionally defined yield strength σ_Y, then $\tau_Y = \sigma_Y/2$.

If stress is directly proportional to load, Eq. 3.3-1 can be expressed in terms of safety factor SF as $(SF)\tau_{max}/\tau_Y = 1$. Since $\tau_{max} = (\sigma_1 - \sigma_3)/2$, the intermediate principal stress σ_2 plays no role in this failure criterion.

Von Mises Criterion. This criterion* postulates that yielding of an isotropic material begins when the strain energy of distortion per unit volume U_{od} reaches a limiting value. Upon comparing Eq. 2.4-5 for τ_{oct}, Eq. 2.6-10 for U_{od}, and Eq. 2.6-12 for σ_e, we see that all these quantities depend on the same function of the state of stress. Therefore the von Mises criterion can be stated in terms of any of these three quantities. We choose to state it in terms of effective stress σ_e. The value of σ_e that defines yielding can be determined from a tension test. Thus, from Eq. 2.6-12, $\sigma_e = \sigma_Y$ when yield strength σ_Y is reached in uniaxial tension, and in any other state of stress,

$$\text{Yielding predicted when} \quad \frac{\sigma_e}{\sigma_Y} \geq 1 \quad (3.3\text{-}2)$$

where the numerator is evaluated from Eq. 2.6-12 for the state of stress in question. Unlike the failure criteria described by Eqs. 3.2-1, 3.2-2, and 3.3-1, Eq. 3.3-2 accounts for *all* components of the state of stress. If all components are directly proportional to load and change by the same factor when the load is changed in intensity, Eq. 3.3-2 can be expressed in terms of safety factor SF as $(SF)\sigma_e/\sigma_Y = 1$.

We could express Eq. 3.3-2 in terms of U_{od}/U_{odY} or in terms of τ_{oct}/τ_{octY}, where U_{odY} and τ_{octY} are respectively the strain energy of distortion per unit volume and the octahedral shear stress when yielding begins in uniaxial tension. In these forms the von Mises criterion would be called the *strain energy of distortion failure criterion* and *the octahedral shear stress failure criterion*. All three forms provide the same results. For example,

---

*What is now known as the von Mises criterion was proposed by Maxwell (in an 1856 letter to Kelvin), but Huber (1904), and by von Mises (1913). The maximum shear stress criterion is also known as the Tresca criterion. It was proposed by Coulomb (1773) and by Tresca (1864). Mohr presented his criterion in 1900. The maximum normal stress criterion is attributed to Rankine.

consider the strain energy of distortion. From Eqs. 2.6-10 and 2.6-12, $U_{od} = \sigma_e^2/6G$. In terms of U_{od} and safety factor SF, the von Mises criterion predicts yielding when

$$\frac{[(SF)\sigma_e]^2/6G}{\sigma_Y^2/6G} = 1 \quad \text{or} \quad \frac{(SF)\sigma_e}{\sigma_Y} = 1 \tag{3.3-3}$$

which is the same equation as seen at the end of the preceding paragraph.

EXAMPLE 1

If a material has yield strength σ_Y in a tension test, what shear stress τ is associated with yielding when a shaft made of this material is loaded by torque?

In the tension test, the maximum shear stress appears on planes inclined at 45° to the load axis and has magnitude $\tau_Y = \sigma_Y/2$. With an equal sign in Eq. 3.3-1 to designate the onset of yielding, the maximum shear stress criterion predicts

$$\frac{\tau}{\sigma_Y/2} = 1 \quad \text{hence} \quad \tau = 0.500\sigma_Y \tag{3.3-4}$$

In the tension test, $\sigma_e = \sigma_Y$. For a state of pure shear τ, Eq. 2.6-12 reduces to $\sigma_e = [6\tau^2]^{1/2}/\sqrt{2} = \sqrt{3}\tau$. With an equal sign in Eq. 3.3-2 to designate the onset of yielding, the von Mises criterion predicts

$$\frac{\sqrt{3}\tau}{\sigma_Y} = 1 \quad \text{hence} \quad \tau = 0.577\sigma_Y \tag{3.3-5}$$

The two criteria do not agree, but there is no requirement that they do so.

EXAMPLE 2

Consider the state of stress examined in the example problem of Section 3.2. Imagine now that the material behaves in a ductile manner and has yield strength $\sigma_Y = 50$ MPa in a tension test. What is the safety factor for the given state of stress?

The maximum shear stress is $\tau_{max} = [\sigma_1 - \sigma_3]/2 = [12.9 - (-30.9)]/2 = 21.9$ MPa. Therefore, according to the maximum shear stress criterion,

$$\frac{(SF)21.9}{50/2} = 1 \quad \text{hence} \quad SF = 1.14 \tag{3.3-6}$$

The effective stress is $\sigma_e = 39.0$ MPa, obtained from Eq. 2.6-12 with principal stresses 12.9 MPa, 0, and -30.9 MPa. Therefore, according to the von Mises criterion,

$$\frac{(SF)39.0}{50} = 1 \quad \text{hence} \quad SF = 1.28 \tag{3.3-7}$$

Remarks. In states of plane stress, the maximum shear stress criterion and the von Mises criterion display the failure envelopes seen in Fig. 3.3-1, where, for this illustration, σ_x and σ_y are principal stresses. A state of plane stress that plots as a point within the respective envelopes is regarded by the respective criteria as safe from yielding.

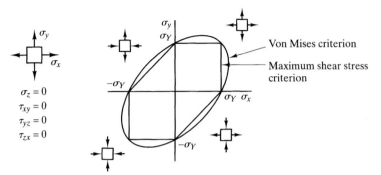

FIGURE 3.3-1 Failure envelopes for plane stress in terms of principal stresses. (Note: Axes are labeled differently than in Figs. 3.2-1 and 3.2-2.)

Which of the two criteria is better? Tests of ductile metals favor the von Mises criterion. Most data points (not shown) fall between the two envelopes in Fig. 3.3-1 but are closer to the von Mises envelope.

When the given state of stress is pure shear, or when $\sigma_x = 2\sigma_y$ or $\sigma_y = 2\sigma_x$ in Fig. 3.3-1, disagreement between safety factors predicted by the two criteria is about 14.5%. For other states of stress, disagreement is less. Accordingly, if we use the maximum shear stress criterion with $\tau_{max} = 0.536\ \sigma_e$, or the von Mises criterion with $\sigma_e = 1.866\ \tau_{max}$, we will not be more than about 7% away from either criterion [3.3]. Because of the closeness of the two criteria and the scatter of experimental data, a choice of one criterion over the other may rest mainly on convenience of use for the application at hand.

3.4 ANISOTROPIC MATERIALS

Anisotropic materials typically do not fail in the same way as isotropic materials and usually be cannot classified as ductile or brittle. Consider wood, for example. Under tensile stress either parallel to the grain or normal to the grain, or under shear stress parallel to the grain, wood fractures. Under compressive stress normal to the grain, wood suffers large permanent deformation. Under compressive stress parallel to the grain, fibers buckle, producing slip-type failure on a plane inclined to the compressive stress direction. For anisotropic materials in general, failure cannot be stated in terms of a single experimentally determined quantity, and a failure criterion usually cannot be described by a single equation. Several failure criteria for anisotropic materials have been proposed [3.4, 3.5]. Many are extensions of failure criteria for isotropic materials. Some may owe their popularity more to simplicity of concept and ease of use than to underlying rigor.

Manufactured composite materials usually are not macroscopically homogeneous. Typically they are laminated, something like plywood, but with orthotropic plies having principal material directions that are not mutually orthogonal. Of course, principal material directions in any ply are in general different from principal stress directions. Fortunately, most manufactured composite materials are made into plates, shells, and pressure vessels so that each ply is loaded in approximately a plane state of stress. Therefore most failure criteria need not have full three-dimensional generality.

Most failure criteria for composite materials address failure of an individual ply. Failure of the laminated structure is more complicated because of unresolved questions about interlaminar stresses, how degraded plies unload, and how failure of some plies influences the remaining intact plies. At present, no failure criterion intended for a laminated structure is reliable enough to be used without experimental confirmation.

3.5 INTRODUCTION TO FRACTURE MECHANICS

Cracks and Brittle Fracture. One expects that materials such as glass and rock will fail in a brittle manner. A normally ductile material such as structural steel may also fail in a brittle manner if it contains a crack in a region of tensile stress. Typically a crack begins at a stress raiser and grows gradually, due to cyclic loading or due to corrosion under steady loading. When a crack reaches a "critical length" it suddenly propagates as a brittle fracture, and the part or structure breaks, perhaps completely in two. Complete separation may be prevented by progagation of the crack into a "crack arrester" such as an existing hole, or by deformations that happen to relieve the mechanism that causes the crack to propagate. Crack propagation speeds may exceed 1000 m/s.

The Liberty cargo ships of World War II are classic examples of this kind of failure. Of some 2700 built, more than 100 broke in two. Part of the trouble was welded construction, in which edges of adjacent plates were welded together. (Previously, ships were constructed of overlapping plates connected by rivets, thus incorporating "crack arresters" in the structure.) Also, the material itself was made more susceptible to brittle fracture by heat of the welding process and by cold conditions in which these ships often operated.

The state of stress at a crack tip causes material there to lose ductility. Consider, for example, a flat plate with a crack oriented perpendicular to the direction of load (Fig. 3.5-1a). Near the crack tip, normal stresses in the plane of the plate are tensile and very large. Consequently, due to the Poisson effect, material around the crack tip tries to contract in the thickness direction (normal to the plate surface). However, the

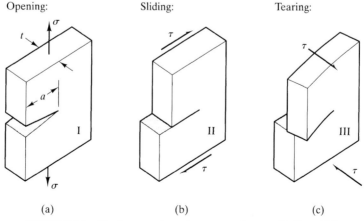

FIGURE 3.5-1 The three crack modes, commonly named I, II, and III.

volume of highly stressed material is very small, so its contraction is largely prevented by adjacent material that has much lower stresses. Thickness-direction tension at the crack tip is the result. Material just ahead of the crack tip is therefore in a state of *triaxial tension,* which favors fracture more than plastic flow.

Why does a crack propagate at great speed after reaching a critical length? The answer is provided by energy considerations, which are summarized as follows. Consider, for example, the geometry of Fig. 3.5-1a, which is shown again in Fig. 3.5-2a. Energy needed to extend the crack an amount da is independent of crack length a, so that energy expended to produce the crack varies linearly with a (Fig. 3.5-2b). As the crack grows, stresses are reduced in material alongside the crack, and stored strain energy is released. Strain energy released varies approximately quadratically with a. (As a conceptual device, imagine that a crack of length a nullifies the uniaxial state of stress in a semicircular disk of radius a, shown dashed in Fig. 3.5-2a. Strain energy released would then be proportional to the volume of this disc, $V = \pi a^2 t/2$.) When a is large enough for the increment of strain energy released to equal the increment of energy needed to extend the crack, sudden fracture impends. Thus $dU_e = dU_s$ in Fig. 3.5-2b identifies the critical crack length. Energy needed to drive the crack is supplied by unloading of stressed material. There is no need (and insufficient time) for energy to be supplied by the work of external forces acting through a distance. The failure load is much less than it would be if the crack were absent because the crack provides a way for intermolecular bonds to be broken sequentially rather than all at once.

Cracks can be categorized as shown in Fig. 3.5-1. In practice, Mode I is most common. Mixed modes are possible. For any mode, one can calculate a *stress intensity factor* for the crack, and compare it with an allowable value to determine whether fracture impends. A stress intensity factor is *not* a stress concentration factor! Indeed, in elementary fracture analysis it is not necessary to use stress concentration data: Peak stresses at a crack tip need not be calculated.

Calculations. Here we consider only isotropic materials, and only Mode I cracks unless stated otherwise. The stress intensity factor for a Mode I crack is denoted by K_I and is given by

$$K_I = \beta \sigma \sqrt{\pi a} \tag{3.5-1}$$

Here σ is the nominal stress that would exist if the crack were absent. Thus, stress σ is independent of crack length. Multiplier β is dimensionless and depends on geometry and type of loading (see Table 3.5-1). Dimension a is defined as either the full crack length or half of it, depending on geometry. Units of K_I are MPa$\sqrt{m}$. Fracture impends if K_I reaches a "critical" value K_{Ic} known as *fracture toughness.* K_{Ic} can be considered a material property, independent of specimen thickness t, if the specimen is sufficiently thick for thickness-direction tension at the crack tip to develop fully. Also, the crack length should exceed a certain minimum. Recommended minimum dimensions are

$$t \geq 2.5 \left(\frac{K_{Ic}}{\sigma_Y} \right)^2 \quad \text{and} \quad a \geq 2.5 \left(\frac{K_{Ic}}{\sigma_Y} \right)^2 \tag{3.5-2}$$

where σ_Y is the yield strength as determined by a tension test of the material, and t is shown in Fig. 3.5-1a. If the actual thickness is less than the value described by Eq. 3.5-2,

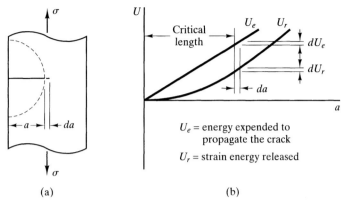

FIGURE 3.5-2 (a) Plate with an edge crack of length a. (b) Energy relations for crack extension.

TABLE 3.5-1 Stress intensity data for flat plates, of isotropic material and uniform thickness, with in-plane loading [3.6].

Tension, central crack of length $2a$

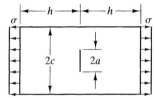

$K_I = \beta \sigma \sqrt{\pi a}$

$$\beta = \frac{1 - 0.5(a/c) + 0.326(a/c)^2}{\sqrt{1 - (a/c)}}$$

Accurate to within 1% for all a/c, provided h/c is "large"

Tension, edge crack of length a

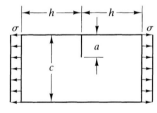

$K_I = \beta \sigma \sqrt{\pi a}$

$\beta = [1.12 - 0.23(a/c) + 10.6(a/c)^2$
$\quad\quad - 21.7(a/c)^3 + 30.4(a/c)^4]$

Accurate to within 1% for $a/c \le 0.6$, provided $h/c > 1$ and sides are free to rotate

Pure bending, edge crack of length a

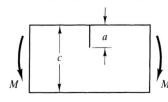

$K_I = \beta \sigma \sqrt{\pi a}$

$$\sigma = \frac{M(c/2)}{I} = \frac{6M}{tc^2}$$

$\beta = [1.12 - 1.39(a/c) + 7.32(a/c)^2$
$\quad\quad - 13.1(a/c)^3 + 14.0(a/c)^4]$

Accurate to within 1% for $a/c \le 0.6$

fracture toughness becomes a function of thickness. Similarly, if a is less than required by Eq. 3.5-2, Eq. 3.5-1 with $K_I = K_{Ic}$ may predict a stress σ greater than the yield strength σ_Y; then one should expect yielding rather than brittle fracture.

With $K_I = K_{Ic}$ and a given σ, Eq. 3.5-1 defines the critical crack dimension a. Note that a is independent of the size of the structure in which the crack appears, except to the extent that β is influenced by larger values of a/c.

Equations 3.5-1 and 3.5-2 result from the study called *linear elastic fracture mechanics* [3.6–3.10]. Actually there is some yielding immediately adjacent to the crack tip; nevertheless, Eq. 3.5-1 is considered sufficiently accurate for practical use if the conditions of Eq. 3.5-2 are met. Formulas for geometries other than those in Table 3.5-1 are available in handbooks or can be calculated by some of the many software packages for numerical stress analysis. Handbooks list K_{Ic} for various materials. K_{Ic} is temperature-dependent. For some steels, K_{Ic} decreases greatly over a small temperature range near 0°C or even above. Thus a structure adequate in a warm climate may become dangerous if moved to a cold climate. In rolled steel plate, K_{Ic} also depends on the orientation of a crack with respect to the direction of rolling [3.9].

Cases in Table 3.5-1 are restricted to in-plane loading. If there is bending that causes out-of-plane deformation, the β factors are different. Thus, for example, the first case in Table 3.5-1 is not applicable to a longitudinal crack in a thin-walled cylindrical pressure vessel because material adjacent to the crack will bulge outward.

In linear elastic fracture mechanics, stress intensity factors for different loads can be superposed, provided the geometry and crack mode are the same for all loads. That is,

$$(K_I)_{net} = (K_I)_1 + (K_I)_2 + (K_I)_3 + \ldots \tag{3.5-3}$$

For example, Eq. 3.5-3 is applicable to the latter two cases in Table 3.5-1. Stress intensity factors from *different* modes do not combine in this way. Consider a central crack of length $2a$ in a very wide plate, with tensile stress σ normal to the crack and shear stress τ parallel to it. Modes I and II are present simultaneously. For this case, the following approximate interaction formula is available [3.8]. Fracture is predicted when

$$\left(\frac{K_I}{K_{Ic}}\right)^2 + \left(\frac{K_{II}}{K_{IIc}}\right)^2 \geq 1 \tag{3.5-4}$$

where $K_I = \sigma\sqrt{\pi a}$, $K_{II} = \tau\sqrt{\pi a}$, and $K_{IIc} \approx 0.75\, K_{Ic}$. Here τ is the *magnitude* of shear stress.

EXAMPLE

A long flat plate, 100 mm wide and 20 mm thick, is loaded axially in tension. A Mode I crack 15 mm long is present on one edge. If the material is Steel C in Table 3.5-2, what axial load P will cause fracture? If this load were applied to an identical plate made of Steel D, what would be the critical crack length?

The crack length and specimen thickness both exceed the minimum of 11 mm for Steel C (Table 3.5-2). Also, $a/c < 0.6$, so formulas for the second case in Table 3.5-1 are applicable. With $K_I = K_{Ic} = 77$ MPa$\sqrt{m}$,

$$77 = [1.12 - 0.23(0.15) + 10.6(0.15)^2 - 21.7(0.15)^3 + 30.4(0.15)^4]\sigma\sqrt{0.015\pi} \tag{3.5-5}$$

TABLE 3.5-2 Ultimate strength σ_u, yield strength σ_Y, fracture toughness K_{Ic}, and minimum crack length and thickness from Eqs. 3.5-2.

Material*	σ_u (MPa)	σ_Y (MPa)	K_{Ic} (MPa$\sqrt{m}$)	Minimum a,t (mm)
Steel A	700	500	175	306
Steel B	900	650	106	66
Steel C	1290	1150	77	11
Steel D	1600	1410	50	3
Aluminum	540	480	30	10
Titanium	900	830	71	18
Glass	80	—	0.25	0

*Data are typical. Materials are not specifically identified because properties depend appreciably on temperature, variations in composition, heat treatment, mechanical working, and testing procedure.

Solving for σ and then for axial load P, we obtain

$$\sigma = 280 \text{ MPa} \quad \text{and} \quad P = (100)(20)\sigma = 560 \text{ kN} \tag{3.5-6}$$

Note that P is calculated from the gross cross section, as if the crack were not present. At fracture, the average nominal stress on the *net* cross section is $P/(20)(100-15) = 329$ MPa, which is well below the yield strength of Steel C. Load P may be called "residual strength," which is what remains of the original strength of a member after a crack has developed.

To answer the second question, we apply axial load $P = 560$ kN to a plate made of Steel D. We are now required to solve a high-order equation for crack length a. In order to obtain an approximate answer easily, we assume that a/c is small, so that terms containing a/c in the expression for β can be neglected. Thus, with $K_I = K_{Ic}$ and Steel D,

$$50 \approx 1.12 \frac{560{,}000}{(100)(20)} \sqrt{\pi a} \tag{3.5-7}$$

from which $a \approx 0.0081$ m $= 8.1$ mm. Thus a/c is indeed small, as assumed. (If all terms are retained in the expression for β, the equation yields the critical crack length $a = 7.6$ mm. This result can be obained by use of mathematical software or a programmable calculator.)

The stronger steel presents a more dangerous condition than the weaker steel because the critical crack length is smaller *for the same load P*. However, the purpose of using a stonger steel is probably to permit a greater load. For a load greater than $P = 560$ kN the critical crack length would be smaller yet. As critical crack lengths become smaller, more costly inspection methods must be used to ensure that a structure has no cracks of critical length.

3.6 INTRODUCTION TO FATIGUE

Fatigue refers to the initiation and gradual propagation of cracks under cyclic loading. Fatigue is the cause of perhaps 80% of the breakage that occurs in machinery. Examples of how stress may vary at a given material point appear in Fig. 3.6-1. Purely alternating stress σ_a would appear in a rotating shaft loaded in pure bending, as the point in question is rotated from the tension side to the compression side. If axial tension is added, mean stress σ_m is superposed, and Fig. 3.6-1b results. The stress history in some part of a vehicle, whose cargo may be light or heavy, is suggested by Fig. 3.6-1c. The randomness of this pattern can be reduced to simpler forms for use in analysis [3.9].

In parts that fail after a great many cycles of loading, cracks begin with yielding on a very small scale. A flaw, inclusion, void, or surface scratch can raise local stresses high enough to produce yielding in a crystal whose planes are oriented parallel to the largest shear stress. With cycling, the crystal strain hardens and cracks. Microcracks grow, join, and eventually produce a macrocrack, whose orientation is usually perpendicular to the maximum principal stress. The rate of crack growth increases with crack length, and can be related to stress intensity factors at the crack tip [3.7–3.9]. When the crack reaches critical length, the part suddenly breaks by brittle fracture.

Figure 3.6-2 shows the typical appearance of a fracture surface after a crack has begun on one side of the cross section and propagated across the member. In the smoother portion, the crack has grown gradually and adjacent surfaces have rubbed against each other. Lines called "beach marks" may be visible. They are produced by temporary cessation of crack growth, due perhaps to changes in operating conditions. Sudden brittle fracture produces the rougher portion. A part that has failed by fatigue may show little or no evidence of ductility, even if its material is normally ductile under static load.

In laboratory specimens, fatigue cracks have been observed at less than 0.1% of the number of load cycles that eventually produce failure. The *rate* of cycling is of no consequence unless temperatures are high. A "size effect" makes small parts endure more cycles. Reasons are as stated for brittle materials at the end of Section 3.2, although with fatigue loading the material may be *ductile* under static load. In cyclic loading, size effect presents another complication; in a rotating shaft with bending moment applied, material on the surface (and any flaw on the surface) is moved in and out of the zone of greatest stress as the shaft rotates. With *static* bending, the flaw may not be in the zone of greatest tensile stress. Most fatigue cracks begin on a surface, because it is usually where stresses are highest. Also, flaws that initiate cracks are more likely to be found on the surface than internally. However, surface treatments that produce unfavorable residual stresses, or contact stresses (Section 2.8), may create a stress field that causes cracks to begin below the surface. Then fatigue damage may not be noticed until pieces of metal flake off. Such "spalling" is seen in ball and roller bearings, gear teeth, and rails.

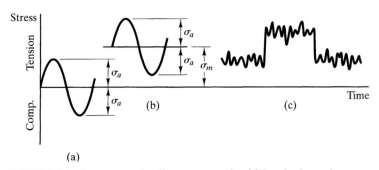

FIGURE 3.6-1 Some types of cyclic stress at a point. (a) Purely alternating stress σ_a. (b) Mean stress σ_m superposed on σ_a. (c) Time-dependent σ_m superposed on a random σ_a.

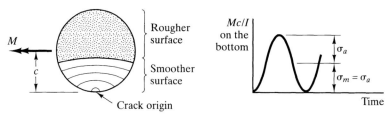

FIGURE 3.6-2 Failure surface in a nonrotating bar of circular cross section, loaded by moment M such that nominal stress Mc/I on the bottom varies with time as shown.

Fatigue data are determined from laboratory tests with specimens of standard dimensions and polished surfaces. In a common test, a bar is rotated with bending moment applied, so that $\sigma_m = 0$. Results of many such tests can be plotted as stress versus number of cycles to failure. Figure 3.6-3 shows three such plots, for metals of different ultimate strengths σ_u. *Fatigue life* is the number of cycles N that produces failure at a given stress level σ_a. *Fatigue strength,* here abbreviated σ_{fs}, is the purely alternating stress σ_a that produces failure in a given number of cycles. Fatigue life depends on the nature of the material, its surface finish and heat treatment, and other factors. For ferrous metals and other metals of similar crystal structure, N appears to be almost infinite if σ_a is small enough. This value of σ_a, where the σ_a versus N curve is horizontal, is called the *endurance limit* or the *fatigue limit.* For materials that do not display an endurance limit, the value of σ_a at $N = 10^8$ may be arbitrarily defined as the endurance limit. Such a definition is analogous to defining a yield strength for materials that display no yield point. Data points, not shown in Fig. 3.6-3, show considerable scatter: For different specimens, the number of cycles to failure at a given stress may vary by a factor of 10 or even 100. Usually the plotted curve represents the median number of cycles. Fatigue life predicted from the curve is only a prediction of probable life, and may be much in error.

The relation between alternating stress σ_a and a superposed tensile mean stress σ_m is shown in Fig. 3.6-4. The solid line from σ_{fs} to σ_u can be regarded as the boundary between failure and no failure, and is seen to be somewhat conservative for the data shown, which are typical of metals that are normally ductile. The data distribution is somewhat different if the material is more nearly brittle, or if alternating shear stress

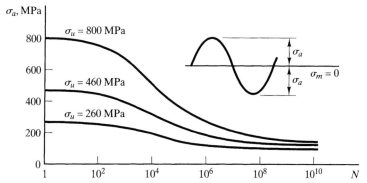

FIGURE 3.6-3 Semilog plots of purely alternating stress (σ_a) versus number of cycles to failure (N) for three aluminum alloys. Data points are not shown.

66 Chapter 3 Failure and Failure Criteria

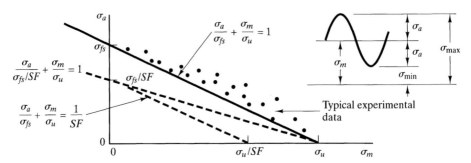

FIGURE 3.6-4 An interaction diagram known as a Goodman diagram (solid line). σ_a = amplitude of alternating stress, σ_m = mean or average stress, σ_{fs} = fatigue strength, σ_u = ultimate strength in tension. Note: $\sigma_m = (\sigma_{max} + \sigma_{min})/2$, $\sigma_a = (\sigma_{max} - \sigma_{min})/2$.

τ_a is superposed on mean shear stress or on mean normal stress. Dashed lines in Fig. 3.6-4 can be used for analysis and design. The upper dashed line, obtained by dividing σ_{fs} by SF (which is equivalent to multiplying σ_a by SF), is usually used with normally ductile materials. The lower dashed line, obtained by dividing both σ_{fs} and σ_u by SF, is usually used with normally brittle materials. For multiaxial states of stress in normally ductile materials, the von Mises failure criterion can be used to calculate "effective" values of σ_a and σ_m [3.9].

Cumulative damage is the name given to cyclic loading in which the load level changes from time to time (Fig. 3.6-1c). The problem is usually treated by the Palmgren-Miner rule [3.9], which states

$$\frac{n_1}{N_1} + \frac{n_2}{N_2} + \dots + \frac{n_n}{N_n} = 1 \quad \text{or} \quad \sum \frac{n_i}{N_i} = 1 \qquad (3.6\text{-}1)$$

where n_i is number of cycles actually applied at load level i, and N_i is the number of cycles required for failure at load level i. Equation 3.6-1 is based on the assumptions that fatigue damage at any stress level is directly proportional to the number of cycles at that level and that the ordering of stress levels is of no consequence. Neither assumption is strictly true. The equation is not a derivable law, but is simple and has adequate accuracy in most cases.

Factors That Affect Fatigue. Several factors, many of them under the control of the designer, can reduce the likelihood that a fatigue crack will begin or can reduce the growth rate of an existing crack. Most are ways to reduce stresses.

- Avoid "fretting," which is surface damage that results from the rubbing of parts against one another. The relative motion may be very small. Fretting is possible in various joints, including shrink fits (Fig. 2.7-3c). In gears and bearings, where some rubbing is inevitable, a lubricant helps by distributing contact forces more evenly, reducing friction forces, and carrying away heat.
- Redesign, by changing dimensions so that nominal stresses are reduced, by rounding internal and re-entrant corners so that stress concentration factors are reduced, or by making changes that reduce the cyclic loads.

- Avoid corrosive environments. Even ordinary moisture can greatly reduce fatigue strength (Fig. 3.6-5).
- A polished surface is most resistant to fatigue (Fig. 3.6-5). Accordingly, remove small surface scratches and tool marks, by *slow* grinding and polishing so as not to introduce tensile residual stress on the surface.
- Cold work the surface, perhaps by shot peening, so as to introduce compressive residual stress on the surface. As suggested by Fig. 3.6-4, a negative σ_m increases the allowable σ_a (or, for a given σ_a, a negative σ_m increases fatigue life).
- Treat the surface. A surface treatment can help or hurt. Helpful treatments include nitriding, to increase surface hardness, and electroplating, to protect against corrosion. However, the plated material should not be less resistant to fatigue than the underlying material and should not be in residual tension after the surface treatment.
- Use heat treatment to introduce residual compressive stress on the surface. A simple rule says "What cools (or solidifies) last is in tension." If there is residual tension on the interior of a shaft, there must be residual compression on the surface in order to preserve static equilibrium. This condition could be accomplished by quenching from a high temperature. In contrast, welding may produce the unfavorable result of residual tension on the surface.
- Use a material that is more resistant to fatigue.

Another reason for the "size effect" can now be noted. The thickness of a cold-worked surface layer in residual compression does not change much with specimen size. Accordingly, when load is applied to a large member, significant tensile stress may yet remain near the base of a cold-worked surface layer. (As an example, for both beams in Fig. 3.2-3, imagine the addition of residual compression of magnitude $\sigma/2$ to top and bottom surface layers of thickness $h_2/2$.)

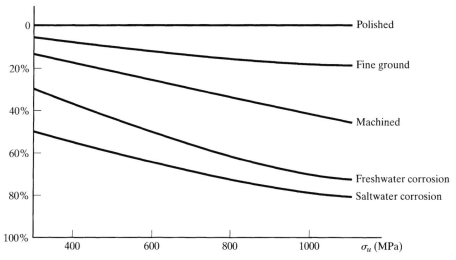

FIGURE 3.6-5 Approximate percentage reduction in fatigue strength, for steels of various static tensile strengths σ_u, due to unfavorable surface finish and environment.

An interesting case related to favorable prestress and fatigue is presented by the Comet aircraft. They were early jet passenger aircraft with a pressurized fuselage. Three disintegrated in flight. The cause was found to be fatigue damage. Yet none of these aircraft had undergone as many cycles of pressurization as the test fuselage, which did not fail after many more than the required number of cycles of pressure had been applied. Investigation showed that the aircraft that disintegrated had been proof tested to 1.3 times the operating gage pressure before being put in service. The test fuselage had been proof tested to twice the operating gage pressure before being fatigue tested by cycling between zero gage pressure and the operating pressure. Highly stressed locations in the test fuselage had yielded during the double-pressure proof test, so that favorable (compressive) residual stresses appeared when pressure was removed. Proof testing the aircraft to 1.3 times the operating gage pressure had not been sufficient to produce the yielding that would have resulted in favorable residual stresses [3.10].

Stress Concentration and Fatigue. A peak stress can be calculated with the assistance of a stress concentration factor K_t, as discussed in Section 2.7. If used with cyclic loading, K_t leads to an underestimate of fatigue strength because of an effect called *notch sensitivity*. For cyclic loading, K_t is modified to produce an *effective* or *fatigue* stress concentration factor K_f, which is less than K_t. Thus arises the somewhat paradoxical situation of K_t factors that for static loading are valid but usually not useful (because local yielding redistributes stresses without failure) and for cyclic loading are useful but not valid (unless they are modified).

Notch sensitivity is at least partly explainable as a size effect that reduces the importance of small-scale stress raisers. For example, a notch of very small radius may raise stresses scarcely more than they are already raised by small-scale flaws inherent in the structure of the material. In this case the material is said to be *notch insensitive*. In contrast, a hypothetical material in which a small notch is fully effective ($K_f = K_t$) is called *notch sensitive*. A definition of K_f is

$$K_f = \frac{\text{Fatigue strength of specimen without notch}}{\text{Fatigue strength of specimen with notch}} \qquad (3.6\text{-}2)$$

K_f may vary with material and the number of cycles for which the fatigue strength is defined. K_f and K_t can be related by the equation

$$K_f = 1 + q(K_t - 1) \qquad (3.6\text{-}3)$$

in which q is called the *notch sensitivity*, where $0 < q < 1$. Factor q approaches zero as the size of the notch approaches zero. Relations among K_t, K_f, and q are plotted qualitatively in Fig. 3.6-6. In practice, numerical values of K_f or q are obtained from tabulations generated by extensive testing of various materials. When alternating stress is superposed on mean stress, as in Fig. 3.6-1b, K_f is applied to the nominal alternating stress and K_t to the nominal mean stress in order to obtain the peak alternating stress σ_a and the peak mean stress σ_m.

Much empirical knowledge and data about fatigue has been accumulated. Serious work with fatigue problems should be preceded by study of a more extensive treatment of the subject than presented here.

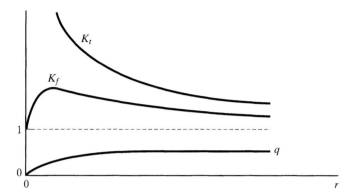

FIGURE 3.6-6 Qualitative variation of K_t, K_f, and q with notch radius r.

EXAMPLE

A flat bar is bent in its plane by a moment M that cycles about a mean value M_m (Fig. 3.6-7). The notch has stress concentration factor $K_t = 2.64$ and notch sensitivity $q = 0.7$. The material is ductile in static tension and has ultimate strength $\sigma_u = 470$ MPa. The fatigue strength is $\sigma_{fs} = 280$ MPa at the required number of loading cycles. If the mean moment is $M_m = 4000$ N·mm and the safety factor is to be $SF = 2$ based on the upper dashed line in Fig. 3.6-4, what alternating moment M_a can be allowed?

The moment of inertia at the net section is $I = 3(10^3)/12 = 250$ mm^4. Values of nominal mean stress and peak mean stress are

$$\sigma_{m(nom)} = \frac{M_m c}{I} = \frac{4000(5)}{250} = 80 \text{ MPa} \quad (3.6\text{-}4a)$$

$$\sigma_m = K_t \sigma_{m(nom)} = 2.64(80) = 211 \text{ MPa} \quad (3.6\text{-}4b)$$

From the upper dashed line in Fig. 3.6-4 we obtain the allowable peak alternating stress σ_a.

$$\frac{\sigma_a}{280/2} + \frac{211}{470} = 1 \quad \text{from which} \quad \sigma_a = 77.1 \text{ MPa} \quad (3.6\text{-}5)$$

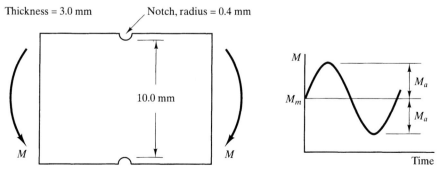

FIGURE 3.6-7 Flat bar of uniform thickness loaded by a cyclic bending moment M.

The allowable nominal alternating stress at the notch is $\sigma_{a(\text{nom})} = \sigma_a/K_f$, where $K_f = 1 + 0.7(2.64 - 1) = 2.15$. Hence $\sigma_{a(\text{nom})} = 77.1/2.15 = 35.9$ MPa. Finally, from the flexure formula, the allowable alternating moment is

$$M_a = \frac{\sigma_{a(\text{nom})} I}{c} = \frac{35.9(250)}{5} = 1794 \text{ N} \cdot \text{mm} \qquad (3.6\text{-}6)$$

If the analysis is repeated, using now the lower dashed line in Fig. 3.6-4, the denominator 470 in Eq. 3.6-5 is replaced by 470/2. This change leads to $\sigma_a = 14.3$ MPa, $\sigma_{a(\text{nom})} = 6.65$ MPa, and $M_a = 333$ N·mm. Clearly the lower dashed line has led to a much more conservative result.

PROBLEMS

3.2-1. (a) Derive Eq. 3.2-2. Suggestions: The radius of an arbitrary Mohr circle tangent to the envelope can be expressed in terms of its extreme principal stresses, σ_1 and σ_3. Combine this radius with the slope of the envelope, which can be expressed in terms of σ_{tf} and σ_{cf}.

(b) Consider a field of pure shear stress τ, as may be produced by twisting a shaft. In terms of σ_{tf} and σ_{cf}, what failure value of τ is predicted by the Mohr criterion?

3.2-2. Let σ_x and σ_y be nonzero principal stresses in a plane state of stress. Sketch a failure envelope, analogous to an envelope in Fig. 3.3-1, for the Mohr failure criterion. Let $\sigma_{cf} = 3\sigma_{tf}$.

3.2-3. A certain isotropic material fails in a brittle manner, at 40 MPa in uniaxial tension and at 240 MPa in uniaxial compression. At a certain point in a structure, one of the principal stresses is known to be 20 MPa tensile and another is known to be 90 MPa compressive. For what range of the remaining principal stress would failure *not* be expected if (a) the maximum normal stress criterion is used, and (b) the Mohr criterion is used?

3.2-4. A solid shaft of circular cross section has radius 40 mm. The material is isotropic and fails in a brittle manner, at 60 MPa in uniaxial tension and at 500 MPa in uniaxial compression. A 1800 kN compressive axial force is applied to the shaft. How large a torque can be added without failure? Use the Mohr failure criterion.

3.2-5. In Problem 3.2-4, let the shaft have radius r. Change the compressive axial force to 1200 kN and let the torque be 10 kN·m. The material is unchanged. For the Mohr failure criterion and a safety factor of 3.0, what radius r is required for the shaft?

3.3-1. Consider the maximum shear stress failure criterion and the states of principal stress treated in Fig. 3.3-1.

(a) For each quadrant (or smaller division) of the plot, describe how the plane of shear failure in the material is oriented with respect to the x, y, and z axes.

(b) In each quadrant (or smaller division) of the plot, σ_x, σ_y, and σ_z can be given appropriate principal stress labels σ_1, σ_2, or σ_3 (greatest to least in algebraic order). In each part of the plot, which labels are appropriate for σ_x, σ_y, and σ_z?

(c) In the fourth quadrant, the failure envelope has the equation $\sigma_x - \sigma_y = \sigma_Y$. Verify this equation. Also state the appropriate equations for other parts of the envelope.

3.3-2. Let σ_Y be the yield strength of a material, as determined by test under uniaxial stress. For each of the following plane states of stress ($\sigma_z = \tau_{yz} = \tau_{zx} = 0$), write an algebraic expression that describes the onset of yielding, in terms of σ_Y and the given stresses.

(a) Given nonzero σ_x and τ_{xy}, with $\sigma_y = 0$. Use Eq. 3.3-1.

(b) Given nonzero σ_x and τ_{xy}, with $\sigma_y = 0$. Use Eq. 3.3-2.

(c) Given nonzero σ_1 and σ_3, with $\sigma_2 = 0$. Use Eq. 3.3-2.
(d) Given nonzero σ_x, σ_y, and τ_{xy}. Use Eq. 3.3-2.

3.3-3. Consider the following three states of principal stresses, in MPa:

(1) $\sigma_1 = 8, \sigma_2 = 3, \sigma_3 = 1$; (2) $\sigma_1 = 6, \sigma_2 = 0, \sigma_3 = -1$; (3) $\sigma_1 = 7.5, \sigma_2 = 1, \sigma_3 = 0$.

Using the following failure criteria, rank these states of stress, from most dangerous to least.
(a) Maximum normal stress criterion.
(b) Mohr criterion, with $\sigma_{tf} = 10$ MPa and $\sigma_{cf} = 180$ MPa.
(c) Maximum shear stress criterion.
(d) Von Mises criterion.

3.3-4. In a certain isotropic material, yielding is observed to begin when stresses $\sigma_x = 90$ MPa, $\sigma_y = 30$ MPa, and $\tau_{xy} = 40$ MPa are reached in a state of plane stress. For what stress σ_Y is yielding predicted in uniaxial tension, according to (a) the maximum shear stress criterion, and (b) the von Mises criterion?

3.3-5. A certain isotropic material yields in uniaxial tension at the stress $\sigma_Y = 140$ MPa. At a certain point in a structure made of this material, two of the three principal stresses are 20 MPa tensile and 90 MPa compressive. If there is to be no yielding, what is the allowable range of the third principal stress according to (a) the maximum shear stress criterion, and (b) the von Mises criterion?

3.3-6. Imagine that yielding is observed in an isotropic material for the circumstances described in (a) and (b), which follow. In each case, what uniaxial yield stress σ_Y does this information predict, according to the maximum shear stress criterion and according to the von Mises criterion?

(a) Torsion of a shaft in which the maximum shear stress is 200 MPa.
(b) A solid bar, 40 mm in diameter, loaded by 480 kN of axial tensile force and by fluid pressure 160 MPa on its cylindrical surface.

3.3-7. A force P is applied to one end of a uniform L-shaped bar of solid circular cross section that is fixed at the other end, as shown on page 72 (Prob. 3.3-7). Force P acts normal to the plane of the bar. If the material is isotropic and yields at 280 MPa in a tension test, what value of P will initiate yielding? Consider the maximum shear stress failure criterion and the von Mises failure criterion.

3.3-8. A solid shaft of circular cross section, 140 mm in diameter, is loaded simultaneously by bending moment 8 kN·m and torque 12 kN·m. The material is isotropic and is known to yield at a stress of 200 MPa in uniaxial tension. What is the safety factor according to (a) the maximum shear stress failure criterion, and (b) the von Mises failure criterion?

3.3-9. A solid steel shaft, 0.020 m in diameter, yields when a torque of 400 N·m is applied. A cylindrical tank, 1.0 m in diameter and made of the same material, is to contain internal pressure $p = 3.0$ MPa. What wall thickness t is required for a safety factor of 2.0? Consider each of two applicable failure criteria.

3.3-10. A shaft of solid circular cross section must carry axial tensile force 30 kN simultaneously with torque 150 N·m. For a safety factor of 1.7, what should be the radius of the cross section? The material is isotropic and yields in uniaxial tension at the stress $\sigma_Y = 400$ MPa. (a) Use the maximum shear stress failure criterion. (b) Use the von Mises failure criterion.

3.3-11. A shaft of solid circular cross section must carry bending moment M simultaneously with torque T. Obtain a formula for the required radius r of the cross section, in terms of M, T,

72   Chapter 3   Failure and Failure Criteria

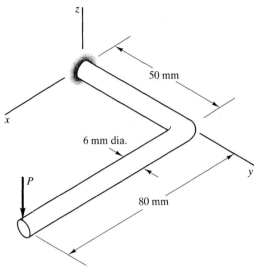

PROBLEM 3.3-7

safety factor SF, and the tensile yield strength σ_Y. (a) Use the maximum shear stress failure criterion. (b) Use the von Mises failure criterion.

3.5-1. In the example problem of Section 3.5, imagine that axial force P is increased by the factor $(\sigma_Y)_D/(\sigma_Y)_C$ when the steel is changed from C to D. What then is the critical crack length, to a good approximation?

3.5-2. The two plates shown are each 20 mm thick and are made of Steel C of Table 3.5-2. One contains a central circular hole, the other a central crack. Stress σ is produced by axial force P.
   (a) What force P will cause yielding in the plate with the hole? (See Eq. 2.7-2.)
   (b) What (approximately) is the force P that will break the plate with the hole?
   (c) What force P will break the plate with the crack?

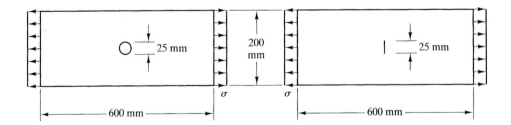

PROBLEM 3.5-2

3.5-3. An aluminum plate, 15 mm thick, has the length and width shown and the material properties stated in Table 3.5-2. For each of the following cases, determine the load (force or moment) that will produce yielding in an uncracked plate, and the load that will fracture a plate having a crack in direction AB.

(a) First case in Table 3.5-1, $2a = 80$ mm, $d = 160$ mm.
(b) Second case in Table 3.5-1, $a = 30$ mm, $d = 80$ mm.
(c) Third case in Table 3.5-1, $a = 30$ mm, $d = 80$ mm.

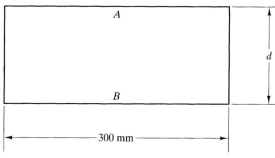

PROBLEM 3.5-3

3.5-4. A steel plate, 20 mm thick, has the dimensions shown and the properties of Steel C in Table 3.5-2. For each of the following cases of cracked plates, apply a load consistent with a safety factor of 3.0, based on the yield strength of the uncracked plate. Then determine the actual safety factor when the crack is taken into account. Each crack is to be located along AB.
(a) First case in Table 3.5-1, $2a = 34$ mm.
(b) Second case in Table 3.5-1, $a = 14$ mm.
(c) Third case in Table 3.5-1, $a = 14$ mm.

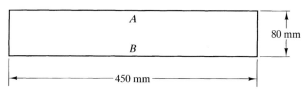

PROBLEM 3.5-4

3.5-5. A titanium plate, 20 mm thick, has the length and width shown and the material properties stated in Table 3.5-2. For each of the following cases determine, to a good approximation, the critical crack length. Each crack is to be located along AB.
(a) First case in Table 3.5-1, $d = 120$ mm, axial load $= 500$ kN.
(b) Second case in Table 3.5-1, $d = 60$ mm, axial load $= 140$ kN.
(c) Third case in Table 3.5-1, $d = 60$ mm, moment load $= 2.9$ kN·m.

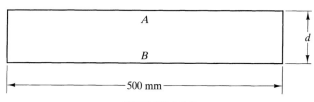

PROBLEM 3.5-5

3.5-6. The plates shown are each 12 mm thick, 36 mm wide, 140 mm long, and made of Steel D in Table 3.5-2. Each edge crack is 18 mm long. Axial load P may act along the top edge, the centerline, or 9 mm from the top edge, as shown. What value of P will produce fracture in each case?

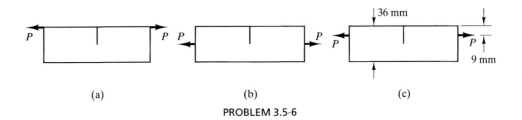

(a)   (b)   (c)

PROBLEM 3.5-6

3.5-7. Plot a failure envelope for Eq. 3.5-4 (change the $\geq$ sign to $=$). Plot K_I/K_{Ic} on the abscissa and K_{II}/K_{Ic} on the ordinate. Use $K_{IIc} = 0.75 K_{Ic}$. Identify the region of the plot for which conditions are considered safe from fracture.

3.5-8. **(a)** The sketch shows the neighborhood of a crack in a wide plate under uniaxial tension σ_x. The angle between the σ_x direction and the normal to the crack is θ. If fracture impends, what equation relates σ_x to angle θ? Let $K_{IIc} = 0.75 K_{Ic}$. [See useful equations stated in part (b).]

(b) Generalize part (a) by adding stress $\sigma_y = k\sigma_x$ as shown, where k is a positive constant. If fracture impends, what equation relates σ_x to k and θ? [Stresses normal and tangent to the crack are $\sigma = \sigma_x \cos^2\theta + \sigma_y \sin^2\theta$ and $\tau = (\sigma_y - \sigma_x)\sin\theta\cos\theta$, respectively.]

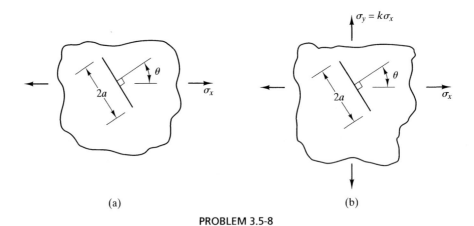

(a)   (b)

PROBLEM 3.5-8

3.5-9. The sketch shows a thin-walled tube of titanium (Table 3.5-2) loaded by torque T. A crack of length $2a = 40$ mm extends through the wall thickness of the tube. Assume that formulas in Section 3.5 are applicable, and determine the approximate torque T that will cause fracture if (a) $\theta = 0$, (b) $\theta = 22.5°$, and (c) $\theta = 45°$.

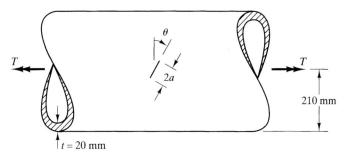

PROBLEM 3.5-9

3.6-1. A cyclic vertical force is applied to the center of the beam shown. The force causes the center to cycle 4.5 mm up and 4.5 mm down relative to the unstressed position. If the material is that of the middle curve in Fig. 3.6-3, what is the expected fatigue life? Assume that $K_f = 1.8$ for the notch.

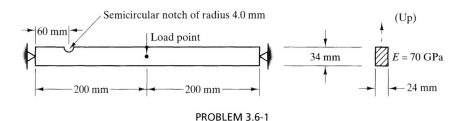

PROBLEM 3.6-1

3.6-2. A prismatic bar is loaded by an axial force that cycles between the values P_{max} and P_{min}. Obtain an expression for the required cross-sectional area in terms of P_{max}, P_{min}, the fatigue strength, the ultimate strength, and the safety factor. The material is ductile in static tension.

3.6-3. (a) A certain ductile material has ultimate tensile strength 1000 MPa and fatigue strength 300 MPa at 10^7 cycles. A part made of this material carries a 350 MPa mean tensile stress, on which is superposed an alternating stress $\sigma_a = 180$ MPa. Will the part survive 10^7 cycles?

(b) A more conservative form of Fig. 3.6-4, known as a Soderberg diagram, uses tensile yield strength σ_Y rather than σ_u for the intercept on the abscissa. If $\sigma_Y = 600$ MPa, what is the safety factor for the conditions stated in part (a)?

3.6-4. A 6000 N load rolls across a simply supported beam that spans 2.0 m. The beam must withstand 10^7 cycles of this loading with a safety factor of 2.4. Use the lowest curve in Fig. 3.6-3 and assume that the material is ductile. What should be the dimensions of a rectangular cross section, if its depth is to be twice its width? Is the design satisfactory if lateral deflection is to be limited to $1/360$ of the span?

3.6-5. The sketch shows a trial design for a bandsaw blade. Load P serves to tension the band. The band has a rectangular cross section 1 mm by 4 mm through the root of the teeth. Properties of the material are $E = 200$ GPa, $\sigma_u = 1000$ MPa, and $\sigma_{fs} = 400$ MPa at the number of cycles required. Assume that the material is so hard it should be regarded as brittle, and that stress concentration factors at roots of the teeth are $K_f = 1.9$ in tension and $K_f = 1.6$ in bending.

(a) What is the safety factor for this design?
(b) Will the likelihood of failure be reduced by increasing the thickness of the blade from 1 mm to 2 mm? Calculate the new safety factor.

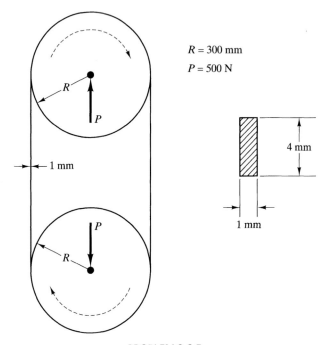

PROBLEM 3.6-5

3.6-6. A cylindrical tank has mean radius 450 mm and wall thickness 8.0 mm. The material has ultimate strength 680 MPa in tension and endurance limit 240 MPa. Internal pressure cycles between 1.7 MPa and 3.4 MPa. What is the safety factor? (Assume that the von Mises failure criterion is applicable, with σ_a and σ_m here regarded as the effective stresses σ_{ae} and σ_{me}.)

3.6-7. Answer the following questions about cumulative damage for the material depicted by the uppermost curve in Fig. 3.6-3. Assume that $\sigma_m = 0$.
(a) If $\sigma_a = 500$ MPa is applied for 8000 cycles, for how many additional cycles can $\sigma_a = 400$ MPa be applied?
(b) A part carries a load that creates the following stress pattern: $\sigma_a = 600$ MPa for 2 cycles, then $\sigma_a = 500$ MPa for 12 cycles. How many repetitions of this pattern will cause failure?

3.6-8. Imagine that because of concern about the safety of a certain design in steel, it is decided to use a steel of higher tensile strength than originally intended. Nothing else is changed. State whether, and why, the likelihood of failure would be decreased by using the stronger steel, if the mode of failure is
(a) Widespread yielding.
(b) Excessive deflection.
(c) Buckling.
(d) Brittle fracture due to an existing crack.
(e) Development of new cracks due to fatigue.

CHAPTER 4

Applications of Energy Methods

4.1 INTRODUCTION

Analytical methods in mechanics can be classified as *vectorial methods,* which are formulated in terms of vector quantities such as force and displacement, and *variational* or *energy methods,* which are formulated in terms of scalar quantities such as work and energy. In some problems an energy-based solution is easier because it does not require a sequence of free-body diagrams or geometric manipulation of force or displacement components. Another merit of some energy methods is their ability to produce an approximate solution for a problem too complicated to be solved exactly. Vectorial methods and energy methods are not mutually exclusive; it is sometimes convenient to use both in a given problem.

Energy methods for solid mechanics are quite general and have been cast in a great variety of forms. In this chapter we consider only two energy methods in some detail: the unit load method (Sections 4.4–4.7) and the method of stationary potential energy (Sections 4.10–4.13). Both methods can be applied whether or not the load versus displacement behavior is linear. However, the difficulties of practical nonlinear problems are usually so great that computer-based methods are more appropriate than pencil-and-paper methods. Accordingly, we will write equations of the unit load method and the stationary potential energy method in forms restricted to linearity. Bars and beams are used to illustrate the methods. Often we seek deflections, or perhaps only redundants if the structure is statically indeterminate.

Terminology. *Degrees of freedom* (d.o.f.) means the number of independent quantities needed to define a displaced configuration of a system. For example, consider a pin-jointed plane truss, built of uniform straight bars that terminate at frictionless joints, loaded only at the joints, and with displacements confined to the plane of the truss. In any bar, displacements vary linearly between the two joints at either end. Displacement of a joint is defined by its two in-plane displacement components. Therefore, a plane truss of n joints has $2n$ d.o.f. These d.o.f., plus knowledge of the original joint locations, are sufficient to define any displaced configuration of the truss.

If there are three support conditions sufficient to prevent rigid body motion of the truss as a whole in its plane, the number of nonzero d.o.f. becomes $2n - 3$. As another example, a straight bar, fixed at end $x = 0$ and carrying arbitrary axial load, requires infinitely many d.o.f. to describe how its axial displacement $u = u(x)$ varies along the length. However, if $u = u(x)$ is idealized as the polynomial $u = a_1 x + a_2 x^2 + \ldots + a_n x^n$, then there are n d.o.f. a_i. The a_i may be called *generalized coordinates*. They control the displacement, but in general no a_i is the displacement of a particular point.

In general statements of energy principles, "displacement" may include rotation as well as linear motion. Similarly, "force" may include moment as well as force. We will use "load" to mean a force that can move as a structure deforms, and "reaction" to mean a force that supports a structure and does not move. Following common practice in mechanics, we often use u, v, and w to denote the x, y, and z components of linear displacement at a point. Components of rotation about the respective coordinate axes are usually called θ_x, θ_y, and θ_z. When it is unnecessary to relate displacement components to a coordinate system, we will often use D and θ as generic symbols for linear displacement and angular displacement respectively.

Work and Energy. Work is done by a force if it has a component parallel to the displacement of the particle to which the force is applied. In Fig. 4.1-1a, work W done by force f as the particle to which it is applied moves from A to B, a distance D, is calculated as follows.

$$dW = (f \cos \beta) du \qquad \text{hence} \qquad W = \int_0^D f \cos \beta \, du \qquad (4.1\text{-}1)$$

where f and β may be functions of u. The component of f normal to the motion, $f \sin \beta$, does no work. Note that W is negative if $\pi/2 < \beta < 3\pi/2$. That is, a force displaced opposite to its direction does negative work. Thus, if a weight W is raised a distance h in a gravitational field, the weight does work $-Wh$. At the same time the external force $F = W$ applied to the weight to raise it does work $+Wh$. Potential energy of the weight increases an amount Wh.

Similar statements apply to work done by a couple: If a particle to which a couple C is applied rotates through an angle $d\theta$ whose vector is parallel to the couple vector, the increment of work done by the couple is $C \, d\theta$. Negative work is done if the couple vector and the rotation vector are oppositely directed.

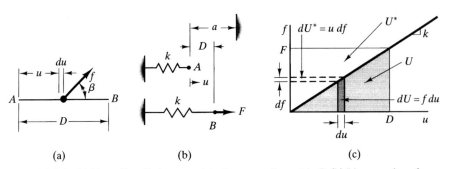

FIGURE 4.1-1 (a) Force F applied to a particle that moves from A to B. (b) Linear spring of stiffness k, before and after stretching. (c) Strain energy U stored in the spring.

When an elastic body is deformed by load, energy is conserved. That is, work done by the load is not lost but is stored as strain energy. Consider the linear spring of stiffness k in Fig. 4.1-1b. Axial load f creates axial displacement u. Let D be the final (equilibrium) value of u when f reaches its final value F. (Use of D in addition to u is not necessary, but may help in understanding this example.) Work done, and consequently strain energy U stored in the spring, is

$$U = \int_0^D f \, du = \int_0^D ku \, du = \frac{kD^2}{2} \qquad (4.1\text{-}2)$$

Since $F = kD$, we can substitute $D = F/k$, and write Eq. 4.1-2 in the form $U = FD/2$, which graphically is the shaded triangular area in Fig. 4.1-1c. Or, we can write Eq. 4.1-2 in the form $U^* = F^2/2k$, where the asterisk is added to U to indicate that strain energy is expressed in terms of force rather than in terms of displacement. Energy U^*, the unshaded triangular area in Fig. 4.1-1c, is called "complementary" strain energy. Clearly $U^* = U$ if the f versus u relation is a straight line.

Virtual Work. Several useful procedures in stress analysis arise from the principle of virtual work. A *virtual displacement* is an imaginary and very small change in configuration of a system. For analysis purposes it is understood that virtual displacements are displacements relative to the equilibrium configuration, when the full loads have been applied. Neither external loads nor internal forces and moments are altered by a virtual displacement. *Virtual work* is work done by all forces, both internal and external, that act on a system during a virtual displacement. Work done by internal forces is equal in magnitude, but opposite in sign, to the change in strain energy. As an example, in Fig. 4.1-1b let there be a virtual displacement du of point B to the right, measured from the equilibrium configuration $u = D$. Load F does positive virtual work in the amount $F \, du$. As in Eq. 4.1-2, the change in strain energy of the spring is $kD \, du$. Total virtual work dV is therefore $dV = (F - kD)du$.

An alternative viewpoint, perhaps less likely to create confusion about algebraic signs, is this: Virtual work is equal in magnitude, but opposite in sign, to the work an observer must do against internal and external forces in creating a virtual displacement. Thus in Fig. 4.1-1b we may isolate point B and consider forces that act on it, namely external load F directed rightward and spring force kD directed leftward. In moving point B an amount du toward the right, measured from the equilibrium configuration $u = D$, we do work $+kD \, du$ against the spring force and work $-F \, du$ against load F. Hence, in Fig. 4.1-1b, increments of total work dW and virtual work dV are

$$dW = kD \, du - F \, du \quad \text{and} \quad dV = -dW = (F - kD)du \qquad (4.1\text{-}3)$$

An *admissible displacement* produces an *admissable configuration,* which is a configuration that does not violate the constraints of internal compatibility and support conditions. That is, material of a continuum is constrained not to split apart or overlap itself; a beam can display no cusp; displacement (but not rotation) must be zero at a simple support; all motion is prevented at a fixed support; and so on.

The *Fourier inequality* states that a mechanical system is in static equilibrium if the virtual work is negative or zero for every admissible virtual displacement [4.1]. Consider, for example, a brick resting on the floor. Admissible displacements include lifting the brick and sliding it, but not lowering it. In either lifting or sliding the brick, an observer must do positive work, against gravity or against friction. Therefore the virtual work is negative, and according to the Fourier inequality the brick is in static equilibrium. If a horizontal load P acts on the brick, it also contributes to virtual work: if P were equal to the weight of the brick times the coefficient of static friction, virtual work associated with a horizontal virtual displacement would be zero. The brick would still be in static equilibrium, although on the verge of sliding. (In this example it does not matter whether displacements are small or large. It *would* matter if we wanted to show that a marble is in equilibrium at the bottom of a shallow dish. Too large a displacement would move the marble out of the dish.)

In the foregoing example, constraints are not workless because energy is dissipated due to friction. In the remainder of this chapter we deal with nondissipative systems. We can therefore apply the *principle of virtual work*, which states: *A system with workless constraints is in static equilibrium if and only if the virtual work is zero for any admissible virtual displacement.* Typically the principle is used to establish the equilibrium configuration. To apply the principle to Fig. 4.1-1b, we set $dV = 0$ in Eq. 4.1-3 and obtain $D = F/k$, which is the correct stretch of the spring due to force F.

Additional examples follow. In Section 4.10 we show that equations obtained in these examples can be produced more easily by the stationary potential energy method.

EXAMPLE 1

The system in Fig. 4.1-2a consists of a linear spring and a weightless bar AB of length L whose ends roll without friction on rigid horizontal and vertical surfaces. The spring is unstretched when $\theta = 0$. Determine the value of θ for equilibrium when load P is applied.

The system has one d.o.f., which we take as θ, although either u or v would also serve. Displacements u and v can be expressed in terms of θ. We determine the work dW an observer must do, against spring force ku and load P, during a virtual displacement $d\theta$, and apply the principle of virtual work. Thus, with $dW = -dV$,

$$-dV = ku\,du - P\,dv \qquad (4.1\text{-}4a)$$

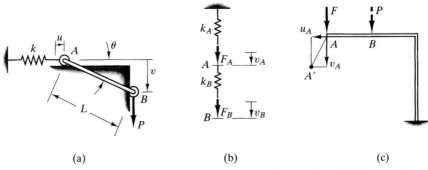

(a) (b) (c)

FIGURE 4.1-2 (a) Linear spring, weightless link, and load P, shown deflected. (b) System of two linear springs loaded by two forces, shown deflected. (c) Frame loaded by one force F or two forces F and P. Deflection components at A are shown.

where

$$u = L(1 - \cos\theta), \quad \text{hence} \quad du = L\sin\theta\, d\theta \quad (4.1\text{-}4b)$$

$$v = L\sin\theta, \quad \text{hence} \quad dv = L\cos\theta\, d\theta \quad (4.1\text{-}4c)$$

$$-dV = kL^2(1 - \cos\theta)\sin\theta\, d\theta - PL\cos\theta\, d\theta \quad (4.1\text{-}4d)$$

$$dV = 0 \quad \text{yields} \quad kL(1 - \cos\theta)\sin\theta - P\cos\theta = 0 \quad (4.1\text{-}4e)$$

The value of θ for equilibrium is obtained by solving the latter equation for θ.

EXAMPLE 2

A system of two linear springs is shown in Fig. 4.1-2b. Springs are unstretched when $v_A = 0$ and $v_B = 0$. Determine the values of v_A and v_B for equilibrium when loads F_A and F_B are applied.

The system has two d.o.f., namely v_A and v_B. The following considerations are not essential but help in obtaining correct algebraic signs. In considering work associated with virtual displacement dv_A, let us imagine that $v_A > v_B$, so that an observer does positive work against both springs. In considering work associated with virtual displacement dv_B, let us imagine that $v_B > v_A$, so that an observer does positive work against spring k_B. Thus

$$-dV = k_A v_A\, dv_A + k_B(v_A - v_B)dv_A + k_B(v_B - v_A)dv_B - F_A dv_A - F_B dv_B \quad (4.1\text{-}5)$$

Or, after gathering terms,

$$-dV = [k_A v_A + k_B(v_A - v_B) - F_A]dv_A + [k_B(v_B - v_A) - F_B]dv_B \quad (4.1\text{-}6)$$

The principle of virtual work states that dV must vanish for *any* admissible virtual displacement. This requirement means that either dv_A or dv_B might be zero while the other is not, or that dv_A might be any multiple of dv_B. For any and all possibilities, dV must vanish. This can happen only if the bracketed expressions in Eq. 4.1-6 vanish separately. Therefore

$$(k_A + k_B)v_A - k_B v_B = F_A \qquad v_A = \frac{1}{k_A}(F_A + F_B)$$

$$\text{hence} \quad (4.1\text{-}7)$$

$$-k_B v_A + k_B v_B = F_B \qquad v_B = \frac{1}{k_A}(F_A + F_B) + \frac{1}{k_B}F_B$$

Conservation of Energy. Work done by loads gradually applied to an elastic structure is stored as strain energy. This concept provides a simple way to determine the displacement or rotation of the load, when an elastic structure is loaded by a single force or a single couple. For example, for the linear spring in Fig. 4.1-1b, conservation of energy requires

$$\frac{1}{2}FD = \frac{1}{2}kD^2 \quad \text{from which} \quad D = \frac{F}{k} \quad (4.1\text{-}8)$$

which is the correct result.

However, the conservation of energy argument has limited usefulness. Consider Fig. 4.1-2c, in which loading causes point A to move to point A', with displacement components u_A and v_A. If a single load F is applied at A, the conservation argument provides v_A but not u_A. Or, if there are loads at point A *and* at point B, the conservation argument provides only one equation, but the equation contains two unknowns, namely v_A and the corresponding displacement v_B at B. Accordingly, for most problems we must use other methods that provide more equations.

Effective Stiffness. A linearly elastic structure loaded by a single force or a single couple can be represented by a linear spring of stiffness k. By definition, k is the force (or couple) divided by the corresponding displacement (or rotation) component of the loaded point. For example, if $P = 0$ in Fig. 4.1-2c, $k = F/v_A$ is the stiffness seen by load F at point A.

4.2 RECIPROCAL THEOREMS

A reciprocal theorem may enable a simple solution for a problem that would be difficult if approached in another way. Reciprocal theorems are restricted to structures whose displacements are directly proportional to load.

Consider a structure that carries two loads, Fig. 4.2-1a. Displacements produced by the two loads P_A and P_B are symbolized as follows.

D_{AA} = displacement component at A parallel to P_A and due to load P_A
D_{AB} = displacement component at A parallel to P_A and due to load P_B
D_{BA} = displacement component at B parallel to P_B and due to load P_A
D_{BB} = displacement component at B parallel to P_B and due to load P_B

Let load P_A be applied first, then load P_B. Work done is stored as strain energy U and is

$$U = \frac{1}{2} P_A D_{AA} + P_A D_{AB} + \frac{1}{2} P_B D_{BB} \qquad (4.2\text{-}1)$$

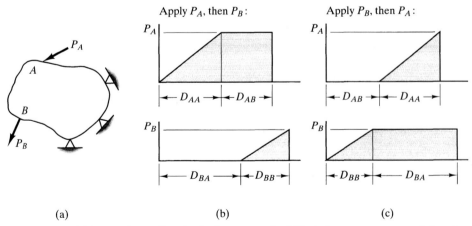

FIGURE 4.2-1 (a) Two loads on a linearly elastic structure. (b, c) Shaded areas represent work done by the loads, applied in different sequence.

Terms that carry the factor 1/2 are associated with loads that increase linearly from zero to their final values, and therefore do work as described in connection with Fig. 4.1-1c. In contrast, load P_A maintains its full value as it is moved through displacement D_{AB} by application of load P_B. Accordingly, there is no factor of 1/2 associated with $P_A D_{AB}$. Work terms in Eq. 4.2-1 correspond to areas in Fig. 4.2-1b. In similar fashion, if load P_B is applied first, then load P_A, work done is shown by areas in Fig. 4.2-1c. This work, stored as strain energy, is

$$U = \frac{1}{2} P_B D_{BB} + P_B D_{BA} + \frac{1}{2} P_A D_{AA} \qquad (4.2\text{-}2)$$

Strain energy U is the same in Eqs. 4.2-1 and 4.2-2 because of the principle of superposition, according to which the final state of deformation is independent of the order in which loads are applied. Therefore, by equating the right-hand sides of Eqs. 4.2-1 and 4.2-2,

$$P_A D_{AB} = P_B D_{BA} \qquad (4.2\text{-}3)$$

Equation 4.2-3 is a statement of *Maxwell's reciprocal theorem.* In words, it states that work done by a first load in moving through displacement produced by a second load is equal to work done by the second load in moving through displacement produced by the first load. It can be shown that the theorem is also valid for groups of loads. Thus, when stating the theorem in words, "load" and "displacement" are replaced by "load system" and "displacements." In the latter form the theorem is known as *Betti's reciprocal theorem* [4.2].

In the foregoing, the words "load" and "displacement" can be given their general meanings. That is, couples and their rotations are included; the theorem is not limited to forces and their displacements. Also, loads may be distributed as well as concentrated.

EXAMPLE 1

Consider cantilever beam AB in Fig. 4.2-2a. Lateral deflection due to loading by couple C alone is a known function of x, specifically $v_C = Cx^2/2EI$. From this information, what is the tip rotation due to uniformly distributed lateral load q?

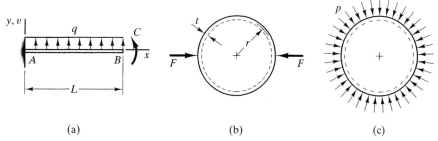

FIGURE 4.2-2 (a) Uniform cantilever beam. (b) Thin-walled spherical tank loaded by diametral pinching forces F. (c) Uniform external pressure p on the tank.

When couple C is applied, each force increment $q\,dx$ acts through displacement v_C, and thus does an increment of work $v_C(q\,dx)$. With θ_{Bq} the desired tip rotation due to load q, Eq. 4.2-3 takes the form

$$C\theta_{Bq} = \int_0^L v_C(q\,dx) \quad \text{where} \quad v_C = \frac{Cx^2}{2EI} \quad (4.2\text{-}4)$$

Integration yields

$$C\theta_{Bq} = q\frac{CL^3}{6EI} \quad \text{from which} \quad \theta_{Bq} = \frac{qL^3}{6EI} \quad (4.2\text{-}5)$$

EXAMPLE 2

A thin-walled spherical tank is pinched by collinear forces F that act along a diameter (Fig. 4.2-2b). What is the change in volume of the tank?

A solution by shell theory would be difficult, but the reciprocal theorem makes the problem easy. Let u represent radial displacement, taken as positive inward in this example. Forces F cause radial displacement u_F, which varies from point to point over the tank. A uniform radial pressure p, Fig. 4.2-2c, produces radial displacement u_p, constant over the tank. According to the reciprocal theorem,

$$2(Fu_p) = \int u_F(p\,dA) \quad \text{or} \quad 2Fu_p = p\,\Delta V_F \quad (4.2\text{-}6)$$

where dA is an increment of surface area of the tank, and the integral of $u_F\,dA$ has been recognized as the desired volume change ΔV_F due to forces F. Displacement u_p is equal to radius r times circumferential strain ϵ_c (Section 1.4). Circumferential strain is given by standard pressure vessel and stress-strain relations (Sections 1.4 and 2.5). Thus

$$u_p = r\epsilon_c = r\frac{1}{E}\left(\frac{pr}{2t} - \nu\frac{pr}{2t}\right) \quad \text{hence} \quad \Delta V_F = \frac{Fr^2(1-\nu)}{Et} \quad (4.2\text{-}7)$$

In this example, ΔV_F is a *decrease* in volume.

4.3 STRAIN ENERGY IN TERMS OF LOADS

Strain energy expressions written in terms of loads are useful for calculating deflections and for solving problems with a low degree of static indeterminacy. Our treatment is restricted to linearly elastic behavior, and to members whose loading can be described as axial force, transverse shear force, bending, and/or twisting.

To begin, consider the prismatic beam in Fig. 4.3-1a and a differential slice of the beam, Fig. 4.3-1b. Bending moment M causes opposite faces of the differential slice to rotate with respect to one another an amount $d\theta$, where, from elementary beam theory, $d\theta = M\,dx/EI$ (see also Eq. 4.8-3). The differential slice responds to load like a linear spring, as described in connection with Fig. 4.1-1c. In the present context, the spring has rotational stiffness. Complementary strain energies dU^* in the differential slice and U^* in the entire beam are

Section 4.3 Strain Energy in Terms of Loads

$$dU^* = \frac{1}{2}M\,d\theta = \frac{1}{2}M\left(\frac{M\,dx}{EI}\right) \quad \text{and} \quad U^* = \int_0^L \frac{M}{2}\frac{M\,dx}{EI} \quad (4.3\text{-}1)$$

In this example, and in general, M is a function of position, $M = M(x)$.

Figure 4.3-1c depicts a differential slice of a beam loaded by axial force N, transverse shear forces V_y and V_z, bending moments M_y and M_z, and torque T. These loads produce axial normal stress σ_x and transverse shear stresses τ_{xy} and τ_{zx}. All three stresses are in general functions of x, y, and z. Point g is the centroid of the cross section. Point S is the shear center, which is the point through which transverse forces V_y and V_z must be directed if they are to produce no tendency for the beam to twist about the x axis (the shear center is discussed in Chapter 10). Axes y and z are principal centroidal axes of the cross section (discussed in Section 1.3). On adjacent cross sections, separated by distance dx, the loads produce the following relative translations and rotations.

- Centroidal axial force N produces x-direction translation $du = N\,dx/EA$, where E is the elastic modulus and A is the area of the cross section.
- Torque T produces rotation $d\theta_x = T\,dx/GK$, where G is the shear modulus and K is a property of cross-sectional area A. For a circular cross section, $K = J$, where J is the centroidal polar moment of area A ($J = \pi r^4/2$ for a solid circular cross section of radius r). For a cross section of general shape, K may be a small fraction of J (see [1.6, 1.7] and Chapter 9).
- Moment M_y produces rotation $d\theta_y = M_y\,dx/EI_y$ about the y axis, where I_y is the moment of inertia of A about the y axis. Because axes yz are principal, M_y produces no rotation about the z axis (this matter is discussed in greater detail in Chapter 10).
- Similarly, moment M_z produces rotation $d\theta_z = M_z\,dx/EI_z$ about the z axis, where I_z is the moment of inertia of A about the z axis. M_z produces no rotation about the y axis.
- Forces V_y and V_z produce translations $dv = k_y V_y\,dx/GA$ and $dw = k_z V_z\,dx/GA$ in y and z directions respectively, where factors k_y and k_z are explained in what follows.

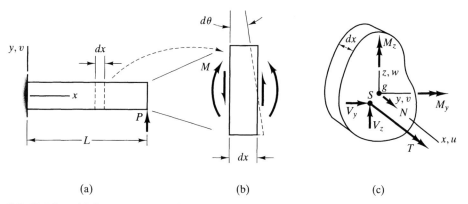

FIGURE 4.3-1 (a) Cantilever beam. (b) Relative rotation $d\theta$ of adjacent cross sections is produced by bending moment M. (c) Differential slice of a beam under a general load system.

Putting all this together, and writing results in the manner of Eq. 4.3-1, we arrive at the following expression for complementary energy U^* in a beam of length L.

$$U^* = \int_0^L \left(\frac{N\,N\,dx}{2\,EA} + \frac{T\,T\,dx}{2\,GK} + \frac{M_y\,M_y\,dx}{2\,EI_y} + \frac{M_z\,M_z\,dx}{2\,EI_z} + \frac{V_y\,k_y V_y\,dx}{2\,GA} + \frac{V_z\,k_z V_z\,dx}{2\,GA} \right) \quad (4.3\text{-}2)$$

This expression contains no product term $M_y M_z$ because y and z are *principal axes of the cross section*. There is no product term such as $V_y M_z$ because the work of M_z in moving through rotation produced by V_y is of higher order than the work of V_y in moving through translation produced by V_y. Similarly, the work of V_y in moving through translation produced by M_z is of higher order than the work of M_z in moving through rotation produced by M_z.

Equation 4.3-2 accounts for shear deflection parallel to a shear force V_y or V_z. If the cross section is unsymmetric, there also exists a shear deflection component directed normal to a shear force, even though y and z are principal axes [4.3]. The normal component is perhaps 20% of the force-parallel component, and terms that account for it are omitted from Eq. 4.3-2.

Most beams are slender enough that T, M_y, and M_z produce much larger stresses and deflections than do N, V_y, and V_z. In such cases, terms containing N, V_y, and V_z can be discarded from Eq. 4.3-2.

Transverse Shear Factors k_y and k_z. In Fig. 4.3-2a, let v_s represent the contribution of transverse shear strain to total tip deflection. An *approximate* value of v_s is

$$v_s = \theta_s L \approx (\gamma_{xy})_{\text{ave}} L = \frac{(\tau_{xy})_{\text{ave}}}{G} L = \frac{V_y L}{GA} \quad (4.3\text{-}3)$$

which implies that $k_y = 1$ in Eq. 4.3-2. To obtain a more accurate value of k_y we must account for the variation of τ_{xy} with y. Using the conservation of energy argument of Eq. 4.1-8, and Eq. 2.6-3 for strain energy, we write

$$\frac{1}{2} V_y v_s = \int \frac{\tau_{xy}^2}{2G} dV \quad \text{or} \quad \frac{1}{2} V_y v_s = \int_0^L \left(\int_{-h/2}^{h/2} \frac{\tau_{xy}^2}{2G} b\,dy \right) dx \quad (4.3\text{-}4)$$

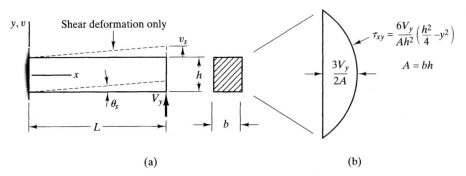

(a) (b)

FIGURE 4.3-2 (a) Cantilever beam, showing deflection v_s produced by transverse shear deformation. (b) Distribution of transverse shear stress in a solid rectangular cross section.

TABLE 4.3-1 Shear factors for use in Eq. 4.3-2.

Shape of Cross Section	k_y	k_z
Solid rectangle	1.20	1.20
Solid circle	≈ 1.11	≈ 1.11
Thin-walled circular cylinder	2.00	2.00
I section, web parallel to z axis	≈ 1.20[a]	≈ 1.00[b]
Closed thin-walled rectangular box	≈ 1.00[b]	≈ 1.00[b]

[a]For A use the combined cross-sectional areas of the flanges.
[b]For A use the cross-sectional area of the web (or webs for a box section).

The expression for τ_{xy} in a solid rectangular cross section is derived in texts on elementary mechanics of materials and is shown in Fig. 4.3-2b. The result of the calculations is

$$\text{Solid rectangular cross section:} \quad v_s = 1.2 \frac{V_y L}{GA} \qquad (4.3\text{-}5)$$

which shows that factor k_y of Eq. 4.3-2 is 1.2 for a solid rectangular cross section. Factors k_y and k_z for various shapes of cross section are stated in Table 4.3-1. All these factors are calculated from transverse shear stress distributions predicted by the elementary beam formula $\tau = VQ/It$.

The beam in Fig. 4.3-2a has tip deflection $v_b = V_y L^3/3EI_z$ due to bending. The ratio v_s/v_b is usually quite small unless the beam is very short.

4.4 UNIT LOAD METHOD

The unit load method is a convenient way of calculating deflections and resolving static indeterminacy in problems having a low degree of static indeterminacy. The present section considers the theory, first in the context of a simple example, then in a more general way. Example applications appear in subsequent sections.

Let us ask for the lateral tip deflection v_o produced by uniformly distributed load q on the cantilever beam in Fig. 4.4-1a. For calculation purposes only, we also apply the unit force shown, at the location where v_o is to be calculated and in the direction of v_o. The unit load is fictitious in the sense that it does not alter v_o. We adopt the following notation for bending moments.

m = bending moment due to the unit load
M = bending moment due to the actual load

These moments are plotted in Fig. 4.4-1b, c. Now imagine that the beam is in static equilibrium under the action of the unit load *alone*. We will apply the virtual work principle to this configuration, and (perhaps surprisingly) come out with v_o due to load q. The virtual work principle says we may consider *any* small admissible displacement relative to the equilibrium configuration. For the present argument, the crucial step lies in making the right choice. We choose the virtual displacement to be equal to the

88 Chapter 4 Applications of Energy Methods

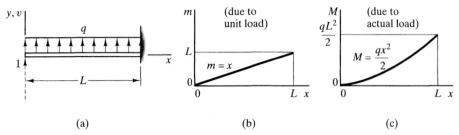

FIGURE 4.4-1 (a) Cantilever beam with uniformly distributed load q. Unit tip load is applied for the purpose of analysis. (b, c) Bending moments produced by the two loads.

displacement function $v = v(x)$ produced by applying the *actual* load q alone to the undeformed beam. It does not matter that the function $v = v(x)$ is not known or that it is superposed on the existing displacement produced by the unit load. During the virtual displacement, moment m on a differential slice of the beam acts through relative rotation $d\theta = M\,dx/EI$ and the unit load acts through translation v_o. Calculating work dW done against internal and external forces, as in Eq. 4.1-3, we have

$$dW = \int_0^L m\frac{M\,dx}{EI} - 1 \cdot v_o \qquad (4.4\text{-}1)$$

(The absence of differential symbols on the right-hand side of Eq. 4.4-1 need cause no alarm. Differential symbols merely indicate "small," and since $v = v(x)$ is a small deflection, we have not departed from the virtual work argument.) Since virtual work must vanish, $dW = 0$, and Eq. 4.4-1 yields

$$v_o = \int_0^L \frac{Mm\,dx}{EI} \qquad (4.4\text{-}2)$$

No factor of 1/2 appears in these terms because the unit load and its bending moment m remain at their full values while moving through deformations produced by the actual load q. With $m = x$ and $M = qx^2/2$, Eq. 4.4-2 yields the expected result, $v_o = qL^4/8EI$. The sign convention for m and M does not matter, except that both must use the *same* sign convention if deflection is to be positive in the direction of the unit load.

Equation 4.4-2 can also be obtained from the reciprocal theorem. Work $1 \cdot v_o$ of the unit load in displacing an amount v_o due to load q is equal to the work of load q in moving through lateral displacement created by the unit load. The latter work is accounted for internally rather than externally by integrating work increment $M\,(m\,dx/EI)$ over length L. Here M is moment due to q and $m\,dx/EI$ is rotation due to the unit load.

Equation 4.4-2 is an instance of the following more general formulation for beams. Let the beam be adequately supported, so that rigid body motion is not possible. Let forces and moments due to *actual* loads on a beam be as shown in Fig. 4.3-1c (all capital letters). Let the corresponding lower case letters represent forces and moments due to a *unit* load, calculated using the same sign convention as used for actual loads. Thus, for example, M_y and m_y are bending moments about the y axis, respectively due to the actual load and the unit load. The unit load must be a force if translation is desired, a couple if rotation is desired, and must be applied at the location and along the line of action of the desired deflection. Let D represent the desired deflection, be it translation or rotation. For a beam of length L,

$$D = \int_0^L \left(\frac{Nn}{EA} + \frac{Tt}{GK} + \frac{M_y m_y}{EI_y} + \frac{M_z m_z}{EI_z} + \frac{k_y V_y v_y}{GA} + \frac{k_z V_z v_z}{GA} \right) dx \quad (4.4\text{-}3)$$

The unit load method is also known as the *dummy load method* and the *Maxwell-Mohr method*. If the point to which the unit load is applied does not deflect parallel to the unit load, D is only the *component* of total deflection parallel to the unit load. If D comes out negative, it is in the direction opposite to the unit load. In particular problems, D may be identified as x-direction displacement u, or as rotation θ_x about the x axis, and so on.

Deflection D due to temperature change can be calculated from Eq. 4.4-3 by replacing $N\,dx/EA$, $M_y\,dx/EI_y$, and so on by deformations produced by temperature change. Thus, for example, if temperature varies with x, $N\,dx/EA$ is replaced by $\alpha\,\Delta T\,dx$, where α is the coefficient of thermal expansion and ΔT is the temperature change.

Castigliano's Second Theorem. This theorem, also known simply as Castigliano's theorem, is a general theorem for linearly elastic structures. It provides Eq. 4.4-3 when applied to beams. The general theorem can be derived as follows.

The arbitrary body in Fig. 4.4-2a is arbitrarily loaded. Deformation produced by loads moves point i to position i'. We propose to calculate D_i, which is the component of displacement of point i in the direction of load P_i. Let U^* be the (complementary) strain energy in the body when loads $P_1, P_2, P_3, \ldots, P_i, \ldots, P_n$ are applied. If load P_i is then incremented an amount dP_i, the total strain energy stored is

$$U^* + \frac{\partial U^*}{\partial P_i} dP_i \quad (4.4\text{-}4)$$

Now let the order of load application be reversed, so that load dP_i is applied first, followed by addition of loads $P_1, P_2, P_3, \ldots, P_i, \ldots, P_n$. Work done, and hence strain energy stored, is

$$\frac{1}{2} dP_i\, dD_i + dP_i\, D_i + U^* \quad (4.4\text{-}5)$$

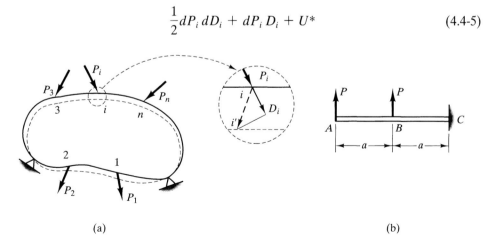

(a) (b)

FIGURE 4.4-2 (a) Linearly elastic body with several applied loads. Inset shows the deflection ii' of the ith load. (b) Cantilever beam loaded by two forces P.

where term $dP_i\, dD_i/2$ represents work done when load dP_i is applied alone and term $dP_i\, D_i$ represents work done by dP_i as it is displaced by other loads while maintaining its full value. Regardless of the order of load application, the same final state of deformation is achieved. Therefore expressions 4.4-4 and 4.4-5 are equal. The U^* terms cancel, and the higher-order term $dP_i\, dD_i/2$ can be discarded. What remains is

$$D_i = \frac{\partial U^*}{\partial P_i} \qquad (4.4\text{-}6)$$

For application to beams, U^* is given by Eq. 4.3-2. As a simple example of how Eq. 4.4-6 may be used, consider the calculation of tip deflection v_L in the beam in Fig. 4.3-1a, with only the bending energy retained. Equations 4.3-2 and 4.4-6 yield

$$v_L = \frac{\partial}{\partial P}\left(\int_0^L \frac{M_z^2\, dx}{2EI_z}\right) \quad \text{or} \quad v_L = \int_0^L \frac{M_z}{EI_z}\frac{\partial M_z}{\partial P}dx \qquad (4.4\text{-}7)$$

In the latter expression, differentiation has been performed before integration, as is usually more convenient. In this example, $M_z = PL - Px$, so $\partial M_z/\partial P = L - x$, and we obtain the expected result, namely $v_L = PL^3/3EI_z$. (As is customary in plane problems, subscripts z on M_z and I_z may be omitted.) Applications of Castigliano's theorem to straight beams are discussed near the end of Section 4.7. An application to sharply curved beams appears in Section 6.7.

The unit load method follows directly from Castigliano's theorem. When Eq. 4.4-6 is applied to Eq. 4.3-2, we obtain the terms $\partial N/\partial P_i$, $\partial T/\partial P_i$, and so on. These quantities are *rates of change* of N, T, etc., with respect to P. They are independent of the magnitude of P. Therefore we may as well use $P = 1$, and we have now arrived at Eq. 4.4-3 where $n = \partial N/\partial P_i$, $t = \partial T/\partial P_i$, and so on.

If we require a displacement where no load is applied, we must use Castigliano's theorem in the following way. Consider, for example, tip deflection of the beam in Fig. 4.4-1. We add a transverse tip load P, so that the bending moment is $M_z = Px + qx^2/2$. After doing the differentiation with respect to P, we set $P = 0$, and obtain the expected result $qL^4/8EI_z$ as the tip deflection due to load q alone.

A pitfall of Castigliano's theorem lies in the treatment of two or more loads that are equal or are proportional to one another. In Fig. 4.4-2b, for example, the calculation $\partial U^*/\partial P$ is found to produce the *sum* of deflections at loaded points rather than deflection at either loaded point. (To circumvent the difficulty, we give one load a different name, say Q for the load at A, then replace Q by P after writing $\partial U^*/\partial Q$ to determine deflection at A.) This pitfall is easy to tumble into, particularly when treating a statically indeterminate problem. Unfortunately, erroneous results may be plausible rather than being so strange that it becomes obvious that something is wrong. Chances for error are reduced if the unit load method is chosen in preference to Castigliano's theorem.

4.5 EXAMPLE SOLUTIONS WITH STRAIGHT MEMBERS

In this section the unit load method is used to determine deflections of statically determinate structures. The discussion continues in Section 4.6, where curved members are treated. Equation 4.4-3 is used frequently, with subscripts omitted if they are not needed in the problem at hand.

EXAMPLE 1

The uniform beam in Fig. 4.5-1a carries uniformly distributed load q. End A is simply supported. End B may deflect vertically but cannot rotate. Determine the vertical deflection at B due to load q.

We will ignore the contribution of transverse shear to deflection. Clearly B will deflect downward, so it would be convenient to direct the unit load at B in the downward direction. However, for consistency with a sign convention in which positive deflections are in positive coordinate directions, we elect to direct the unit load upward at B. Bending moments due to load q and due to unit upward load at B are respectively

$$M = qLx - \frac{qx^2}{2} \quad \text{and} \quad m = -x \tag{4.5-1}$$

In M, the term qL is the reaction at A. In m, the reaction at A is unity (downward), so $m = -1 \cdot x$. We need only one moment term from Eq. 4.4-3 to obtain vertical deflection v_B at B. Thus

$$v_B = \int_0^L \frac{Mm}{EI} dx \quad \text{hence} \quad v_B = -\frac{5qL^4}{24EI} \tag{4.5-2}$$

That v_B is negative means it is opposite in direction to the unit load at B.

EXAMPLE 2

In the structure of Fig. 4.5-1b, assume that connections at A, C, and D are frictionless. Consider only the bending deformation of beam ABC and the stretching deformation of spring CD, and determine the vertical deflection at B due to load P.

If axial force is constant over length L of a uniform straight bar, the axial-load term in Eq. 4.4-3 yields NnL/EA. But EA/L is recognized as axial stiffness k of the bar, where k is load divided by the deflection it produces. Accordingly $NnL/EA = Nn/k$, and in this form the axial-load term applies directly to spring CD of Fig. 4-5-1b. Elementary statics shows that force in the spring is $N = \sqrt{2}P/2$ due to load P and $n = -\sqrt{2}/2$ due to a unit upward load at B. As for beam ABC, bending moments from A to B, due respectively to load P and to a unit upward load at B, are

$$M = \frac{Px}{2} \quad \text{and} \quad m = -\frac{x}{2} \quad \text{from } A \text{ to } B \tag{4.5-3}$$

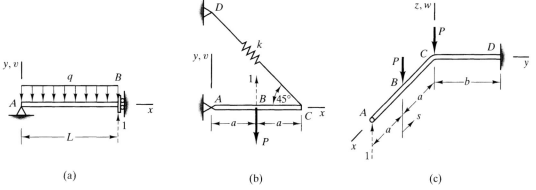

FIGURE 4.5-1 (a) Uniformly loaded beam. (b) Beam with one end supported by a linear spring. (c) Bent bar with out-of-plane loading.

Bending deformation in the beam is symmetric about B, hence strain energy due to bending is the same in each half of the beam. We may therefore integrate only from A to B and double the result. Vertical deflection at B is therefore

$$v_B = \frac{Nn}{k} + 2\int_0^a \frac{Mm}{EI}dx \quad \text{hence} \quad v_B = -\frac{P}{2k} - \frac{Pa^3}{6EI} \quad (4.5\text{-}4)$$

Negative signs indicate that v_B is downward.

An alternative method of solution would be to add the center deflection of a simply supported beam to the deflection at B produced by stretching of the spring. This solution method would involve geometric manipulation of deflection components, which is avoided by the unit load method.

EXAMPLE 3

A slender uniform bar $ABCD$ lies in the xy plane (Fig. 4.5-1c). The cross section is circular, so I is its moment of inertia about any centroidal axis, and constant K for twisting is $K = 2I$. Calculate the vertical deflection at A due to the z-parallel loads P.

Because the bar is slender, we will include only the contributions of bending and twisting. We apply a unit load in the z direction at A. It is convenient to introduce an auxiliary coordinate s that originates at B, as shown. From B to C there is only bending about the y axis.

$$M_y = Ps \quad \text{and} \quad m_y = -a - s \quad \text{from } B \text{ to } C \quad (4.5\text{-}5)$$

From C to D there is bending about the x axis and torque.

$$M_x = 2Py \qquad m_x = -y$$
$$\text{and} \qquad \text{from } C \text{ to } D \quad (4.5\text{-}6)$$
$$T = Pa \qquad t = -2a$$

Vertical deflection at A, positive in the z direction, is

$$w_A = \int_0^a \frac{M_y m_y}{EI}ds + \int_0^b \frac{M_x m_x}{EI}dy + \int_0^b \frac{Tt}{GK}dy \quad (4.5\text{-}7a)$$

$$w_A = -\left(\frac{5a^3 + 4b^3}{6EI} + \frac{2a^2 b}{GK}\right)P \quad (4.5\text{-}7b)$$

Integration over AB is not required because loads P create no moment or torque in AB. We see that point A has no deflection components in the xy plane because an x- or y-direction unit load at A produces zero values of m_x, m_y, and t throughout $ABCD$, so that products $M_x m_x$, $M_y m_y$, and Tt are all zero.

EXAMPLE 4

The plane truss in Fig. 4.5-2a consists of uniform two-force members connected by frictionless pins. Determine the vertical deflection at joint B due to loads P.

Forces in members due to the actual loads and due to a unit upward load at B are shown in Fig. 4.5-2b, c. The only term needed from Eq. 4.4-3 is the first. Since axial force is constant along

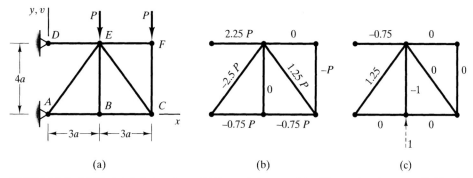

FIGURE 4.5-2 (a) Pin-jointed plane truss. (b) Forces in bars, positive if in tension, due to applied loads. (c) Forces in bars, positive if in tension, due to upward unit load at B.

the length of each member, and each member is uniform, this term becomes the sum of NnL/EA contributions from all eight bars of the truss.

$$D_B = \sum_{i=1}^{8} \frac{N_i n_i L_i}{E_i A_i} \qquad (4.5\text{-}8)$$

However, in the calculation of vertical deflection at B the product $N_i n_i$ is nonzero only for members AE and DE. Vertical deflection at B is therefore, for the case of E and A the same in all bars,

$$D_B = \frac{(-2.5P)(1.25)(5a)}{EA} + \frac{(2.25P)(-0.75)(3a)}{EA} = -\frac{20.7Pa}{EA} \qquad (4.5\text{-}9)$$

D_B is negative because it is opposite in direction to the unit load applied.

What if relative motion between points B and F were required? We could apply a unit load at F in direction BF and calculate the BF-parallel component of deflection at F, then apply a unit load at B in direction BF and calculate the BF-parallel component of deflection at B. The former deflection minus the latter is the BF-parallel deflection component of F relative to B, positive in the direction B to F. Upon writing out terms of the calculations we would discover that the same result is obtained by applying oppositely directed unit loads at B and F *simultaneously*, so that n in Eq. 4.5-8 represents the combined effect of two BF-parallel unit loads. The summation then extends over only the five bars stressed by simultaneous application of the two unit loads, namely BC, CF, FE, EB, and CE. Since $N_i = 0$ for bars FE and EB, the final summation includes only three nonzero terms.

4.6 CURVED MEMBERS

The unit load method is particularly well suited to problems of rings and arches. Emphasis is placed on circular rings in what follows. By "circular ring" we mean a member whose centerline forms a plane circular arc. Usually the arc is less than a complete circle. In applying Eq. 4.4-3 we assume that the cross section is solid and its dimensions are small in comparison with the radius of the arc. Otherwise corrections may be needed (see Sections 6.1, 6.5–6.7).

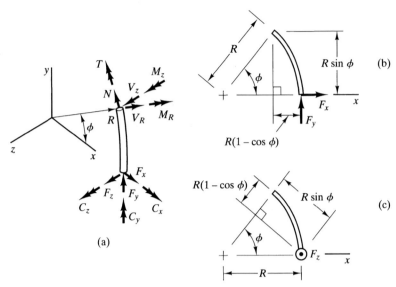

FIGURE 4.6-1 (a) Internal forces and moments in a circular ring due to loads applied at $\phi = 0$. Moment vectors are directed according to the right-hand rule. (b, c) Moment arms used to calculate internal moments.

Circular Ring Formulas. Figure 4.6-1a shows a portion of a ring, with the xy plane placed in the plane of the ring. The portion shown is a free body diagram, with externally applied forces and couples at end $\phi = 0$ and internal forces and couples at the other end, where the portion shown has been cut free of the remainder of the ring. Internal actions are determined from conditions of static equilibrium, by summing forces and moments with reference to radial, axial, and z directions at the cut end. Figure 4.6-1b shows moment arms used to calculate the contributions to M_z of in-plane forces F_x and F_y. Figure 4.6-1c shows moment arms used to calculate the contributions to M_R and T of out-of-plane force F_z. Results of all such calculations are

$$\begin{aligned}
N &= F_x \sin \phi - F_y \cos \phi \\
V_R &= -F_x \cos \phi - F_y \sin \phi \\
V_z &= -F_z \\
M_R &= F_z R \sin \phi - C_x \cos \phi - C_y \sin \phi \\
M_z &= -F_x R \sin \phi - F_y R(1 - \cos \phi) - C_z \\
T &= -F_z R(1 - \cos \phi) + C_x \sin \phi - C_y \cos \phi
\end{aligned} \quad (4.6\text{-}1)$$

EXAMPLE 1

Slender uniform half-ring ABC in Fig. 4.6-2a is loaded at B by force P, which is directed normal to the plane of the ring and into the paper. Determine the deflection at C.

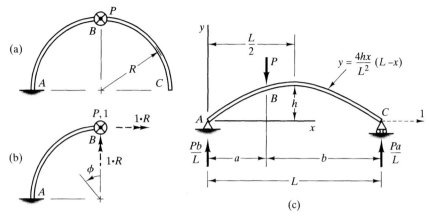

FIGURE 4.6-2 (a) Plane half-ring loaded by force P directed normal to the plane of the ring. (b) Loads used to calculate deflection components at C. (c) Parabolic arch loaded by in-plane force P.

When integrated from B to C, all terms in Eq. 4.4-3 yield zero because BC carries no force or moment due to applied load P. We may therefore treat only portion AB, taking care that a unit load applied at C is properly transferred to B. Due to applied load P in Fig. 4.6-2b, M_z is zero and

$$M_R = -PR \sin \phi \quad \text{and} \quad T = PR(1 - \cos \phi) \quad (4.6\text{-}2)$$

A unit load at C, parallel to P and in the same direction, produces unit force, moment $1 \cdot R$, and torque $1 \cdot R$ at B (Fig. 4.6-2b). Hence, from B to A,

$$m_R = (-1)R \sin \phi - (1 \cdot R) \cos \phi - (-1 \cdot R) \sin \phi = -R \cos \phi \quad (4.6\text{-}3a)$$

$$t = -(-1)R(1 - \cos \phi) + (1 \cdot R) \sin \phi - (-1 \cdot R) \cos \phi = R(1 + \sin \phi) \quad (4.6\text{-}3b)$$

The ring is slender, so we ignore transverse shear deformation. For a ring, length increment dx becomes $R \, d\phi$. Ring-normal deflection at C is therefore

$$D_C = \int_0^{\pi/2} \frac{M_R m_R}{EI} R \, d\phi + \int_0^{\pi/2} \frac{Tt}{GK} R \, d\phi \quad (4.6\text{-}4)$$

Integration is facilitated by Table 4.6-1. Equations 4.6-2, 4.6-3, and 4.6-4 yield

$$D_C = \frac{PR^3}{2EI} + \left(\frac{\pi - 1}{2}\right) \frac{PR^3}{GK} \quad (4.6\text{-}5)$$

In-plane unit loads at C produce nonzero m_z but zero m_R and t; accordingly, there is no in-plane deflection at C due to load P.

EXAMPLE 2

The slender parabolic arch in Fig. 4.6-2c is pinned at A and roller-supported at C. Vertical load P at B is in the plane of the arch. Explain in detail the calculations needed to determine the horizontal displacement at C.

TABLE 4.6-1 Selected integration formulas

No.	$f(\phi)$	$\int f(\phi)\, d\phi$	$\int_0^{\pi/2} f(\phi)\, d\phi$	$\int_0^{\pi} f(\phi)\, d\phi$	$\int_0^{3\pi/2} f(\phi)\, d\phi$	$\int_0^{2\pi} f(\phi)\, d\phi$
1	$\sin\phi$	$-\cos\phi$	1	2	1	0
2	$\cos\phi$	$\sin\phi$	1	0	-1	0
3	$\sin^2\phi$	$\frac{1}{2}\left(\phi - \frac{1}{2}\sin 2\phi\right)$	$\frac{\pi}{4}$	$\frac{\pi}{2}$	$\frac{3\pi}{4}$	π
4	$\cos^2\phi$	$\frac{1}{2}\left(\phi + \frac{1}{2}\sin 2\phi\right)$	$\frac{\pi}{4}$	$\frac{\pi}{2}$	$\frac{3\pi}{4}$	π
5	$\sin\phi\cos\phi$	$\frac{1}{2}\sin^2\phi$	$\frac{1}{2}$	0	$\frac{1}{2}$	0
6	$1 - \cos\phi$	$\phi - \sin\phi$	$\frac{\pi}{2} - 1$	π	$\frac{3\pi}{2} + 1$	2π
7	$(1 - \cos\phi)^2$	$\frac{3}{2}\phi - 2\sin\phi + \frac{1}{4}\sin 2\phi$	$\frac{3\pi}{4} - 2$	$\frac{3\pi}{2}$	$\frac{9\pi}{4} + 2$	3π
8	$(1 - \cos\phi)\sin\phi$	$-\cos\phi - \frac{1}{2}\sin^2\phi$	$\frac{1}{2}$	2	$\frac{1}{2}$	0
9	$(1 - \cos\phi)\cos\phi$	$\sin\phi - \frac{1}{2}\left(\phi + \frac{1}{2}\sin 2\phi\right)$	$1 - \frac{\pi}{4}$	$-\frac{\pi}{2}$	$-1 - \frac{3\pi}{4}$	$-\pi$
10	$\phi\sin\phi$	$\sin\phi - \phi\cos\phi$	1	π	-1	-2π
11	$\phi\cos\phi$	$\cos\phi + \phi\sin\phi$	$-1 + \frac{\pi}{2}$	-2	$-1 - \frac{3\pi}{2}$	0
12	$(1 - \cos\phi)(1 + \sin\phi)$	$\phi - \sin\phi - \cos\phi - \frac{1}{2}\sin^2\phi$	$\frac{\pi}{2} - \frac{1}{2}$	$2 + \pi$	$\frac{3\pi}{2} + \frac{3}{2}$	2π

Because the arch is slender, we need consider only bending moment. Due to load P, bending moment is

$$M_a = \frac{Pb}{L} x \qquad \text{for } 0 < x < a \qquad (4.6\text{-}6a)$$

$$M_b = \frac{Pb}{L} x - P(x - a) \qquad \text{for } a < x < L \qquad (4.6\text{-}6b)$$

where Pb/L is the support reaction at A. Due to a unit horizontal load at C, bending moment is

$$m = 1 \cdot y \qquad \text{which is} \qquad m = \frac{4hx}{L^2}(L - x) \qquad (4.6\text{-}7)$$

Horizontal displacement at C, positive to the right, is

$$u_C = \int \frac{Mm\,ds}{EI} \qquad \text{or} \qquad u_C = \int_0^a \frac{M_a m}{EI} f\,dx + \int_a^L \frac{M_b m}{EI} f\,dx \qquad (4.6\text{-}8)$$

where ds is an increment of length along the arch. The first integral spans the curved length of the arch; each of the latter two integrals spans a horizontal distance. Factor f in the latter two integrals relates ds to dx, and is determined as follows.

$$ds = (dx^2 + dy^2)^{1/2} = \left[1 + \left(\frac{dy}{dx}\right)^2\right]^{1/2} dx \approx \left[1 + \frac{1}{2}\left(\frac{dy}{dx}\right)^2\right] dx \qquad (4.6\text{-}9)$$

Factor f is given by either of the two bracketed expressions. The former expression is exact; the latter is approximate. The approximation is valid if the arch is shallow, so that slope dy/dx is small. However, the approximation for ds is only 6% high for $dy/dx = 1$. Its error is greatest at A and C, whereas the greatest contribution to u_C comes from the portion of the arch nearest B, where bending moments are greatest and the approximation for ds is more accurate. Therefore the error in u_C is appreciably less than the maximum error in the approximation for ds.

A simple numerical solution may also be considered, for use with either pencil and paper calculation or computer spreadsheet. The length of the arch can be divided into a finite number of length increments Δs. At the center of each increment, numerical values of $M, m, E,$ and I can be computed and assumed constant over the span of the increment. Thus, the first integral in Eq. 4.6-8 becomes a sum over the number of length increments Δs (or, a sum over length increments Δx when using forms that include factor f). This kind of numerical solution is convenient if flexural stiffness EI is not constant throughout the arch.

4.7 STATICALLY INDETERMINATE PROBLEMS

In this section we show by example how the unit load method can be used to solve statically indeterminate problems. The unit load method is suited to hand calculation and to problems that are indeterminate to the first, second, or possibly third degree. It is awkward to program the method for computer solution.

A *redundant* is a force or a moment that can be removed without making a structure a mechanism and thus destroying its stability. A statically indeterminate structure may have external and/or internal redundants. If external, there are more support reactions than needed for equilibrium; if internal, there are excess connections (for example, one or more bars of a truss that might be removed without destroying stability). The number of redundants is equal to the degree of static indeterminacy. In a given problem the choice of redundants is not unique, and some choices may make calculations easier than others.

In general, when using the unit load method, one calculates redundants in terms of applied loads as follows. First, redundants are chosen, equal in number to the degree of static indeterminancy. Then redundants are regarded as loads, applied to a structure that is now statically determinate. Values of redundants are determined by enforcing conditions of compatibility. Typically this means stating in equation form that displacement is zero at certain locations. The number of such equations is equal to the number of redundants.

In the following examples we assume that deformations due to axial force and transverse shear force can be neglected in members that also carry bending moment or torque.

EXAMPLE 1

The left end of the uniform cantilever beam in Fig. 4.7-1a is supported at B by a linear spring of stiffness k, which is unstressed when uniformly distributed load q is absent. Determine the force in the spring and deflection v_B at B due to application of load q.

The structure is statically indeterminate to the first degree. Any one of support reactions R, Q, or M_C can be used as the redundant. Here we elect to use R. The spring can be treated in the manner shown by Eq. 4.5-4. Force in the spring and bending moment in the beam, due to actual loads and the unit load, are

$$N = -R \quad \text{and} \quad M = Rx - \frac{qx^2}{2} \tag{4.7-1a}$$

$$n = -1 \quad \text{and} \quad m = x \tag{4.7-1b}$$

where N and M are force and bending moment produced by actual loads, and n and m are force and bending moment produced by a unit load acting upward at A. The base of the spring at A does not displace vertically, so Eq. 4.4-3 yields

$$0 = \frac{Nn}{k} + \int_0^L \frac{Mm}{EI} dx \tag{4.7-2}$$

where EA/L in Eq. 4.4-3 becomes k in the present problem. Redundant R is obtained from Eqs. 4.7-1 and 4.7-2. Finally, v_B can be calculated from the compression of the spring.

$$R = \frac{3kqL^4}{8(3EI + kL^3)} \quad \text{and} \quad v_B = -\frac{R}{k} = -\frac{3qL^4}{8(3EI + kL^3)} \tag{4.7-3}$$

One can check that this solution yields the correct v_B for the limiting case of no spring ($k=0$) and the correct R for the limiting case of a simple support at B (infinite k).

An alternative choice of redundant is force Q. Thus, end C is regarded as having the support condition seen at B in Fig. 4.5-1a, which allows deflection but not rotation, and Eqs. 4.7-1 are replaced by

$$N = -qL + Q \quad \text{and} \quad M = (qL - Q)x - \frac{qx^2}{2} \tag{4.7-4a}$$

$$n = 1 \quad \text{and} \quad m = -x \tag{4.7-4b}$$

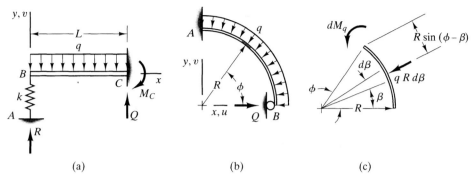

FIGURE 4.7-1 (a) Cantilever beam with a spring support. (b) Quarter-ring with a roller support. (c) Moment dM_q produced by increment $qR\,d\beta$ of distributed load.

when n and m are produced by a unit load acting upward at C. Equation 4.7-2 now states that point C has no vertical deflection, and yields reaction force Q.

Yet another choice of redundant is moment M_C. Thus, end C is regarded as simply supported, and Eqs. 4.7-1 are replaced by

$$N = -\frac{qL}{2} + \frac{M_C}{L} \quad \text{and} \quad M = \left(\frac{qL}{2} - \frac{M_C}{L}\right)x - \frac{qx^2}{2} \tag{4.7-5a}$$

$$n = \frac{1}{L} \quad \text{and} \quad m = -\frac{x}{L} \tag{4.7-5b}$$

where n and m are produced by a clockwise unit couple load at C. Equation 4.7-2 now states that the beam has no rotation at C, and yields reaction moment M_C.

Remarks. Note that internal forces and moments due to a unit load can be calculated using statics: the unit load is applied to a structure that we have made *statically determinate* by regarding excess supports as loads (that is, as redundants). Deformations of this structure are identical to those of the original statically indeterminate structure because calculated values of redundants are such as to enforce the original constraints.

As a calculation device, when the unit load is merely a unit value of a redundant, one can obtain internal forces and moment by differentiation. For example, in Eqs. 4.7-5, $n = \partial N/\partial M_C$ and $m = \partial M/\partial M_C$. Here N and M must be the original expressions, not modified by substitution of redundants subsequently determined.

EXAMPLE 2

The uniform quarter-ring in Fig. 4.7-1b is roller-supported at B and is loaded by a uniformly distributed radial load q. Determine the vertical deflection at B.

Before calculating deflection we must resolve the indeterminacy so that reaction force Q becomes known in terms of distributed load q. Bending moment due to q is not obvious and must be determined by integration. Auxilliary angle β is introduced, as shown in Fig. 4.7-1c. An increment of load $qR\,d\beta$, Fig. 4.7-1c, has moment arm $R\sin(\phi-\beta)$. Therefore at angle ϕ the bending moment due to q is

$$M_q = \int_0^\phi [R\sin(\phi-\beta)](qR\,d\beta) = qR^2(1-\cos\phi) \tag{4.7-6}$$

which, incidentally, is the moment that would be associated with a concentrated force qR acting vertically downward at B. We elect to use reaction Q as the redundant. Bending moments due to actual loads and due to a unit load acting rightward at B are respectively

$$M = qR^2(1-\cos\phi) - QR\sin\phi \quad \text{and} \quad m = -R\sin\phi \tag{4.7-7}$$

The sign convention for bending moment is arbitrary, but is the same for M and m. Here M and m are regarded as positive when they act to decrease the radius of curvature of the ring. Point B on the ring has zero horizontal displacement. Therefore

$$0 = \int_0^{\pi/2} \frac{Mm}{EI} R\,d\phi \quad \text{yields} \quad Q = \frac{2qR}{\pi} \tag{4.7-8}$$

Vertical deflection at B may now be calculated. For this purpose the quarter-ring can be regarded as statically determinate and subjected to known loads q and Q. M remains as stated in Eq. 4.7-7. Bending moment m due to unit vertical load at B, directed upward, is $m = -R(1-\cos\phi)$. Hence

$$v_B = \int_0^{\pi/2} \frac{M[-R(1-\cos\phi)]}{EI} R\,d\phi = \left(\frac{1}{\pi} + 2 - \frac{3\pi}{4}\right)\frac{qR^4}{EI} = -0.0379\frac{qR^4}{EI} \tag{4.7-9}$$

That v_B is negative means that point B on the ring moves opposite to the direction of the unit load; that is, B moves downward, as is reasonable.

EXAMPLE 3

In the arch in Fig. 4.6-2c, make the support condition at C the same as at A, so that the arch is attached to the ground by hinges at A and C. Establish the equation that serves to determine the horizontal component of support reaction at A and at C.

Each of the bending moments in Eqs. 4.6-6 must be augmented by Hy, where H is the desired horizontal force, considered positive when acting to the right on the arch at C. Equation 4.6-7 is unchanged. As C does not move relative to A, the present problem requires that $u_C = 0$ in Eq. 4.6-8. If EI is constant, we obtain

$$H = -\int_0^L M_P\,y\,ds \bigg/ \int_0^L y^2\,ds \tag{4.7-10}$$

where M_P is the bending moment due to load P alone, as stated by Eqs. 4.6-6, and $y = y(x)$ is the known shape of the arch. (If EI were unity and there were a roller support at C, each integral in Eq. 4.7-10 would represent horizontal displacement at C; the numerator for load P, the denominator for a unit horizontal load at C.) Axial deformation has not been included in Eq. 4.7-10. The error of this omission is small unless the arch is quite shallow and not slender [4.4].

EXAMPLE 4

A uniform semicircular arch is clamped at both ends and centrally loaded, as shown in Fig. 4.7-2a. Explain and compare strategies for determining the support reactions.

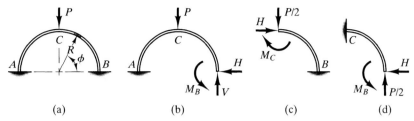

FIGURE 4.7-2 (a) Semicircular arch with fixed ends. (b–d) Diagrams used in various methods of determining redundants.

The problem is statically indeterminate to the third degree. If reactions at B in Fig. 4.7-2b are regarded as the redundants, we must impose at B the conditions of zero horizontal displacement, zero vertical displacement, and zero rotation. There are two expressions for bending moment, one valid from B to C, the other from C to A. Three simultaneous equations must be solved to determine H, V, and M_B in terms of P, R, E, and I.

Considerable effort is saved if symmetry is exploited. We imagine that load P is applied as two loads, each $P/2$, which straddle point C an infinitesimal distance apart. A vertical section through point C exposes internal force H and internal moment M_C, but no transverse shear force (because of symmetry). We now regard H and M_C as redundants, and determine them from two simultaneous equations, which say that at point C the quarter-circle arch in Fig. 4.7-2c has no horizontal displacement and no rotation.

Figure 4.7-2d shows another alternative, similar to that of Fig. 4.7-2c. Vertical reaction $P/2$ is known because of symmetry, and H and M_B are determined by stating that there is no horizontal displacement and no rotation at B. The quarter-rings in Figs. 4.7-2c and 4.7-2d have the same deformation. The latter quarter-ring is translated upward relative to the former, an amount equal to the magnitude of the (downward) displacement at C in the arch. Rigid body motion has no effect on the state of deformation, and consequently has no effect on internal forces or support reactions.

EXAMPLE 5

The plane truss in Fig. 4.7-3a has six uniform bars. The two diagonal bars are not connected where they cross. The truss is internally indeterminate to the first degree. Explain how the indeterminacy can be resolved.

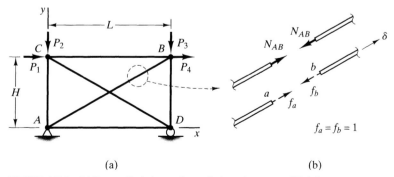

FIGURE 4.7-3 (a) Internally indeterminate six-bar plane truss. (b) A bar cut open, showing internal force N_{AB} and unit loads f_a and f_b applied for the purpose of analysis.

We arbitrarily choose force N_{AB} in diagonal bar AB as the redundant, and cut the bar open (Fig. 4.7-3b). The truss is now statically determinate, and axial forces N_i in all bars ($i = 1, 2, \ldots, 6$) can be stated in terms of external loads P_1 through P_4 and force N_{AB}, which is to be determined. Unit loads f_a and f_b at the cut are shown in Fig. 4.7-3b. We now write displacements δ_a and δ_b at cut ends a and b as if unit loads f_a and f_b were applied separately rather than simultaneously. With n_{ai} and n_{bi} the force bar in i due to f_a and f_b respectively, Eq. 4.5-8 yields

$$\delta_a = \sum_{i=1}^{5} \frac{N_i n_{ai} L_i}{E_i A_i} + \frac{N_{AB} f_a L_{Aa}}{E_{AB} A_{AB}} \quad \text{where} \quad f_a = 1 \quad (4.7\text{-}11a)$$

$$\delta_b = \sum_{i=1}^{5} \frac{N_i n_{bi} L_i}{E_i A_i} + \frac{N_{AB} f_b L_{Bb}}{E_{AB} A_{AB}} \quad \text{where} \quad f_b = 1 \quad (4.7\text{-}11b)$$

Each summation extends over all bars except bar AB. No gap or overlap occurs across ab, therefore $\delta_a = -\delta_b$, or $\delta_a + \delta_b = 0$. Hence

$$0 = \sum_{i=1}^{5} \frac{N_i (n_{ai} + n_{bi}) L_i}{E_i A_i} + \frac{N_{AB} n_{AB} (L_{Aa} + L_{Bb})}{E_{AB} A_{AB}} \quad \text{where} \quad n_{AB} = 1 \quad (4.7\text{-}12)$$

Now $n_{ai} + n_{bi}$ is simply n_i, the force in any bar i due to unit axial force in bar AB. Also, $L_{Aa} + L_{Bb} = L_{AB}$. Accordingly, Eq. 4.7-12 is simply stated as

$$0 = \sum_{i=1}^{6} \frac{N_i n_i L_i}{E_i A_i} \quad (4.7\text{-}13)$$

where summation extends over all bars of the truss. This equation contains redundant N_{AB} as the only unknown. Equation 4.7-13 states that at the imagined cut, the relative approach or separation of cut ends a and b is zero. This result may be made more plausible by imagining that the cut in AB is made at A. Then $\delta_a = 0$, and δ_b becomes the absolute displacement of end b, which clearly must be zero in order to preserve compatibility.

For a truss with more bars and greater static indeterminacy, one can write as many equations as there are redundants. Each equation has the form of Eq. 4.7-13. However, the method quickly becomes laborious. Computer programs, usually based on a different analytical method, are well suited to truss analysis and are recommended for all but the simplest truss problems.

The argument that produces Eqs. 4.7-11 to 4.7-13 is not limited to truss analysis. The essential statement is that a continuous member does not break apart or develop a cusp at an imaginary cut. Equations that express this statement serve to determine the internal forces and moments at any location in a member.

Castigliano's Second Theorem. In statically indeterminate problems, there is an appreciable chance that Castigliano's theorem (Eq. 4.4-6) will be misapplied. For example, imagine that loading on the beam in Fig. 4.7-1a consists of a couple M_o at B, rather than distributed loading q, and that rotation θ_B of the beam at B is required. One first calculates reaction R in terms of M_o, by any method. A correct result for θ_B is obtained by expressing bending moment M in terms of R and M_o, forming U^* in terms of R and M_o, writing $\theta_B = \partial U^* / \partial M_o$, and finally substituting for R its value in terms of M_o. An *incorrect* result for θ_B is obtained by substituting for R in terms of M_o before differentiation, so that M and U^* are expressed in terms of M_o alone, then writing $\theta_B = \partial U^* / \partial M_o$. This pitfall is the same as described in the last paragraph of Section 4.4. The unit load method avoids this chance for error because the basic equation, Eq. 4.4-3, is a post-differentiation result into which redundants may be substituted after they have been determined.

As another example of how Castigliano's theorem may be misapplied, imagine that the spring in Fig. 4.7-1a is infinitely stiff, so that the beam acts as if it were simply supported at B. End C remains fixed. If we regard both Q and M_C as redundants, write bending moment M and energy U^* in terms of q, Q, and M_C, and attempt to impose fixity at C by writing $\partial U^*/\partial Q = 0$ and $\partial U^*/\partial M_C = 0$, we obtain values of Q and M_C appropriate to a beam fixed at *both* ends. This happens because Eq. 4.4-6 presumes that equilibrium is satisfied when Q and M_C can vary independently [4.2]. But if Q and M_C are independent, equilibrium cannot be satisfied if end B is simply supported. To treat end B as simply supported we would be obliged to relate Q and M_C by the equation of moment equilibrium about B, then use either $\partial U^*/\partial Q = 0$ or $\partial U^*/\partial M_C = 0$ (we cannot use both; the problem is statically indeterminate to only the first degree). With the unit load method, it may be more obvious that there is something wrong with regarding Q and M_C as independent: the beam, supported only by the spring at its left end, would not be in equilibrium under a unit load corresponding to Q or M_C.

Castigliano's theorem can be made to work for the cantilever beam example of the preceding paragraph by satisfying equilibrium implicitly as part of the solution, rather than explicitly at the outset. The trick is to add λg to U^*, where λ is a Lagrange multiplier and $g = QL - M_C - qL^2/2$. In general, $g = 0$ is an equation of constraint; here it is an equilibrium equation stating that moments about B sum to zero. Unknowns now include λ. Partial derivatives of U^* with respect to Q, M_C, and λ are each set to zero, and correct results for Q and M_C are obtained. More generally, in order to implicitly satisfy the n equations of equilibrium $g_i = 0$, U^* must be augmented by terms $\lambda_i g_i$, where $i = 1, 2, \ldots, n$. Derivatives with respect to each of the original variables and with respect to each multiplier λ_i are set to zero. Thus, if there are m original variables, $m + n$ simultaneous equations must be solved.

4.8 FORMULAS FROM GEOMETRIC CONSIDERATIONS

Formulas that result from the unit load method can also be obtained directly by considering the geometry of deformation. Although the same equations are obtained, geometric arguments are worthy of study in order to improve physical understanding and to make the equations more than merely a manipulation device.

Consider a plane bar with arbitrary in-plane loads, Fig. 4.8-1a. The bar need not be uniform, but we will assume that all cross sections have the xy plane as a plane of symmetry so that all displacements are in the xy plane. Deformations due to axial force are considered first. At an arbitrary point C, axial force N can be determined from the known loading and geometry. Length increment ds elongates an amount $d\delta$, and $d\delta$ can be resolved into its x and y components du and dv.

$$d\delta = \frac{N\,ds}{EA} \qquad du = d\delta \sin\alpha \qquad dv = -d\delta \cos\alpha \qquad (4.8\text{-}1)$$

Displacement components at an arbitrarily located point B due to axial deformation of the bar are each the sum of all contributions from B to D.

$$u_B = \int_B^D \frac{N \sin\alpha}{EA}\,ds \qquad v_B = \int_B^D \frac{N(-\cos\alpha)}{EA}\,ds \qquad (4.8\text{-}2)$$

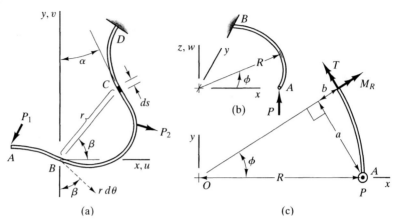

FIGURE 4.8-1 (a) Curved bar in the xy plane loaded by arbitrary forces in its plane. (b) Quarter-circle bar with out-of-plane load P. (c) Internal moment and torque in the quarter-circle bar due to load P.

Terms $\sin \alpha$ and $-\cos \alpha$ can be regarded as axial forces at C due, respectively, to unit x-direction load at B and unit y-direction load at B. Thus, $\sin \alpha$ or $-\cos \alpha$ is the form of n in the first term of Eq. 4.4-3 appropriate to calculation of u_B or v_B. As usual, expressions for N and n must observe the same sign convention.

Displacement at the arbitrary point B due to bending can be formulated as follows. First we must write an expression for relative rotation between the two ends of length increment ds. Let y' be an axis defined by the intersection of the xy plane with the plane of a cross section, and let $y' = 0$ at the neutral axis of the cross section. Then

$$\left. \begin{array}{l} \text{From Eq. 1.6-1:} \quad d\theta = -\dfrac{\epsilon\, ds}{y'} \\[6pt] \text{Also:} \quad \epsilon = \dfrac{\sigma}{E} = \dfrac{1}{E}\left(-\dfrac{My'}{I}\right) = -\dfrac{My'}{EI} \end{array} \right\} \quad \text{hence} \quad d\theta = \dfrac{M}{EI}\, ds \quad (4.8\text{-}3)$$

If we temporarily imagine that the bar is rigid except for length increment ds, and that the end of ds farthest from B is fixed, then the rotation at B is also $d\theta$, and the displacement at B is equal to $r\, d\theta$. This displacement can be resolved into its x and y components. Thus

$$d\theta_B = d\theta \qquad du_B = (r\, d\theta) \sin \beta \qquad dv_B = -(r\, d\theta) \cos \beta \qquad (4.8\text{-}4)$$

But $r \sin \beta = y$ and $r \cos \beta = x$, where x and y are coordinates of C with respect to B. Letting all length increments between B and D deform, and adding their contributions at B, we obtain from Eqs. 4.8-3 and 4.8-4

$$\theta_B = \int_B^D \frac{M}{EI}\, ds \qquad u_B = \int_B^D \frac{My}{EI}\, ds \qquad v_B = -\int_B^D \frac{Mx}{EI}\, ds \qquad (4.8\text{-}5)$$

Now x and y can be regarded as bending moments at C due, respectively, to a unit y-direction load at B and a unit x-direction load at B. Thus, x and y are the forms of m in

Eq. 4.4-3 appropriate to calculation of u_B or v_B due to bending. In the first of Eqs. 4.8-5, $m = 1$. Signs in Eqs. 4.8-5 correspond to M positive for the directions of P_1 and P_2 shown in Fig. 4.8-1a, and m positive for unit loads at B in the positive x and positive y directions.

Similar arguments apply to out-of-plane displacements. Consider, for example, the z-direction deflection w_A at A in Fig. 4.8-1b. The z-direction motion dw_A at A produced by bending and twisting of a length increment $ds = R\, d\phi$ that carries bending moment M_R and torque T is

$$dw_A = \frac{M_R\, ds}{EI} a + \frac{T\, ds}{GK} b \quad \text{hence} \quad w_A = \int_A^B dw_A \quad (4.8\text{-}6)$$

where, as shown in Fig. 4.8-1c,

$$a = R \sin \phi \quad \text{and} \quad b = R(1 - \cos \phi) \quad (4.8\text{-}7)$$

A unit load at A, upward and parallel to load P, produces bending moment $m_R = R \sin \phi$ and torque $t = R(1 - \cos \phi)$ at angle ϕ. Therefore Eqs. 4.8-6 are also equations of the unit load method.

While helping with physical understanding, geometric considerations require careful attention to algebraic signs. For example, no negative sign is attached to T in Eq. 4.8-6, despite the incorrect assumed direction for T shown in Fig. 4.8-1b. With the unit load method, confusion with signs is reduced merely by using the same convention for T and t.

For the special case of a *straight* beam with loading in a single plane, Eqs. 4.8-5 can be interpreted as follows. Let the beam lie along the x axis, so that $ds = dx$. Also let B lie at $x = 0$, and D lie at positive x. Then the first of Eqs. 4.8-5 says that relative rotation between points B and D is equal to the area of the M/EI diagram between B and D. The last of Eqs. 4.8-5 says that lateral deflection at B relative to a tangent to the beam axis at D is equal to the first moment of the M/EI diagram about B. (If D is not a fixed support, this relative deflection is in general not the deflection relative to a fixed reference frame.) These interpretations are known as the *moment-area method*. Details may be found in elementary texts about mechanics of materials.

4.9 STRAIN ENERGY IN TERMS OF DISPLACEMENTS

Strain energy expressions are particularly useful in approximate solution methods. As an example, the popular finite element method for general stress analysis can be described as a method that uses piecewise-continuous displacement fields to satisfy the virtual work principle. In following sections, beam problems serve as vehicles to explain approximate solutions based on displacement fields. Energy expressions needed for these solutions are described in the present section.

Straight Members. Figure 4.9-1 shows a typical cross section of a beam, with axial force N, bending moments M_y and M_z, and torque T. Transverse shear forces are not shown; we will consider only bars under axial load and slender members, for which transverse shear deformation may be neglected. Point O is the centroid of the cross

Chapter 4 Applications of Energy Methods

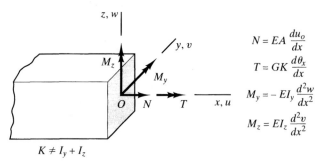

$K \neq I_y + I_z$

FIGURE 4.9-1 Axial force, torque, and bending moments in a straight beam.

section, and u_o is its axial displacement. In the expression for N in Fig. 4.9-1, du_o/dx represents axial strain ϵ_x along the line of centroidal axes of cross sections. This term is the usual definition of normal strain as change in length divided by original length, applied to a length dx that changes in length an amount du_o. Thus, an expression for $u_o = u_o(x)$ yields ϵ_x as a function of x by differentiation. The expression for N in Fig. 4.9-1 may be substituted into the first term in Eq. 4.3-2, with U^* replaced by U to indicate that strain energy is expressed in terms of displacement quantities rather than force quantities. Thus, for a straight bar of length L under axial load,

$$\text{Bar, axial load:} \quad U = \int_0^L \frac{EA}{2}\left(\frac{du_o}{dx}\right)^2 dx \quad (4.9\text{-}1)$$

In the expression for T in Fig. 4.9-1, the deformation term is rate of twist $d\theta_x/dx$ about the x axis. We use K, rather than polar moment of inertia J of the cross-sectional area, to permit treatment of noncircular cross sections (see [1.6] and Chapter 9). Expressions for M_y and M_z in Fig. 4.9-1 are the familiar moment-curvature relations from elementary beam theory (see Eq. 1.7-1). Substituting torque and moment terms into Eq. 4.3-2, and assuming that terms containing N, V_y, and V_z are negligible, we obtain

$$\text{Slender beam:} \quad U = \int_0^L \left[\frac{GK}{2}\left(\frac{d\theta_x}{dx}\right)^2 + \frac{EI_y}{2}\left(\frac{d^2w}{dx^2}\right)^2 + \frac{EI_z}{2}\left(\frac{d^2v}{dx^2}\right)^2\right] dx \quad (4.9\text{-}2)$$

Circular Rings. Consider a circular ring of mean radius R, loaded in its plane. Let us assume that the ring is sufficiently slender that axial deformation can be neglected in comparison with bending deformation. Then deformation in the plane of the ring can be described entirely by radial displacement v, which is a function of angular coordinate ϕ. It can be shown [4.5] that the strain energy of flexure of a slender ring bent in its own plane is

$$U = \int \frac{EI}{2}\left(\frac{1}{R^2}\frac{d^2v}{d\phi^2} + \frac{v}{R^2}\right)^2 R\, d\phi \quad (4.9\text{-}3)$$

The second-derivative term is analogous to d^2v/dx^2 in Fig. 4.9-1. The term v/R^2 is the change in curvature produced by radial displacement $v = v(\phi)$, assuming the "inextensibility condition"—that axial strain is zero, so that v produces bending but not axial strain. Thus, the term in parentheses is curvature κ, and bending moment is $M = EI\kappa$. Length increment dx has become arc length $R\, d\phi$. Because of its two-term form, Eq. 4.9-3 is not so readily used as Eq. 4.9-2 if an approximate solution is to be obtained by hand calculation. Computer solutions may be preferred.

4.10 PRINCIPLE OF STATIONARY POTENTIAL ENERGY

The following arguments are closely related to arguments regarding work, energy, and virtual work in Section 4.1, which the reader may wish to review. Equations produced by the principle of stationary potential energy can also be produced by the principle of virtual work. Most readers will probably conclude that the potential energy approach is more straightforward.

Consider the system of two springs shown in Fig. 4.10-1a. The potential energy of this system, given the symbol Π, is

$$\Pi = \frac{1}{2}k_A v_A^2 + \frac{1}{2}k_B(v_B - v_A)^2 - F_A v_A - F_B v_B \qquad (4.10\text{-}1)$$

Terms that contain k_A and k_B represent strain energy U in the springs. These terms come from Eq. 4.1-2, which reads $U = kD^2/2$, where D is the stretch (or shortening), relative to the unstressed configuration, of a spring of stiffness k. In the present example, $D = v_A$ for the upper spring and $D = v_B - v_A$ for the lower spring. Since D is squared, writing $D = v_A - v_B$ produces the same result. The latter two terms in Eq. 4.10-1 represent potential energy of the loads. These terms are negative because the same direction is taken as positive for load and displacement. Thus, if a constant force F_A displaces a distance v_A, F_A does positive work $F_A v_A$ and thereby loses potential in this amount. (If a weight W were raised an amount v in a gravitational field, with v positive upward, the weight would *gain* potential in the amount Wv.) The datum for potential energy of a load is arbitrary. For example, in Fig. 4.1-1b we could write either $-Fu$ or $F(a - u)$ as the potential of load F; in subsequent differentiation with respect to u, the added constant Fa would disappear. One can regard Π as the work an observer must do against all forces, internal and external, in displacing the system from the unstressed configuration $v_A = v_B = 0$.

The principle of stationary potential energy applies to a conservative system, which is a system in which work needed to alter the system from one configuration to another is independent of the path taken. Thus, for example, a system in which energy

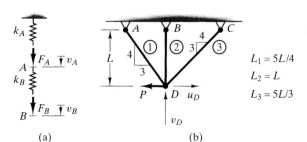

FIGURE 4.10-1 (a) System of two linear springs loaded by two forces, shown deflected. (b) Three-bar plane truss.

$L_1 = 5L/4$
$L_2 = L$
$L_3 = 5L/3$

is dissipated by friction is *not* conservative. The principle of stationary potential energy may be stated as follows: *Among all admissible displaced configurations of a conservative system, those that satisfy the equations of equilibrium make the potential energy stationary with respect to small admissible variations of displacement.* If the stationary condition is a relative minimum, the equilibrium state is stable. For derivation and extensive discussion of these statements, see [4.1, 4.2].

For the system of Fig. 4.10-1a, all values of v_A and v_B are admissible. Values of v_A and v_B that actually occur are those that place the system in static equilibrium under loads F_A and F_B. According to the principle of stationary potential energy, v_A and v_B are determined from Eq. 4.10-1 as follows.

$$\frac{\partial \Pi}{\partial v_A} = 0 \quad \text{yields} \quad k_A v_A - k_B(v_B - v_A) = F_A \tag{4.10-2a}$$

$$\frac{\partial \Pi}{\partial v_B} = 0 \quad \text{yields} \quad k_B(v_B - v_A) = F_B \tag{4.10-2b}$$

The same system is analyzed by virtual work in Section 4.1. The results, Eqs. 4.1-7, agree with Eqs. 4.10-2. The connection of the present method with virtual work is apparent if we note that work dW done by an observer in creating a small admissible displacement is equal to the change $d\Pi$ in potential energy, and is also equal in magnitude to virtual work dV. For the system of Fig. 4.10-1a,

$$d\Pi = \frac{\partial \Pi}{\partial v_A} dv_A + \frac{\partial \Pi}{\partial v_B} dv_B \tag{4.10-3}$$

If v_A and v_B satisfy static equilibrium conditions, virtual work must vanish. Therefore $d\Pi = 0$. Since dv_A and dv_B are arbitrary and independent displacements, $d\Pi$ can vanish only if $\partial\Pi/\partial v_A$ and $\partial\Pi/\partial v_B$ vanish separately. Thus we obtain Eqs. 4.10-2.

As another example, consider the plane truss of Fig. 4.10-1b. We take as d.o.f. (degrees of freedom) the displacement components u_D and v_D of joint D. The change in length of any bar is determined by the components of u_D and v_D parallel to the bar. Hence, dividing change in length by total length, we obtain the axial strain in each bar.

$$\epsilon_1 = \frac{1}{L_1}\left(\frac{3}{5}u_D - \frac{4}{5}v_D\right) \quad \epsilon_2 = -\frac{1}{L_2}v_D \quad \epsilon_3 = \frac{1}{L_3}\left(-\frac{4}{5}u_D - \frac{3}{5}v_D\right) \tag{4.10-4}$$

Strain energy in a bar is given by Eq. 4.9-1, in which axial strain is the gradient of axial displacement, $\epsilon = du_o/dx$. If each bar is uniform, the potential energy of the system is

$$\Pi = \sum_{i=1}^{3} \frac{E_i A_i}{2} \epsilon_i^2 L_i + P u_D \tag{4.10-5}$$

Here potential energy of the load is $+Pu_D$ rather than $-Pu_D$ because P and u_D are oppositely directed: A positive u_D moves the load opposite to its direction and therefore increases its potential energy. Stationary potential energy requires that

$$\frac{\partial \Pi}{\partial u_D} = 0 \quad \text{and} \quad \frac{\partial \Pi}{\partial v_D} = 0 \tag{4.10-6}$$

These two equations can be solved for u_D and v_D. Strains in the bars are determined by substituting these displacements into Eqs. 4.10-4. Finally, stresses in the bars are

$$\sigma_1 = E_1 \epsilon_1 \quad \sigma_2 = E_2 \epsilon_2 \quad \sigma_3 = E_3 \epsilon_3 \quad (4.10\text{-}7)$$

Static indeterminacy has no effect on the solution procedure. Attaching one bar less to joint D, or several bars more, would change only the number of terms in the summation in Eq. 4.10-5. There would still be only two d.o.f., which would still be determined by Eqs. 4.10-6.

Thermal Loads. With the potential energy method, as with any numerical method in which displacement quantities are the d.o.f., calculation of response to heating or cooling proceeds as follows.

1. With all d.o.f. maintained at zero, determine loads that arise from heating or cooling. Also determine "initial" stresses, which are stresses due to temperature change while all d.o.f. are maintained at zero.
2. Free the d.o.f. not prescribed by support conditions, and calculate displacements produced by loads determined in Step 1.
3. Calculate stresses produced by displacements calculated in Step 2, and superpose them on the initial stresses of Step 1.

As an example, imagine that only bar 2 in Fig. 4.10-1b is heated, an amount ΔT. Let α be the coefficient of thermal expansion. Downward force $F = \alpha EA \, \Delta T$ appears at joint D. Initial stress in bar 2 is $\sigma_o = -E\alpha \, \Delta T$ (compressive). Bars 1 and 3 are free of initial stress. Next, displacements u_D and v_D at joint D due to force F are calculated. In this example, v_D is negative (downward). The final stress in bar 2 is $\sigma = \sigma_o - E(v_D/L)$. Thus, σ in bar 2 is a compressive stress, smaller in magnitude than σ_o. Final stresses in bars 1 and 3 come only from u_D and v_D; for example, in bar 1 the final stress is $E(0.6u_D - 0.8v_D)/(1.25L)$. If the truss were made statically *determinate*, say by omitting bar 3, initial stresses due to any temperature changes applied to bars 1 and 2 would be equal in magnitude but opposite in sign to stresses produced in these bars by displacements. In general, temperature change in a statically determinate truss produces displacement but not stress.

Comparison with Other Energy Methods. With the principle of stationary potential energy, strain energy U is expressed in terms of displacements. Compatibility of displacements is ensured by allowing only admissible displacements. The number of equations generated is equal to the number of displacement d.o.f., and is independent of the degree of static indeterminacy. Equations generated are equilibrium equations, to be solved for displacement d.o.f. Internal forces or stresses, if desired, are determined last, by differentiation of the displacement field.

The unit load method is a particular result of the principle of stationary complementary energy [4.1, 4.2]. Complementary strain energy U^* is expressed in terms of forces. Equilibrium is ensured by providing supports sufficient to prevent rigid body motion. For a statically indeterminate structure, the number of equations generated is equal to the degree of static indeterminacy. Equations generated are compatibility equations, to be solved for redundant forces. Displacements, if desired, are determined last, by integration of strain and/or curvature fields.

Despite a superficial similarity, the principle of stationary potential energy is unrelated to the conservation of energy argument. With d.o.f. v_i (as for example in Eq. 4.10-2), the stationary principle has the form, for n d.o.f.,

$$\frac{\partial}{\partial v_i}\left(U - \sum F_i v_i\right) = 0 \qquad i = 1, 2, \ldots, n \qquad (4.10\text{-}8)$$

while the conservation of energy argument can be stated as

$$U = \frac{1}{2}\sum F_i v_i \qquad i = 1, 2, \ldots, n \qquad (4.10\text{-}9)$$

The stationary principle pertains to derivatives of potential, not the potential itself; the conservation argument pertains to energy quantities, not their derivatives. The factor of 1/2 in Eq. 4.10-9 arises because force increases linearly when stretching a linear spring. Potential energy of a load has nothing to do with linear properties of a spring. There is no factor of 1/2 in Eq. 4.10-8 because each load is regarded as always acting at its full value rather than gradually increasing. Equation 4.10-8 yields as many equations as there are d.o.f. Equation 4.10-9 remains a single equation, regardless of the number of d.o.f., and therefore has limited usefulness.

4.11 RAYLEIGH-RITZ METHOD

The Rayleigh-Ritz method was suggested by Lord Rayleigh in the 1870s and generalized by W. Ritz some 35 years later. The method is applicable to many problems of continuous media. As applied to stress analysis, it is a way of using the principle of stationary potential energy to obtain an approximate solution of a problem too difficult to solve exactly. Usually, the source of difficulty is that the problem is described by partial differential equations that cannot be solved because the geometry, loading, or support conditions are not simple enough. To obtain the mathematical model that constitutes a Rayleigh-Ritz approximation, one defines a set of displacement modes, each admissible, and each multiplied by a generalized coordinate. The calculation process selects values of generalized coordinates such that potential energy is made stationary. Thus the equilibrium configuration of the mathematical model is obtained. Strains are obtained by differentiation of the displacement field, and stresses by applying a stress-strain relation to the strains. In what follows we consider details of the process only as they apply to beam problems.

As described in Section 4.1, an *admissible configuration* is a displacement state that violates neither support conditions nor internal compatibility of the material. The term *kinematically admissible configuration* has the same meaning. Figure 4.11-1a shows *in*admissible lateral displacement fields of a beam: They violate the conditions that deflections must be zero at supports, that rotation must be zero at the wall, or that the elastic curve must display no break and no cusp. Figure 4.11-1b shows three of infinitely many admissible configurations. Only the lowest of the curves shown seems physically likely, and it is such a curve that is selected in preference to other admissible curves by the principle of stationary potential energy.

Some details of the Rayleigh-Ritz procedure may be explained with reference to the cantilever beam problem of Fig. 4.11-2a. For this problem, admissible polynomial fields for lateral displacement include

$$v = a_2 x^2 \quad\quad v = a_2 x^2 + a_3 x^3 \quad\quad v = a_2 x^2 + a_3 x^3 + a_4 x^4 \quad\quad (4.11\text{-}1)$$

and so on. All are continuous and all satisfy kinematic boundary conditions, which are $v = 0$ and $dv/dx = 0$ at $x = 0$. Fields $v = a_0$ and $v = a_0 + a_1 x$ are *not* admissible for this problem because they violate kinematic boundary conditions. Each d.o.f. a_i in Eqs. 4.11-1 is a generalized coordinate. Here a_2, a_3, and a_4 describe amplitudes of the respective displacement modes x^2, x^3, and x^4.

Let us consider the two-term approximation in Eq. 4.11-1. Strain energy in the beam is calculated by using the last term in Eq. 4.9-2, which involves curvature d^2v/dx^2. Potential energy of the load is obtained by summing potential energies of force increments $q\, dx$ that move through distances v. Thus, potential energy of the system is

$$\Pi = \int_0^L \frac{EI_z}{2}(2a_2 + 6a_3 x)^2\, dx + \int_0^L (a_2 x^2 + a_3 x^3) q\, dx \quad\quad (4.11\text{-}2)$$

in which the latter integral is positive because q and v are oppositely directed. To make potential energy stationary with respect to the d.o.f., we write $\partial\Pi/\partial a_2 = 0$ and $\partial\Pi/\partial a_3 = 0$. Thus

$$4EI_z L a_2 + 6EI_z L^2 a_3 + \frac{qL^3}{3} = 0 \quad\quad (4.11\text{-}3\text{a})$$

$$6EI_z L^2 a_2 + 12EI_z L^3 a_3 + \frac{qL^4}{4} = 0 \quad\quad (4.11\text{-}3\text{b})$$

Equations 4.11-3 yield

$$a_2 = -\frac{5qL^2}{24EI_z} \quad\quad a_3 = \frac{qL}{12EI_z} \quad\quad (4.11\text{-}4)$$

The largest deflection appears at $x = L$. As predicted by the approximation, it is

$$v_L = a_2 L^2 + a_3 L^3 \quad\text{or}\quad v_L = -\frac{qL^4}{8EI_z} \quad\quad (4.11\text{-}5)$$

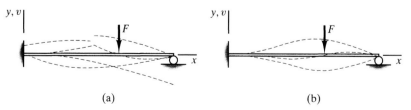

(a)   (b)

FIGURE 4.11-1 Dashed lines show deflections of a propped cantilever beam, greatly exaggerated. (a) Inadmissible configurations. (b) Admissible configurations.

Bending moment is calculated by multiplying curvature d^2v/dx^2 by flexural stiffness EI_z. The largest bending moment appears at $x=0$. As predicted by the approximation, it is

$$M_o = EI_z\left(\frac{d^2v}{dx^2}\right)_{x=0} = EI_z(2a_2) = -0.417\, qL^2 \qquad (4.11\text{-}6)$$

Tip deflection v_L happens to agree with elementary beam theory, but deflections in the range $0 < x < L$ do not. Bending moment M_o is about 17% low in magnitude.

As seen in Fig. 4.11-2, the one-term approximation $v = a_2 x^2$ has greater error than the two-term approximation. (The value of a_2 in the one-term approximation can be obtained from Eq. 4.11-3a after setting $a_3 = 0$.) The three-term polynomial $v = a_2 x^2 + a_3 x^3 + a_4 x^4$ yields a solution that agrees completely with elementary beam theory, because the solution from beam theory, obtained by integration of $EI(d^2v/dx^2) = M(x)$, is a polynomial of this form. Adding the next term $a_5 x^5$ would not change anything: we would obtain $a_5 = 0$ and the same solution as before from the remaining three terms. *If the assumed field is capable of representing the exact solution, the Rayleigh-Ritz method yields exact results.*

Remarks. Curvature of a beam, and strains in general, are calculated by differentiation of the displacement field, and therefore have greater error than an approximate displacement field. For example, consider the two lateral displacement functions $v = 4a_1 x(L-x)/L^2$ and $v = a_1 \sin(\pi x/L)$ over the span $0 < x < L$. The latter function is exact for distributed lateral load of the form $q = q_o \sin(\pi x/L)$ on a uniform simply supported beam. While the two functions $v = v(x)$ look similar, their derivatives dv/dx look different, and each successive differentiation produces greater differences. Accordingly, strains (and therefore stresses) at an arbitrary location are likely to be less accurate than displacements, although at some points displacements may be less accurate than strains and stresses.

A more general explanation of the Rayleigh-Ritz method is as follows. Consider an elastic body with known supports and loads, whose static deflections and stresses are required. Let u, v, and w be displacement components of an arbitrary material point in x, y, and z directions. The displacement field of the body is approximated as

$$u = \sum_{i=1}^{\ell} a_i f_i \qquad v = \sum_{i=\ell+1}^{m} a_i f_i \qquad w = \sum_{i=m+1}^{n} a_i f_i \qquad (4.11\text{-}7)$$

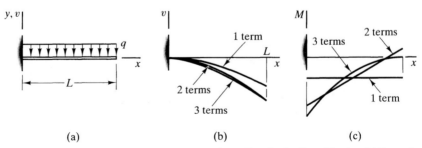

FIGURE 4.11-2 (a) Uniform cantilever beam with uniformly distributed load q. (b) Lateral deflection $v = v(x)$, calculated by polynomial approximation. (c) Bending moment $M = M(x)$, calculated as $M = EI_z(d^2v/dx^2)$.

Each of the functions $f_i = f_i(x,y,z)$ must be an admissible displacement field. Strains dictated by Eqs. 4.11-7 are determined by differentiation ($\epsilon_x = \partial u/\partial x$, and so on; see Section 7.2). Hence, expressions for stresses in terms of the a_i can be written, as well as an expression for potential energy Π, which is composed of strain energy in the elastic body and potential energy of applied loads. The n d.o.f. a_i are determined by solving the n algebraic equations

$$\frac{\partial \Pi}{\partial a_i} = 0, \quad i = 1, 2, \ldots, n \tag{4.11-8}$$

Finally, knowing the a_i, we return to Eqs. 4.11-7. Thus we obtain displacements at points of interest, evaluate strains $\epsilon_x = \partial u/\partial x$ and so on, then use stress-strain relations to obtain stresses.

The f_i in Eqs. 4.11-7 are modes of deformation, and are often taken as polynomials or trigonometric functions because these forms are comparatively easy to manipulate. The a_i are amplitudes of the deformation modes. The solution is exact if Eqs. 4.11-7 are capable of representing the exact state of deformation. An exact representation rarely happens, because we have used a finite number of degrees of freedom to approximate a displacement field that in general has infinitely many d.o.f. The approximation constrains the displacement field to a form obtainable by superposition of modes f_i, which, being inexact, is a form that the structure does not prefer. Thus the effect of constraints is to stiffen the mathematical model relative to a model without such constraints, so that approximate displacements are too small in an average sense. Also, since differential equations of equilibrium are not used in a Rayleigh-Ritz analysis, these equations are in general not satisfied at any particular location by an approximate solution. Instead, equilibrium equations are satisfied over the volume of the structure in an average sense.

Skill is needed in order to select appropriate f_i. Fields must be admissible and should not omit terms of low order. For example, *poor* displacement fields for the beam problem of Fig. 4.11-2a include $v = a_3 x^3$ and $v = a_2 x^2 + a_4 x^4$. Omission of a term renders approximate series incomplete and prevents convergence to correct results as the number of terms approaches infinity. It is not necessary that displacement fields satisfy "natural" boundary conditions, and doing so is usually difficult, but if it can be done, accuracy will be improved for a given number of a_i. (Natural boundary conditions for Fig. 4.11-2a are that transverse shear force and bending moment both vanish at $x = L$, which implies that $d^2v/dx^2 = 0$ and $d^3v/dx^3 = 0$ at $x = L$.)

Engineering judgment is needed in order to assess the quality of results. There is no easy way to check the accuracy of the approximation. However, one can check how well natural boundary conditions are satisfied by the approximate solution. Also, the solution can be repeated with another term added to the series. If the additional term does little to alter results, we have evidence (but not proof) that results are good.

A Rayleigh-Ritz solution, such as the one just described, has the disadvantages of providing no error estimate, often requiring tedious calculations, and not having a form suited to computer implementation. The finite element method, summarized in Section 4.13, can be regarded as a form of the Rayleigh-Ritz method that overcomes these disadvantages.

4.12 TRIGONOMETRIC SERIES

Approximating series used in the Rayleigh-Ritz method need not be restricted to polynomial forms. Trigonometric series are better suited to some problems. Consider the uniform fixed-fixed beam of Fig. 4.12-1a and the following series for lateral displacement v.

$$v = \sum_{i \text{ even}} a_i \left(1 - \cos \frac{i\pi x}{L}\right) \quad (4.12\text{-}1)$$

For all even integer values of i, the mode in parentheses is admissible because it yields $v = 0$ and $dv/dx = 0$ at each end of the beam. Curvature is

$$\frac{d^2 v}{dx^2} = \sum_{i \text{ even}} a_i \left(\frac{i\pi}{L}\right)^2 \cos \frac{i\pi x}{L} \quad (4.12\text{-}2)$$

In squaring the curvature to evaluate U by the last term in Eq. 4.9-2, we obtain all squared terms $\cos^2(i\pi x/L)$ and all cross-product terms $\cos(i\pi x/L)\cos(j\pi x/L)$, where i and j are different integers. These terms must be integrated over the length of the beam. Integrals of this type are evaluated by the following formulas, in which i and j are integers.

$$\int_0^\pi \cos i\theta \cos j\theta \, d\theta = \begin{cases} \pi & \text{for } i = j = 0 \\ \dfrac{\pi}{2} & \text{for } i = j \neq 0 \\ 0 & \text{for } i \neq j \end{cases}$$

$$\int_0^\pi \sin i\theta \sin j\theta \, d\theta = \begin{cases} \dfrac{\pi}{2} & \text{for } i = j \neq 0 \\ 0 & \text{for all other } i \text{ and } j \end{cases} \quad (4.12\text{-}3)$$

These results are halved if the upper limit of integration is changed to $\pi/2$ and are doubled if the upper limit of integration is changed to 2π.

With the substitution $x = L\theta/\pi$, the integral of $(d^2v/dx^2)^2$ in Eq. 4.9-2 is converted to the form of Eq. 4.12-3, and an expression for strain energy U is obtained. With v positive upward in Fig. 4.12-1a, potential energy of the load is Fv_c, where $v_c =$

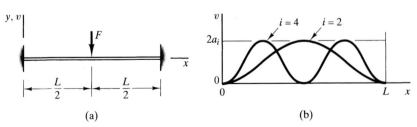

FIGURE 4.12-1 (a) Uniform fixed-fixed beam with a central lateral load. (b) Plots of each of the first two series terms of Eq. 4.12-1.

$2(a_2 + a_6 + a_{10} + \ldots)$ is the value of v at $x = L/2$. Potential energy of the system is $\Pi = U + Fv_c$, which is

$$\Pi = \frac{EI_z}{2} \frac{L}{2} \sum_{i \text{ even}} a_i^2 \left(\frac{i\pi}{L}\right)^4 + 2F \sum_{i=2,6,10,\ldots} a_i \qquad (4.12\text{-}4)$$

The latter summation does not include $i = 4$, $i = 8$, and so on because $v = 0$ at midspan for these values of i. The stationary condition is achieved by writing $\partial \Pi / \partial a_i = 0$ for all values of i used in Eq. 4.12-4. Each such equation has the same form, namely

$$EI_z \frac{L}{2} a_i \left(\frac{i\pi}{L}\right)^4 + 2F = 0 \qquad \text{for } i = 2, 6, 10, \ldots \qquad (4.12\text{-}5)$$

from which

$$a_i = -\frac{4FL^3}{EI_z \pi^4 i^4} \qquad \text{for } i = 2, 6, 10, \ldots \qquad (4.12\text{-}6)$$

Now that a_i is known, deflection v and bending moment $M = EI_z(d^2v/dx^2)$ can be calculated at any x. At midspan,

$$\text{At } x = \frac{L}{2}: \quad v = -\frac{8FL^3}{EI_z \pi^4}\left(\frac{1}{2^4} + \frac{1}{6^4} + \frac{1}{10^4} + \cdots\right) \qquad (4.12\text{-}7)$$

$$\text{At } x = \frac{L}{2}: \quad M = \frac{4FL}{\pi^2}\left(\frac{1}{2^2} + \frac{1}{6^2} + \frac{1}{10^2} + \cdots\right) \qquad (4.12\text{-}8)$$

The negative sign in Eq. 4.12-7 means that v is downward at midspan. Complete series converge to exact results, namely $v = -FL^3/192EI_z$ and $M = FL/8$ at midspan. Truncated series yield the following percentage errors.

	One Term	Two Terms	Three Terms	Four Terms
v at $x = L/2$:	-1.45%	-0.23%	-0.07%	-0.03%
M at $x = L/2$:	-18.94%	-9.94%	-6.69%	-5.04%

As expected, approximate deflections obtained by truncating the series are too small in magnitude. Also, the expression for v is more accurate than that for M and converges at a faster rate.

A different location of the origin $x = 0$, or different support conditions, would require a series that differs from Eq. 4.12-1. For example, if ends of the beam in Fig. 4.12-1a were simply supported rather than clamped, a sine series for lateral deflection would be admissible:

$$v = \sum a_i \sin \frac{i\pi x}{L} \qquad i = 1, 2, 3, \ldots \qquad (4.12\text{-}9)$$

This series also satisfies the natural boundary condition of zero curvature at each end, which is not necessary but improves accuracy for a given number of terms. If loads are symmetric with respect to the middle of the beam, only odd terms of the sine series are needed.

116　Chapter 4　Applications of Energy Methods

The advantage of trigonometric series is seen in Eq. 4.12-6: Although there are infinitely many a_i, they "decouple" so that each is independent of all others, and a single generic equation suffices to determine all of them.

4.13　FINITE ELEMENT METHOD

In the classical Rayleigh-Ritz method, each approximating function f_i in Eq. 4.11-7 spans the entire structure, and generalized coordinates a_i usually have no direct physical meaning. The finite element method can be described as a form of the Rayleigh-Ritz method in which there are many approximating functions, each comparatively simple and each spanning a limited region of the structure. Also, the d.o.f. are actual displacements of specific points instead of generalized coordinates a_i. These changes make the method physically appealing and well suited to computer implementation. In the present section we illustrate the finite element method as applied to a simple beam problem, then remark on more general aspects of the method.

Beam Example. Consider the uniform beam of Fig. 4.13-1a. Its d.o.f. are lateral displacement and rotation at each end ($v_1, \theta_{z1}, v_2, \theta_{z2}$). In terms of these d.o.f., an admissible lateral displacement field $v = v(x)$ is

$$v = \left(1 - \frac{3x^2}{L^2} + \frac{2x^3}{L^3}\right)v_1 + \left(x - \frac{2x^2}{L} + \frac{x^3}{L^2}\right)\theta_{z1}$$
$$+ \left(\frac{3x^2}{L^2} - \frac{2x^3}{L^3}\right)v_2 + \left(-\frac{x^2}{L} + \frac{x^3}{L^2}\right)\theta_{z2} \quad (4.13\text{-}1)$$

This cubic interpolation polynomial is in full agreement with the lateral displacement field $v = v(x)$ predicted by elementary beam theory when end displacements and rotations are $v_1, \theta_{z1}, v_2,$ and θ_{z2}. One may easily check that Eq. 4.13-1 yields $v = v_1$ at $x = 0$, $dv/dx = \theta_{z1}$ at $x = 0$, and so on. With Eq. 4.13-1, the last term in Eq. 4.9-2 gives the strain energy of the beam as [4.1, 4.2]

$$U = \frac{2EI_z}{L}\left[\theta_{z1}^2 + \theta_{z1}\theta_{z2} + \theta_{z2}^2 - 3\frac{v_2 - v_1}{L}(\theta_{z1} + \theta_{z2}) + 3\left(\frac{v_2 - v_1}{L}\right)^2\right] \quad (4.13\text{-}2)$$

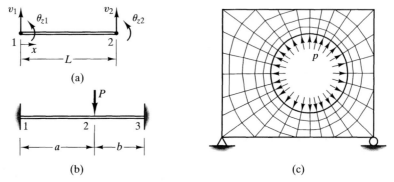

FIGURE 4.13-1　(a) Degrees of freedom of a plane beam element. (b) Uniform fixed-fixed beam with an off-center lateral load. (c) Plane body with a pressurized hole.

The beam may be called an "element" and its end points 1 and 2 called "nodes." Elements can be connected together at nodes. Thus, for example, a beam continuous across several supports and carrying several loads can be modeled by several beam elements connected end to end.

If two beam elements are used to model the uniform beam in Fig. 4.13-1b, lateral displacement $v = v(x)$ is represented by two separate cubic polynomials. This happens to be the correct form of displacement field for this problem and therefore provides exact results. Nodes 1, 2, and 3 in Fig. 4.13-1b are numbered at the "structure level," so that the rightmost element has node numbers 2 and 3. Numerical subscripts in Eq. 4.13-2 become 2 and 3 to obtain U for the rightmost element. The displacement field is made admissible by setting d.o.f. v_1, θ_{z1}, v_3, and θ_{z3} to zero. Potential energy Π of the structure is the strain energy U of each element plus potential energy Pv_2 of load P at node 2.

$$\Pi = \frac{2EI_z}{a}\left[\theta_{z2}^2 - 3\frac{v_2}{a}\theta_{z2} + 3\left(\frac{v_2}{a}\right)^2\right]$$
$$+ \frac{2EI_z}{b}\left[\theta_{z2}^2 - 3\frac{-v_2}{b}\theta_{z2} + 3\left(\frac{-v_2}{b}\right)^2\right] + Pv_2 \quad (4.13\text{-}3)$$

The stationary condition is $\partial\Pi/\partial v_2 = 0$ and $\partial\Pi/\partial\theta_{z2} = 0$. Thus we obtain the equations

$$12EI_z\left(\frac{1}{b^3} + \frac{1}{a^3}\right)v_2 + 6EI_z\left(\frac{1}{b^2} - \frac{1}{a^2}\right)\theta_{z2} = -P$$
$$6EI_z\left(\frac{1}{b^2} - \frac{1}{a^2}\right)v_2 + 4EI_z\left(\frac{1}{b} + \frac{1}{a}\right)\theta_{z2} = 0 \quad (4.13\text{-}4)$$

which have the solution

$$v_2 = -\frac{Pa^3b^3}{3EI_z(a+b)^3} \quad \text{and} \quad \theta_{z2} = \frac{P(a-b)a^2b^2}{2EI_z(a+b)^3} \quad (4.13\text{-}5)$$

Bending moment is computed as $M = EI_z(d^2v/dx^2)$. For example, at node 1 of the structure, from the leftmost element,

$$M_1 = EI_z\left(\frac{d^2v}{dx^2}\right)_{x=0} \quad \text{yields} \quad M_1 = -\frac{Pab^2}{(a+b)^2} \quad (4.13\text{-}6)$$

For this particular problem, the methods of Sections 4.11 and 4.12, which use generalized coordinates a_i, would not be as simple as the foregoing method because an off-center load makes it difficult to contrive an appropriate expression for $v = v(x)$ that spans the entire beam. Equation 4.12-1, for example, imposes symmetry about the center, and therefore will not produce correct results for an off-center load, regardless of the number of terms used.

Remarks. Figure 4.13-1c shows a more representative finite element structure. Elements are plane quadrilaterals and a few plane triangles. Nodes that connect elements together are located at corners of elements. D.o.f. of this problem are horizontal and vertical displacement components at each node. The displacement field adopted for each element may be a simple polynomial of such form that elements do not overlap, separate, or generate strain concentrations at nodes. In practice, an expression for Π is not constructed. Instead, simultaneous algebraic equations analogous to Eqs. 4.13-4 are written directly, using numerical coefficients, and in matrix format. These equations are solved for the unknown d.o.f. In practical problems, the simultaneous equations analogous to Eqs. 4.13-4 usually number in the thousands, so that computer implementation is mandatory. Computed results are not exact, but exact results can be approached by using more and more elements.

The finite element method is extremely versatile. There are no restrictions as to structure type, shape, loading, or support condition. Problems of stress analysis, buckling, vibration, and dynamic response can be solved. Nonlinearities, such as plastic flow, can be included. The method is not restricted to structural mechanics; it is also applicable to heat transfer, magnetic fields, fluid flow, and other problem areas. Computer software is commercially available in all these areas. A typical software user deals mainly with preparing input and inspecting output. Graphics-oriented "preprocessors" facilitate preparation of the finite element model. "Postprocessors" can plot stress contours, and display or even animate displaced shapes.

Such versatility and ease of use comes at a price. All stress analysis algorithms, finite element algorithms included, are based on assumptions and procedures that have limited ranges of applicability. If limitations are forgotten, or never learned, software may be pushed beyond its range of validity. Also, software has no way of knowing if the software user has modeled the actual problem satisfactorily: Geometry, material properties, loads, supports, and other things may be misunderstood by the user or not entered properly as data. Whether good or bad, results are displayed with great polish by the software, and many users are entirely too willing to accept computed results at face value. An approximate pencil-and-paper solution, even if quite rough, may show that a computer solution is not to be believed. Good finite element analysis relies more on physical grasp of the problem and a commitment to check for errors than on knowledge of finite element theory [4.6].

PROBLEMS

4.1-1. The sketch shows a weightless beam supported by three linear springs, each of stiffness k. In this problem assume that the beam is rigid (therefore, as is easily shown, when the rigid beam is loaded by weight W at A, the leftmost spring carries compressive force $5W/6$). If W is the weight of a small self-propelled cart, how much energy must the cart expend to move from A to B? Express the answer in terms of W and k.

4.1-2. The structure shown consists of two uniform bars, each of weight W. Connections, at each end of each bar, are frictionless. Couple C is applied to the end of one bar where the bar

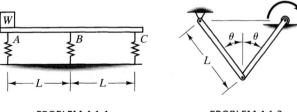

PROBLEM 4.1-1   PROBLEM 4.1-2

rolls on a horizontal rigid surface. Determine angle θ for static equilibrium by the virtual work method. Use equilibrium considerations to check the result.

4.1-3. A weightless rigid bar is supported by a frictionless hinge and two linear springs, as shown. Springs are unstressed when the bar is horizontal. Use the virtual work method to determine the force in each spring due to application of couple C to the right end of the bar. Use equilibrium considerations to check results.

4.1-4. The uniform bar shown has weight W and slides without friction on the horizontal and vertical rigid surfaces. The linear spring is unstressed when $\theta = \pi/2$. Determine the value of θ for equilibrium by the virtual work method. Use equilibrium considerations to check the result.

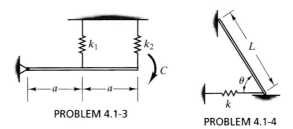

PROBLEM 4.1-3

PROBLEM 4.1-4

4.1-5. A weightless rigid bar is suspended by three identical linear springs, as shown. Initially the bar is horizontal and springs are unstressed. Use the virtual work method to determine the force in each spring due to application of load P. Use equilibrium considerations to check results.

4.1-6. As shown, a weightless rigid bar of length $2a$ is hinged at its left end and carries a weight W at three-quarters span. A second weight W is attached to the bar by linear springs of stiffnesses k_1 and k_2 and an inextensible cord that passes over frictionless pulleys. As d.o.f., use rotation θ of the bar and vertical displacement v of the second weight. Springs are unstressed when $\theta = v = 0$. Use the virtual work method to set up equations for the equilibrium values of θ and v. Solve these equations for the case $k_1 = k_2$ and use equilibrium considerations to check results.

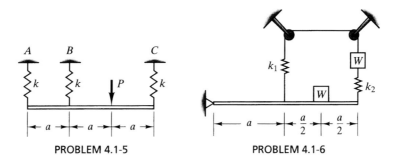

PROBLEM 4.1-5

PROBLEM 4.1-6

4.1-7. A uniform and linearly elastic bar is fixed to a rigid wall at one end, as shown. A small gap g exists between the other end of the bar and a second rigid wall. A force P, more than adequate to close the gap, causes the center of the bar to move an amount u_c. Why is strain energy in the bar not $Pu_c/2$? What is the correct strain energy in terms of $P, L, A, E,$ and g?

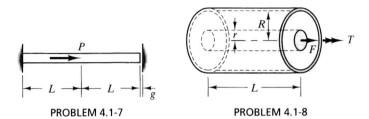

PROBLEM 4.1-7

PROBLEM 4.1-8

4.1-8. The sketch represents a solid bar of radius r within a cylinder of inner radius R. The space between r and R is filled with a linearly elastic material of low shear modulus G. Assume that the cylinder is supported so that it does not move, and that the cylinder and the bar are rigid.

  (a) Let $T = 0$. Determine the displacement of the bar due to loading by force F.

  (b) Let $F = 0$. Determine the rotation of the bar due to loading by torque T.

4.1-9. Solve Problem 1.8-2 by the conservation of energy method.

4.2-1. Consider a uniform cantilever beam of length L, loaded by a lateral tip force P and a tip couple C. If tip lateral deflection due to C is $CL^2/2EI$, what is tip rotation due to P?

4.2-2. A solid prismatic bar of arbitrary cross section is pinched by two equal and opposite lateral forces P. Forces P act in the plane of a cross section and are applied to points a distance d apart. Determine the change in length of the bar.

4.2-3. Alter Example 2 of Section 4.2 in the following ways. In each case obtain an expression for the change in volume of the tank produced by collinear pinching forces F.

  (a) Apply collinear pinching forces to any two points on a spherical tank rather than to two points at opposite ends of a diameter. Let the points be separated by a distance h, where $0 < h < 2r$.

  (b) Apply pinching forces along a diameter near the middle of a long thin-walled cylindrical tank.

4.2-4. A uniform slender circular ring is loaded by couples C at diametrically opposite points A and B, as shown. Demonstrate that diameter AB does not change in length.

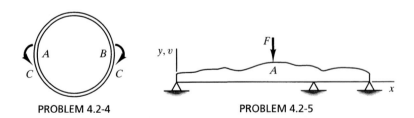

PROBLEM 4.2-4

PROBLEM 4.2-5

4.2-5. A linearly elastic beam of nonuniform cross section is to be loaded by lateral force F at point A (see sketch). The lateral deflection curve $v = v(x)$ is to be determined experimentally. Explain how (or why) $v = v(x)$ can be determined by measuring v at a single point as force F moves slowly along the beam.

4.2-6. A uniform circular ring is loaded in its plane by a mixture of forces and couples. The ring is slender, and is therefore assumed to be inextensible; that is, the ring flexes but does not stretch or compress circumferentially. Show that the area enclosed by the ring does not change when loads are applied.

4.2-7. A solid body, isotropic and of arbitrary shape, is pinched by two collinear forces F, applied to points a distance h apart. What volume change is produced by forces F?

4.3-1. Verify the assertion about higher-order terms made following Eq. 4.3-2, with specific reference to V_y and M_z.

4.3-2. Axial stress σ_x in Fig. 4.3-1c has the form

$$\sigma_x = \frac{N}{A} + \frac{M_y z}{I_y} - \frac{M_z y}{I_z}$$

Use this expression and Eq. 2.6-2 to obtain the terms in Eq. 4.3-2 that contain N, M_y, and M_z.

4.3-3. By calculation of the kind shown in Eq. 4.3-4, verify the following factors in Table 4.3-1: (a) 1.20, (b) 1.11, and (c) 2.00.

4.4-1. In Fig. 4.4-1a, determine rotation at $x = 0$ and lateral deflection at $x = L/2$ due to load q, by means of (a) the unit load method, and (b) Castigliano's theorem.

4.5-1. The beam shown has a rectangular cross section. Shear load q on its lower surface is uniform and has dimensions [force/length]. Use the unit load method to determine the deflection components of point C, which is at the middle of the right end, in terms of q, L, b, h, and E.

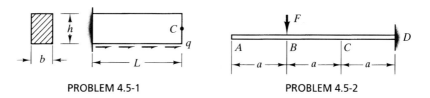

PROBLEM 4.5-1   PROBLEM 4.5-2

4.5-2. Determine the following for the slender uniform cantilever beam shown, in terms of F, a, E, and I. (a) Deflection and rotation at A. (b) Deflection and rotation at C. (c) Rotation of a straight line (such as an imaginary taught string) that connects A and C.

4.5-3. Distributed load on the slender uniform beam shown varies linearly, reaching intensity q_L at the right end. Determine the displacement and rotation at the center of the beam.

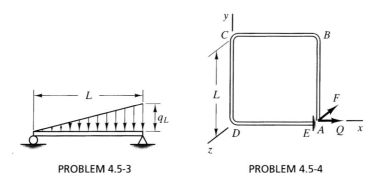

PROBLEM 4.5-3   PROBLEM 4.5-4

4.5-4. A slender uniform bar of circular cross section is bent to form a square in the xy plane, as shown. One corner is left open, where end E is fixed and end A is loaded by force Q in the x direction and force F in the negative z direction. Determine the x, y, and z components of displacement at A if (a) $F=0$, and (b) $Q=0$.

4.5-5. The slender uniform bent bar shown is fixed at A and has legs of lengths a, b, and c. Load q on leg BC is uniform and acts in the y direction.
  (a) Determine the y-direction deflection at corner C.
  (b) Determine the z-direction deflection at corner D.
  (c) At corner C, determine the rotation of the bar about the x axis.

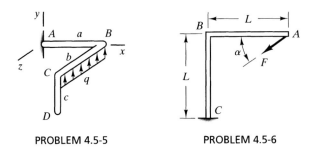

PROBLEM 4.5-5   PROBLEM 4.5-6

4.5-6. The angled frame shown is slender and uniform. Load F acts in the plane of the frame. What should be angle α if the displacement of point A is to be parallel to load F?

4.5-7. Let all bars of the plane truss shown have the same elastic modulus E and the same cross-sectional area A. Determine the following in terms of P, L, E, and A.
  (a) Horizontal deflection at B.
  (b) Vertical deflection at B.
  (c) Horizontal deflection at C.
  (d) Vertical deflection at C.
  (e) Relative motion of B and F along a line between B and F.

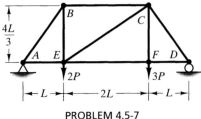

PROBLEM 4.5-7

4.6-1. In Fig. 4.6-2a, determine the resultant rotation at C in terms of P, R, E, and I. Also determine the orientation of this resultant. For convenience, let $EI = GK$. (For what kind of material and cross section is it true that $EI = GK$?)

4.6-2. In Fig. 4.6-2c, let $a = L/2$ and $h = L/2$. Also let EI be constant. Divide the arch from $x = 0$ to $x = L/2$ into five equal parts, each of length $\Delta x = L/10$. Use numerical integration to determine the horizontal deflection at C in terms of P, L, E, and I. Calculate three results: in Eq. 4.6-9, use both exact and approximate forms for ds in terms of dx (here Δs in terms of Δx); also use the simpler approximation $\Delta s = \Delta x$. Compare the two approximate values of deflection at C with that based on the exact form of ds.

4.6-3. The slender bent bar shown lies in the xy plane and has a uniform circular cross section. Determine the x, y, and z components of deflection at the end due to the two couple-loads C, whose vectors are y-parallel and z-parallel.

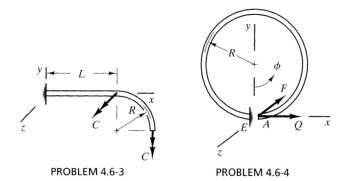

PROBLEM 4.6-3   PROBLEM 4.6-4

4.6-4. Replace the square in Problem 4.5-4 by a uniform slender circular ring of circular cross section and mean radius R, as shown. Determine the same results as are requested in Problem 4.5-4.

4.6-5. The slender semicircular bar BC shown is uniform. Regard the straight attachment AB as rigid. Load F and structure ABC lie in the xy plane. Show that the deflection of point A is parallel to F, regardless of how F is oriented in the xy plane. Also determine the effective spring constant k seen by load F.

4.6-6. In Problem 4.6-5, replace force F by a z-parallel force P, applied at A. Determine all components of displacement at A, including rotations.

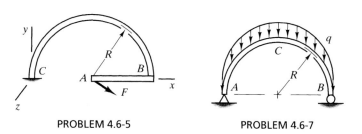

PROBLEM 4.6-5   PROBLEM 4.6-7

4.6-7. The slender and uniform semicircular arch shown is supported by a hinge at A and by a roller at B. Distributed load q is constant per unit of *horizontal* distance. Determine (a) the horizontal displacement at end B, and (b) the vertical displacement at crown C.

4.6-8. Remove forces F and Q in Problem 4.6-4. Instead let the ring be loaded by its own weight, which may be taken as q units of force per unit of arc length. Determine the x, y, and z components of displacement at A. Consider that gravity acts in (a) the x direction, (b) the y direction, and (c) the z direction.

4.6-9. For the respective directions of gravity loading in Problem 4.6-8, determine the components of rotation at A.

4.6-10. A uniformly distributed load q acts tangent to the uniform quarter-ring shown.
  (a) Determine the horizontal and vertical components of deflection at $\phi = 0$ due to bending.
  (b) Determine the vertical deflection at $\phi = 0$ due to axial strain.

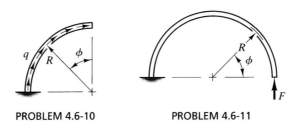

PROBLEM 4.6-10   PROBLEM 4.6-11

4.6-11. For the slender uniform half-ring shown, what is the radial displacement at an arbitrary angle ϕ, expressed as a function of F, R, E, I, and ϕ? Suggestion: Apply a unit radial force at the arbitrary angle ϕ.

4.6-12. The sketch shows the cross section of a helical spring, fixed to a support at the base. The cutting plane contains the axis of the helix. Load P is collinear with z, the axis of the helix. Let the spring be "closely coiled," which means that angle α is small. Thus the total length of all coils may be taken as $2\pi R n$ for a spring of n coils. Assume that $c \ll R$.

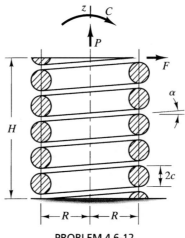

PROBLEM 4.6-12

(a) Show that under stretching load P, a typical cross section of a coil carries torque of magnitude $T = PR$ and bending moment $M_R = 0$ (see Fig. 4.6-1a).

(b) Use the results of part (a) to determine the z-direction elongation of the spring due to P, in terms of P, R, c, n, and G.

(c) Determine the angle of rotation at the top of the spring due to couple C, in terms of C, R, c, n, E, and Poisson's ratio ν.

(d) The result of part (c) yields an expression for the effective flexural stiffness of the spring. Use this result to determine the lateral displacement at the top of the spring due to loading by lateral force F. Assume that $\tan \alpha = \alpha$, and express the result in terms of F, α, c, E, ν, and H.

4.7-1. Solve Problem 1.7-7 by use of the unit load method.

4.7-2. In Problem 4.1-1, let the beam be flexible rather than rigid, with uniform flexural stiffness EI, and let $EI = kL^3$. Determine the fraction of W carried by the left spring when weight W is placed above it on the beam.

4.7-3. The uniform rectangular frame shown is plane, and is loaded in its plane by forces P. Consider bending only, and determine the separation of the points to which forces P are applied.

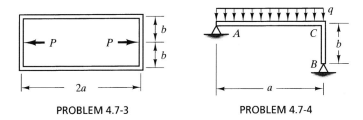

PROBLEM 4.7-3   PROBLEM 4.7-4

4.7-4. Ends A and B of the slender frame shown are hinged, EI is constant throughout, and load q is uniform. Determine the horizontal component of support reaction at B.

4.7-5. The uniform simply supported beam shown makes contact with the central support only after gap g is closed by application of the uniformly distributed load q. What then is the reaction force at the central support?

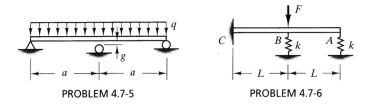

PROBLEM 4.7-5   PROBLEM 4.7-6

4.7-6. Flexural stiffness EI is constant throughout the cantilever beam shown. Each of the two supporting springs has stiffness $k = 3EI/L^3$. Determine the deflection at A, in terms of F, L, E, and I.

4.7-7. Determine the tension in the wire in the structure shown, in terms of load F, length L, cross-sectional area A of the wire, and moment of inertia I of members of the plane frame. All parts have the same elastic modulus.

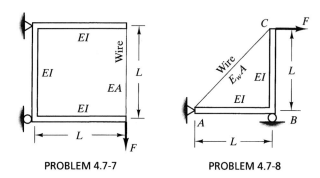

PROBLEM 4.7-7

PROBLEM 4.7-8

4.7-8. Frame ABC in the structure shown has a rectangular cross section, $0.10L$ units on a side. The cross-sectional area and elastic modulus of wire AC are each one-tenth the corresponding values for the frame. Consider bending of the frame and stretching of the wire, and determine the horizontal component of displacement at C, in terms of F, E, and L.

4.7-9. The queen-post truss shown consists of beam $ABCD$, compression posts BE and CF, and cable $AEFD$ that is attached to ends A and D of the beam. All dimensions and properties are known. Stress analysis under uniformly distributed load q is required. Outline an analysis procedure in detail. State what effects can be ignored, and why.

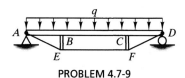

PROBLEM 4.7-9

4.7-10. The sketch shows a slender plane frame with in-plane loading. Supports at A and B are hinged. Flexural stiffness EI is uniform throughout the frame. Displacements are confined to the plane of the frame. Obtain an expression for the bending moment where force F is applied.

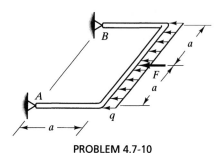

PROBLEM 4.7-10

4.7-11. The sketch shows a slender plane frame with out-of-plane loading. Supports at A and B are fixed. Flexural stiffness EI and torsional stiffness GK are uniform throughout the frame. Obtain an expression for the bending moment where force F is applied.

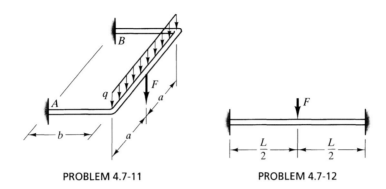

PROBLEM 4.7-11   PROBLEM 4.7-12

4.7-12. Supporting walls for the uniform beam shown are not completely rigid: When load F is applied, each end of the beam rotates through a small angle β. Determine the bending moments at the walls and the center deflection in terms of F, L, E, I, and β.

4.7-13. Consider the parabolic arch discussed in Example 3 of Section 4.7. Each end of the arch is hinged at its support. Assume that EI is constant and that deformations due to axial force N can be neglected.

  (a) Let the arch be uniformly heated an amount ΔT. Write an expression for the horizontal component of support reaction, analogous to Eq. 4.7-10.

  (b) Let the arch be loaded by a downward distributed load q, where q is constant per unit of *horizontal* distance. Show that the horizontal component of support reaction is $H = qL^2/8h$. What then is the net bending moment in the arch?

4.7-14. The ring shown is slender and symmetric about the axis of loading, but has an offset hole. The maximum tensile stress is required. For approximate analysis, half the ring is divided into 10 segments. Explain in detail the analysis steps that will lead to a solution.

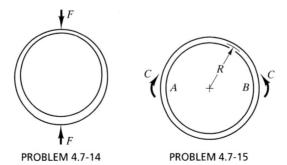

PROBLEM 4.7-14   PROBLEM 4.7-15

4.7-15. A slender uniform ring is loaded by couples C at diametrically opposite points A and B, as shown. Determine the amount of rotation at A or B. Suggestion: What does the reciprocal theorem say about radial displacement at A and B?

4.7-16. Flexural stiffness EI is constant throughout the slender plane frame shown. (a) Determine the bending moment at the point where load F is applied. (b) Determine the vertical displacement at this point.

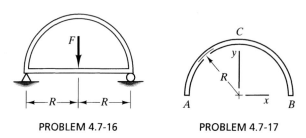

PROBLEM 4.7-16

PROBLEM 4.7-17

4.7-17. A slender uniform half-ring is shown in the sketch. Various in-plane loadings and supports are posed in the following exercises. Express answers in terms of the load, R, E, and I.

(a) A and B are hinged. A force P in the negative y direction is applied at C. Determine the bending moment at C.

(b) A is fixed and B can slide on a frictionless horizontal surface. A force P is applied in the negative y direction at C. Determine the bending moment at A.

(c) A and B are hinged. A force P in the x direction is applied at C. Determine the x-direction displacement at C.

(d) A and B are fixed. A force P in the negative y direction is applied at C. Determine the bending moment at C.

(e) A is fixed and B can slide on a frictionless horizontal surface. A force P in the negative x direction is applied at B. Determine the x-direction displacement at B.

(f) A is fixed and B can slide on a frictionless horizontal surface. A clockwise couple M_C is applied at C. Determine the y-direction displacement at C.

4.7-18. The ring shown is slender and uniform. At A it is welded to a support post. Axes x and y are in the plane of the ring; axis z is normal to this plane. Various loadings and supports are posed in the following exercises. Express answers in terms of the load, R, EI, and GK.

(a) A force P acts in the y direction at C. Determine the change in length of diameter AC.

(b) A force P acts in the y direction at C. Determine the change in length of diameter BD.

(c) A couple M_o about the x axis is applied at C. At C, determine the bending moment in the ring and the z-direction deflection.

(d) A couple M_o about the y axis is applied at C. Let $EI = GK$, and determine the torque in the ring at C.

(e) A force P is applied in the z direction at C. Determine the bending moment at C.

(f) The entire circumference is loaded by a uniformly distributed load q that acts tangent to the ring in the clockwise direction. Determine the bending moment at B.

(g) The ring has mass ρ per unit of arc length, and spins with constant angular velocity ω about the y axis. Determine the bending moment at C and the change in diameter along the axis of spin.

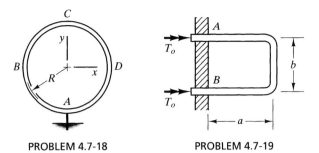

PROBLEM 4.7-18   PROBLEM 4.7-19

4.7-19. A slender uniform bar is bent into a U-shape, as shown. Ends A and B, to which torques T_o are applied, are held by frictionless bearings that permit only rotation about the axis of torque. What is the bending moment at A?

4.7-20. Verify that the Lagrange multiplier method, as described for an example problem in the final paragraph of Section 4.7, leads to correct values of support reactions M_C and Q in that example problem.

4.8-1. Consider the helical spring of Problem 4.6-12. Use geometric arguments, of the kind suggested by Fig. 4.8-1b, to express the z-direction elongation of the spring due to load P in terms of P, R, c, n, and G. (The geometric argument suggests that arc lengths ds of coils produce lateral displacement increments as well as z-direction displacement increments. What becomes of the lateral increments?)

4.10-1. Let all three bars in Fig. 4.10-1b have the same cross-sectional area and the same elastic modulus. Use the stationary potential energy principle to determine the force in each bar in terms of load P.

4.10-2. A weightless rigid bar is hinged so that it can rotate without friction about its base in the xy plane. The top is connected to a support by a horizontal linear spring, as shown. For what value of load P is the potential energy stationary? Show that a larger value of P makes the system unstable when the bar is in the vertical position shown.

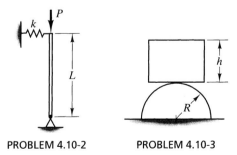

PROBLEM 4.10-2   PROBLEM 4.10-3

4.10-3. A homogeneous, rigid rectangular block rests on a rigid cylindrical surface, as shown. Show that this configuration is stable for $h < 2R$.

4.10-4. Discard bar 3 from the truss of Fig. 4.10-1b. Heat bar 1 by ΔT degrees. Determine displacement components of joint D and stresses in bars 1 and 2, by use of the method described in Section 4.10.

4.11-1. For the uniform cantilever beam of Fig. 4.11-2a, use natural boundary conditions at $x = L$ to reduce the three-term approximation in Eq. 4.11-1 to a single degree of freedom. Then use the Rayleigh-Ritz method to determine lateral deflection at $x = L$ and bending moment at $x = 0$.

4.11-2. A uniform cantilever beam carries a transverse tip load F, as shown. Determine deflection v_L at the tip and bending moment M_o at the support, using the following polynomials for lateral deflection. Also determine the percentage errors of v_L and M_o.

(a) $v = a_2 x^2$
(b) $v = a_2 x^2 + a_3 x^3$
(c) $v = a_3 x^3$ (which is a defective approximation)
(d) $v = a_2 x^2 + a_4 x^4$ (which is a defective approximation)

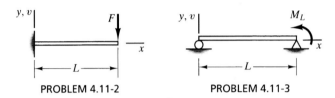

PROBLEM 4.11-2   PROBLEM 4.11-3

4.11-3. Moment M_L is applied to end $x = L$ of the uniform simply supported beam shown. Determine the rotation at $x = L$. Use the Rayleigh-Ritz method with a single generalized coordinate, which multiplies a polynomial for $v = v(x)$ that is not only admissible but also yields zero bending moment at $x = 0$.

4.11-4. Consider a uniform beam that is simply supported at end $x = 0$ and at end $x = L$. For the following loadings, determine the lateral deflection and the bending moment at midspan. Compare exact and approximate results. Use the approximate lateral displacement field $v = ax(L - x)$, where a is a generalized coordinate.

(a) Concentrated lateral force F at midspan.
(b) Uniformly distributed lateral load q.
(c) Distributed lateral load that varies linearly from zero at $x = 0$ to intensity q_L at $x = L$.

4.11-5. Use the Rayleigh-Ritz method to obtain an approximation for the deflection of load F shown in the sketch. Let $k = EI/L^3$, where EI is the flexural stiffness of the uniform beam. Use a polynomial that contains two generalized coordinates. Express the answer in terms of F and k.

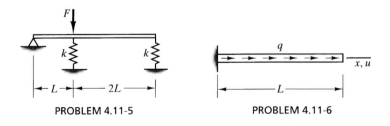

PROBLEM 4.11-5   PROBLEM 4.11-6

4.11-6. A uniform bar is fixed at one end and carries uniformly distributed axial load q, as shown. Compare exact and Rayleigh-Ritz solutions for axial displacement and axial stress, if the approximate axial displacement field $u = u(x)$ has (a) one generalized coordinate, and (b) two generalized coordinates.

4.11-7. (a) For the uniform propped cantilever beam shown, write an admissible polynomial for $v = v(x)$ that contains a single generalized coordinate. (It will also contain x, x^2, and L.)

(b) Assume that a cylinder of weight W can roll freely along the beam. Ends of the beam are at the same elevation. Use the result of part (a) to determine the lateral deflection of the beam at the rest position of W, in terms of W, L, E, and I.

(c) Let the beam have imaginary sides, each parallel to the plane of the paper, so that water poured on the beam can "pond" and create the deflected shape shown. Use the result of part (a), and obtain the equation that makes the potential energy of the system stationary. Solve for EI, and interpret the meaning of this result.

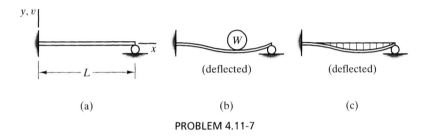

PROBLEM 4.11-7

4.11-8. Cross-sectional area A of the tapered bar shown varies linearly from A_0 at $x = 0$ to $A_0/2$ at $x = L$. The end at $x = L$ is forced to displace u_L units to the right by a rightward force P applied at $x = L$. Use a two-term polynomial approximation for $u = u(x)$, and determine the magnitude of P. What is the percentage error of this result?

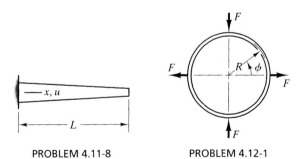

PROBLEM 4.11-8   PROBLEM 4.12-1

4.12-1. A slender, uniform circular ring is loaded by four radial forces F, as shown. Write an expression for radial displacement v that has a single generalized coordinate. Hence, determine the radial displacement and bending moment at $\phi = 0$.

4.12-2. Repeat Problem 4.12-1 with a different loading: Remove forces F, and let the ring spin with constant angular velocity ω about its vertical diameter.

4.12-3. Solve for tip deflection and root bending moment in the cantilever beam of Fig. 4.11-2a by means of a trigonometric series based on cosine terms. Calculate the percentage error of results predicted by use of one, two, three, and four series terms.

4.12-4. Solve Problem 4.11-4(a, b, c) by use of trigonometric series. When comparing exact and approximate results at midspan, use one, two, three, and finally four terms of the series approximation.

4.12-5. Solve Problem 4.11-6 by means of a sine series for $u = u(x)$. Let the series have a form that satisfies the natural boundary condition $\epsilon_x = 0$ at $x = L$, where $\epsilon_x = du/dx$.

4.13-1. Apply the procedure described in the beam example of Section 4.13 to the following problems of uniform beams. Verify that correct results are obtained for bending moment at $x = 0$ and for deflection and/or rotation at $x = L$. In each case use as few beam elements as possible.

 (a) Cantilever, fixed at $x = 0$, loaded by transverse tip force P at $x = L$.
 (b) Cantilever, fixed at $x = 0$, loaded by tip moment M_L at $x = L$.
 (c) Simply supported at $x = 0$ and at $x = L$, loaded by moment M_L at $x = L$.
 (d) Simply supported at $x = 0$ and at $x = L$, loaded by transverse force P at $x = L/2$. Exploit the symmetry of the problem.

CHAPTER 5

Beams on an Elastic Foundation

5.1 INTRODUCTION

The study of beams on an elastic foundation arose from the practical need to analyze railroad track. A railroad rail acts as a beam, whose supports (ties and the ballast below them) are not rigid but deflect when load is applied. A rail could be analyzed as a beam supported by discrete elastic springs, but this approach is analytically cumbersome. It is usually easier, and sufficiently accurate, to idealize the supports as a continuous elastic foundation. The idealization is applicable to more than just railroad rails. For example, it can be applied to a beam supported by other beams, to a pier built on pilings and loaded by horizontal force, to a slender structure that floats on water, and to the rim of a spoked bicycle wheel (the rim acts as a beam, the spokes act as elastic springs [5.1]). Also, differential equations for a cylindrical shell under axisymmetric loading have the same form as differential equations for a beam on elastic foundation. Accordingly, solutions of one set of equations become solutions of the other by a change of physical constants, and an understanding of one problem is readily adapted to the other.

The simplest analytical model of a continuous elastic foundation is the Winkler model, which dates from Winkler's work in the 1860s. This model states that if deflection w is imposed on the foundation, it resists with pressure $k_o w$, where k_o is the foundation modulus. Values of k_o for soils often lie in the range $20(10^6)$ N/m^2/m to $200(10^6)$ N/m^2/m. Large values are best: If k_o were infinite, a beam supported by the foundation would have no deflection and no flexural stress. In calculations it is convenient to use $k = k_o b$ rather than k_o, where b is the width of the beam in contact with the foundation. Dimensions of k are [force/length/length]. Lateral deflection w causes a Winkler foundation to apply distributed force kw to a beam of uniform width, where kw has dimensions [force/length].

The Winkler foundation model is *exact* for a prismatic beam that floats on liquid with no part submerged. For solid foundation materials, or for a beam supported by discrete springs, the Winkler model is approximate. However, our intent is to analyze the beam, not the foundation; what is needed is a way to represent the *effect* of the foundation on the beam. Since most foundation models are approximate, replacing the

Winkler model by a more sophisticated alternative may increase mathematical complexity but add little accuracy to calculated deflections and stresses in the beam.

Many problems of prismatic beams on a Winkler foundation have simple analytical solutions. If a beam has complexities of loading or support conditions, changes in flexural stiffness EI, or discrete supporting springs too widely spaced to be "smeared" (Section 5.5), then a computational solution is preferable. A beam on elastic foundation has a simple finite element model. (However, particular care is appropriate: A rather fine discretization may be needed in order to obtain accurate stresses, but a very fine finite element discretization may provoke numerical error associated with ill-conditioned equations.) Similarly, plates and *curved* beams on an elastic foundation are analytically complicated, and a computational solution may be preferred.

In this chapter we consider only straight beams, linearly elastic conditions, and static loading. A static load on a long rail creates a symmetric valley. A *moving* load creates an asymmetric valley, such that the load must be continually driven uphill by an energy input, even when there is no damping. When damping is small and a critical velocity is approached, deflections and bending moments are greatly amplified [5.2].

It is perhaps obvious that a discrete spring need not be a helical spring. Any linearly elastic support acts as a linear spring, whose stiffness K is by definition equal to applied load divided by the corresponding displacement. For example, $K = 3EI/L^3$ for a cantilever beam of length L loaded by a transverse tip force.

Limitations. Regardless of the foundation model adopted, its behavior should be clearly understood. A Winkler foundation model resists only force normal to its surface. It does not resist force directed parallel to its surface. Also, it deflects only where there is load. Adjacent material is utterly unaffected (Fig. 5.1-1b). Such is not the case for a true elastic solid: As shown in Fig. 5.1-1c, adjacent material deflects even though it is not loaded normal to its surface, and uniform deflection does not produce uniform pressure. Soil and railroad ballast are not accurately represented by either the Winkler model or an elastic solid; nevertheless, the Winkler model may be adequate. Railroad ballast is often loose until compacted by increasing load, so there is a nonlinear response [5.3].

The Winkler model assumes that contact is never broken between beam and foundation. Thus, where a beam deflects upward, it is assumed that the foundation pulls down on the beam. If this does not happen, the problem becomes nonlinear. In solving such a problem one must locate the zone of contact as well as determine the beam deflection function $w = w(x)$.

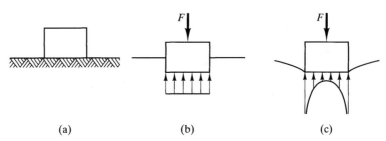

FIGURE 5.1-1 Rigid block on an elastic foundation. (a) No load applied. (b) Load applied, Winkler foundation. (c) Load applied, elastic solid foundation.

One should also ask if secondary effects may be important. For example, foundation pressure on the surface of a beam of thin-walled cross section causes cross sections to deform. Thus, in a wide-flange beam, a flange in contact with the foundation sustains flexural stresses directed normal to the web. Also, foundation pressure $k_o w$ becomes nonuniform across the flange, which may alter the foundation force per unit length along the beam.

5.2 BASIC EQUATIONS, WINKLER FOUNDATION

Figure 5.2-1a shows an arbitrary loading q and the resulting foundation reaction kw near the end of a beam. In subsequent sections we consider particular cases of loading and support conditions. Following custom, we define downward as positive for external load and for deflection.

We begin by isolating a differential element, Fig. 5.2-1b, and writing two equilibrium equations, one for vertical forces and another for moments about an axis normal to the paper. These equations, and the result of combining them, are

$$\left. \begin{array}{l} \dfrac{dV}{dx} = kw - q \\[2mm] \dfrac{dM}{dx} = V \end{array} \right\} \quad \dfrac{d^2 M}{dx^2} = kw - q \qquad (5.2\text{-}1)$$

The moment-curvature relation, from elementary beam theory, is

$$M = -EI \frac{d^2 w}{dx^2} \qquad (5.2\text{-}2)$$

The negative sign is needed because positive bending moment M is associated with negative curvature $d^2 w/dx^2$. Combination of Eq. 5.2-2 with the equilibrium equation $dM/dx = V$ yields, for a uniform beam (constant EI),

$$V = -EI \frac{d^3 w}{dx^3} \qquad (5.2\text{-}3)$$

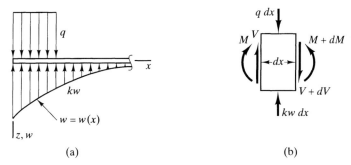

FIGURE 5.2-1 (a) Distributed load q and foundation reaction kw near end $x = 0$ of a beam. (b) Forces and moments that act on a differential element of the beam.

Combination of Eqs. 5.2-1 and 5.2-2 yields the fourth-order governing equation for a uniform beam on a Winkler foundation:

$$EI\frac{d^4w}{dx^4} + kw = q \quad (5.2\text{-}4)$$

The form of Eq. 5.2-4 is that of Eq. 1.7-2, but here the net distributed lateral load is $q - kw$ rather than only q. If the foundation were absent ($k = 0$), one could repeatedly integrate both sides of Eq. 5.2-4, as in elementary beam theory, and obtain a solution for lateral deflection $w = w(x)$. With $k \neq 0$, the solution must be obtained by methods for solving differential equations. Here we skip the details, noting only that for $q = 0$ Eq. 5.2-4 is solved by taking w as the product of $e^{\beta x}$ or $e^{-\beta x}$ times $\sin \beta x$ or $\cos \beta x$. Four such products may be written, each associated with a different constant of integration. For convenience, we have introduced β, where

$$\beta = \left[\frac{k}{4EI}\right]^{1/4} \quad (5.2\text{-}5)$$

We recall from Section 5.1 that k has dimensions [force/length/length] and that $k = k_o b$. Dimensions of β are [1/length]. The solution of Eq. 5.2-4 for constant k, constant EI, and $q = 0$ is

$$w = e^{\beta x}(C_1 \sin \beta x + C_2 \cos \beta x) + e^{-\beta x}(C_3 \sin \beta x + C_4 \cos \beta x) \quad (5.2\text{-}6)$$

where C_1 through C_4 are constants of integration. For $q \neq 0$, the general solution of Eq. 5.2-4 consists of Eq. 5.2-6 plus a particular solution. In this chapter we discuss concentrated loads, and also uniformly distributed loads q, for which the particular solution is $w = q/k$. In a typical analysis problem, one prescribes the elastic properties of beam and foundation, the loading, and boundary conditions. With this information, one can evaluate β and C_1 through C_4. Thus $w = w(x)$ is determined. To obtain flexural stress in the beam, one uses the flexure formula $\sigma = Mc/I$, where M is obtained from Eq. 5.2-2.

Many loadings and boundary conditions are possible. For each different case there is a different set of integration constants. As will be shown, many useful cases, including those where $q \neq 0$, are easy to treat by superposition of results for a few simple cases, thus avoiding the need to evaluate integration constants for each different situation.

In subsequent discussion it is convenient to use the functions here defined:

$$A_{\beta x} = e^{-\beta x}(\cos \beta x + \sin \beta x) \qquad B_{\beta x} = e^{-\beta x} \sin \beta x$$
$$C_{\beta x} = e^{-\beta x}(\cos \beta x - \sin \beta x) \qquad D_{\beta x} = e^{-\beta x} \cos \beta x \quad (5.2\text{-}7)$$

Useful relations among these functions, valid for constant β, are

$$A_{\beta x} = -\frac{1}{\beta}\frac{dD_{\beta x}}{dx} = \frac{1}{2\beta^2}\frac{d^2 C_{\beta x}}{dx^2} = \frac{1}{2\beta^3}\frac{d^3 B_{\beta x}}{dx^3} \quad (5.2\text{-}8\text{a})$$

$$B_{\beta x} = -\frac{1}{2\beta}\frac{dA_{\beta x}}{dx} = \frac{1}{2\beta^2}\frac{d^2 D_{\beta x}}{dx^2} = -\frac{1}{4\beta^3}\frac{d^3 C_{\beta x}}{dx^3} \quad (5.2\text{-}8\text{b})$$

$$C_{\beta x} = \frac{1}{\beta}\frac{dB_{\beta x}}{dx} = -\frac{1}{2\beta^2}\frac{d^2 A_{\beta x}}{dx^2} = \frac{1}{2\beta^3}\frac{d^3 D_{\beta x}}{dx^3} \quad (5.2\text{-}8\text{c})$$

Section 5.2 Basic Equations, Winkler Foundation

$$D_{\beta x} = -\frac{1}{2\beta}\frac{dC_{\beta x}}{dx} = -\frac{1}{2\beta^2}\frac{d^2 B_{\beta x}}{dx^2} = \frac{1}{4\beta^3}\frac{d^3 A_{\beta x}}{dx^3} \qquad (5.2\text{-}8\text{d})$$

Note that βx is a dimensionless quantity; *radians* when used in $\sin \beta x$ and $\cos \beta x$. A numerical tabulation of functions appears in Table 5.2-1. Values not tabulated can be found by interpolation or directly from Eqs. 5.2-7.

TABLE 5.2-1 Selected values of terms defined by Eqs. 5.2-7

βx	$A_{\beta x}$	$B_{\beta x}$	$C_{\beta x}$	$D_{\beta x}$
0	1	0	1	1
0.02	0.9996	0.0196	0.9604	0.9800
0.04	0.9984	0.0384	0.9216	0.9600
0.10	0.9907	0.0903	0.8100	0.9003
0.20	0.9651	0.1627	0.6398	0.8024
0.30	0.9267	0.2189	0.4888	0.7077
0.40	0.8784	0.2610	0.3564	0.6174
0.50	0.8231	0.2908	0.2415	0.5323
0.60	0.7628	0.3099	0.1431	0.4530
0.70	0.6997	0.3199	0.0599	0.3798
$\pi/4$	0.6448	0.3224	0	0.3224
0.80	0.6354	0.3223	−0.0093	0.3131
0.90	0.5712	0.3185	−0.0657	0.2527
1.00	0.5083	0.3096	−0.1108	0.1988
1.10	0.4476	0.2967	−0.1457	0.1510
1.20	0.3899	0.2807	−0.1716	0.1091
1.30	0.3355	0.2626	−0.1897	0.0729
1.40	0.2849	0.2430	−0.2011	0.0419
1.50	0.2384	0.2226	−0.2068	0.0158
$\pi/2$	0.2079	0.2079	−0.2079	0
1.60	0.1959	0.2018	−0.2077	−0.0059
1.70	0.1576	0.1812	−0.2047	−0.0235
1.80	0.1234	0.1610	−0.1985	−0.0376
1.90	0.0932	0.1415	−0.1899	−0.0484
2.00	0.0667	0.1231	−0.1794	−0.0563
2.20	0.0244	0.0896	−0.1548	−0.0652
$3\pi/4$	0	0.0670	−0.1340	−0.0670
2.40	−0.0056	0.0613	−0.1282	−0.0669
2.60	−0.0254	0.0383	−0.1019	−0.0636
2.80	−0.0369	0.0204	−0.0777	−0.0573
3.00	−0.0423	0.0070	−0.0563	−0.0493
π	−0.0432	0	−0.0432	−0.0432
3.20	−0.0431	−0.0024	−0.0383	−0.0407
3.40	−0.0408	−0.0085	−0.0237	−0.0323
3.60	−0.0366	−0.0121	−0.0124	−0.0245
3.80	−0.0314	−0.0137	−0.0040	−0.0177
$5\pi/4$	−0.0279	−0.0139	0	−0.0139
4.00	−0.0258	−0.0139	0.0019	−0.0120
$3\pi/2$	−0.0090	−0.0090	0.0090	0
2π	0.0019	0	0.0019	0.0019

Equations 5.2-8 are also useful when integration is required. As an example, from Eq. 5.2-8a,

$$\int A_{\beta x}\, dx = -\int \frac{1}{\beta}\, dD_{\beta x} \quad \text{hence} \quad \int A_{\beta x}\, dx = -\frac{1}{\beta} D_{\beta x} + C_5 \quad (5.2\text{-}9)$$

where C_5 is a constant of integration.

5.3 SEMI-INFINITE BEAMS WITH CONCENTRATED END LOADS

The uniform beam shown in Fig. 5.3-1a lies on the x axis and extends from $x = 0$ to positive infinity. Such a beam is called "semi-infinite" or perhaps "one-way infinite." (An actual beam has finite length; remarks on when it may be idealized as infinitely long appear in Section 5.7.) At $x = 0$, loads are force P_o and/or moment M_o. Deflection and rotation at $x = 0$ are w_o and θ_o, both shown in the positive sense in Fig. 5.3-1b. In general, one seeks w, θ, M, and V in the beam as functions of x.

The problem is solved by appropriate specialization of Eq. 5.2-6. Saint-Venant's principle requires that w vanish as x approaches infinity. Accordingly, terms associated with $e^{\beta x}$ in Eq. 5.2-6 must vanish. We therefore must have $C_1 = 0$ and $C_2 = 0$. What remains, in terms of functions in Eqs. 5.2-7, is

$$w = C_3 B_{\beta x} + C_4 D_{\beta x} \quad (5.3\text{-}1)$$

Boundary conditions needed to determine C_3 and C_4 are, from Eqs. 5.2-2 and 5.2-3,

$$M_o = -EI\left(\frac{d^2w}{dx^2}\right)_{x=0} \quad \text{and} \quad -P_o = -EI\left(\frac{d^3w}{dx^3}\right)_{x=0} \quad (5.3\text{-}2)$$

P_o is a negative V because the positive sense of V, defined by Fig. 5.2-1b, is upward on material to the right of a cut. Differentiation of w to obtain d^2w/dx^2 and d^3w/dx^3 is assisted by Eqs. 5.2-8. Constant C_4 is found to drop out of the first of Eqs. 5.3-2. This equation therefore yields C_3, which can be written in different forms by substitution from Eq. 5.2-5.

$$C_3 = \frac{M_o}{2EI\beta^2} \quad \text{or} \quad C_3 = \frac{2\beta^2 M_o}{k} \quad (5.3\text{-}3)$$

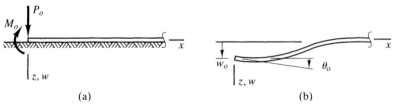

FIGURE 5.3-1 (a) Concentrated loads P_o and M_o at the end of a semi-infinite beam on a Winkler foundation. (b) End deflection w_o and end rotation $\theta_o = (dw/dx)_{x=0}$, both shown in the positive sense.

Section 5.3 Semi-Infinite Beams with Concentrated End Loads

Knowing C_3 and proceeding in similar fashion, we obtain C_4 from the second of Eqs. 5.3-2.

$$C_4 = \frac{2\beta P_o}{k} - \frac{2\beta^2 M_o}{k} \tag{5.3-4}$$

Now Eqs. 5.3-3 and 5.3-4 can be substituted into Eq. 5.3-1, whereupon we find that the substitution $B_{\beta x} - D_{\beta x} = -C_{\beta x}$ is appropriate. Thus we obtain Eq. 5.3-5 as shown. Hence, again with the assistance of Eqs. 5.2-8, we obtain expressions for θ, M, and V.

$$w = w(x) \qquad w = \frac{2\beta P_o}{k} D_{\beta x} - \frac{2\beta^2 M_o}{k} C_{\beta x} \tag{5.3-5}$$

$$\theta = \frac{dw}{dx} \qquad \theta = -\frac{2\beta^2 P_o}{k} A_{\beta x} + \frac{4\beta^3 M_o}{k} D_{\beta x} \tag{5.3-6}$$

$$M = -EI\frac{d^2w}{dx^2} \qquad M = -\frac{P_o}{\beta} B_{\beta x} + M_o A_{\beta x} \tag{5.3-7}$$

$$V = -EI\frac{d^3w}{dx^3} \qquad V = -P_o C_{\beta x} - 2\beta M_o B_{\beta x} \tag{5.3-8}$$

As seen by consulting Eqs. 5.2-7, variations with x of w, θ, M, and V are represented by damped sine and cosine waves. Plots of w and M appear in Fig. 5.3-2.

For equilibrium of vertical forces, the total foundation reaction must balance load P_o. Accordingly, as an alternative to the second of Eqs. 5.3-2, we can equate P_o to the integral of $kw\,dx$ from zero to infinity. The same results as already presented are of course obtained.

Note that P_o and/or M_o may be unknown in a particular application. For example, if end $x = 0$ in Fig. 5.3-1a were simply supported and loaded by prescribed moment M_o, unknowns would be end rotation θ_o and support reaction P_o. A solvable problem is posed when any two of the four end quantities in Fig. 5.3-1 are prescribed.

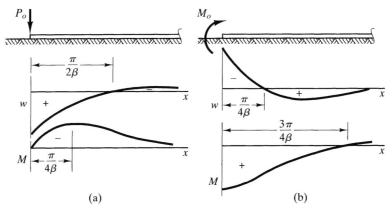

FIGURE 5.3-2 Variation with x of lateral deflection w and bending moment M in end-loaded, uniform semi-infinite beams on a Winkler foundation. (a) End force load. (b) End moment load.

EXAMPLE

A semi-infinite steel beam ($E = 200$ GPa) has a square cross section 80 mm on a side and rests on a Winkler foundation of modulus $k_o = 0.25$ N/mm²/mm. A downward load of 50 kN is applied to the end, at $x = 0$. Determine the largest upward and downward deflections and their locations. Also determine the largest flexural stress and its location.

Necessary constants are

$$\left. \begin{array}{l} EI = 200{,}000 \dfrac{80^4}{12} = 6.827(10^{11}) \text{ N} \cdot \text{mm}^2 \\ k = 80k_o = 20 \text{ N/mm/mm} \end{array} \right\} \quad \begin{array}{l} \beta = \left[\dfrac{k}{4EI}\right]^{1/4} \\ \beta = 0.001645/\text{mm} \end{array} \quad (5.3\text{-}9)$$

Equation 5.3-5 becomes $w = 2\beta P_o D_{\beta x}/k$ and yields the largest downward deflection at $x = 0$, where $D_{\beta x} = 1$.

$$w_{\max} = w_o = \frac{2\beta P_o}{k} = \frac{2(0.001645)50{,}000}{20} = 8.23 \text{ mm} \quad (5.3\text{-}10)$$

The largest upward deflection appears at the smallest x for which $\theta = 0$. From Eq. 5.3-6, this is where $A_{\beta x} = 0$. Scanning Table 5.2-1, we see that $A_{\beta x} = 0$ at $\beta x = 3\pi/4$, or $x = 1432$ mm. Here $D_{\beta x} = -0.0670$, a value that could have been obtained by simply scanning the table for the largest negative value of $D_{\beta x}$. The largest negative (upward) deflection is

$$w_{\min} = \frac{2\beta P_o}{k}(-0.0670) = -0.0670 w_{\max} = -0.55 \text{ mm} \quad (5.3\text{-}11)$$

Remember that our analysis presumes that beam and foundation do not separate from one another where w is upward. If separation occurs in the actual structure, our solution may yet be a good approximation because $|w_{\min}| \ll |w_{\max}|$ in this particular example. Bending moment is $M = -P_o B_{\beta x}/\beta$. We scan Table 5.2-1 and find that the largest magnitude of $B_{\beta x}$ is 0.3224 and appears at $\beta x = \pi/4$. Hence the bending moment of largest magnitude, and the associated flexural stress, are

$$M = |M_{\min}| = \frac{(50{,}000)(0.3224)}{0.001645} = 9.80(10^6) \text{ N} \cdot \text{mm} \quad (5.3\text{-}12)$$

$$\sigma = \frac{Mc}{I} = \frac{9.80(10^6)(40)}{80^4/12} = 115 \text{ MPa} \quad (5.3\text{-}13)$$

This stress is tensile on top of the beam and appears at $x = \pi/4\beta = 477$ mm from the end.

5.4 INFINITE BEAMS WITH CONCENTRATED LOADS

We consider a uniform beam that extends to infinity in both directions from the point where loads are applied. An infinite beam may be thought of as a "both-way infinite" beam. (An actual beam can often be idealized as infinitely long; see Section 5.7.) Let an x coordinate originate at the point of load and be positive rightward. Equation 5.2-6 applies for $+x$ or for $-x$, but not for both, because $w = w(x)$ cannot be represented as a single function for the entire beam. We elect to represent the $+x$ side. Accordingly, C_1 and C_2 must vanish, for reasons stated above Eq. 5.3-1, and Eq. 5.3-1 is applicable to cases treated in the present section. Constants C_3 and C_4 need not be determined explicitly for the present case because we can use results already developed in Section 5.3, as follows.

Section 5.4 Infinite Beams with Concentrated Loads

Concentrated Force Loading. Deflections are symmetric with respect to the load (Fig. 5.4-1b). Therefore $\theta = 0$ at $x = 0$. Imagine two vertical cuts in the beam, straddling load P_o and an infinitesimal distance dx apart. Thus we isolate a beam fragment of length dx, which carries a central downward load P_o, upward shear force $P_o/2$ on each side, and positive bending moment M_o on each side. Hence, the beam extending rightward from P_o carries downward end force $P_o/2$ and clockwise moment M_o. Equations of Section 5.3 are applicable to this problem if we replace P_o by $P_o/2$ and use a value of M_o such that $\theta = 0$ at $x = 0$. From Eq. 5.3-6,

$$0 = -\frac{2\beta^2(P_o/2)}{k} + \frac{4\beta^3 M_o}{k} \qquad \text{hence} \qquad M_o = \frac{P_o}{4\beta} \tag{5.4-1}$$

Substituting for P_o and M_o in Eq. 5.3-5, noting that $2D_{\beta x} - C_{\beta x} = A_{\beta x}$, and using Eqs. 5.2-5 and 5.2-8, we obtain

$$w = w(x) \qquad w = \frac{\beta P_o}{2k} A_{\beta x} \tag{5.4-2}$$

$$\theta = \frac{dw}{dx} \qquad \theta = -\frac{\beta^2 P_o}{k} B_{\beta x} \tag{5.4-3}$$

$$M = -EI\frac{d^2w}{dx^2} \qquad M = \frac{P_o}{4\beta} C_{\beta x} \tag{5.4-4}$$

$$V = -EI\frac{d^3w}{dx^3} \qquad V = -\frac{P_o}{2} D_{\beta x} \tag{5.4-5}$$

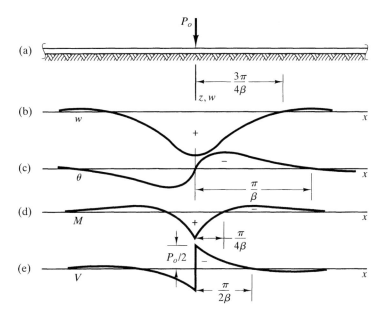

FIGURE 5.4-1 (a) Concentrated load P_o at $x = 0$ on a uniform infinite beam on a Winkler foundation. (b–e) Plots of deflection, rotation, bending moment, and transverse shear force in the beam.

Recall that these results are valid only for the right half of the beam. Results in the left half ($x < 0$) can be obtained from results in the right half ($x > 0$) and the following conditions of symmetry and antisymmetry.

$$w(x) = w(-x) \qquad \theta(x) = -\theta(-x)$$
$$M(x) = M(-x) \qquad V(x) = -V(-x) \qquad (5.4\text{-}6)$$

If k or EI has a step change at $x = 0$, none of the foregoing equations in the present section is valid. Such a problem can be solved by use of equations in Section 5.3 and careful attention to algebraic signs (see Example 4 of Section 5.5).

Concentrated Moment Loading. Results for concentrated moment can be obtained in much the same way as for concentrated force; that is, by adapting results developed in Section 5.3. The right half of an infinite beam, Fig. 5.4-2, is loaded by moment $M_o/2$ and by end force P_o such that $w = 0$ at $x = 0$. From Eq. 5.3-5,

$$0 = \frac{2\beta P_o}{k} - \frac{2\beta^2 (M_o/2)}{k} \qquad \text{hence} \qquad P_o = \frac{M_o \beta}{2} \qquad (5.4\text{-}7)$$

Substituting for P_o and M_o in Eq. 5.3-5, noting that $D_{\beta x} - C_{\beta x} = B_{\beta x}$, and using Eqs. 5.2-5 and 5.2-8, we obtain

$$w = w(x) \qquad w = \frac{\beta^2 M_o}{k} B_{\beta x} \qquad (5.4\text{-}8)$$

$$\theta = \frac{dw}{dx} \qquad \theta = \frac{\beta^3 M_o}{k} C_{\beta x} \qquad (5.4\text{-}9)$$

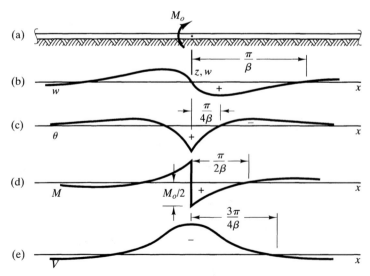

FIGURE 5.4-2 (a) Concentrated moment load M_o at $x = 0$ on a uniform infinite beam on a Winkler foundation. (b–e) Plots of deflection, rotation, bending moment, and transverse shear force in the beam.

$$M = -EI\frac{d^2w}{dx^2} \qquad M = \frac{M_o}{2} D_{\beta x} \qquad (5.4\text{-}10)$$

$$V = -EI\frac{d^3w}{dx^3} \qquad V = -\frac{\beta M_o}{2} A_{\beta x} \qquad (5.4\text{-}11)$$

Recall that these results are valid only for the right half of the beam. Results in the left half ($x < 0$) can be obtained from results in the right half ($x > 0$) and the following conditions of symmetry and antisymmetry.

$$\begin{array}{ll} w(x) = -w(-x) & \theta(x) = \theta(-x) \\ M(x) = -M(-x) & V(x) = V(-x) \end{array} \qquad (5.4\text{-}12)$$

5.5 EXAMPLE PROBLEMS

EXAMPLE 1

In Fig. 5.4-1a, let $k = 0.25$ N/mm/mm for the foundation and $EI = 441(10^9)$ N·mm² for the beam. Determine the deflection and bending moment at the location where a load $P_o = 18$ kN acts on the beam.

From Eq. 5.2-5,

$$\beta = \left[\frac{k}{4EI}\right]^{1/4} = \left[\frac{0.25}{4(441)10^9}\right]^{1/4} = 6.136(10^{-4})/\text{mm} \qquad (5.5\text{-}1)$$

From Eqs. 5.4-2 and 5.4-4, at $x = 0$,

$$w_{max} = \frac{\beta P_o}{2k} A_{\beta x} = \frac{6.136(10^{-4})(18{,}000)}{2(0.25)}(1.0) = 22.1 \text{ mm} \qquad (5.5\text{-}2)$$

$$M_{max} = \frac{P_o}{4\beta} C_{\beta x} = \frac{18{,}000}{4(0.0006136)}(1.0) = 7.33 \text{ kN·m} \qquad (5.5\text{-}3)$$

EXAMPLE 2

For the beam in Fig. 5.5-1, determine the deflection and bending moment at the center and the largest spring force. Do so by "smearing" the springs and assuming that the beam is infinitely long. (Thus we approximate the actual situation, and must ask: How accurate are the results?)

Springs can be "smeared" if they are of equal stiffness, equal spacing, and the spacing is sufficiently small (as defined following Eq. 5.5-5). A spring with stiffness K and deflection w applies force Kw. If s is the spring spacing, force Kw can be idealized as uniformly distributed over a total span s ($s/2$ on either side of the spring). A Winkler foundation of modulus k and deflection w exerts force kws over span s. If the two forces are to be the same, then $Kw = kws$, so

$$k = \frac{K}{s} \quad \text{here} \quad k = \frac{275}{1100} = 0.25 \text{ N/mm/mm} \qquad (5.5\text{-}4)$$

Because of the choice of numerical constants in the present example, we obtain the same numerical values of k and β as in Example 1. And, if the beam is idealized as infinitely long, we obtain the same numerical results for w_{max} and M_{max} as in Eqs. 5.5-2 and 5.5-3. At a spring

Chapter 5 Beams on an Elastic Foundation

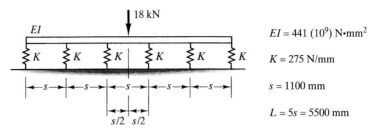

FIGURE 5.5-1 Uniform beam supported by equally spaced springs, each of stiffness K.

nearest the load, $x = 550$ mm, $\beta x = 0.337$, and $A_{\beta x} = 0.9096$. Hence, from Eqs. 5.4-2 and 5.5-2 the spring force Kw is

$$Kw = K(w_{max} A_{\beta x}) = 275(22.1)(0.9096) = 5530 \text{ N} \quad (5.5\text{-}5)$$

How accurate are these approximate results? First, we note that the following ad hoc guidelines have been met. Experience indicates that it is acceptable to "smear" springs if there are at least three [1.6] or four [5.4] springs per half wave. A half wave is $\beta x = \pi$. Thus, if we adopt the more stringent guideline of four springs per half wave, we require that

$$s \leq \frac{\pi}{4\beta} \quad (5.5\text{-}6)$$

In the present example, $\pi/4\beta = 1280$ mm, so the spacing guideline has been met. Another ad hoc guideline is that a centrally loaded beam of length L can be treated as if it were infinitely long if $\beta L > 3$ (see Section 5.7). In the present example, $\beta L = 3.37$, which satisfies the guideline.

A definitive assessment of accuracy requires that the actual problem be solved. A finite element model that is exact according to elementary beam theory was constructed as follows. The six spans in Fig. 5.5-1 (the outer four of length s, the inner two of length $s/2$) were each modeled by a single beam element. Springs were each represented by a single spring element. Symmetry was exploited by modeling only the right half of the structure, with rotation prevented at the left node. This node had no attached spring and was loaded by a 9 kN force. The model had seven degrees of freedom and yielded seven simultaneous equations. Computed results are deflection $w_{max} = 24.1$ mm and bending moment $M_{max} = 8.11$ kN·m, both beneath the load, and compressive force 6010 N in each of the two springs nearest the load. Thus Eqs. 5.5-2, 5.5-3, and 5.5-5 have errors of -8.3% for w_{max}, -9.6% for M_{max}, and -8.0% for the largest spring force. These errors are nearly "worst case" values, as we are near recommended limits for spring spacing and beam length, and the load is midway between springs rather than over a spring. If the physical problem is altered so that the two springs adjacent to the load are replaced by three springs, one beneath the load and two at distances $s/2$ away, each of stiffness $2K/3$ so as to maintain the same total support stiffness, and the finite element model is similarly revised, we then obtain $w_{max} = 23.1$ mm and $M_{max} = 7.03$ kN·m from the finite element model. Equations 5.5-2 and 5.5-3 still provide w_{max} and M_{max} for a Winkler foundation model of this problem, but these quantities now have the respective errors -4.3% and $+4.3\%$.

With current finite element software, an analysis such as that just described is not difficult, and may be the preferred method if constants of the structure are accurately known, accurate results are required, and the work is done carefully.

EXAMPLE 3

An infinitely long rail on a Winkler foundation has the following properties: $EI = 441(10^9)$ N·mm², $k = 0.25$ N/mm/mm, and $\beta = 6.136(10^{-4})$/mm. Two downward wheel loads, 18 kN each and 2.6 m apart, are applied to the rail (Fig. 5.5-2a). What are the maximum deflection and the maximum bending moment?

Values of w_{max} and M_{max} for a *single* 18 kN load on this structure have aready been calculated (Eqs. 5.5-2 and 5.5-3). In the present problem we must superpose results for two loads at different locations. Let us first calculate deflection and bending moment at C due to load at A (Fig. 5.5-2). For these quantities we adopt the notation $w_{C(A)}$ and $M_{C(A)}$. With $x = 0$ at A, x at C is 2600 mm, and $\beta x = 1.595$. From Eqs. 5.4-2 and 5.4-4,

$$w_{C(A)} = \frac{\beta P_o}{2k} A_{\beta x} = \frac{6.136(10^{-4})(18,000)}{2(0.25)}(0.1978) = 4.4 \text{ mm} \quad (5.5\text{-}7)$$

$$M_{C(A)} = \frac{P_o}{4\beta} C_{\beta x} = \frac{18,000}{4(6.136)10^{-4}}(-0.2078) = -1.52 \text{ kN·m} \quad (5.5\text{-}8)$$

At C, and due to load at C, deflection and bending moment are $w_{C(C)} = 22.1$ mm and $M_{C(C)} = 7.33$ kN·m (from Eqs. 5.5-2 and 5.5-3). Superposing results at C, we obtain

$$w_C = w_{C(C)} + w_{C(A)} = 22.1 + 4.4 = 26.5 \text{ mm} \quad (5.5\text{-}9)$$

$$M_C = M_{C(C)} + M_{C(A)} = 7.33 - 1.52 = 5.81 \text{ kN·m} \quad (5.5\text{-}10)$$

The same results prevail at A because of symmetry about midpoint B; for example, $w_A = w_{A(A)} + w_{A(C)} = 26.5$ mm. Note, however, that if $w_{A(C)}$ were to be numerically evaluated, x would have to be positive because our formulas are not valid for negative x.

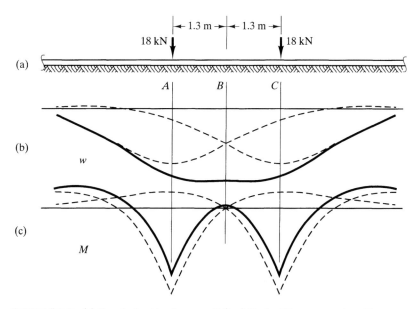

FIGURE 5.5-2 (a) Equal wheel loads on a rail. (b, c) Dashed lines are produced by component loads. Solid lines are superposed results, for $w = w(x)$ and $M = M(x)$.

It is possible that w_{max} will appear at B, but unlikely that M_{max} will appear at B. Nevertheless we will calculate both quantities at B. Because of symmetry about B, we can obtain results at B by simply doubling the effect of either load. For this calculation, $x = 1300$ mm and $\beta x = 0.798$. From Eqs. 5.4-2 and 5.4-4,

$$w_B = 2\frac{\beta P_o}{2k}A_{\beta x} = 2\frac{6.136(10^{-4})(18,000)}{2(0.25)}(0.6369) = 28.1 \text{ mm} \qquad (5.5\text{-}11)$$

$$M_B = 2\frac{P_o}{4\beta}C_{\beta x} = 2\frac{18,000}{4(6.136)10^{-4}}(-0.0078) = -0.11 \text{ kN·m} \qquad (5.5\text{-}12)$$

Because M_B is negative, we know that the rail is concave down at B. Therefore, although $w_B > w_C$, the maximum deflection does not appear at B; rather, there must be two points between A and C, symmetrically located with respect to B, that have the same value of w_{max}. However, since M_B is small and $w_B \approx w_C$, we conclude that the rail is almost flat in the neighborhood of point B. As a practical matter, it is probably not necessary to know w_{max} exactly, and the Winkler foundation model is probably an approximation of actual foundation characteristics. Accordingly, for two equal or unequal loads on a rail, one might accept the largest of w_A, w_B, and w_C as a satisfactory approximation of w_{max}.

EXAMPLE 4

An infinite beam is constructed by welding together two semi-infinite beams having the same width, one of flexural stiffness EI, the other of flexural stiffness $2EI$. The infinite beam is placed on a uniform Winkler foundation and moment load M_o is applied at the weld (Fig. 5.5-3a). What are the lateral deflection, rotation, and bending moments at the weld?

Equations of Section 5.4 are not directly applicable because they apply to a *uniform* infinite beam. To solve the present problem we treat each half as a semi-infinite beam, and enforce compatibility of displacement and rotation at the weld. Thus we determine the internal shear force and internal moment at the weld, and can then determine the required information.

Figure 5.5-3b shows the beam cut apart at the weld, with internal force V and internal couple C thus exposed. V and C are action-reaction loads that each half of the beam applies to the other. Arbitrarily, external load M_o has been associated with the right half. (The same final

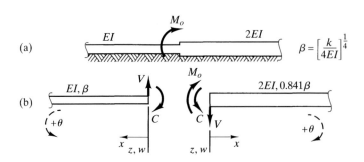

FIGURE 5.5-3 (a) Stepped beam on a Winkler foundation, with moment load M_o applied at the step. (b) Internal force V and internal couple C exposed by a cut at the step. The foundation reaction is not shown.

results would be obtained if M_o were associated with the left half instead.) Two equations of compatibility can be written. One says that both halves have the same deflection w at $x=0$. From Eq. 5.3-5,

$$-\frac{2\beta V}{k} + \frac{2\beta^2 C}{k} = \frac{2(0.841\beta)V}{k} - \frac{2(0.841\beta)^2(M_o - C)}{k} \qquad (5.5\text{-}13)$$

The other equation says that both halves have the same rotation θ at $x=0$. In writing this equation we must recognize that since x must be positive when directed away from the end in each half, equal positive rotations $\theta = dw/dx$ have opposite sign at $x=0$. Arbitrarily choosing clockwise as positive, and adjusting algebraic signs accordingly, we obtain from Eq. 5.3-6

$$-\frac{2\beta^2 V}{k} + \frac{4\beta^3 C}{k} = -\frac{2(0.841\beta)^2 V}{k} + \frac{4(0.841\beta)^3(M_o - C)}{k} \qquad (5.5\text{-}14)$$

Equations 5.5-13 and 5.5-14 yield

$$V = 0.450\beta M_o \quad \text{and} \quad C = 0.414 M_o \qquad (5.5\text{-}15)$$

Bending moments are $C = 0.414 M_o$ immediately left of the weld and $M_o - C = 0.586 M_o$ immediately right of the weld, both directed clockwise. Deflection and rotation at the weld can be obtained from either the left-hand sides or the right-hand sides of Eqs. 5.5-13 and 5.5-14, and are

$$w = -0.072\beta^2 M_o/k \quad \text{and} \quad \theta = 0.757\beta^3 M_o/k \qquad (5.5\text{-}16)$$

The deflection is upward and the rotation is clockwise.

For a *uniform* beam of the average stiffness $1.5EI$, Eq. 5.4-9 yields $\theta = 0.738\,\beta^3 M_o/k$. Why is this value so close to that of Eq. 5.5-16? The beam is only part of the elastic system; the foundation is the remainder. Hence, the influence of flexural stiffness EI is not as great as simple beam theory might suggest.

5.6 UNIFORMLY DISTRIBUTED LOAD

Semi-Infinite Beam. If a beam carries uniformly distributed load q over its entire span and a Winkler foundation provides the only support, the beam displaces without bending. Displacement w is independent of x, and foundation reaction kw is uniform and equal to q. Thus the deflection is $w = q/k$ (Fig. 5.6-1a). Such is the case whether the beam is finite, semi-infinte, or infinite.

Superposition can be used to solve simple problems in which additional support is provided. Consider Fig. 5.6-1b. If the simple support were absent, deflection would be $w = q/k$ for all x. Force R applied by the simple support can be obtained from Eq. 5.3-5 by setting $M_o = 0$ and $w = -q/k$ at $x=0$. Thus $P_o = -q/2\beta$, where $P_o = -R$. Knowing P_o, one can use Eq. 5.3-7 to obtain $M = M(x)$ and hence calculate flexural stresses.

In less straightforward problems, supports may be placed at locations other than $x=0$, load q may cover a small span, or load q may not be uniform. Analytical solutions are possible for such problems, but they are sufficiently tedious that computational solutions may be preferred.

148   Chapter 5   Beams on an Elastic Foundation

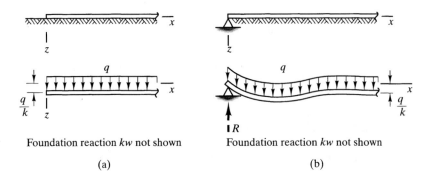

Foundation reaction kw not shown   Foundation reaction kw not shown

(a)   (b)

FIGURE 5.6-1   (a) Beam on Winkler foundation with uniformly distributed load q over its entire span, shown deflected. (b) The same beam, now with a simple support at the left end, also shown deflected.

Infinite Beam.   Figure 5.6-2 shows an infinite beam that carries uniformly distributed load q over a finite span ℓ. In what follows we obtain expressions for deflection and bending moment at an arbitrary point Q that lies within span ℓ.

Deflection can be calculated by superposing deflection increments produced by infinitely many load increments $q\,dx$. From Eq. 5.4-2, the deflection increment at point Q produced by load increment $q\,dx$ at point O is

$$dw_Q = \frac{\beta q\,dx}{2k} A_{\beta x} \tag{5.6-1}$$

The next contribution to dw_Q at point Q is obtained by conceptually moving point O a distance dx along the beam, and writing Eq. 5.6-1 again. Summing contributions as point O is moved across the entire span $\ell = a + b$, and using Eq. 5.2-9 for integration, we obtain the total deflection w_Q at point Q.

$$w_Q = \frac{\beta q}{2k}\int_0^\ell A_{\beta x}\,dx = \frac{\beta q}{2k}\left[\left(-\frac{1}{\beta}D_{\beta x}\right)_0^a + \left(-\frac{1}{\beta}D_{\beta x}\right)_0^b\right] \tag{5.6-2a}$$

$$w_Q = \frac{q}{2k}(2 - D_{\beta a} - D_{\beta b}) \tag{5.6-2b}$$

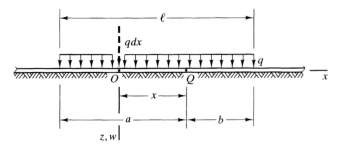

FIGURE 5.6-2   Uniformly distributed load q over a length ℓ of an infinite beam on a Winkler foundation.

In doing the integration one imagines that x is positive to the left when point O is right of point Q because our formulas require that x be positive. As partial checks on the result, we note that Eq. 5.6-2b reduces to $w_Q = q/k$ when a and b are both large, and to $w_Q = 0$ when a and b are both small.

Rotation at point Q is $\theta_Q = dw_Q/dx$, where $dx = da$. From Eq. 5.6-2b,

$$\theta_Q = \frac{q}{2k}\left(-\frac{dD_{\beta a}}{da} - \frac{dD_{\beta b}}{db}\frac{db}{da}\right) \tag{5.6-3}$$

Using Eq. 5.2-8a and noting that $db/da = -1$, we obtain

$$\theta_Q = \frac{\beta q}{2k}(A_{\beta a} - A_{\beta b}) \tag{5.6-4}$$

Another differentiation yields curvature d^2w_Q/dx^2, and with Eqs. 5.2-2, 5.2-5, and 5.2-8b we obtain M_Q, the bending moment at point Q.

$$M_Q = \frac{q}{4\beta^2}(B_{\beta a} + B_{\beta b}) \tag{5.6-5}$$

Finally, transverse shear force at point Q is $V_Q = dM_Q/dx$.

$$V_Q = \frac{q}{4\beta}(C_{\beta a} - C_{\beta b}) \tag{5.6-6}$$

The foregoing expressions pertain to any point Q *within* span ℓ. Depending on physical constants, M_Q may have greatest magnitude at the center of span ℓ or near ends of span ℓ. And, as noted below, it is possible for bending moment *outside* span ℓ to have slightly larger magnitude than given by Eq. 5.6-5.

Infinite Beam, $\beta\ell \leq \pi$. The case $\beta\ell \leq \pi$ may be called the case of "small $\beta\ell$." It arises when ℓ is small, the foundation is soft, or the beam is stiff. It is the case for which the largest bending moment appears at the middle of span ℓ, where M_Q has the value

$$M_Q = \frac{q}{2\beta^2} B_{\beta\ell/2} \qquad \left(\text{for } a = b = \frac{\ell}{2}\right) \tag{5.6-7}$$

The argument for having a maximum at midspan when $\beta\ell \leq \pi$ is as follows. Using Eq. 5.6-5, M_Q at midspan ($a = b = \ell/2$) can be compared with M_Q at any other point (a arbitrary, $b = \ell - a$). The midspan value of M_Q is largest if the condition $2B_{\beta\ell/2} \geq B_{\beta a} + B_{\beta b}$ is satisfied. Using Table 5.2-1, and trying several values of βa for each of several trial values of $\beta\ell$, one concludes that the condition is satisfied when $\beta\ell \leq \pi$.

Infinite Beam, $\beta\ell$ "Large." By "$\beta\ell$ large" we mean that $\beta\ell$ is sufficiently large that bending moment almost vanishes near the center of span ℓ. For this case the largest bending moments appear near ends of span ℓ, and are easily calculated as follows.

Imagine that the beam is cut apart at the left end of the loaded span. Then the portion not loaded does not displace, while the adjacent loaded portion displaces without bending, an amount q/k (Fig. 5.6-3). To restore the continuity present in the actual beam, cut ends must apply transverse shear forces V to one another, of such magnitude that

150   Chapter 5   Beams on an Elastic Foundation

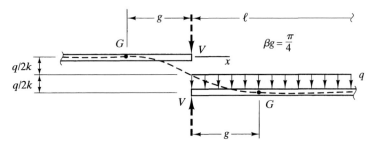

Foundation reaction kw not shown

FIGURE 5.6-3   Reconnecting at an imagined cut in the beam of Fig. 5.6-2 at the left end of span ℓ, when $\beta\ell$ is large. Dashed line shows the actual displaced shape of the beam axis.

each end displaces an amount $q/2k$. No moments are needed to restore continuity because forces V alone produce equal rotations at the imagined cut. Thus, with $w = q/2k$, Eq. 5.3-5 yields $V = q/4\beta$. Hence, Eq. 5.3-7 yields the bending moment at any location. Bending moments of largest magnitude appear at points labeled G in Fig. 5.6-3, and are

$$|M_G| = \frac{q}{4\beta^2} B_{\beta g} = 0.0806 \frac{q}{\beta^2} \quad \left(\text{for } g = \frac{\pi}{4\beta}\right) \tag{5.6-8}$$

(This result can also be obtained from Eq. 5.6-5 by letting b become large and setting $a = \pi/4\beta$.) The same situation prevails at the right end of span ℓ. Thus, a bending moment of magnitude M_G is found at a total of four points.

When may $\beta\ell$ be considered large? Several finite element solutions, each using a different value of $\beta\ell$, suggest that if $\beta\ell > 4$, the actual bending moment of greatest magnitude, whether inside or outside of span ℓ, is less than 3% greater than given by Eq. 5.6-8.

Infinite Beam, $\beta\ell$ "Intermediate."   In this case, $\beta\ell > \pi$, so that the largest bending moment is not given by Eq. 5.6-7, yet $\beta\ell$ is not large enough for Eq. 5.6-8 to be valid. The bending moment of largest magnitude may appear *outside* of span ℓ. Accurate bending moments for this "intermediate" case may be obtained computationally or from lengthy formulas [5.4]. Finite element analyses suggest the following procedure, which may be adequate for most problems of uniformly distributed load on an infinite uniform beam. Calculate M_Q from Eq. 5.6-7 and M_G from Eq. 5.6-8. The actual bending moment of largest magnitude is either M_Q (if $\beta\ell \leq \pi$) or will exceed the larger of M_Q and M_G by less than 10% (if $\beta\ell > \pi$). The largest error occurs when $\beta\ell = 3.6$, which is in the "intermediate" range (and for which, incidentally, M_Q and M_G are equal).

5.7   BEAMS OF FINITE LENGTH

For a beam of finite length, all four constants C_1 through C_4 in Eq. 5.2-6 must be retained because the argument that precedes Eq. 5.3-1 does not apply to a finite beam. There are sufficient conditions to determine all four constants. For example, in Fig. 5.7-1a, we can state that at $x = 0$, $\theta = 0$ and $V = -P/2$; and at $x = L/2$, $M = 0$ and $V = 0$. Thus $w = w(x)$ becomes known for $x > 0$, Eq. 5.2-2 can be used to determine $M = M(x)$, and so on. For this and other problems of finite beams, the algebra is considerable and final expressions are often lengthy. Nevertheless, results are known and are tabulated for several cases [1.6, 5.4].

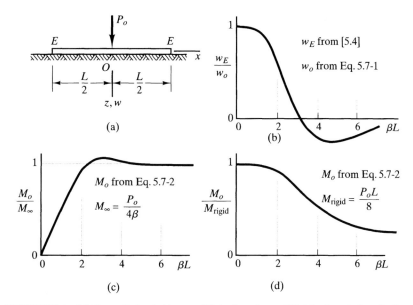

FIGURE 5.7-1 (a) Centrally loaded beam of finite length on a Winkler foundation. (b-d) Plots of deflection ratio and bending moment ratios versus βL.

The case depicted by Fig. 5.7-1a is among the simplest. Consider first the case of a very short beam (βL small). Then the beam can be idealized as *rigid*, so that the foundation reaction on the beam is uniformly distributed and of intensity P/L. The entire beam deflects an amount $w = P/kL$, and it is a simple matter of statics to show that the maximum bending moment is $M_{\text{rigid}} = PL/8$, at center point O.

For larger values of βL, deformation of the beam must be taken into account. General expressions for central deflection and bending moment, at $x = 0$ in Fig. 5.7-1a, valid for all values of βL, are [5.4]

$$w_o = \frac{\beta P_o}{2k} \frac{2 + \cosh \beta L + \cos \beta L}{\sinh \beta L + \sin \beta L} \tag{5.7-1}$$

$$M_o = \frac{P_o}{4\beta} \frac{\cosh \beta L - \cos \beta L}{\sinh \beta L + \sin \beta L} \tag{5.7-2}$$

As βL becomes large, w_o approaches $\beta P_o/2k$ and M_o approaches $P_o/4\beta$, which are the results for an infinitely long beam. For the following discussion, let us symbolize this bending moment result as $M_\infty = P_o/4\beta$.

The ratios M_o/M_∞ and M_o/M_{rigid} are plotted in Fig. 5.7-1. Also plotted is w_E/w_o, which is the ratio of deflection at ends of the beam to deflection at the center. These plots suggest the following grouping.

1. Short beams, for which $\beta L < 1$
2. Intermediate beams, for which $1 < \beta L < 3$
3. Long beams, for which $\beta L > 3$

The bounds $\beta L = 1$ and $\beta L = 3$ are of course somewhat arbitrary. A beam in group 1 may be analyzed as if it were rigid, with only about 1% error in w_o and M_o. A beam in group 2 can be analyzed by Eqs. 5.7-1 and 5.7-2. A beam in group 3 may be considered infinitely long, with less than 10% error in either w_o and M_o for any value of $\beta L > 3$.

Similar considerations apply to other loadings. For example, let load P_o in Fig. 5.7-1a be applied at an end E rather than at center O. Then a rigid-beam analysis yields deflection $4P_o/kL$ at the loaded end and maximum bending moment $4PL/27$ at $x = L/3$. As compared with exact results [5.4], these rigid-beam values are less than 1% in error if $\beta L < 1$. For a long end-loaded beam one can use formulas for a semi-infinite beam, with less than 2% error in either end deflection or maximum bending moment if $\beta L > 3$.

Generalizing, we conclude that finite beams under various loadings can often be analyzed as if they are either rigid or long and flexible, with intermediate cases treated either by available formulas [1.6, 5.4] or by numerical analysis with existing software.

PROBLEMS

5.2-1. A flat plate has specific gravity slightly greater than unity. Explain how the plate may yet be made to float on still water by placing a weight near the center of the plate.

5.2-2. A uniform straight pipe has mass ρ per unit length. The pipe is hung from a ceiling by means of several cables, each of length ℓ. Imagine that loads are applied in a *horizontal* plane and normal to the pipe axis. What is the effective Winkler foundation modulus for small lateral displacements of the pipe in the horizontal plane? Express the result in terms of ρ, ℓ, and g (the acceleration of gravity).

5.2-3. Verify that if $q = 0$ in Eq. 5.2-4, the form $w = C_1 e^{\beta x} \sin \beta x$ is a solution of this equation. (Other terms in Eq. 5.2-6 will check in similar fashion.)

5.2-4. A long cable rests on a Winkler foundation. The cable undergoes *small* lateral displacements, so its axial tension T may be taken as constant. Assume that the cable has no bending stiffness. At a point $x = 0$, far from end supports that maintain tension T, the cable is pushed down an amount w_o. Obtain the expression for the deflected shape $w = w(x)$, in terms of w_o, k, T, and x.

5.2-5. (a) A very long, prismatic elastic shaft of circular cross section is embedded in an elastic medium. When a length dx of the shaft rotates an amount θ, the medium applies torque $k\theta \, dx$ to the length dx. Torque T_o is applied to the end of the shaft at $x = 0$. Obtain expressions for rotation $\theta = \theta(x)$ and torque $T = T(x)$ in terms of T_o, k, G, J, and x.

(b) Revise part (a) so that the medium resists axial displacement rather than rotation, with force $ku \, dx$ for axial displacement u of a length dx. The bar resists stretching rather than twisting, and the loading at $x = 0$ is axial force P_o rather than torque T_o. Obtain expressions for axial displacement $u = u(x)$ and axial force $P = P(x)$ in terms of P_o, k, E, A, and x.

5.3-1. (a) Equate P_o to the total foundation reaction (as suggested in the paragraph that follows Eq. 5.3-8). Also use the first of Eqs. 5.3-2. Hence, verify Eqs. 5.3-3 and 5.3-4.

(b) Verify that Eqs. 5.3-6 through 5.3-8 follow from Eq. 5.3-5.

5.3-2. (a) At end $x = 0$ of a semi-infinite beam on a Winkler foundation, deflection w_o and end rotation θ_o are prescribed. Express the force and moment applied at $x = 0$ in terms of w_o, θ_o, k, and β.

(b) Hence, for $x > 0$, express w, θ, M and V as functions of x.

5.3-3. In the example problem of Section 5.3, how much force is applied to the beam by the portion of the foundation that deflects upward? Ignore values of βx greater than $3\pi/2$ in the calculation.

5.3-4. A semi-infinite beam, supported only by a Winkler foundation, is loaded by moment M_o at end $x = 0$.
  (a) Determine the ratio w_{max}/w_{min} and the ratio M_{max}/M_{min}.
  (b) In the range $0 < \beta x < 5\pi/4$, calculate the upward force applied to the beam (where $w > 0$) and the downward force applied to the beam (where $w < 0$). Explain any discrepancy between the forces.

5.3-5. A long wooden plank has specific gravity 0.5, elastic modulus $E = 13$ GPa, and a cross section 200 mm by 40 mm. The plank floats on still water. How large a weight can be placed at midwidth on one end if the upper surface is to stay dry? And, for those portions of the plank that deflect upward, how does the "foundation" apply downward pressure, as is assumed by our theory?

5.3-6. A semi-infinite beam of elastic modulus $E = 200$ GPa rests on a Winkler foundation of modulus $k_o = 80$ N/mm²/mm. The beam has a rectangular cross section, 42 mm by 68 mm, with the 42 mm dimension in contact with the foundation. An end deflection of 0.7 mm is produced by a vertical force applied at the end. What, and where, is the largest flexural stress in the beam?

5.3-7. The end of a semi-infinite beam on a Winkler foundation is simply supported and loaded by moment M_o, as shown. Obtain expressions for w, θ, M, and V as functions of x. Qualitatively sketch the deflected shape.

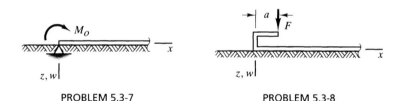

PROBLEM 5.3-7   PROBLEM 5.3-8

5.3-8. An angle bracket is attached to the end of a semi-infinite beam on a Winkler foundation, as shown. Load F is applied to the bracket.
  (a) In terms of β, what should be dimension a if the end of the beam is not to rotate?
  (b) For this value of dimension a, at what values of βx does one find the largest values of tensile stress and compressive stress on the top surface of the beam? What are the bending moments at these locations?

5.3-9. In the sketch for Problem 5.3-8, let the foundation have modulus $k_o = 0.018$ N/mm²/mm. Let the beam have a 10 mm by 25 mm rectangular cross section, with the 10 mm dimension in contact with the foundation. Use $E = 200$ GPa for the beam.
  (a) What should be dimension a if there is to be no deflection w at $x = z = 0$?
  (b) For this value of dimension a, what value of load F will produce flexural stress 230 MPa in the beam?

5.3-10. Bar AB in the sketch is rigid. It has long deformable beams welded to either side. If flexural stress in the beams must be limited to 200 MPa, what central load Q may be applied? Data are as follows. $EI = 6.83(10^{11})$ N·mm² for each beam, $k = 20$ N/mm/mm for the foundation, and each beam has a square cross section. For the entire structure, an 80 mm dimension is in contact with the foundation.

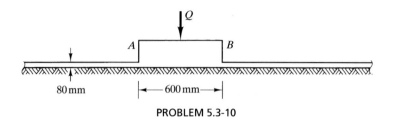

PROBLEM 5.3-10

5.3-11. Two uniform beams, cantilever beam AB and semi-infinite beam BC, are connected by a hinge at B, as shown. Beam BC is supported by a Winkler foundation. Load P is applied at the hinge. Derive an expression for deflection w_B at B.

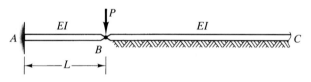

PROBLEM 5.3-11

5.3-12. A vertical load $P_o = 120$ kN is applied to end $x = 0$ of a semi-infinite beam of modulus $E = 204$ GPa on a Winkler foundation of modulus $k_o = 0.04$ N/mm^2/mm. If the allowable flexural stress in the beam is 240 MPa, what should be the side length of a square cross section?

5.3-13. An infinite beam is fabricated with a cusp, as shown in part (a) of the sketch. Then the beam is everywhere bonded to a Winkler foundation that was initially flat, and released. The result is shown in part (b) of the sketch. If $k = 0.25$ N/mm/mm and $EI = 441(10^9)$ N·mm^2, what is the resulting bending moment at the cusp?

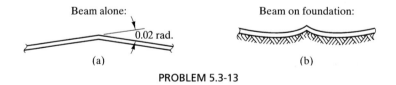

PROBLEM 5.3-13

5.3-14. A very long uniform beam is supported by a Winkler foundation, but the foundation is absent over a span $2a$ centered about load P, as shown. What is the bending moment in the beam at A and at B, in terms of P, a, and β? And for what value of βa does span AB act as a simply supported beam?

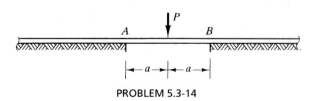

PROBLEM 5.3-14

5.4-1. (a) Verify that Eqs. 5.4-3 through 5.4-5 follow from Eq. 5.4-2.
(b) Verify that Eqs. 5.4-9 through 5.4-11 follow from Eq. 5.4-8.

5.4-2. In Section 5.4, results from Section 5.3 are used to solve the problems of Figs. 5.4-1 and 5.4-2. In this exercise work the other way: Derive Eq. 5.3-5 by using formulas for infinite beams in Section 5.4.

5.4-3. Imagine that an infinite beam on a Winkler foundation is loaded by two forces P, one up and the other down, a small distance Δx apart. Use Eq. 5.4-2, recognize that $P\,\Delta x$ is a moment, and in this way derive Eq. 5.4-8.

5.4-4. Use Maxwell's reciprocal theorem to obtain Eq. 5.4-8 from Eq. 5.4-3. Suggestion: Place loads M_o and P_o a distance x apart, as shown.

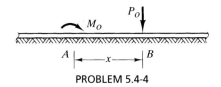

PROBLEM 5.4-4

5.4-5. A long uniform beam rests on many identical linear springs, each of stiffness K and uniformly spaced a distance s apart. Downward loads, each of magnitude P and spaced a distance s apart, are applied to the beam midway between springs. Thus, each load P is a distance $s/2$ from springs on either side. In terms of P, K, s, E, and I, what are the greatest deflection and the greatest bending moment in the beam? Suggestion: Sketch the deformed shape before doing an analysis.

5.4-6. A semi-infinite beam is loaded by force P at distance a from its end, as shown. Explain how formulas for this problem can be obtained from formulas in Sections 5.3 and 5.4.

PROBLEM 5.4-6   PROBLEM 5.4-7

5.4-7. In the sketch, flexural stiffness EI is the same for both the vertical beam and the horizontal beam on a Winkler foundation. Determine the rotation at the intersection, where moment M_o is applied, in terms of M_o, h, β, E, and I. Assume that horizontal restraint prevents the horizontal beam from moving parallel to the foundation.

5.4-8. A long beam is loaded near its center by a concentrated force. What is the effect on maximum deflection and maximum stress of overestimating the Winkler foundation modulus k by 100%? Repeat the exercise, now using a moment rather than a force to load the beam.

5.4-9. Estimate the maximum tensile stress in the sandwich construction shown in the sketch.

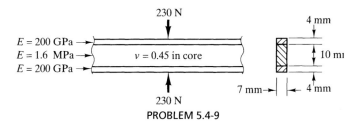

PROBLEM 5.4-9

5.5-1. Let an idealized bicycle wheel consist of 36 radial spokes, all in the midwidth plane of the rim, with data as follows. Spokes: diameter 2.1 mm, $E = 210$ GPa, length $= 309.4$ mm from the center of the wheel to the axis of the rim. Rim: $E = 70$ GPa, $I = 1469$ mm^4. Apply to the rim a 1000 N radially directed force, and assume that this force is not sufficient to overcome prestress in any spoke. Use formulas for a straight beam on a Winkler foundation to approximate the largest bending moment in the rim and the largest change in stress in a spoke [5.1].

5.5-2. A 25 kN load is applied to an infinite beam on a Winkler foundation. The beam has a rectangular cross section, with a 20 mm width in contact with the foundation. For the beam, $E = 200$ GPa; for the foundation, $k_o = 0.015$ N/mm^2/mm. What should be the depth of the cross section so that the maximum flexural stress does not exceed 180 MPa?

5.5-3. The sketch shows the top view and the end view of a long beam that is fixed at end A and free at end B. Beam AB is supported by several identical and equally spaced cross-beams CD, which are 0.5 m apart and are simply supported at their ends. For all beams, $E = 200$ GPa, $I = 25(10^6)$ mm^4, and depth $2c = 180$ mm. A load of 70 kN is applied to the middle of beam AB, directed normal to directions AB and CD. Determine the largest flexural stress in beam AB and the largest flexural stress in a cross beam.

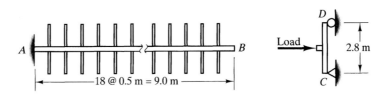

PROBLEM 5.5-3

5.5-4. A long rail on a Winkler foundation is loaded by two equal wheel forces. How far apart should the wheels be in order to minimize flexural stress? Express the answer in terms of β.

5.5-5. A long rail, initially straight, rests on a Winkler foundation. After loading by a downward force P applied near the center of the rail, an optical technique is used to measure the vertical distance between the highest and lowest points on top surface of the rail. If this distance is called g, what is the foundation modulus k, in terms of P, g, E, and I?

5.5-6. Two loads P are applied to a uniform infinite beam on a Winkler foundation, as shown. For a simpler but approximate analysis, the two loads can be replaced by a single load $2P$, as shown. In terms of β, how large can distance s be if the magnitude of error of the approximate analysis must be limited to (a) 5% on the maximum deflection, and (b) 5% on the maximum bending moment?

PROBLEM 5.5-6

5.5-7. In terms of β, what should be the distance between two equal forces on an infinite beam if the deflection midway between loads is to be the same as the deflection at each load?

5.5-8. In Example 2 of Section 5.5, let the loads applied at A and C be 6 kN and 12 kN respectively. Leave other data of the problem unchanged. Determine the location and magnitude of (a) the largest deflection, and (b) the largest bending moment.

5.5-9. In the sketch for Problem 5.4-4, let us make rotation θ vanish at A, and let us do so by keeping M_o constant, using the smallest possible value of load P_o, and choosing the "best" value of distance x. Determine x (in terms of β) and P_o (in terms of β and M_o). Maintain the directions shown for M_o and P_o and keep B to the right of A.

5.5-10. A uniform beam rests on a Winkler foundation and a simple support, as shown. Let $\beta = 6.136(10^{-4})$/mm and $P = 9$ kN. For what value of spacing a does the bending moment over the support have greatest magnitude, and what is this bending moment?

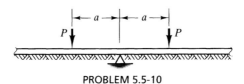

PROBLEM 5.5-10

5.5-11. An infinite beam, for which $E = 200$ GPa, $I = 28.4(10^6)$ mm^4, and depth $2c = 180$ mm, rests on a Winkler foundation of modulus $k = 10$ N/mm/mm.

(a) Determine the maximum deflection and maximum flexural stress when a single load $P = 100$ kN is applied to the beam.

(b) Change the loading to three equal loads, each 100 kN and 1.7 m apart. Determine the maximum deflection and maximum flexural stress in the beam.

5.5-12. An infinite beam rests on a Winkler foundation. For the foundation, $k_o = 0.0025$ N/mm^2/mm; for the beam, $E = 70$ GPa. Loads $2P$, $3P$, and P are applied as shown. Determine, in terms of P, the flexural stress beneath (a) load $2P$, (b) load $3P$, and (c) load P.

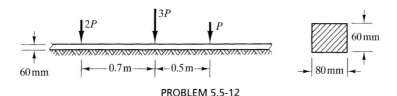

PROBLEM 5.5-12

5.5-13. The railroad rail shown can be considered infinitely long. Each of the many ties that support the rail has spring constant $K = 5$ kN/mm.

(a) Load P is directly over a tie. What are the deflection and bending moment in the rail directly beneath load P?

(b) What are the deflection and bending moment in the rail at a position two ties away from load P?

(c) During a test, one of the ties is replaced by a special support of spring constant $K = 30$ kN/mm. If load P is directly over the special support, what is the deflection of the rail beneath the load?

(d) If P is applied two ties away from the special support, what is the deflection of the rail beneath the load? What and where is the bending moment of largest magnitude?

158 Chapter 5 Beams on an Elastic Foundation

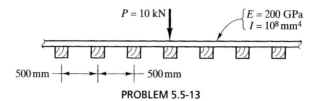

PROBLEM 5.5-13

5.5-14. A rigid bar is connected to a long beam on a Winkler foundation by three linear springs of known stiffnesses K_1, K_2, and K_1, as shown. Load P is applied to the middle of the rigid bar. It is required to determine forces P_1, P_2, and P_1 in the respective springs and deflection w_b of the rigid bar. Assume that P, a, β, and k are known. Write, but do not solve, the necessary equations (in terms of the aforementioned quantities, $A_{\beta a}$, and $A_{2\beta a}$).

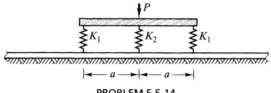

PROBLEM 5.5-14

5.5-15. In Fig. 5.5-3b, apply moment load M_o to the left side instead of the right side, and show that the same final results are obtained.

5.5-16. In Fig. 5.5-3, replace moment M_o by vertical force F applied at the step in the beam. Write equations analogous to Eqs. 5.5-13 and 5.5-14, and determine the bending moment at the step in terms of force F and β.

5.6-1. (a) Determine the deflection equation $w = w(x)$ of the beam in Fig. 5.6-1b. Express w in terms of q, k, and $D_{\beta x}$.
  (b) Similarly, for the same beam, establish the bending moment equation $M = M(x)$.
  (c) Imagine that the left end of the beam in Fig. 5.6-1b is fully fixed rather than simply supported. Express deflection w in terms of q, k, and $A_{\beta x}$.
  (d) What are the maximum and minimum bending moments for the beam of part (c)?

5.6-2. Qualitatively sketch $w = w(x)$ and $M = M(x)$ for the beam of Fig. 5.6-2. Consider the cases (a) $\beta \ell < \pi$, and (b) $\beta \ell$ "large".

5.6-3. Derive the equation $M_Q = (q/4\beta^2)(B_{\beta a} - B_{\beta b})$, which pertains to a point Q that is b units to the right of the loaded zone in Fig. 5.6-2. Suggestion: Add self-equilibrating loads q, as shown in the second part of the sketch for the present problem.

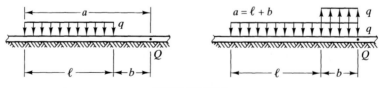

PROBLEM 5.6-3

5.6-4. A long beam carries uniformly distributed load q over a span $\ell = 3000$ mm. Let $k = 0.10$ N/mm/mm for the foundation and $EI = 3240$ N·m² for the beam. Determine the following bending moments, in terms of q.

(a) M at the center of span ℓ.

(b) M at the ends of span ℓ.

(c) M within span ℓ, a distance $\pi/4\beta$ from either end of span ℓ.

(d) The M of largest magnitude within span ℓ.

(e) The M of largest magnitude outside of span ℓ (see Problem 5.6-3).

5.6-5. A long beam rests on a Winkler foundation and on two simple supports, as shown. Determine the bending moments at the supports and at the center of span L, in terms of q. Let $L = 3.61$ m, $k = 4.0$ N/mm/mm, and $\beta = 0.001107$/mm.

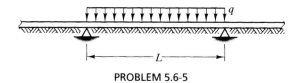

PROBLEM 5.6-5

5.6-6. A long beam on a Winkler foundation carries a distributed load that has a step change from q_1 to q_2, as shown. What is the maximum bending moment, in terms of q_1, q_2, and β?

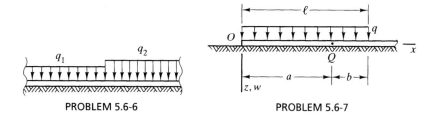

PROBLEM 5.6-6   PROBLEM 5.6-7

5.6-7. A semi-infinite beam on a Winkler foundation carries uniformly distributed load q over span ℓ, as shown. Apply superposition concepts and formulas in Sections 5.3 and 5.6 to develop expressions for deflection w and bending moment M at an arbitrary point Q within span ℓ. Verify that these expressions yield the expected result for M at $x = 0$ and the expected result for w at $x = 0$ when $\beta\ell$ is large.

5.7-1. The beam shown is simply supported at $x = 0$, where it is loaded by moment M_o. The beam rests on a Winkler foundation and may *not* be considered infinitely long. What boundary conditions may be used to evaluate constants C_1 through C_4 in Eq. 5.2-6?

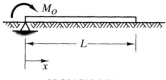

PROBLEM 5.7-1

5.7-2. Show that Eqs. 5.7-1 and 5.7-2 reduce to the expected rigid beam formulas as βL approaches zero.

5.7-3. The short beam shown rests on a Winkler foundation and can be regarded as rigid. Obtain expressions for deflection w and foundation pressure p in terms of P, L, k_o, a, b, and x.

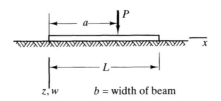

PROBLEM 5.7-3

5.7-4. (a) For what range of dimension a in Problem 5.7-3 does the foundation exert no downward pressure on the rigid beam?
(b) What are the largest and smallest deflections w of load P as it moves along the rigid beam through this range?

5.7-5. Let $a = L/4$ in Problem 5.7-3, and let the foundation be unable to pull down on the rigid beam. If the maximum deflection of the beam is w_o, what is P in terms of w_o, k, and L?

5.7-6. (a) Let $a = L/2$ in Problem 5.7-3. Show that the bending moment at the middle of the rigid beam is $PL/8$.
(b) Let $a = 0$ in Problem 5.7-3. As usual, the foundation can pull as well as push on the beam. Derive the location and value of the bending moment of largest magnitude in the rigid beam.

5.7-7. Consider a beam supported entirely by a Winkler foundation and short enough that it may be regarded as rigid. Obtain an expression for the energy that must be expended by a small self-propelled cart of weight W in traveling from one end of the beam to its center.

5.7-8. The sketch represents the top view of a rigid circular plate of radius R on a Winkler foundation. A downward force P is applied to the plate at radius a.
(a) Express the maximum deflection of the plate in terms of k_o, P, R, and a.
(b) If the foundation is not allowed to pull on the plate or separate from it, what is the maximum permissible value of a/R?

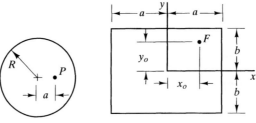

PROBLEM 5.7-8   PROBLEM 5.7-9

5.7-9. The sketch represents the top view of a rigid rectangular plate on a Winkler foundation. A downward force F is applied to the plate at coordinates x_o and y_o, measured from the center of the plate.
(a) Express the foundation pressure in terms of F, a, b, x_o, y_o, x, and y.
(b) Assume that the foundation must not pull on the plate or separate from it. Determine, and sketch, the region of the plate to which force F must be restricted.

CHAPTER 6

Curved Beams

6.1 INTRODUCTION. STRAIN DISTRIBUTION

A *curved beam* is a member whose loading produces bending moment and whose axis is curved before load is applied. In this chapter we consider loads that bend a curved beam in its own plane. Examples appear in Fig. 6.1-1. Additional examples include chain links and frames of industrial presses. A curved beam is called *slender* if $h \ll R$, where h is the depth of the cross section and R is the radius of curvature at the centroid of the cross section. The beam is called *sharply curved* if h is comparable to R.

The elementary flexure formula $\sigma = Mc/I$ is intended for straight beams, and underestimates circumferential stress in a curved beam. Error is greatest for a sharply curved beam. Also, if the section is thin-walled, the elementary formula overlooks stresses that become important when a beam is slender but curved. These stresses are associated with distortion of the cross section, which is depicted for a wide-flange section in Fig. 6.1-1c. The "flapping" distortion of the flanges causes them to have z-direction flexural stresses, which may exceed circumferential stresses. This action is not present in a straight beam and can be greatly reduced in a curved beam by welding stiffening plates between flanges (typical stiffening plates would be rectangular, in spaces $ABED$ and $CBEF$ in Fig. 6.1-1c).

Assumptions. Curved beam formulas considered in this chapter are based on the following assumptions. The material is homogeneous, isotropic, and linearly elastic. In geometric terms, the beam is a solid of revolution, but it is usually not a complete ring. The beam axis, and inner and outer edges of the beam, all form circular arcs about the axis of revolution. The beam axis lies in a plane; it is not helical as in a coil spring. Loads act in the plane of curvature. The cross section is uniform. The cross section is assumed to have symmetry about a central radial line. As usual, minor departures from ideal geometry often cause only minor errors.

Subject to the foregoing assumptions, theory of elasticity [6.1] provides exact solutions for certain loadings on a curved beam, but only if the cross section is a solid circle or a solid rectangle. For general loadings and general shapes of cross section,

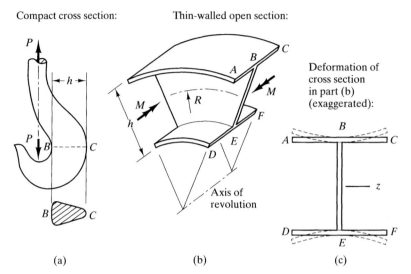

FIGURE 6.1-1 (a) Typical shape and loading of a crane hook. (b, c) Bending of a wide-flange curved beam.

formulas are obtained from mechanics of materials analysis. The original such analysis was done by Winkler (1858). The restriction to symmetry of the cross section is made so that a cross section will have no tendency to rotate in its own plane. Asymmetric cross sections are considered in [4.2]. Note, however, that if every cross section is restrained from rotating in its own plane, then asymmetry does not affect the state of deformation, and equations presented in this chapter remain applicable.

Unless distortion of the cross section is considered (in Fig.6.3-2b and in Sections 6.5 and 6.6), formulas developed in this chapter are based on the assumption that stresses do not vary in a direction parallel to the axis of revolution.

Deformations and Strains. As is the case for a straight beam, for mechanics of materials analysis we assume that plane cross sections remain plane after load is applied. Without writing equations, we can see that circumferential strain ϵ_ϕ does *not* vary linearly with distance y from the neutral axis (Fig. 6.1-2). Although displacements such as CC' are directly proportional to y, circumferential strain $\epsilon_\phi = CC'/BC$ involves an initial length BC that is smaller toward the center of curvature. The result is the strain distribution shown in Fig. 6.1-2a. For linearly elastic conditions, the associated stress σ_ϕ is similarly distributed. Accordingly, the neutral axis of pure bending does *not* pass through the centroid of the cross section; rather, the neutral axis is shifted toward the center of curvature.

A formula for circumferential strain ϵ_ϕ is as follows. With y the distance from the neutral axis and r_n the radius of curvature at the neutral axis, from Fig. 6.1-2b,

$$\epsilon_\phi = \frac{CC'}{BC} = \frac{y\,d\theta}{r\,d\phi} = \frac{y\,d\theta}{(r_n - y)\,d\phi} \tag{6.1-1}$$

This formula confirms the preceding verbal argument: ϵ_ϕ has greater magnitude when y is positive (inward) than when y is negative (outward).

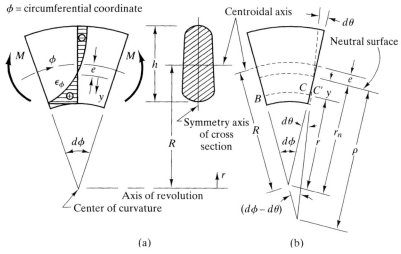

FIGURE 6.1-2 (a) Distribution of circumferential strain ϵ_ϕ in a curved beam bent by moments M. (b) Geometry of deformation in bending. Point C moves to C'.

An alternative expression for ϵ_ϕ is obtained as follows. For arc length of the neutral surface over arc $d\phi$, we can write $r_n d\phi = \rho(d\phi - d\theta)$, where ρ is the radius of curvature at the neutral surface after moment M is applied. Solving this expression for $d\theta/d\phi$ and substituting into Eq. 6.1-1, we obtain

$$\epsilon_\phi = \frac{y}{r_n - y} r_n \left(\frac{1}{r_n} - \frac{1}{\rho} \right) \tag{6.1-2}$$

The expression in parentheses is the change in curvature between unloaded and loaded states. In the limiting case of a straight beam, r_n approaches infinity, and Eq. 6.1-2 reduces to the expected expression $\epsilon_\phi = -y/\rho$. Thus we see that curved beam theory includes straight beam theory as a limiting case. Whether a beam is initially curved or initially straight, ϵ_ϕ is proportional to *change* in curvature.

Related Problems in Springs. Consider a helical spring, close-coiled so that coils almost touch one another before loading. Let z represent the axis of the helix (see the figure for Problem 4.6-12). If the spring is torsionally loaded, by a torque vector along the z axis, coils act as curved beams in almost pure bending. If instead the spring is stretched by force P along the z axis, coils are loaded in direct shear by force P and in torsion by torque $T = PR$, where R is the coil radius. Consider a small slice of the coil, bounded by two radial cutting planes that intersect along the z axis and include angle $d\phi$. Adjacent cut faces of the slice rotate relative to each other, thus producing shear strain equal to the relative displacement of the cut faces divided by the distance between them. This distance is $R\,d\phi$ on the coil axis, but smaller for points nearer the z axis and larger for points farther away. Accordingly, like ϵ_ϕ in Eq. 6.1-1, shear strain due to $T = PR$ is largest on the inside of a coil. If coils are slender, the difference is negligible. For massive coils, correction formulas are available [1.6, 6.2].

6.2 CIRCUMFERENTIAL STRESS

In this section we assume that circumferential stress σ_ϕ is the only nonzero normal stress. Actually a radial stress σ_r is also present, so the assumption is inconsistent with elasticity equations of compatibility and radial equilibrium. However, σ_ϕ and σ_r have their largest magnitudes at different locations in the cross section, so the error of ignoring their interaction is small. We also assume that the cross section has no distortion in its own plane. This assumption is reasonable if the cross section is compact rather than thin-walled, although appreciable error is still possible, as described in Section 6.3.

With σ_ϕ the only nonzero normal stress, the stress-strain relation is $\sigma_\phi = E\epsilon_\phi$, and with Eq. 6.1-1 we obtain

$$\sigma_\phi = E\frac{y\,d\theta}{(r_n - y)\,d\phi} \quad \text{or} \quad \sigma_\phi = E\frac{d\theta}{d\phi}\left(\frac{r_n}{r} - 1\right) \tag{6.2-1}$$

where the latter form comes from the substitution

$$y = r_n - r \tag{6.2-2}$$

We want σ_ϕ to be expressed in terms of loading and dimensions of the cross section, and so must express $d\theta/d\phi$ and r_n in terms of these quantities. Two equilibrium equations are available:

$$N = \int_A \sigma_\phi \, dA \quad \text{and} \quad M = \int_A (e + y)\sigma_\phi \, dA \tag{6.2-3}$$

where A is the area of the cross section. The first equation says that force increments $\sigma_\phi \, dA$ sum to the total circumferential force N. The second equation says that moment increments about the centroidal axis of the cross section sum to the bending moment M (see Fig. 6.2-1).

Pure Bending. With $N = 0$, Eq. 6.2-1 and the first of Eqs. 6.2-3 yield

$$E\frac{d\theta}{d\phi}\left[r_n \int_A \frac{dA}{r} - A\right] = 0 \tag{6.2-4}$$

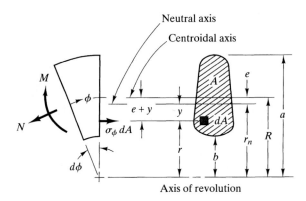

FIGURE 6.2-1 Moment load M and circumferential force load N are both shown in the positive sense. Coordinate y is measured from the neutral axis, positive inward.

Since neither E nor $d\theta/d\phi$ is zero, the expression in brackets must vanish. Therefore

$$r_n = \frac{A}{\int_A dA/r} \quad \text{or} \quad \frac{A}{r_n} = \int_A \frac{dA}{r} \tag{6.2-5}$$

The first form defines the location of the neutral axis of pure bending. So does the second form, but it may be easier to remember. The eccentricity of the neutral axis of pure bending,

$$e = R - r_n \tag{6.2-6}$$

is now established in terms of the geometry of the beam.

In the second of Eqs. 6.2-3, the integral of $e\sigma_\phi$ vanishes when $N=0$. What remains is the integral of $y\sigma_\phi$, which yields, after substitution from Eqs. 6.2-1 and 6.2-2,

$$M = E\frac{d\theta}{d\phi}\left[r_n \int_A \left(\frac{r_n}{r} - 1\right)dA - r_n \int_A dA + \int_A r\,dA\right] \tag{6.2-7}$$

The first integral within brackets vanishes according to Eq. 6.2-4. The second integral is simply A. The third integral is RA, which is the first moment of area A about the center of curvature. Thus Eq. 6.2-7 yields

$$M = E\frac{d\theta}{d\phi} A(R - r_n) \quad \text{from which} \quad \frac{d\theta}{d\phi} = \frac{M}{EAe} \tag{6.2-8}$$

Substituting the expression for $d\theta/d\phi$ and Eq. 6.2-2 into Eq. 6.2-1, we obtain, for pure bending,

$$\sigma_\phi = \frac{M(r_n - r)}{Aer} \quad \text{or} \quad \sigma_\phi = \frac{My}{Aer} \tag{6.2-9}$$

In the first form, coordinate r locates the radial position at which σ_ϕ is calculated. The second form is easier to remember but contains *two* coordinates, r and y, which must satisfy Eq. 6.2-2. Note that y is negative when $r > r_n$. When M and y have correct signs, so does σ_ϕ.

Bending and Direct Force. When $N \neq 0$, Eqs. 6.2-1 and 6.2-3 remain applicable, but the algebra needed to obtain σ_ϕ becomes tedious (see Problem 6.2-1). The result is

$$\sigma_\phi = \frac{N}{A} + \frac{M(r_n - r)}{Aer} \tag{6.2-10}$$

The direct-stress term N/A is the same form as used for a straight beam. As an ad hoc modification [6.3], one can replace N/A by Nr_n/Ar, which makes the formula more accurate at locations where $r < r_n$.

When bending moment M is produced by an externally applied force, as for example in Fig. 6.2-2, M is calculated as moment about the *centroidal* axis of the cross section, as shown (not as moment about the neutral axis).

Aids to Calculation. One must determine A, R, e, and r_n in order to calculate σ_ϕ from Eq. 6.2-9 or 6.2-10. A sampling of useful formulas appears in Table 6.2-1. More extensive tabulations appear in [1.6, 1.7]. If a cross section is of such shape that

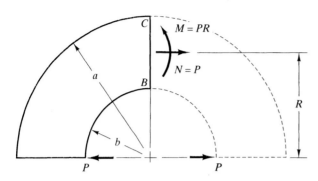

FIGURE 6.2-2 Internal force N and moment M at the midpoint of a half-ring, produced by external loads P.

tabulated formulas do not readily supply what is needed, a simple numerical integration process can be used instead. As shown in Fig. 6.2-3, one divides a cross section into rectangular strips parallel to the axis of revolution and replaces integration by summation. In many problems, roughly ten strips provide quite accurate results.

Calculated stress σ_ϕ has considerable error if e is inaccurate. If a curved beam is slender, then R and r_n are almost equal, and their difference $e = R - r_n$ has few correct digits unless R and r_n are each calculated with great accuracy. The need for great accuracy can be avoided by using the following approximation (for a derivation, see the first edition of this book or [6.4]).

$$e \approx \frac{I}{RA} \quad \text{for} \quad \frac{R}{h} > 8 \tag{6.2-11}$$

TABLE 6.2-1 Selected data for curved beams of trapezoidal, rectangular, triangular, and circular cross sections.

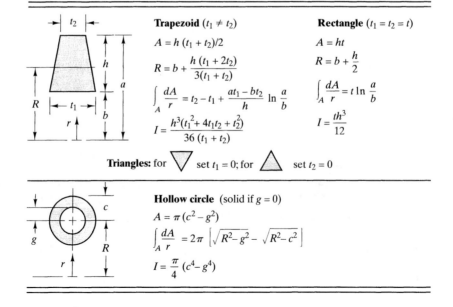

Trapezoid ($t_1 \neq t_2$)

$A = h(t_1 + t_2)/2$

$R = b + \dfrac{h(t_1 + 2t_2)}{3(t_1 + t_2)}$

$\displaystyle\int_A \frac{dA}{r} = t_2 - t_1 + \frac{at_1 - bt_2}{h} \ln \frac{a}{b}$

$I = \dfrac{h^3(t_1^2 + 4t_1 t_2 + t_2^2)}{36(t_1 + t_2)}$

Rectangle ($t_1 = t_2 = t$)

$A = ht$

$R = b + \dfrac{h}{2}$

$\displaystyle\int_A \frac{dA}{r} = t \ln \frac{a}{b}$

$I = \dfrac{th^3}{12}$

Triangles: for ▽ set $t_1 = 0$; for △ set $t_2 = 0$

Hollow circle (solid if $g = 0$)

$A = \pi(c^2 - g^2)$

$\displaystyle\int_A \frac{dA}{r} = 2\pi \left[\sqrt{R^2 - g^2} - \sqrt{R^2 - c^2} \right]$

$I = \dfrac{\pi}{4}(c^4 - g^4)$

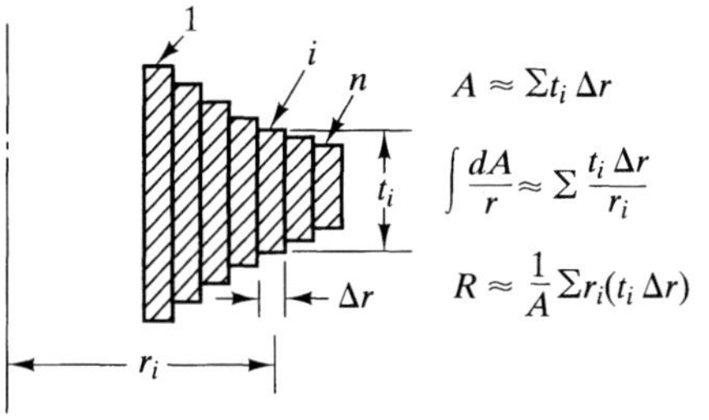

FIGURE 6.2-3 Approximate calculation of cross-sectional properties by means of strips of thickness Δr. Summations run from 1 to n.

$$A \approx \Sigma t_i \, \Delta r$$

$$\int \frac{dA}{r} \approx \Sigma \frac{t_i \, \Delta r}{r_i}$$

$$R \approx \frac{1}{A} \Sigma r_i (t_i \, \Delta r)$$

Here h is the depth of the cross section, shown in Fig. 6.1-2, and I is the moment of inertia of cross-sectional area A about its centroidal axis parallel to the axis of revolution. Indeed for $R/h > 8$ the curved-beam stress formula may not be needed; the straight-beam formula $\sigma = Mc/I$ may be sufficiently accurate. For example, in a solid rectangular cross section with $R/h = 8$, e from Eq. 6.2-11 is 0.1% low, and flexural stress magnitudes calculated by the formula $\sigma = Mc/I$ are 4.2% low at $r = b$ and 4.1% high at $r = a$.

If supplied with a correction factor, the straight-beam flexure formula can be used to calculate flexural stress at $r = b$ or at $r = a$ in a curved beam. We write

$$\sigma_\phi = K \frac{Mc}{I} \qquad (6.2\text{-}12)$$

Factor K depends on the geometry of the cross section, its centroidal radius of curvature R, and whether c represents distance from the centroidal axis to the inside of the cross section or distance to the outside. Equation 6.2-12 is convenient if K is already available for the geometry being analyzed. Tabulated values of K [1.6] are obtained by computing the ratio of stress σ_ϕ from Eq. 6.2-9 to stress from the straight-beam flexure formula $\sigma = Mc/I$. Thus, for a solid rectangular cross section with $r/h = 8$, $K_b = 1.043$ at $r = b$ (inside) and $K_a = 0.960$ at $r = a$ (outside). For other shapes of cross section, K_b may be larger. An approximate rule says that if c/R is less than 0.1, K_b is usually less than 1.1 [2.8].

6.3 NUMERICAL EXAMPLES

EXAMPLE 1

The crane hook of Fig. 6.1-1a is shown again in Fig. 6.3-1a. For a load $P = 20$ kN, circumferential stresses at points B and C on the horizontal cross section BC are desired.

Curvature and cross section vary around the hook. We will assume that these departures from the ideal can be ignored. For a simple but approximate analysis, we idealize the cross section as two trapezoids, BD and DC in Fig. 6.3-1b, and apply formulas from Table 6.2-1. Initially we carry enough digits to guarantee accuracy in the calculation $e = R - r_n$. The cross-sectional area is

$$A = A_{BD} + A_{DC} = 340 + 1800 = 2140 \text{ mm}^2 \qquad (6.3\text{-}1)$$

The centroid of inner trapezoid BD has radial coordinate

$$R_{BD} = 30 + \frac{10(22 + 92)}{3(22 + 46)} = 35.588 \text{ mm} \qquad (6.3\text{-}2)$$

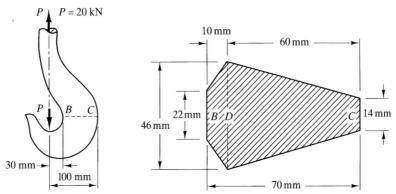

FIGURE 6.3-1 (a) Crane hook. (b) Cross section, shown enlarged, and idealized as two trapezoids.

Similarly, $R_{DC} = 64.667$ mm for outer trapezoid DC. The radius to the centroid of the entire cross section BDC is

$$R = \frac{1}{A}(R_{BD}A_{BD} + R_{DC}A_{DC}) = 60.047 \text{ mm} \quad (6.3\text{-}3)$$

For inner trapezoid BD,

$$\left(\int_A \frac{dA}{r}\right)_{BD} = 46 - 22 + \frac{40(22) - 30(46)}{10}\ln\frac{40}{30} = 9.616 \text{ mm} \quad (6.3\text{-}4)$$

Similarly, for outer trapezoid DC, $\int dA/r = 29.697$ mm. For the entire cross section,

$$\int_A \frac{dA}{r} = \left(\int_A \frac{dA}{r}\right)_{BD} + \left(\int_A \frac{dA}{r}\right)_{DC} = 39.313 \text{ mm} \quad (6.3\text{-}5)$$

Radius r_n to the neutral axis of pure bending, and its eccentricity e, are

$$r_n = \frac{A}{\int_A dA/r} = \frac{2140}{39.313} = 54.435 \text{ mm} \quad (6.3\text{-}6)$$

$$e = R - r_n = 60.047 - 54.435 = 5.612 \text{ mm} \quad (6.3\text{-}7)$$

Henceforth we need not carry so many digits. The bending moment is

$$M = PR = 20{,}000(60.0) = 1.20(10^6) \text{ N}\cdot\text{mm} \quad (6.3\text{-}8)$$

Equation 6.2-10 yields the circumferential stresses at B and C.

$$(\sigma_\phi)_B = \frac{20{,}000}{2140} + \frac{1.2(10^6)(54.4 - 30.0)}{2140(5.61)(30.0)} = 9.3 + 81.3 = 90.6 \text{ MPa} \quad (6.3\text{-}9)$$

$$(\sigma_\phi)_C = \frac{20{,}000}{2140} + \frac{1.2(10^6)(54.4 - 100.0)}{2140(5.61)(100.0)} = 9.3 - 45.6 = -36.3 \text{ MPa} \quad (6.3\text{-}10)$$

What flexural stresses are predicted by the straight-beam formula $\sigma = Mc/I$? Moment of inertia I is calculated by use of the transfer theorem, $I = (I_{cg} + Ad^2)_{BD} + (I_{cg} + Ad^2)_{DC}$, where d represents distance from the centroid of the entire cross section to the centroid of trapezoid BD or DC. Here

$I = 733{,}000$ mm^4. Distance c is $R - b$ at B and $R - a$ at C, and Mc/I yields 49.2 MPa at B and -65.4 MPa at C. It is clear that if a beam is curved, Mc/I may yield a large and nonconservative error at the inner edge of the cross section.

EXAMPLE 2

For a rectangular cross section loaded by either pure bending or the "shear loading" of Fig. 6.2-2, exact solutions are provided by theory of elasticity [6.1]. Using these solutions, we can evaluate the accuracy of curved-beam and straight-beam formulas when applied to solid rectangular cross sections of various a/b ratios (Table 6.3-1). We see that the curved-beam formula is much more accurate than the straight-beam formula. (Also, for $a/b = 16$ with shear loading, the ad hoc modification noted following Eq. 6.2-10 changes the entry 0.785 to 1.049.) That $|\sigma_b/\sigma_a| = a/b$ for shear loading is fortuitous; $|\sigma_b/\sigma_a|$ will not equal a/b for other loadings or other shapes of cross section. No exact solution is available for a general shape of cross section. One must then either use the curved-beam formula, conduct experiments, or use numerical tools such as the finite element method.

EXAMPLE 3

We consider pure bending of a trapezoidal cross section, Fig. 6.3-2. For a given bending moment M, we calculate circumferential stress σ_ϕ by curved beam analysis, by Eq. 6.2-9, and by finite element analysis. Two different finite element algorithms agree well [4.6, 6.5]; the finite element results presented in Fig. 6.3-2 are considered reasonably accurate. We see that curved beam analysis may seriously underestimate the largest circumferential stress. (For the same cross section, but with $b = 17.6$ mm so that $a/b = 6$, if M is such that curved beam analysis yields $\sigma_\phi = 100$ MPa along ABC, finite element analysis yields $\sigma_{\phi B} = 140$ MPa and $\sigma_{\phi A} = \sigma_{\phi C} = 23$ MPa.) Curved-beam analysis is in error because it neglects distortion of the cross section, which has appreciable effect even though this trapezoidal cross section is reasonably compact. Sections 6.5

TABLE 6.3-1 Largest circumferential stress σ_ϕ in a rectangular cross section. Except for $|\sigma_b/\sigma_a|$, stress ratios pertain to the inside edge $r = b$. Sources: σ_{curved} from Eq. 6.2-10, $\sigma_{\text{straight}} = N/A \pm Mc/I$, $\sigma_{\text{elasticity}}$ from [6.1]. Pure bending: see Fig. 6.1-2. Shear loading: see Fig. 6.2-2.

a/b	Pure Bending ($M \neq 0, N = 0$)			Shear Loading ($M = PR, N = P$)		
	$\|\sigma_b/\sigma_a\|$ (elasticity)	$\sigma_{\text{curved}} / \sigma_{\text{elasticity}}$	$\sigma_{\text{straight}} / \sigma_{\text{elasticity}}$	$\|\sigma_b/\sigma_a\|$ (elasticity)	$\sigma_{\text{curved}} / \sigma_{\text{elasticity}}$	$\sigma_{\text{straight}} / \sigma_{\text{elasticity}}$
1.05	1.03	1.000	0.984	1.05	1.000	0.984
1.1	1.07	1.000	0.968	1.10	1.000	0.968
1.2	1.13	1.000	0.939	1.20	0.998	0.940
1.4	1.25	0.999	0.888	1.40	0.994	0.890
2.0	1.58	0.996	0.774	2.00	0.977	0.776
3.5	2.22	1.000	0.613	3.50	0.932	0.609
6.0	2.99	1.024	0.487	6.00	0.879	0.465
11.0	4.00	1.093	0.379	11.00	0.819	0.331
16.0	4.67	1.167	0.327	16.00	0.785	0.263

170 Chapter 6 Curved Beams

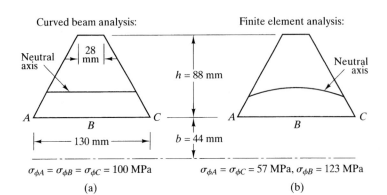

FIGURE 6.3-2 Pure bending of a curved beam of trapezoidal cross section, for which $a/b = 3$ and Poisson's ratio $\nu = 0.3$.

and 6.6 discuss corrections for distortion, but formulas that include corrections are available only for thin-walled T or wide-flange sections and thin-walled pipes.

Unfortunately, finite element analysis is not easily applied to curved beams. The difficulty is that the problem is not axisymmetric because points displace circumferentially as well as radially and axially. The method of [6.5] requires only a two-dimensional mesh, but is limited to pure bending and uses an algorithm not found in commercial software. With pure bending and commercial software, three-dimensional elements need occupy only a single layer in the ϕ direction, but one must locate the neutral axis by two trial calculations, then impose displacements such that plane sections remain plane. The software computes σ_ϕ, then the applied moment M by integration (Eq. 6.2-3). Finally, computed stresses σ_ϕ are scaled by the ratio of actual moment to computed moment [4.6]. Alternatively, a full three-dimensional finite element analysis can be used, and this appears to be the only correct option if the loading is not pure bending.

6.4 RADIAL STRESS AND SHEAR STRESS

Figure 6.4-1a shows forces that act on a differential slice of a curved beam. For some shapes of cross section, such as a wide-flange section, radial stress σ_r may be a significant stress. The physical action that produces σ_r also acts to distort the shape of some cross sections, particularly those of thin-walled members.

A Simple Approximation. In Fig. 6.4-1, let $V = 0$, and imagine that $(r_1 - b) \ll R$ and that σ_ϕ has the constant value $(\sigma_\phi)_{\text{ave}}$ over distance $r_1 - b$. Then a force $(\sigma_\phi)_{\text{ave}} A_1$ acts on each side of the slice, where area A_1 is the portion of A where $r < r_1$ (Fig. 6.4-1c). The inward component of this force, acting parallel to a radial line through the center of the slice, is $[(\sigma_\phi)_{\text{ave}} A_1] \sin (d\phi/2)$. Let σ_r be assumed constant over the area $t_1(r_1 d\phi)$ on which it acts. Then, for equilibrium of radial forces on the slice,

$$\sigma_r t_1(r_1 d\phi) - 2[(\sigma_\phi)_{\text{ave}} A_1] \sin \frac{d\phi}{2} = 0 \qquad (6.4\text{-}1)$$

Section 6.4 Radial Stress and Shear Stress

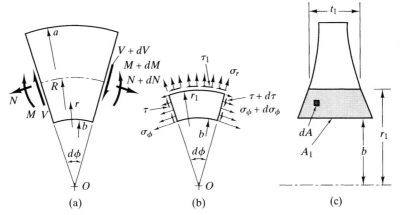

FIGURE 6.4-1 (a) Forces that act on an infinitesimal slice of a curved beam. (b) Stresses that act on the inner portion of the infinitesimal slice. (c) Geometry of an arbitrary cross section.

Since $d\phi$ is infinitesimal, $\sin(d\phi/2) = d\phi/2$, and Eq. 6.4-1 yields

$$\sigma_r = \frac{(\sigma_\phi)_{\text{ave}} A_1}{t_1 r_1} \quad \text{for} \quad (r_1 - b) \ll R \tag{6.4-2}$$

where t_1 is the thickness of the cross section at radius r_1. It should be no surprise that if Eq. 6.4-2 is solved for $(\sigma_\phi)_{\text{ave}}$ the resulting expression is consistent with the formula for circumferential stress in a thin-walled cylindrical pressure vessel, $\sigma = pr/t$.

The foregoing argument is elaborated upon in what follows, to account for the variation with r of σ_ϕ and the presence of shear stress τ. We continue to assume that stresses do not vary in the direction parallel to dimension t_1 in Fig. 6.4-1c.

Equilibrium Equations. In Fig. 6.4-1a, summation of forces along the radial centerline, circumferential forces (normal to the radial centerline), and moments about center of curvature O yields, after neglecting higher-order terms,

$$\frac{dV}{d\phi} = -N \qquad \frac{dN}{d\phi} = V \qquad \frac{dM}{d\phi} = VR \tag{6.4-3}$$

The same equilibrium considerations can be expressed in terms of stresses shown in Fig. 6.4-1b. With τ the shear stress at arbitrary r and τ_1 the shear stress at $r = r_1$,

$$-d\phi \int_b^{r_1} \sigma_\phi \, dA - \int_b^{r_1} d\tau \, dA + \sigma_r t_1 r_1 \, d\phi = 0 \tag{6.4-4}$$

$$\int_b^{r_1} d\sigma_\phi \, dA - d\phi \int_b^{r_1} \tau \, dA - \tau_1 t_1 r_1 \, d\phi = 0 \tag{6.4-5}$$

$$\int_b^{r_1} d\sigma_\phi \, r \, dA - \tau_1 t_1 r_1^2 \, d\phi = 0 \tag{6.4-6}$$

Shear Stress. From Eqs. 6.2-10, 6.4-6, and the second and third of Eqs. 6.4-3, we obtain a formula for transverse shear stress at radius r_1 [6.6, 6.7].

$$\tau_1 = \frac{V r_n}{A e \, t_1 r_1^2} [R A_1 - Q_1] \tag{6.4-7}$$

where

$$A_1 = \int_b^{r_1} dA \quad \text{and} \quad Q_1 = \int_b^{r_1} r \, dA \tag{6.4-8}$$

In a rectangular cross section of thickness t, for example, $A_1 = (r_1 - b)t$ and $Q_1 = A_1(b + r_1)/2$.

Often the shear stress is not large, and a satisfactory approximation is provided by the usual straight-beam formula, $\tau = VQ/It$, in which τ is maximum at the centroidal axis unless t is unusually large there.

Radial Stress. From Eqs. 6.2-10, 6.4-5, 6.4-7, and the second and third of Eqs. 6.4-3, we obtain an expression for the second integral in Eq. 6.4-5. Substituting this result into Eq. 6.4-4 and using the first of Eqs. 6.4-3, we obtain a formula for radial stress at radius r_1 [6.7].

$$\sigma_r = \frac{r_n}{A e \, t_1 r_1} \left[(M - NR) \left(\int_b^{r_1} \frac{dA}{r} - \frac{A_1}{r_n} \right) + \frac{N}{r_1} (R A_1 - Q_1) \right] \tag{6.4-9}$$

A simplification appears for the special case $M = NR$, which arises when $M_o = 0$ in Fig. 6.4-2. Then, on their right-hand sides, the formulas for τ_1 and σ_r have identical multipliers for V and N respectively. And, for a rectangular cross section, τ_1 and σ_r then have greatest magnitude for $r_1 = 2ab/(a+b)$. Note, however, that τ_1 is largest along DE and FG in Fig. 6.4-2, while σ_r is largest along BC.

EXAMPLE 1

In Fig. 6.4-2, let $P = 2000$ N and $M_o = 0$. Let the cross section be rectangular, with $a = 30$ mm, $b = 10$ mm, and $t = 12$ mm. Determine the largest values of the circumferential, radial, and shear stresses.

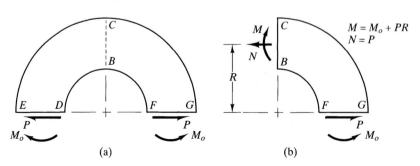

FIGURE 6.4-2 (a) External loads P and M_o on a half-ring. (b) Internal force N and moment M on cross section BC.

Straightforward application of foregoing formulas yields $R = 20$ mm, $r_n = 18.2$ mm, $e = 1.80$ mm, $r_1 = 2ab/(a + b) = 15$ mm, $A = 240$ mm², $A_1 = 60$ mm², and $Q_1 = 750$ mm³. Hence, from Eqs. 6.2-10, 6.4-7, and 6.4-9,

$(\sigma_\phi)_{max} = 84.3$ MPa   (at B in Fig. 6.4-2, at $r = 10$ mm)

$(\sigma_r)_{max} = 14.0$ MPa   (along BC in Fig. 6.4-2, at $r_1 = 15$ mm)

$(\tau_1)_{max} = 14.0$ MPa   (along DE and FG in Fig. 6.4-2, at $r_1 = 15$ mm)

Strictly, this value of $(\tau_1)_{max}$ is valid only if forces P are applied as shear stress distributions that vary with r_1 as described by Eq. 6.4-7.

An exact solution for this problem, by theory of elasticity [6.1], yields maximum radial and shear stresses of 13.9 MPa, both at $r_1 = 15$ mm. The elementary formula for a straight beam, $\tau = VQ/It$, yields a maximum shear stress of 12.5 MPa at $r = 20$ mm. If we consider bending only, so that $M = PR$ and $N = 0$ in Eq. 6.4-9, this equation shows a maximum σ_r of 14.8 MPa, at $r_1 = 15.7$ mm. It is reassuring that these various methods all yield similar results.

EXAMPLE 2

Determine the radial stress along the dashed line in Fig. 6.4-3a, at $r = 100$ mm.

Straightforward calculation yields $R = 111.82$ mm and $A = 6600$ mm². Also

$$\int_A \frac{dA}{r} = 90 \ln \frac{100}{60} + 30 \ln \frac{200}{100} = 66.77 \text{ mm} \qquad (6.4\text{-}10)$$

Hence $r_n = 6600/66.77 = 98.85$ mm and $e = R - r_n = 13.0$ mm. Also $r_1 = 100$ mm, $t_1 = 30$ mm, $A_1 = 3600$ mm², and $Q_1 = 3600(80) = 288{,}000$ mm³. Next

$$\int_b^{r_1} \frac{dA}{r} = 90 \ln \frac{100}{60} = 45.97 \text{ mm} \qquad (6.4\text{-}11)$$

On the cross section identified by the dashed line in Fig. 6.4-3a, $N = -80{,}000$ N and $M = -80{,}000(130 + R) = -19.35(10^6)$ N·mm. Equation 6.4-9 yields, after partially combining terms, $\sigma_r = 384(10^{-9})[-99.4(10^6) - 92.0(10^6)]$, from which

$$\sigma_r = -73.5 \text{ MPa} \qquad (6.4\text{-}12)$$

This result is a radial compressive stress at the intersection of stem and flange. It is an average value because stress concentration effects near re-entrant corners have been neglected. If the cross section were rectangular, we would expect to find the largest magnitude of σ_r at a radius

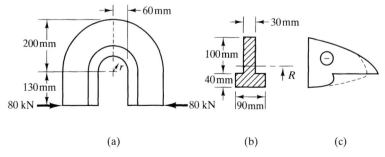

FIGURE 6.4-3 Machine frame having a T cross section. Part (c) shows the approximate distribution of radial stress along the symmetry axis of the cross section.

174 Chapter 6 Curved Beams

less than r_n. In the present problem r_n locates an axis within the flange, whose width t is large. Accordingly we accept Eq. 6.4-12 as a satisfactory estimate of the largest average σ_r. The dashed line in Fig. 6.4-3c represents an estimate of the actual variation of σ_r along the symmetry axis of the cross section. (For comparison, where the dashed line in Fig. 6.4-3a meets the inner edge of the flange, Eq. 6.2-10 yields circumferential stress $\sigma_\phi = -159$ MPa.)

6.5 SECTIONS HAVING THIN FLANGES

Deformation in the plane of the cross section can have an appreciable effect on stiffness and stress. Formulas presented thus far have not taken this deformation into account. The effect may be very significant for any thin-walled section, even if the beam is not sharply curved. The present discussion is concerned with open flanged sections, such as wide-flange and T sections. Closed sections such as pipes are discussed in Section 6.6.

Physical Behavior. Imagine that we prescribe a certain relative rotation $d\theta$ between adjacent cross sections, and apply whatever bending moment M is needed to maintain this relative rotation (Fig. 6.5-1a). Also imagine that initially all deformation of the cross section in its own plane is somehow prevented. We next describe how the cross section tends to deform, and consequences of allowing it to do so.

Circumferential stress σ_ϕ generates radial force (see, for example, the second group of terms in Eq. 6.4-1). Radial force on a flange not allowed to deform can be represented by radially directed pressure p (Fig. 6.5-1b). If deformation of the cross section is now allowed, pressure p causes the "flapping" deformation of flanges shown in

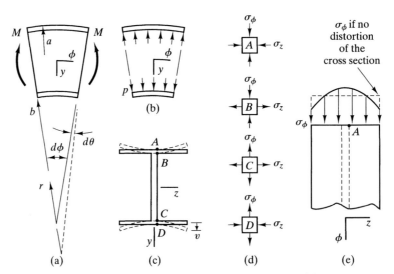

FIGURE 6.5-1 (a) Bending of a curved beam having wide flanges. (b) Effective pressure that deforms flanges. (c) Deformation of the cross section (dashed lines). (d) Normal stresses in ϕ and z directions at selected points. (e) Final circumferential stress distribution in the outer flange (solid line).

Fig. 6.5-1c. Thus there is flexure of flanges in the yz plane, which is accompanied by flexural stress σ_z. We will see that σ_z may be larger in magnitude than circumferential stress σ_ϕ. Figure 6.5-1d shows the directions of σ_ϕ and σ_z at selected points in the cross section. At points B and D these stresses may combine to produce large shear stress. At points B and C, radial stress in the web (not shown) is tensile for the direction of M depicted. Thus, a state of triaxial tension exists at C. Triaxial tension tends to promote fracture. (Reversal of the direction of moment M would be accompanied by reversal of all deformations and all stresses.)

Not only does deformation of a cross section in its own plane produce stress σ_z, it also alters circumferential stress σ_ϕ. Again imagine that angle $d\theta$ of Fig. 6.5-1a is maintained while imaginary constraints that prevent deformation of the cross section are removed. On (say) the midsurface of the inner flange, a typical circumferential fiber moves inward an amount v, thus acquiring a circumferential compressive strain, $\epsilon_\phi = -v/r_m$, where r_m is the radius to the flange midsurface. (The basis of this formula is explained in connection with Fig. 1.4-2b.) The associated circumferential stress, $-Ev/r_m$, is superposed on the existing circumferential stress, which is tensile on the inner flange. Thus, circumferential tensile stress is reduced, most greatly near flange tips where v is largest. Similarly, the magnitude of circumferential compressive stress is reduced in the outer flange. Accordingly, the moment M required to maintain a prescribed angle $d\theta$ is reduced by deformation of the cross section. Or, if we prescribe M instead of $d\theta$, we conclude that $d\theta$ is increased by deformation of the cross section; that is, flexural stiffness seen by moment M is reduced. Also, to sustain a given moment M, flange circumferential stress σ_ϕ must be larger near the web and smaller near flange tips than in a nondeforming cross section (Fig. 6.5-1e). As noted in Section 6.1, deformation of the cross section and its undesirable consequences can be greatly reduced by the addition of radial stiffening plates between flanges.

Analysis. The standard method of accounting for the foregoing effects is to use a substitute cross section [6.8, 6.9], which differs from the actual cross section by having reduced flange dimensions in the z direction. Stresses in the actual cross section are obtained by applying previously derived formulas to the substitute cross section, which is regarded as having no in-plane distortion. The substitute cross section is also used in calculating deflections of a curved beam.

We summarize the derivation rather than giving full details. Consider the typical cross section shown in Fig. 6.5-2a. Let $\overline{\sigma}_\phi = \overline{\sigma}_\phi(z)$ be circumferential stress at a flange midsurface, where $r = r_m$. Let $\overline{\sigma}_{\phi f}$ be the value of $\overline{\sigma}_\phi$ at $z = 0$, where a flange midsurface meets the web (point A or point B in Fig. 6.5-2a). Elsewhere along a flange midsurface,

$$\overline{\sigma}_\phi = \overline{\sigma}_{\phi f} - E\frac{v}{r_m} \qquad (6.5\text{-}1)$$

where v is the flange deflection, positive inward, at $r = r_m$. The Ev/r_m term is not inconsequential; for example, if $v/r_m = 0.001$, then $Ev/r_m \approx 200$ MPa for steel. Therefore it is necessary to determine $v = v(z)$, as follows.

An expression for pressure p on an undeformed flange (Fig. 6.5-1b) can be obtained from Eq. 6.4-2 with the substitutions $\sigma_r = p$, $(\sigma_\phi)_{\text{ave}} = \overline{\sigma}_{\phi f}$, $A_1 = \ell t_f$, and $t_1 r_1 = \ell r_m$.

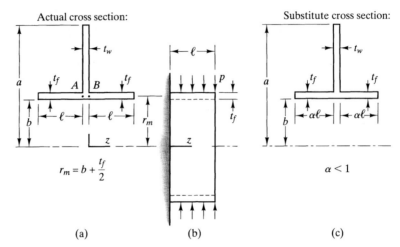

FIGURE 6.5-2 (a) T-shaped cross section. (b) Cylindrical shell model used in developing formulas for the substitute cross section. (c) The substitute cross section.

Thus $p = (t_f/r_m)\bar{\sigma}_{\phi f}$. A flange is now regarded as a thin cylindrical shell of mean radius r_m, fixed at one end and free at the other, and loaded by pressure p (Fig. 6.5-2b). From Eq. 13.6-4, the governing differential equation for this shell problem is

$$\frac{Et_f^3}{12(1-\nu^2)} \frac{d^4v}{dz^4} + \frac{Et_f}{r_m^2} v = \frac{t_f}{r_m} \bar{\sigma}_{\phi f} \tag{6.5-2}$$

where ν is Poisson's ratio. This equation is solved for $v = v(z)$ and the results substituted into Eq. 6.5-1, whereupon we have an expression for $\bar{\sigma}_\phi = \bar{\sigma}_\phi(z)$. This stress, in acting on the actual flange width ℓ, produces a circumferential force. We require that the same circumferential force be produced by constant stress $\bar{\sigma}_{\phi f}$ in acting on the substitute flange width $\alpha\ell$. We assume that a flange is thin enough that σ_ϕ may be regarded as varying linearly across dimension t_f. Accordingly the midsurface value $\bar{\sigma}_\phi$ may be used to calculate circumferential force. Equality of the two circumferential forces therefore requires

$$\int_0^\ell \bar{\sigma}_\phi t_f \, dz = \bar{\sigma}_{\phi f} (\alpha \ell t_f) \tag{6.5-3}$$

from which we solve for α. Finally, flexural stress σ_z in a flange can be determined from the elastic properties, thickness t_f, and curvature d^2v/dz^2. The σ_z stresses of largest magnitude appear on flange surfaces where a flange meets the web, tensile on one flange surface and compressive on the other. These stresses are written in terms of $\bar{\sigma}_{\phi f}$:

$$(\sigma_z)_{\max} = \pm \beta \bar{\sigma}_{\phi f} \tag{6.5-4}$$

Manipulation of the foregoing equations yields the following:

$$\alpha = \frac{1}{\lambda \ell} \frac{\sinh 2\lambda\ell + \sin 2\lambda\ell}{2 + \cosh 2\lambda\ell + \cos 2\lambda\ell} \tag{6.5-5a}$$

$$\beta = \sqrt{3} \frac{\cosh 2\lambda\ell - \cos 2\lambda\ell}{2 + \cosh 2\lambda\ell + \cos 2\lambda\ell} \tag{6.5-5b}$$

Section 6.5 Sections Having Thin Flanges

TABLE 6.5-1 Factors α and β from Eqs. 6.5-5, for Poisson's ratio $\nu = 0.3$.

$\dfrac{\ell^2}{rt}$	0	0.1	0.2	0.3	0.4	0.5	0.6	0.7	0.8	0.9
α	1.000	0.995	0.979	0.955	0.923	0.888	0.851	0.813	0.776	0.741
β	0	0.284	0.555	0.802	1.018	1.200	1.347	1.462	1.550	1.615
$\dfrac{\ell^2}{rt}$	1.0	1.1	1.2	1.4	1.6	2.0	3.0	4.0	6.0	10.0
α	0.708	0.678	0.651	0.603	0.564	0.506	0.422	0.374	0.314	0.246
β	1.661	1.693	1.713	1.731	1.731	1.711	1.674	1.676	1.707	1.731

where

$$\lambda = \left[\frac{3(1-\nu^2)}{r_m^2 t_f^2} \right]^{1/4} \tag{6.5-5c}$$

For $\nu = 0.3$, Table 6.5-1 provides a short compilation of factors α and β. If the actual flange width ℓ is very large, Eqs. 6.5-5 yield $\alpha \approx 1/\lambda\ell$ and $\beta \approx \sqrt{3}$. For $\nu = 0.3$ and very large ℓ in the actual cross section, the flange in the substitute cross section has width $\alpha\ell = 1/\lambda = 0.778\sqrt{r_m t_f}$. Or, for two identical flanges as in Fig. 6.5-1c, the total z-parallel dimension of the substitute cross section becomes $1.556\sqrt{r_m t_f} + t_w$. Web thickness t_w and all radius-parallel dimensions are the same as in the actual cross section.

EXAMPLE

A long straight steel pipe, of inner radius 1.500 m and wall thickness 3.12 mm, is reinforced at midlength by a circumferential T-section ring welded to the outside of the pipe as shown in Fig. 6.5-3a. The pipe is lifted by a single hook at the top of the reinforcing ring. At this location, circumferential tensile force is estimated as 4200 N and bending moment is estimated as 6.7 kN·m. (This moment acts opposite to the direction shown in Fig. 6.5-1a.) Significant values of stress are required. Let $\nu = 0.3$.

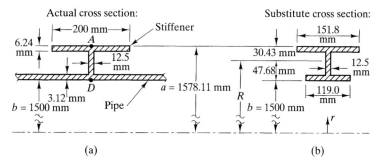

FIGURE 6.5-3 Cross section of one side of a long, thin-walled pipe with an external T-section ring stiffener. Actual and substitute cross sections are shown.

First we must calculate dimensions of the substitute cross section. The inner flange of the curved beam is the pipe wall, so that ℓ is very large. As noted in the preceding paragraph, the total width of the substitute inner flange is

$$2\alpha\ell + t_w = 1.556 \sqrt{r_m t_f} + t_w$$
$$= 1.556 \sqrt{1501.56(3.12)} + 12.5 = 119.0 \text{ mm} \qquad (6.5\text{-}6)$$

For the outer flange, $r_m = 1575.0$ mm, $\ell = (200 - 12.5)/2 = 93.75$ mm, and $\ell^2/r_m t_f = 0.894$. Interpolating in Table 6.5-1, we obtain $\alpha = 0.743$. The total width of the substitute outer flange is

$$2\alpha\ell + t_w = 2(0.743)(93.75) + 12.5 = 151.8 \text{ mm} \qquad (6.5\text{-}7)$$

The substitute cross section is shown in Fig. 6.5-3b. Its cross-sectional area, centroidal radius, and centroidal moment of inertia are

$$A = 2178 \text{ mm}^2 \qquad R = 1547.68 \text{ mm} \qquad I = 1.927(10^6) \text{ mm}^4 \qquad (6.5\text{-}8)$$

The ratio of R to depth h of the cross section is $1547.68/78.11 = 19.8$. This R/h ratio is sufficiently large that slender-beam theory is adequate for calculation of circumferential stresses; that is, $\sigma_\phi = N/A \pm My/I$. Bending moment M is negative in this problem. Thus

$$\text{At } r = b: \quad \sigma_\phi = \frac{4200}{2178} - \frac{6.7(10^6)(47.68)}{1.927(10^6)} = -164 \text{ MPa} \qquad (6.5\text{-}9\text{a})$$

$$\text{At } r = a: \quad \sigma_\phi = \frac{4200}{2178} + \frac{6.7(10^6)(30.43)}{1.927(10^6)} = 108 \text{ MPa} \qquad (6.5\text{-}9\text{b})$$

In the actual cross section, these σ_ϕ stresses appear at points D and A in Fig. 6.5-3a. For inner and outer flanges respectively, $\beta = 1.732$ and $\beta = 1.611$. Using these β values and circumferential stresses $\overline{\sigma}_{\phi f}$ at flange midsurfaces from M alone, Eq. 6.5-4 yields the following flexural stresses. In the actual cross section, these σ_z stresses appear at surfaces of inner and outer flanges where flanges meet the web.

$$\text{Inner:} \quad (\sigma_z)_{\max} = \pm 1.732 \frac{6.7(10^6)(46.12)}{1.927(10^6)} = \pm 278 \text{ MPa} \qquad (6.5\text{-}10\text{a})$$

$$\text{Outer:} \quad (\sigma_z)_{\max} = \pm 1.611 \frac{6.7(10^6)(27.31)}{1.927(10^6)} = \pm 153 \text{ MPa} \qquad (6.5\text{-}10\text{b})$$

At the inner surface of the inner flange, point D in Fig. 6.5-3a, σ_ϕ is compressive and $(\sigma_z)_{\max}$ is tensile. Here the maximum shear stress is

$$\tau_{\max} = \frac{278 - (-164)}{2} = 221 \text{ MPa} \qquad (6.5\text{-}11)$$

Such a large shear stress may initiate yielding. With yielding, flanges deflect more, thus reducing circumferential bending stiffness and increasing the magnitudes of both σ_ϕ and $(\sigma_z)_{\max}$. Thus, in this problem yielding has a destabilizing effect, possibly leading to plastic collapse.

In summary, because the cross section is thin-walled and the beam is curved, a substitute cross section is needed, yet in this example curvature of the substitute beam can be ignored in the calculation of circumferential stresses. This procedure may at first seem strange, but it is not in error. It is probably more significant that we have ignored the effect of lengthwise flexural stresses in the pipe that arise when it is lifted by the hook at midlength. On the upper portion of the pipe, these stresses are tensile, and act to reduce radial deformation of the inner flange. Our formulas make no provision for this effect. (A similar problem, the effect of axial force on beam bending, is discussed in Sections 14.3 and 14.4.)

6.6 CLOSED SECTIONS WITH THIN WALLS

Bending of a curved pipe causes its cross section to deform. The physical mechanism, which involves radial components of circumferential forces, is as explained for open sections in Section 6.5. If the closed cross section is initially circular, the deformation is called "ovalization" (Fig. 6.6-1b). The effect of ovalization is to decrease the stiffness seen by moment M, increase the largest circumferential stress σ_ϕ, and produce "hoop" flexural stresses σ_η directed around the originally circular cross section. In the upper part of Fig. 6.6-1b, σ_η on the inner surface of the pipe is tensile at $\eta = 0$, compressive at $\eta = 90°$, and so on. Stress σ_η is of opposite sign on the outer surface. Deformations, and signs of all stresses, are reversed if M is reversed. Pipes of noncircular cross section behave similarly (Fig. 6.6-1c).

Circumferential and hoop stresses present in a curved pipe of circular cross section are shown in Fig. 6.6-2 [6.10]. In this example, dimensions are such that $R/t = 64.0$, $c/t = 21.3$, and $R/c = 3.0$. Stresses are reported as the *ratio* of actual stress to the maximum flexural stress that would be present if the pipe were straight, namely $(\sigma_\phi)_{st} = Mc/I$. The σ_ϕ distribution is striking in that the maximum σ_ϕ is about three times as large as $(\sigma_\phi)_{st}$ and appears well away from the inner and outer extremes of the cross section. At inner and outer extremes of the cross section, σ_ϕ is small and even opposite in sign to what one would expect from elementary beam theory. The σ_η plot

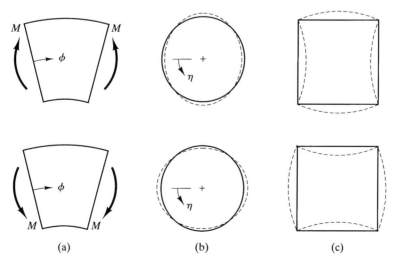

FIGURE 6.6-1 Distortion of circular and rectangular thin-walled cross sections when a curved beam is bent in its plane of curvature.

180 Chapter 6 Curved Beams

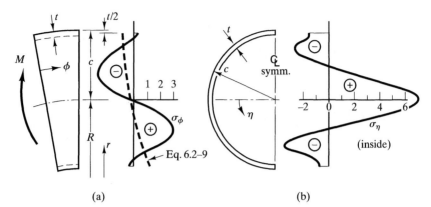

FIGURE 6.6-2 Bending of a curved beam of thin-walled circular cross section. Plots show circumferential stress σ_ϕ at the midsurface and hoop flexural stress σ_η on the inner surface, plotted versus radial coordinate r. Here $R/c = 3.0$ and $R/t = 64$.

shows that σ_η on the inner surface of the pipe is more than double the largest σ_ϕ and is about seven times the value of $(\sigma_\phi)_{st}$.

Analysis procedures for curved pipes are beyond the scope of this book. The following summary and additional references are offered. Reference 6.11 considers pipes of circular cross section. Peak stresses and flexural stiffness are related to E, ν, R, c, t, external loads, and internal pressure in the pipe. External loads considered include moment M in Fig. 6.6-2, moment about the $\eta = \pi/2$ axis, and torque T about the curved axis of the pipe. Maximum shearing stress is determined by locating peak combinations of σ_ϕ and σ_η (taking membrane and bending actions into account) and torsional shear stress associated with T. Reference 6.12 presents formulas for pipes of rectangular cross section. Finite element analysis can be performed using shell elements rather than three-dimensional elements; otherwise, general finite element analysis has the same awkward features as noted for solid curved beams at the end of Section 6.3. However, software adapted to piping problems often includes a pipe-bend element that takes ovalization into account.

6.7 DEFLECTIONS OF SHARPLY CURVED BEAMS

We consider deflections of a sharply curved beam loaded in its own plane. Assumptions stated in Section 6.1 remain in force. For a cross section of wide-flange or T shape, properties $A, e,$ and R should refer to the substitute cross section discussed in Section 6.5. Analogous adjustments for other shapes of thin-walled cross section may be found in the literature.

Let $d\theta$ represent the relative rotation between two cross sections separated by distance ds (for a straight beam) or by angle $d\phi$ (for a curved beam). Then, from Eqs. 4.8-3 and 6.2-8,

$$\text{Straight beam:} \quad d\theta = \frac{M}{EI} ds \quad (6.7\text{-}1a)$$

Section 6.7   Deflections of Sharply Curved Beams   181

$$\text{Sharply curved beam:} \quad d\theta = \frac{M}{EAe} d\phi \tag{6.7-1b}$$

Now let us substitute into Eq. 6.7-1b: $e \approx I/RA$ (from Eq. 6.2-11) and $R\,d\phi = ds$. Then Eqs. 6.7-1a and 6.7-1b become the same. Thus we see that when a curved beam is loaded in pure bending, the error of applying the straight-beam formula for $d\theta$ is proportional to the difference between exact and approximate values of $1/e$. For a rectangular cross section with $R = h$, which is the same as $a/b = 3$, the straight-beam formula yields a value of $d\theta$ that is 7.7% high. This error is rather small for so sharp a curvature.

However, when normal force N and transverse shear force V are taken into account, error may be much greater, and the theory described next should be used in preference to the simpler deflection formulas used in Chapter 4. The following theory pertains to displacements at radius $r = R$.

Theory. A customary and worthwhile simplification states that the contribution to strain energy of radial stress σ_r is negligible in comparison with the contribution of circumferential stress σ_ϕ. We will not ignore the effect of shear stress, but will assume that the straight-beam formula $\tau = VQ/It$ is adequate, so as to avoid the more complicated Eq. 6.4-7. Complementary strain energy in the beam is the volume integral of strain energy densities. From Eqs. 2.6-2 and 2.6-3,

$$U^* = \iint \left(\frac{\sigma_\phi^2}{2E} + \frac{\tau^2}{2G} \right) r\,dA\,d\phi \tag{6.7-2}$$

where the volume element is $r\,dA\,d\phi$. We substitute for σ_ϕ from Eq. 6.2-10 and note that area integrals of r, $(r_n - r)^2/r$ and $(r_n - r)$ have the respective values AR, Ae, and $-Ae$. The shear term is integrated as described in Section 4.3. Thus Eq. 6.7-2 becomes

$$U^* = \int \left(\frac{N^2 R}{2EA} + \frac{M^2}{2EAe} - \frac{MN}{EA} + \frac{kV^2 R}{2GA} \right) d\phi \tag{6.7-3}$$

A physical derivation of the MN coupling term is as follows. Force resultant N acts at the centroid of a cross section and is therefore a distance e from the neutral axis of bending. In Fig. 6.1-2b we see that rotation $d\theta$ associated with positive M will displace a positive (tensile) N a distance $e\,d\theta$ in the negative direction. Work done by N is therefore $-Ne\,d\theta$. Substituting for $d\theta$ from the second of Eqs. 6.7-1, we obtain the MN coupling term in Eq. 6.7-3, with its negative sign. Because MN is a product term rather than a squared term, algebraic signs of M and N must be observed.

Equation 6.7-3 can be manipulated by Castigliano's second theorem to obtain deflections and solve statically indeterminate problems, with due attention to cautions noted in Section 4.4.

EXAMPLE 1

The sharply curved quarter-ring in Fig. 6.7-1a is loaded in pure bending. Determine the horizontal component of deflection at radius R on the lower end due to applied moment load M_o.

As required by Castigliano's second theorem, a fictitious force F is applied at the location and in the direction of the displacement to be determined (Fig. 6.7-1b). The required horizontal

182 Chapter 6 Curved Beams

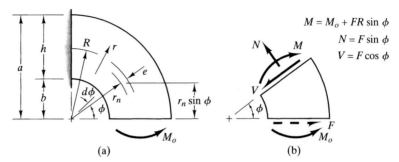

FIGURE 6.7-1 (a) External moment load M_o applied to a quarter-ring. (b) Force F is a fictitious force used in the calculation of horizontal displacement at $\phi = 0$.

displacement is $u_o = \partial U^*/\partial F$, where U^* comes from Eq. 6.7-3 because the ring is sharply curved. Thus

$$u_o = \int_0^{\pi/2} \left(\frac{NR}{EA}\frac{\partial N}{\partial F} + \frac{M}{EAe}\frac{\partial M}{\partial F} - \frac{M}{EA}\frac{\partial N}{\partial F} - \frac{N}{EA}\frac{\partial M}{\partial F} + \frac{kVR}{GA}\frac{\partial V}{\partial F} \right) d\phi \qquad (6.7\text{-}4)$$

where $\partial N/\partial F = \sin \phi$ and $\partial M/\partial F = R \sin \phi$. Having differentiated with respect to F, we can now set F to zero. Thus $M = M_o$, $N = 0$, and $V = 0$. Equation 6.7-4 yields

$$u_o = \int_0^{\pi/2} \left(\frac{M_o R}{EAe} - \frac{M_o}{EA} \right) \sin \phi \, d\phi = \frac{M_o}{EA}\left(\frac{R}{e} - 1\right) \qquad (6.7\text{-}5)$$

A more physical derivation of this result is as follows. Consider du_o, an increment of horizontal displacement at the lower end of the ring associated with deformation of a typical slice spanned by angle $d\phi$ (Fig. 6.7-1). Since the loading is pure bending, adjacent faces of the slice rotate about the neutral axis at $r = r_n$. The amount of relative rotation, $d\theta$, is given by Eq. 6.7-1b. Hence

$$du_o = (r_n \sin \phi)\, d\theta \qquad \text{hence} \qquad u_o = \int_0^{\pi/2} (r_n \sin \phi)\frac{M_o}{EAe}\, d\phi \qquad (6.7\text{-}6)$$

With the substitution $r_n = R - e$, Eq. 6.7-6 becomes identical to Eq. 6.7-5. For a slender ring, $R \gg e$, and the substitution $e \approx I/RA$ from Eq. 6.2-11 is appropriate. Thus u_o from Eq. 6.7-5 approaches $u_o = M_o R^2/EI$, which is the slender-beam result predicted by methods of Section 4.6.

For a numerical example, let the cross section be rectangular and let $a/b = 3.5$. Then, in comparison with the exact value of u_o from theory of elasticity [6.1], the value of u_o given by Eq. 6.7-5 is 11.4% low and the slender-beam result $u_o = M_o R^2/EI$ is 9.8% high. In *stress* calculation for these dimensions, the straight-beam formula $\sigma = Mc/I$ yields a flexural stress at $r = b$ that is 38.7% low. The implication is that effects of initial curvature are much more important for stress calculation than for deflection calculation.

EXAMPLE 2

Replace moment M_o in Fig. 6.7-1a by the rightward force F. Again determine the horizontal displacement component at radius R on the lower end of the quarter-ring.

As in Example 1, $\partial N/\partial F = \sin\phi$ and $\partial M/\partial F = R\sin\phi$, but now

$$N = F\sin\phi \qquad M = FR\sin\phi \qquad V = F\cos\phi \tag{6.7-7}$$

Equation 6.7-4 remains applicable, and becomes

$$u_o = \int_0^{\pi/2}\left(\frac{FR}{EA}\sin^2\phi + \frac{FR^2}{EAe}\sin^2\phi - 2\frac{FR}{EA}\sin^2\phi + \frac{kFR}{GA}\cos^2\phi\right)d\phi \tag{6.7-8}$$

Integrals of $\sin^2\phi$ and $\cos^2\phi$ from 0 to $\pi/2$ are both $\pi/4$. To introduce additional numbers, let us say that the cross section is rectangular, $a/b = 3.5$, $k = 1.2$, and Poisson's ratio $\nu = 0.3$. Thus $G = 0.385E$. Equation 6.7-8 becomes, with terms written in the same order,

$$u_o = \frac{\pi FR}{4EA}(1.00 + 8.84 - 2.00 + 3.12) = 8.61\frac{FR}{EA} \tag{6.7-9}$$

We see that the contribution of shear deformation is considerable, and that the MN coupling term contributes twice as much to u_o as the direct stress term, but with opposite sign. The slender-beam result, predicted by methods of Section 4.6, is $u_o = \pi FR^3/4EI$. In comparison with the exact result [6.1] for the numerical properties chosen here, u_o from Eq. 6.7-9 is only 0.5% low, while the slender-beam value of u_o is 11% low.

PROBLEMS

6.1-1. The sketch shows approximately one-quarter of a coil of a massive helical spring. The spring is close-coiled and is stretched by force P along the axis of the helix. Torque $T = PR$ in a coil produces shear stress τ, whose distribution is shown along the y axis of the coil cross section. (The direct shear contribution $\tau = VQ/It$ is not shown.) If the cross section rotates about point Q, show that τ due to T varies in the same manner as σ_ϕ in Eq. 6.2-1.

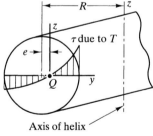

PROBLEM 6.1-1

6.2-1. Show that Eq. 6.2-10 is obtained from Eqs. 6.2-1 and 6.2-3 with $N \neq 0$. Suggestions: Obtain an expression for $Er_n(d\theta/d\phi)$ from the moment equation. Eliminate r_n between this result and the equation for N; thus obtain an expression for $E(d\theta/d\phi)$. Substitute the expressions for $Er_n(d\theta/d\phi)$ and $E(d\theta/d\phi)$ into Eq. 6.2-1, and use Eqs. 6.2-2, 6.2-5, and 6.2-6.

6.2-2. Substitute Eq. 6.2-11 into Eq. 6.2-9, and show how the resulting expression reduces to the flexure formula $\sigma = Mc/I$ if the beam is slender.

6.2-3. Show that, for the loading of Fig. 6.2-2, $\sigma_\phi = 0$ on the centroidal axis of cross section BC, regardless of the shape of the cross section.

Chapter 6 Curved Beams

6.2-4. For a trapezoidal cross section, derive the formula for $\int dA/r$ (see Table 6.2-1).

6.2-5. Apply the calculation method of Fig. 6.2-3 to the cross section in Fig. 6.3-1b. Calculate percentage errors of the resulting values of A, $\int dA/r$, R, r_n, and e.
   (a) Use 7 strips, with $\Delta r = 10$ mm for each strip.
   (b) Use 14 strips, with $\Delta r = 5$ mm for each strip.

6.2-6. For a solid circular cross section, use Eq. 6.2-11 to evaluate e in terms of c and R. Then show that for $R \gg c$, formulas in Table 6.2-1 lead to the same expression for e. Suggestion: In the formula for $\int dA/r$, write the radical expression as a series.

6.2-7. For a triangular cross section with apex pointed outward ($t_2 = 0$ in Table 6.2-1), and for a/b ratios of 1.2, 1.6, 3.0, and 8.0, calculate the following quantities.
   (a) e/h according to Eq. 6.2-6.
   (b) e/h according to Eq. 6.2-11.
   (c) K in Eq. 6.2-12 for circumferential stress at $r = b$.
   (d) K in Eq. 6.2-12 for circumferential stress at $r = a$.

6.2-8. In Problem 6.2-7, for what approximate value of a/b do circumferential stresses at $r = b$ and at $r = a$ have the same magnitude?

6.2-9. Consider pure bending of a curved beam of rectangular cross section. Let bending moment M, thickness t, and outer radius a be prescribed.
   (a) Provide a verbal argument that there must be an optimum value of inner radius b, such that circumferential stress given by Eq. 6.2-9 is minimized.
   (b) Solve analytically and/or numerically for the optimum value of b/a.

6.3-1. A positive bending moment $M = 2.0$ kN·m is applied to a curved beam having the T section shown. Determine circumferential stresses at $r = b$ and at $r = a$. Do you think proportions of the cross section are well chosen? (In this exercise, ignore adjustments discussed in Section 6.5.)

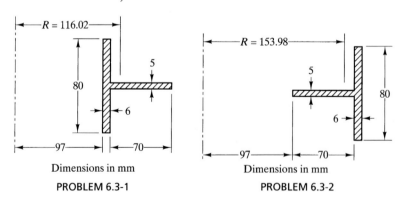

PROBLEM 6.3-1 PROBLEM 6.3-2

6.3-2. Repeat Problem 6.3-1 for the cross section shown.

6.3-3. (a) In Problem 6.3-1, determine the percentage error of the circumferential stress at $r = b$ if this stress is calculated by the flexure formula $\sigma = Mc/I$.
   (b) In Problem 6.3-1, imagine that an error in calculation yields a value of r_n that is 2% smaller than the correct value. What is the resulting error in the circumferential stress at $r = b$?

6.3-4. The frame shown is made of a material that yields in uniaxial tension at 400 MPa. If the safety factor against yielding must be 2.5, what is the allowable load P?

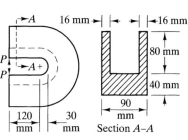
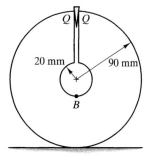

PROBLEM 6.3-4   PROBLEM 6.3-5

6.3-5. The circular ring shown has a solid circular cross section and a narrow radial cut on one side. A wedge is used to pry the cut open a small amount. A gage at point B shows that the resulting circumferential strain is $800(10^{-6})$. What horizontal forces are applied by the wedge to the contact points Q? Let $E = 70$ GPa.

6.3-6. For the T section shown, what should be dimension t_i if pure bending is to create circumferential stresses of the same magnitude on inside and outside edges of the cross section? (In this exercise, ignore adjustments discussed in Section 6.5.)

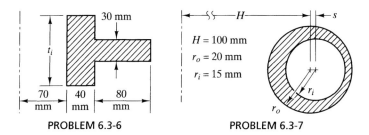

PROBLEM 6.3-6   PROBLEM 6.3-7

6.3-7. A curved beam is subjected to pure bending. The cross section is circular, of outer radius r_o, and with a hole of radius r_i offset a distance s as shown. Distance s is to be such that maximum and minimum circumferential stresses have the same magnitude. Determine s for the dimensions shown.

6.3-8. (a) The bracket shown has a rectangular cross section. Estimate the maximum tensile stress. Approximate as necessary. Make "pessimistic" approximations, so as to overestimate the maximum stress rather than underestimate it.

(b) At what points between fillet B and corner C are stresses particularly small?

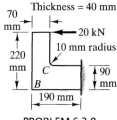

PROBLEM 6.3-8

6.4-1. (a) Derive Eqs. 6.4-3.
(b) Derive Eqs. 6.4-4, 6.4-5, and 6.4-6.
(c) Show that Eq. 6.4-7 is obtained as described in the text.
(d) Show that Eq. 6.4-9 is obtained as described in the text.

6.4-2. (a) For a beam loaded in pure bending, at what value of r_1 in Fig. 6.4-1b is radial force provided by the radial component of σ_ϕ a maximum? Is this where σ_r is maximum? Why or why not?
(b) Show that, for the loading in Fig. 6.2-2, σ_r in a rectangular cross section is maximum at $r_1 = 2ab/(a+b)$.
(c) For pure bending and a rectangular cross section, determine the value of r_1 for which σ_r is maximum, in terms of b and r_n.

6.4-3. In Example 1 of Section 6.4, ignore force N on cross section BC of Fig. 6.4-2 so that the loading on cross section BC is regarded as being only the bending moment $M = PR$. What radial stress σ_r at $r_1 = 15$ mm is now predicted by Eq. 6.4-9?

6.4-4. In Fig. 6.4-3, determine τ_1 at $r_1 = R$ using Eq. 6.4-7. Also determine $\tau = VQ/It$ at the same radius. At what locations along the U-shaped beam axis are these results applicable?

6.4-5. (a) For the geometry and loading of Example 1 of Section 6.4, use Eq. 6.4-2 to estimate the radial stress at $r_1 = R$. To estimate $(\sigma_\phi)_{ave}$, make the simple assumption that tensile (or compressive) stress varies linearly from the value calculated at $r = b$ (or at $r = a$) to zero at the neutral axis. For what reasons is the calculated σ_r not exact?
(b) Repeat part (a), with reference to Example 2 of Section 6.4 and the radial stress at $r_1 = 100$ mm.
(c) Let pure bending be applied to a curved beam of rectangular cross section. If σ_ϕ is obtained from the flexure formula $\sigma = Mc/I$, what σ_r is predicted by Eq. 6.4-2 at $r_1 = R$, in terms of M, t, h, and R?
(d) For $a/b = 3$, what is the percentage error of the result in part (c) at $r = R$?

6.4-6. Let a curved beam have a rectangular cross section with $a = 40$ mm, $b = 10$ mm, and uniform thickness t. Also let $M_o = -PR$ in Fig. 6.4-2, so that cross section BC carries tensile force $N = P$ but no bending moment. Theory of elasticity predicts $\sigma_r = -0.00980P/t$ at $r_1 = 24$ mm. What σ_r does Eq. 6.4-9 predict at this location?

6.4-7. In this exercise, ignore adjustments discussed in Section 6.5.
(a) Use Eq. 6.4-9 to determine σ_r at $r_1 = 103$ mm in Problem 6.3-1.
(b) Use Eq. 6.4-9 to determine σ_r at $r_1 = 167$ mm in Problem 6.3-2.

6.4-8. (a) Using cross section A for the curved beam shown, calculate the maximum circumferential stress. Also calculate the radial stress at $r_1 = 400$ mm.
(b) Using cross section B for the curved beam shown, calculate the required thickness t of weld material, based on an allowable average radial tensile stress of 200 MPa in the weld.

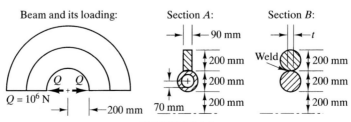

PROBLEM 6.4-8

6.4-9. A ring of rectangular cross section is loaded by diametral force P, as shown. Sketch the approximate variation of stresses σ_ϕ and σ_r, with identification of algebraic signs, along sections AB and CD.

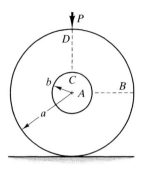

PROBLEM 6.4-9

6.4-10. In the sandwich beam shown, $E_f \gg E_c$ and $h \gg t$. Determine σ_y, the y-direction normal stress in the core, in terms of $P, x, E_f, b, t,$ and h. Assume that load P produces negligible change in dimension h and that σ_x is constant through thickness t of each facing. Suggestion: The beam acquires radius of curvature ρ; consider the radial component of force associated with σ_x in a facing. Do not expect σ_y to be directly proportional to P.

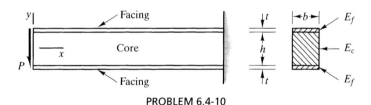

PROBLEM 6.4-10

6.5-1. The sketch shows two thin-walled open cross sections, used for curved beams loaded by moment M as in Fig. 6.5-1. In each case, qualitatively show the deformation of the cross section by sketching its deformed midline on top of the undeformed midline.

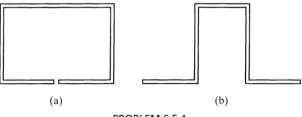

PROBLEM 6.5-1

6.5-2. Revise the example problem in Section 6.5 by attaching the stem of the T-section stiffener to the inside of the pipe rather than to the outside. Thus, $a = 1503.12$ mm, $b = 1425.01$ mm, and the inner flange has width 200 mm and thickness 6.24 mm. Calculate the same quantities calculated in the example problem.

6.5-3. A curved beam of the wide-flange section shown is loaded by a bending moment M that increases its radius of curvature. Calculate the following stresses in terms of M and indicate exactly where they appear on the original cross section. Let $\nu = 0.3$.

(a) The σ_ϕ and σ_z stresses of largest magnitude.

(b) The average radial stress in the web at $r = 255$ mm.

(c) The shear stress of largest magnitude.

(d) Suggest how the proportions of the cross section might be improved.

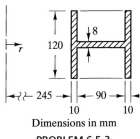

Dimensions in mm
PROBLEM 6.5-3

6.5-4. Do the following for the cross section used in Problem 6.3-1, now taking in-plane deformation of the cross section into account. Let $\nu = 0.3$.

(a) Calculate the tensile and compressive circumferential stresses of largest magnitude.

(b) Calculate the largest flexural stresses σ_z in a flange.

(c) Express the flexural stiffness as a percentage of the flexural stiffness of a non-deforming cross section. (For a given $d\phi$, use Eq. 6.2-8 to obtain stiffness $M/d\theta$.)

(d) Calculate the average radial stress in the web where it meets the flange.

(e) Calculate the largest shear stress in the cross section.

6.5-5. Repeat Problem 6.5-4 for the cross section used in Problem 6.3-2.

6.6-1. Let bending moment M have the direction shown in the upper part of Fig. 6.6-1a, and consider the rectangular cross section in Fig. 6.6-1c. Qualitatively plot circumferential stress σ_ϕ and hoop flexural stress σ_η around the outside of the left half of the cross section.

6.6-2. In the present exercise, assume that numbers on the stress plots in Fig. 6.6-2 represent stresses in MPa.

(a) Use Eq. 6.4-2 to approximate the radial stress at $r_1 = R$. (This σ_r is a membrane stress, constant through thickness t, that is superposed on hoop flexural stress σ_η at $\eta = 0$ and $\eta = 180°$.)

(b) Use the plots of σ_ϕ and σ_η to approximate the largest shear stress on the inner surface of the pipe bend. For the geometry of the pipe, use proportions stated in the text of Section 6.6.

6.7-1. Show that Eq. 6.7-3 is indeed obtained from Eq. 6.7-2 in the manner described in the text.

6.7-2. In Fig. 6.7-1a, replace moment M_o by an upward force F, applied at $r = R$ on the lower end of the quarter-ring. Obtain an expression for the vertical deflection of force F. Then, for a solid circular cross section with $\nu = 0.25$ and $a/b = 3$, evaluate the numerical components of this result (in the manner of Eq. 6.7-9). What is the ratio of this result to the result given by the slender-beam theory of Section 4.6?

6.7-3. The half-ring shown is loaded by uniform radial pressure p on its outer surface. The cross section is rectangular, with dimensions h by t. Obtain an expression for vertical deflection at the free end. Evaluate this expression in terms of p and E for the case $h = 20$ mm, $t = 10$ mm, $R = 20$ mm, and $G = 0.4E$. For the same case, what deflection is predicted by the slender-beam theory of Section 4.6?

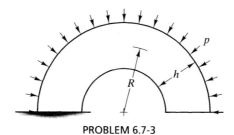

PROBLEM 6.7-3

6.7-4. A complete ring is supported by a rigid surface and loaded by a force P on top, as shown in the sketch for Problem 6.4-9.
  (a) Show that the bending moment on cross section AB is $(PR/2)(1 - 2/\pi + 2e/\pi R)$.
  (b) Obtain an expression for the vertical displacement v_T of cross section CD.
  (c) For a rectangular cross section with $\nu = 0.3$ and $a/b = 3.5$, determine the ratio of v_T from part (b) to v_T from slender-beam theory.

CHAPTER 7

Elements of Theory of Elasticity

7.1 INTRODUCTION

Theory of elasticity treats the usual questions of static stress analysis: Given the geometry and elastic properties of a body, and how it is supported and loaded, what are the stresses and what are the deflections? Within the framework of restrictions cited below, an elasticity solution is exact. A mechanics of materials solution is usually approximate, although it may be quite satisfactory in practice. The two methods are compared in a descriptive way in Section 1.1, which the reader may wish to review.

Theory of elasticity is more complicated than mechanics of materials. However, it can solve problems that mechanics of materials cannot. Several useful elasticity solutions are known. An understanding of the subject makes it possible to use the solutions correctly, and clarifies conditions that must be met by a satisfactory solution, however obtained. Theory of elasticity augments other analysis methods by providing insights that might otherwise be missed. Advanced analysis methods often rely on concepts from theory of elasticity. Concepts and equations developed in this chapter are also used in subsequent chapters.

Unlike a mechanics of materials solution, an elasticity solution is not based on an assumed mode of deformation. Instead, an elasticity solution must satisfy the following requirements.

1. *Compatibility of deformations:* The material must nowhere split apart or overlap itself. That is, it must have no displacement discontinuities.
2. *Equilibrium:* Every differential element of the body must be in static equilibrium.
3. *Boundary conditions:* At the boundary of the body, the solution must exhibit displacements that agree with support conditions, and stresses that agree with loads (for example, if a surface at right angles to the x axis is loaded by pressure p, the solution must yield $\sigma_x = -p$ on that surface).

Restrictions. Solutions discussed in this chapter are based on the assumptions that the material is homogeneous, isotropic, and linearly elastic, and that displacements (including rotations) are small. In thermal problems we assume that the coefficient of thermal expansion is independent of temperature. Nonlinear materials and large displacements are not considered here. They usually make a problem more complicated, and raise the possibility that a solution is not unique. That is, a nonlinear problem may have two or more solutions for a given set of loads and support conditions.

It is not always possible to obtain a correct stress field without explicit consideration of displacements. For example, in polar coordinates $r\theta$, an elasticity solution may contain ln r terms in the stress field. For the problem of a curved beam of rectangular cross section loaded by bending moment, and the problem of a thick cylinder loaded by internal pressure, ln r terms satisfy the equations of equilibrium, stress boundary conditions, and compatibility expressed in terms of stress. The ln r terms are appropriate for a curved beam. But ln r terms must be discarded from the pressurized cylinder solution because they lead to displacements that are not "single valued"; that is, ln r terms lead to displacements that are not the same at $\theta = 0$ and $\theta = 2\pi$. In summary, solutions based on stress equations alone are inadequate for some "multiply connected" bodies, which include plane bodies with one or more holes. Such solutions are not considered in this chapter.

One-Dimensional Example. The following problem illustrates in a simple way some arguments of elasticity theory that are also used in two- and three-dimensional problems. The bar shown in Fig. 7.1-1a is assumed to be in a state of uniaxial stress. This idealization is not exact because the fixed support at $x = 0$ would in reality involve three-dimensional effects. However, such effects are localized near $x = 0$.

In Fig. 7.1-1c, forces must sum to zero on the typical differential element shown. Load $q = q(x)$ has dimensions [force/length]. Therefore

$$-A\sigma_x + A(\sigma_x + d\sigma_x) + q\,dx = 0 \quad \text{or} \quad \frac{d\sigma_x}{dx} + \frac{q}{A} = 0 \quad (7.1\text{-}1)$$

The latter equation is a differential equation of equilibrium. It must be satisfied for all x between 0 and L. It can be integrated when q is a known function of x. There will be

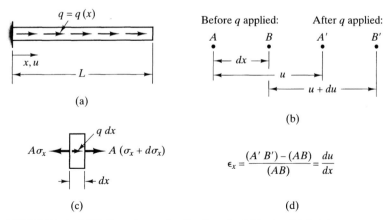

FIGURE 7.1-1 (a) Uniform bar under distributed axial load q. (b) Axial displacements. (c) Forces that act on a differential element, shown enlarged. (d) Axial strain-displacement relation.

one integration constant, which can be evaluated from the boundary condition $\sigma_x = 0$ at $x = L$. Thus $\sigma_x = \sigma_x(x)$ is known at any location along the bar.

Axial displacement at any location is $u = u(x)$. In Fig. 7.1-1b we see that typical points A and B both displace, and their separation changes an amount du. Following the usual definition of normal strain as change in length divided by original length, we obtain the strain-displacement relation $\epsilon_x = du/dx$ shown in Fig. 7.1-1d. Thus we see that a strain field can be obtained from a displacement field by differentiation. If we substitute $\epsilon_x = du/dx$ into the uniaxial stress-strain relation $\sigma_x = E\epsilon_x$, and substitute the result into Eq. 7.1-1, we obtain

$$E\frac{d^2u}{dx^2} + \frac{q}{A} = 0 \qquad (7.1\text{-}2)$$

Integration of Eq. 7.1-2 produces two integration constants, which can be evaluated from two boundary conditions: $u = 0$ at $x = 0$, and $du/dx = 0$ at $x = L$ (since $\sigma_x = 0$ at $x = L$ and $du/dx = \sigma_x/E$). Thus $u = u(x)$ is known, and $\sigma_x = \sigma_x(x)$ is then given by $\sigma_x = E(du/dx)$.

Load $q = q(x)$ in Fig. 7.1-1a can be used to describe either an axial load on the surface of the bar or an axial body force B_x, which is a force per unit volume. For axial body force loading, $q\,dx$ in Fig. 7.1-1c would be replaced by $B_x A\,dx$. Possible causes of B_x include x-direction acceleration and gravity acting parallel to the x axis.

2-D and 3-D Problems. Practical elasticity problems are two- or three-dimensional. Governing equations of a three-dimensional problem are direct extensions of governing equations in two dimensions. However, *solving* the governing equations usually becomes much more difficult with each increase in dimensionality. Accordingly, this chapter emphasizes two-dimensional problems.

Particular attention will be given to the plane stress problem. Thus, we will consider various geometries and in-plane loadings of a flat plate or sheet of material of uniform thickness. The plate, and loads applied to it, are parallel to the xy plane, and the plate does not bend out of its original plane. It is assumed that σ_x, σ_y, and τ_{xy} are the only nonzero stresses. This assumption violates three-dimensional compatibility conditions unless lateral strain ϵ_z is a linear function of x and y. In reality such linearity usually does not prevail [6.1]. However, the assumption is *almost* exact if the plate thickness is small in comparison with other dimensions. For example, although the assumption is not exact near a hole of diameter less than the plate thickness, the resulting error may be insignificant.

Stress-strain relations in Cartesian coordinates xy for an isotropic material in a state of plane stress, in forms explicit in strains and explicit in stresses, are

$$\epsilon_x = \frac{1}{E}(\sigma_x - \nu\sigma_y) + \alpha T \qquad \sigma_x = \frac{E}{1-\nu^2}(\epsilon_x + \nu\epsilon_y) - \frac{E\alpha T}{1-\nu}$$

$$\epsilon_y = \frac{1}{E}(\sigma_y - \nu\sigma_x) + \alpha T \qquad \sigma_y = \frac{E}{1-\nu^2}(\epsilon_y + \nu\epsilon_x) - \frac{E\alpha T}{1-\nu} \qquad (7.1\text{-}3)$$

$$\gamma_{xy} = \frac{1}{G}\tau_{xy} \qquad \tau_{xy} = G\gamma_{xy}$$

where E is the elastic modulus, ν is Poisson's ratio, α is the coefficient of thermal expansion, and $T = T(x,y)$ is the temperature relative to an initial temperature state in which thermal stress is zero. Equations 7.1-3 are the same as Eqs. 2.5-2, where ΔT is used to designate temperature change (the convention in elasticity is to use the symbol T instead). Recall also that $G = 0.5E/(1+\nu)$ for an isotropic material. In this chapter we assume that material properties are not temperature dependent. Thus, when there is a temperature gradient, derivatives of material properties with respect to the coordinates are zero (see also remarks that follow Eqs. 2.5-1).

7.2 DISPLACEMENTS, STRAINS, AND COMPATIBILITY

Two Dimensions. A displacement field in the xy plane consists of two components, $u = u(x,y)$ parallel to the x axis and $v = v(x,y)$ parallel to the y axis. This field simultaneously accounts for rigid body motion and deformation. For small deformations, strains ϵ_x, ϵ_y, and γ_{xy} are expressed in terms of u and v as follows.

In Fig. 7.2-1, imagine that points B, O, and A are marked on the plane body prior to loading, with BOA a right angle. Loading causes these points to move to B', O', and A'. Normal strain in the x direction is the x-direction motion of A relative to O, divided by the original separation OA. Thus

$$\epsilon_x = \frac{(O'A') - (OA)}{(OA)} = \frac{(\partial u/\partial x)dx}{dx} = \frac{\partial u}{\partial x} \quad (7.2\text{-}1)$$

Here $\partial u/\partial x$ is the rate of change of u with respect to x and $(\partial u/\partial x)dx$ is the small *amount* of change of u over distance dx; that is, $(\partial u/\partial x)dx$ is the elongation of an x-parallel line of original length dx. In similar fashion we obtain $\epsilon_y = \partial v/\partial y$. Shear strain is defined as the amount of change of a right angle such as BOA in Fig. 7.2-1. Angles α and β are very small, so that $\tan \alpha \approx \alpha$ and $\tan \beta \approx \beta$. Thus

$$\gamma_{xy} = \alpha + \beta = \frac{u + \dfrac{\partial u}{\partial y}dy - u}{dy} + \frac{v + \dfrac{\partial v}{\partial x}dx - v}{dx} = \frac{\partial u}{\partial y} + \frac{\partial v}{\partial x} \quad (7.2\text{-}2)$$

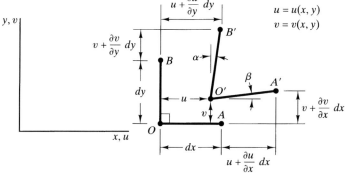

FIGURE 7.2-1 Displacement and deformation in the xy plane move points B, O, and A to locations B', O', and A'.

Collecting results, we have the *plane strain-displacement relations*, here expressed in rectangular Cartesian coordinates x and y:

$$\epsilon_x = \frac{\partial u}{\partial x} \qquad \epsilon_y = \frac{\partial v}{\partial y} \qquad \gamma_{xy} = \frac{\partial u}{\partial y} + \frac{\partial v}{\partial x} \qquad (7.2\text{-}3)$$

Three Dimensions. In rectangular Cartesian coordinates xyz, displacement components parallel to the respective axes are $u = u(x,y,z)$, $v = v(x,y,z)$, and $w = w(x,y,z)$. Strain-displacement relations consist of Eqs. 7.2-3 plus the following equations:

$$\epsilon_z = \frac{\partial w}{\partial z} \qquad \gamma_{yz} = \frac{\partial v}{\partial z} + \frac{\partial w}{\partial y} \qquad \gamma_{zx} = \frac{\partial w}{\partial x} + \frac{\partial u}{\partial z} \qquad (7.2\text{-}4)$$

Strain-displacement relations may be stated in various coordinate systems; for example, cylindrical coordinates are appropriate for a solid of revolution.

Compatibility. Compatibility exists if the displacement field is continuous and single-valued. Thus, no cracks appear and the material does not overlap itself. Compatibility is guaranteed by a displacement field that is single-valued, continuous, and has continuous derivatives. Because there are more strain components than displacement components, compatibility may not be satisfied by arbitrarily devised strain expressions, even if each strain expression is continuous and single-valued. Strains must also satisfy *compatibility equations* if they are to describe a physically possible deformation field.

In two dimensions, the two displacement components u and v define the three strains ϵ_x, ϵ_y, and γ_{xy}. Differentiating the first of Eqs. 7.2-3 twice with respect to y, the second twice with respect to x, and the third with respect to x and y, we observe that

$$\frac{\partial^2 \epsilon_x}{\partial y^2} + \frac{\partial^2 \epsilon_y}{\partial x^2} = \frac{\partial^2 \gamma_{xy}}{\partial x \partial y} \qquad (7.2\text{-}5)$$

This equation is called the *compatibility equation* for a plane problem. A strain field that does not satisfy this equation is not valid. In three dimensions, three displacement components define six strains, and there are five additional compatibility equations [6.1]. They are not written here because we will not use them in this book.

Thus far in this section we have dealt entirely with matters of geometry and have not invoked any kind of stress-strain relation.

Application of Concept. Figure 7.2-2 shows a threaded connection. Load P is transferred from bolt to sleeve and then to the wall on the right. In elementary mechanics of materials it would probably be assumed that each thread from A to B carries the same load. Is this result possible, in view of compatibility requirements?

If all threads are equally loaded, axial force is $F_b = P(s/L)$ in the bolt and $F_s = P(x/L)$ in the sleeve. These forces dictate the distributions of axial strain shown in Fig. 7.2-2. Thus, at A the bolt elongates but the sleeve does not; at B the sleeve elongates but the bolt does not. The differing elongations imply that threads, which fit together before loading, do not fit afterward. Also, somewhere between A and B, bolt and sleeve must have the same elongation; there the threads carry no load at all. These

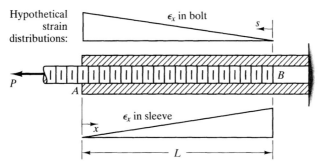

FIGURE 7.2-2 Threaded connection betweeen a bolt and a sleeve, seen in cross section. Strain distributions correspond to hypothetical (but impossible) uniform loading of all threads.

contradictions with the assumption of uniform thread load compel us to conclude that the assumption cannot be true.

In such a connection, the actual stress distribution involves complicated local patterns near A and B. The two or three threads closest to A and B (four to six threads altogether) carry most of the load. Similar considerations apply to a nut on a bolt, and show that stresses in threads would not be reduced much if the nut were made thicker so that it extended a considerable distance along the bolt, because most threads would then carry almost no load.

7.3 EQUILIBRIUM EQUATIONS AND BOUNDARY CONDITIONS

Equilibrium Equations. Consider first a two-dimensional state of stress. Stresses that act on a differential element are shown in Fig. 7.3-1a. A stress derivative such as $\partial \tau_{xy}/\partial y$ is the rate of change of τ_{xy} with respect to y, and $(\partial \tau_{xy}/\partial y)dy$ is the small *amount* of change in τ_{xy} over distance dy. Stresses σ_x, σ_y, and τ_{xy} may each be functions of x and y. Body forces B_x and B_y are forces per unit volume, which can be used to account for acceleration, self-weight, or a magnetic field.

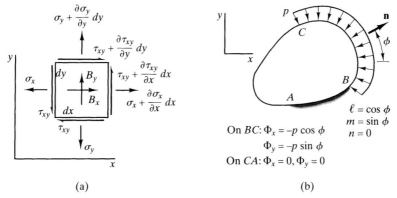

FIGURE 7.3-1 (a) Stresses on a differential element in a plane problem. (b) Boundary conditions on a plane body.

The thickness of a plane body does not matter if it is uniform. Accordingly, following common practice, we take the thickness as unity. Areas on which stresses act are therefore $(1)dx$ and $(1)dy$, and body forces act on volume $(1)dx\,dy$. For static equilibrium to prevail, forces acting on the differential element in Fig. 7.3-1a must sum to zero in both x and y directions. Thus

$$-\sigma_x dy - \tau_{xy} dx + \left(\sigma_x + \frac{\partial \sigma_x}{\partial x} dx\right) dy + \left(\tau_{xy} + \frac{\partial \tau_{xy}}{\partial y} dy\right) dx + B_x dx\,dy = 0 \tag{7.3-1a}$$

$$-\tau_{xy} dy - \sigma_y dx + \left(\tau_{xy} + \frac{\partial \tau_{xy}}{\partial x} dx\right) dy + \left(\sigma_y + \frac{\partial \sigma_y}{\partial y} dy\right) dx + B_y dx\,dy = 0 \tag{7.3-1b}$$

These equations reduce to the following, which are the *differential equations of equilibrium* for a plane body of uniform thickness.

$$x \text{ direction:} \quad \frac{\partial \sigma_x}{\partial x} + \frac{\partial \tau_{xy}}{\partial y} + B_x = 0 \tag{7.3-2a}$$

$$y \text{ direction:} \quad \frac{\partial \tau_{xy}}{\partial x} + \frac{\partial \sigma_y}{\partial y} + B_y = 0 \tag{7.3-2b}$$

These equations are also valid for a state of plane strain, which is a state in which ϵ_z, γ_{yz}, and γ_{zx} are all zero everywhere in the body.

In three dimensions, we consider a volume element $dx\,dy\,dz$, all six stresses and their variations with respect to x, y, and z, and body forces per unit volume in the x, y, and z directions. The differential equations of equilibrium are

$$x \text{ direction:} \quad \frac{\partial \sigma_x}{\partial x} + \frac{\partial \tau_{xy}}{\partial y} + \frac{\partial \tau_{zx}}{\partial z} + B_x = 0 \tag{7.3-3a}$$

$$y \text{ direction:} \quad \frac{\partial \tau_{xy}}{\partial x} + \frac{\partial \sigma_y}{\partial y} + \frac{\partial \tau_{yz}}{\partial z} + B_y = 0 \tag{7.3-3b}$$

$$z \text{ direction:} \quad \frac{\partial \tau_{zx}}{\partial x} + \frac{\partial \tau_{yz}}{\partial y} + \frac{\partial \sigma_z}{\partial z} + B_z = 0 \tag{7.3-3c}$$

For a state of plane stress, σ_z, τ_{yz}, τ_{zx}, and B_z are all zero, and Eqs. 7.3-3 reduce to Eqs. 7.3-2. Equilibrium equations can be written using other coordinate systems, and are independent of the stress-strain relation of the material.

A physically possible state of stress satisfies the differential equations of equilibrium at every point. Some approximate solutions, such as those based on displacement fields and the Rayleigh-Ritz method, satisfy equilibrium equations only in an average or integral sense.

Boundary Conditions. Prescribed conditions on the boundary of a body include displacement quantities and stress quantities. Typically, displacements are prescribed on part of the boundary to provide support, and stresses or "tractions" are prescribed on another part to provide loads. Equations for stress boundary conditions are

Section 7.3 Equilibrium Equations and Boundary Conditions

obtained from Eqs. 2.2-7: We define "surface tractions" as $\Phi_x = dR_x/dA$, $\Phi_y = dR_y/dA$, and $\Phi_z = dR_z/dA$, where Φ_x, Φ_y, and Φ_z are prescribed forces per unit of boundary area dA. Thus

x direction: $\quad \Phi_x = \sigma_x \ell + \tau_{xy} m + \tau_{zx} n$ \hfill (7.3-4a)

y direction: $\quad \Phi_y = \tau_{xy} \ell + \sigma_y m + \tau_{yz} n$ \hfill (7.3-4b)

z direction: $\quad \Phi_z = \tau_{zx} \ell + \tau_{yz} m + \sigma_z n$ \hfill (7.3-4c)

where ℓ, m, and n are direction cosines of an outward normal **n** on the boundary (see Fig. 2.2-2).

An example of boundary conditions for a plane stress problem appears in Fig. 7.3-1b. Along AB, displacements u and v must be zero because of the fixed support. Along BC and CA there are stress boundary conditions. From Eqs. 7.3-4,

$$\Phi_x = \sigma_x \cos \phi + \tau_{xy} \sin \phi \quad \text{and} \quad \Phi_y = \tau_{xy} \cos \phi + \sigma_y \sin \phi \quad (7.3\text{-}5)$$

in which Φ_x and Φ_y are obtained from Eqs. 2.2-7: For example, along BC, $dR = -p\, dA$ and $dR_x = \ell dR$, hence $\Phi_x = dR_x/dA = -p\ell = -p \cos \phi$. Note that along BCA Eqs. 7.3-5 yield expected results such as $\sigma_x = -p$ and $\tau_{xy} = 0$ at $\phi = 0$, $\sigma_y = -p$ and $\tau_{xy} = 0$ at $\phi = \pi/2$, and $\sigma_x = 0$ and $\tau_{xy} = 0$ at $\phi = \pi$. The elasticity solution of this example problem, if known, would display zero displacements along AB and have a stress field that satisfies Eqs. 7.3-5 along BC and CA.

In order to simplify a problem, a load is sometimes replaced by a statically equivalent load. Consider a cantilever beam problem. A concentrated tip force P, Fig. 7.3-2a, makes the problem difficult. A simple elasticity solution is available for the problem of Fig. 7.3-2b, where tip force P is applied as a quadratic distribution of surface traction $\Phi_y = -\tau_{xy}$ that is zero at $y = \pm c$ and maximum at $y = 0$. According to Saint-Venant's principle, stresses in the two problems are practically indistinguishable at a distance of $2c$ or more from the left end. Therefore one may regard the elasticity solution as applicable to either problem. This solution is discussed in Section 7.5. It confirms the results of elementary beam theory for a beam of unit thickness: $\sigma_x = 3Pxy/2c^3$, $\sigma_y = 0$, and $\tau_{xy} = 3P(c^2 - y^2)/4c^3$.

Figure 7.3-2b implies that the displacement boundary condition at the right end of the beam is $u = 0$ and $v = 0$ for $-c < y < c$. This condition makes the elasticity problem difficult. It is much easier if an initially vertical line at $x = L$ is allowed to warp into

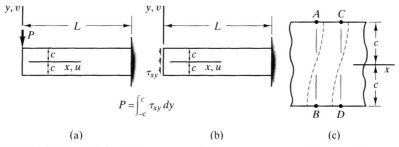

FIGURE 7.3-2 (a, b) Statically equivalent tip forces on a cantilever beam of unit thickness. (c) The portion of deformation associated with shear stress τ_{xy}.

the shapes shown for sections AB and CD in Fig. 7.3-2c. This deformation is associated with shear stress τ_{xy}. But now the "fixed end" support condition admits many interpretations. We might set $\partial v/\partial x = 0$ at $x = L$ and $y = 0$ to keep the beam axis horizontal at $x = L$, or set $\partial u/\partial y = 0$ at $x = L$ and $y = 0$ to prevent rotation of a vertical line at $x = L$, or set to zero the average x-direction displacement along the end at $x = L$, and so on. The various choices are associated with the same stress field but different rotations about end $x = L$. If $c \ll L$ so that transverse shear deformation becomes negligible, the various choices all yield lateral tip deflection $PL^3/3EI$, as expected from elementary beam theory.

Real supports are not rigid, and their elasticity contributes to the total deformation. This effect has been studied for a beam that terminates in a plane support of the same modulus [1.6, 6.1].

EXAMPLE

Investigate the displacement field

$$u = a_1(x^2 - y^2) - a_2 y + a_3 \quad \text{and} \quad v = 2a_1 xy + a_4 \qquad (7.3\text{-}6)$$

where the a_i are constants. Can it be the solution of a static plane stress problem if there are no body forces?

Applying Eqs. 7.2-3, we obtain the strains

$$\epsilon_x = 2a_1 x \qquad \epsilon_y = 2a_1 x \qquad \gamma_{xy} = -a_2 \qquad (7.3\text{-}7)$$

These strains satisfy the compatibility equation, Eq. 7.2-5. This result should be anticipated because Eqs. 7.3-6 are continuous and have continuous derivatives. Equations 7.1-3 and 7.3-7 yield the stresses

$$\sigma_x = \frac{2Ea_1 x}{1 - \nu} \qquad \sigma_y = \frac{2Ea_1 x}{1 - \nu} \qquad \tau_{xy} = -\frac{Ea_2}{2(1 + \nu)} \qquad (7.3\text{-}8)$$

When Eqs. 7.3-2 are applied to these stresses, with $B_x = 0$ and $B_y = 0$, we observe that the x-direction equilibrium equation does not become the identity $0 = 0$. Therefore Eqs. 7.3-6 are not a possible solution.

7.4 STRESS FIELD SOLUTION FOR PLANE STRESS PROBLEMS

Stresses can often be obtained without explicitly considering displacements, if one is willing to accept ambiguities in displacement boundary conditions of the sort described in connection with Fig. 7.3-2. In the present section we use rectangular Cartesian coordinates. Analogous equations in polar coordinates appear in Section 7.7.

As a convenience, body forces per unit volume are defined as derivatives of a "potential function" $V = V(x,y)$.

$$B_x = -\frac{\partial V}{\partial x} \quad \text{and} \quad B_y = -\frac{\partial V}{\partial y} \qquad (7.4\text{-}1)$$

This formulation is adequate for most problems. If body forces are to be ignored, one sets $V = 0$.

Section 7.4 Stress Field Solutions for Plane Stress Problems

Airy Stress Function. In 1863, G. B. Airy suggested that stresses in a plane problem be defined as derivatives of a "stress function" $F = F(x,y)$.

$$\sigma_x = \frac{\partial^2 F}{\partial y^2} + V \qquad \sigma_y = \frac{\partial^2 F}{\partial x^2} + V \qquad \tau_{xy} = -\frac{\partial^2 F}{\partial x \, \partial y} \qquad (7.4\text{-}2)$$

The merit of these definitions is that the equilibrium equations, Eqs. 7.3-2, are identically satisfied by Eqs. 7.4-1 and 7.4-2. If compatibility is also satisfied, F is the solution of a plane problem in elasticity. We express the compatibility condition in terms of F by substitution of Eqs. 7.4-2 into Eqs. 7.1-3 to obtain strains, and substitution of the strains into the compatibility equation, Eq. 7.2-5. Thus, F is required to satisfy the equation

$$\nabla^4 F + (1 - \nu) \nabla^2 V + E\alpha \nabla^2 T = 0 \qquad (7.4\text{-}3)$$

where ∇^2 is called the *harmonic* differential operator and $\nabla^4 = \nabla^2(\nabla^2)$ is called the *biharmonic* differential operator. For example, for the function $F = F(x,y)$,

$$\nabla^2 F = \frac{\partial^2 F}{\partial x^2} + \frac{\partial^2 F}{\partial y^2} \qquad \text{and} \qquad \nabla^4 F = \frac{\partial^4 F}{\partial x^4} + 2\frac{\partial^4 F}{\partial x^2 \partial y^2} + \frac{\partial^4 F}{\partial y^4} \qquad (7.4\text{-}4)$$

In Eq. 7.4-3, if α were temperature-dependent and T were a function of the coordinates, the term $E\alpha\nabla^2 T$ would become $E\nabla^2(\alpha T)$ (see also remarks that follow Eqs. 2.5-1.) If there is no body force and no temperature change, Eq. 7.4-3 reduces to

$$\nabla^4 F = 0 \qquad (7.4\text{-}5)$$

Equation 7.4-5 is valid for plane strain problems as well as plane stress problems. A function F that satisfies Eq. 7.4-5 is called *biharmonic*. A biharmonic F solves a plane elasticity problem because a biharmonic F satisfies equilibrium and compatibility conditions. The particular problem solved depends on boundary conditions. Examples appear in Section 7.5.

EXAMPLE

A rectangular block stands on a rigid horizontal support and is loaded by its own weight (Fig. 7.4-1a). Let ρ be the mass density and g the acceleration of gravity. As a proposed solution of this problem, investigate the equations

$$F = -\rho g \left(\frac{y^3}{6} + \frac{Lx^2}{2} \right) \qquad \text{and} \qquad V = \rho g y \qquad (7.4\text{-}6)$$

Equations 7.4-2 and 7.4-6 yield

$$\sigma_x = 0 \qquad \sigma_y = \rho g(y - L) \qquad \tau_{xy} = 0 \qquad (7.4\text{-}7)$$

These stresses satisfy free-surface boundary conditions on the top edge and the vertical edges. On the bottom edge, $\sigma_y = -\rho g L$. This stress could also be obtained from statics by dividing the weight of the block by the area of its base. For these stress boundary conditions, Eqs. 7.4-7 are an exact solution. It remains to consider whether the associated displacements agree with the support condition shown in Fig. 7.4-1a.

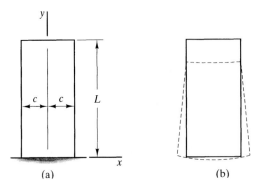

FIGURE 7.4-1 (a) Rectangular block loaded by its own weight. (b) Deformation of the block is shown by dashed lines.

Displacements $u = u(x,y)$ and $v = v(x,y)$ can be obtained by substituting Eqs. 7.4-7 into Eqs. 7.1-3 to obtain strains, substituting strains into Eqs. 7.2-3, and integrating the partial differential equations (see Section 7.6). Without undertaking this calculation we can see that there is trouble with the support condition at $y = 0$. Equations 7.4-7 require that the block have the deformed shape shown in Fig. 7.4-1b: The block shortens vertically, and becomes wider toward the base due to the Poisson effect. Right angles are preserved because shear strain is zero. But a curved shape at the base is incompatible with a rigid horizontal support. Also, if the block is bonded to the support, sideways expansion is prevented at the base. Despite these inconsistencies, according to Saint-Venant's principle Eqs. 7.4-7 should be quite accurate for $y > 2c$.

Solution Methods. An elasticity solution to a given problem is obtained if one devises a stress field that satisfies equilibrium, compatibility, and the boundary conditions of the given problem. For most problems this task is so difficult that no one can do it. Elasticity solutions are available only for special combinations of geometry, boundary conditions, and loading. Nevertheless, many problems of practical interest have been solved.

Several solutions have been obtained via the Airy stress function. Any function F that has continuous derivatives through fourth order and satisfies Eq. 7.4-3 or Eq. 7.4-5 is the solution of some elasticity problem. This observation suggests the *inverse method*, in which we choose a biharmonic function F, then investigate what problem we have solved. The investigation includes inspecting stresses on the boundary of a chosen region. Unfortunate choices of location, shape, and orientation of the region to inspect may make a useful solution look useless. By trial and combination of simple biharmonic functions it is possible to obtain solutions of practical interest. The inverse method is used in Sections 7.5 and 7.9. There is also a *semi-inverse method*, in which the form of the solution is partially defined or assumed at the outset, and the remainder contrived so that the complete solution satisfies all necessary conditions (see Sections 7.8 and 7.11).

Without using a stress function, it is sometimes possible to obtain a stress field that satisfies all necessary conditions. The following equation can be obtained by substituting Eqs. 7.1-3 into Eqs. 7.2-5, then substituting for the shear stress terms from Eqs. 7.3-2. Thus

$$\nabla^2(\sigma_x + \sigma_y) + (1 + \nu)\left(\frac{\partial B_x}{\partial x} + \frac{\partial B_y}{\partial y}\right) + E\alpha\nabla^2 T = 0 \qquad (7.4\text{-}8)$$

This equation is known as the *compatibility equation in terms of stress.* Actually, it combines compatibility, equilibrium, and stress-strain relations. Normal stresses that satisfy Eq. 7.4-8 are a valid solution of some problem (however, note restrictions listed in Section 7.1).

It is also possible to start with a displacement field instead of a stress field or a stress function. This procedure is often the basis of an approximate method. In particular, it is the basis of many finite element procedures, although this fact is not apparent to the user of commercial software.

7.5 POLYNOMIAL SOLUTIONS IN CARTESIAN COORDINATES

In the present section we ignore body forces and thermal stresses. Therefore Eq. 7.4-5 is the governing equation. Also, in this section we use the inverse method; that is, we examine candidate biharmonic functions F and ask what problem is solved.

Quadratic and Cubic Stress Functions. Quadratic terms are the lowest-order terms that yield nonzero stresses from an Airy stress function. Consider

$$F = a_1 x^2 + a_2 xy + a_3 y^2 \qquad (7.5\text{-}1)$$

where the a_i are constants. Equations 7.4-2 and 7.5-1 yield the stresses

$$\sigma_x = 2a_3 \qquad \sigma_y = 2a_1 \qquad \tau_{xy} = -a_2 \qquad (7.5\text{-}2)$$

Taken separately, these are states of x-direction normal stress, y-direction normal stress, and pure shear, each uniform over the entire plane.

Consider next the following cubic function, where a_1 is a constant.

$$F = a_1 y^3 \quad \text{yields} \quad \sigma_x = 6a_1 y \quad \text{and} \quad \sigma_y = \tau_{xy} = 0 \qquad (7.5\text{-}3)$$

We can display this solution by choosing a region of the xy plane and showing the stress boundary conditions, which are determined from Eqs. 7.3-5 and the stress field evaluated at the boundary. Whether or not the solution appears useful depends on the region chosen. If we choose the region shown in Fig. 7.5-1a, we recognize a state of pure bending; if we choose the region in Fig. 7.5-1b, we recognize a state of bending plus axial load. The region chosen in Fig. 7.5-1c makes the solution appear to have no practical interest. In parts (a) and (b) of Fig. 7.5-1, edges $y = \text{constant}$ are free of stress,

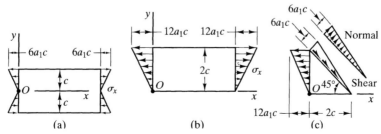

FIGURE 7.5-1 Stress field associated with $F = a_1 y^3$, illustrated as stress boundary conditions on three different regions in the xy plane.

and stresses are independent of the length of the beam. These results are in complete agreement with stresses calculated by elementary beam theory.

A Quartic Stress Function. With a_4 a constant, consider the biharmonic function

$$F = a_4 x y^3 \quad \text{which yields} \tag{7.5-4a}$$

$$\sigma_x = 6a_4 xy \qquad \sigma_y = 0 \qquad \tau_{xy} = -3a_4 y^2 \tag{7.5-4b}$$

Figure 7.5-2a shows stress boundary conditions associated with this solution on the region $0 \leq x \leq L$, $-c \leq y \leq c$. The solution appears to lack practical interest. But now let us remove shear stress on $y = \pm c$ by superposing on Eq. 7.5-4a the stress function $F = -3a_4 c^2 xy$, which is obtained from Eq. 7.5-1 by setting $a_1 = 0$, $a_3 = 0$, and $a_2 = -3a_4 c^2$. Thus we obtain

$$F = a_4(xy^3 - 3c^2 xy) \quad \text{which yields} \tag{7.5-5a}$$

$$\sigma_x = 6a_4 xy \qquad \sigma_y = 0 \qquad \tau_{xy} = 3a_4(c^2 - y^2) \tag{7.5-5b}$$

This stress distribution corresponds to a cantilever beam problem, Fig. 7.5-2b. The elasticity solution is exact—provided that the load at $x = 0$ and the support at $x = L$ both apply the parabolic distributions of τ_{xy} shown, and provided that the support at $x = L$ applies the linear distribution of σ_x shown. All these conditions are not likely to happen in a real beam, so the elasticity solution must be regarded as approximate near the ends, but according to Saint-Venant's principle the solution is essentially exact for $2c < x < (L - 2c)$. The elasticity solution agrees completely with stresses of elementary beam theory. For example, if we calculate the transverse tip load from τ_{xy} in Eq. 7.5-5b on a beam of unit thickness and then apply the flexure formula, we obtain

$$P = \int_{-c}^{c} \tau_{xy} \, dy = 4a_4 c^3 \quad \text{hence} \quad \sigma_x = \frac{My}{I} = \frac{Pxy}{I} = 6a_4 xy \tag{7.5-6}$$

Also, one can easily show that the beam formula $\tau = VQ/It$ with $V = P$ yields a shear stress in agreement with τ_{xy} of Eq. 7.5-5b.

A Quintic Stress Function. We consider an Airy stress function that solves the problem of a simply supported and uniformly loaded beam, Fig. 7.5-3a. Again, support con-

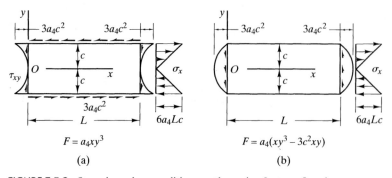

FIGURE 7.5-2 Stress boundary conditions on the region $0 \leq x \leq L$ and $-c \leq y \leq c$, associated with the stress functions of Eqs. 7.5-4 and 7.5-5.

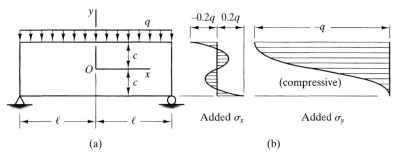

FIGURE 7.5-3 (a) Simply supported beam of unit thickness with uniformly distributed load q. (b) "Added" stresses that appear in the elasticity solution but not in beam theory.

ditions shown are at odds with boundary tractions of the elasticity solution on edges $x = \pm \ell$, and Saint-Venant's principle must be invoked. Load q has dimensions of pressure, or dimensions [force/length] for a beam of unit thickness. The Airy stress function F can be devised by clever use of superposition, but we will simply present it.

$$F = -\frac{q}{40c^3}(y^5 - 5x^2y^3 + 15c^2x^2y + 5\ell^2 y^3 - 2c^2 y^3 + 10c^3 x^2) \quad (7.5\text{-}7)$$

It is easy to show that $\nabla^4 F = 0$. Equations 7.4-2 and 7.5-7 yield the following stresses. In the format shown, the initial group of terms agrees with the formulas $\sigma = My/I$ and $\tau = VQ/It$ of elementary beam theory, while the second group of terms (in brackets) appears only in the elasticity solution.

$$\sigma_x = \frac{3qy}{4c^3}(x^2 - \ell^2) + \left[\frac{qy}{c^3}(0.3c^2 - 0.5y^2)\right] \quad (7.5\text{-}8a)$$

$$\sigma_y = 0 + \left[\frac{q}{4c^3}(y^3 - 3yc^2 - 2c^3)\right] \quad (7.5\text{-}8b)$$

$$\tau_{xy} = \frac{3qx}{4c^3}(c^2 - y^2) \quad (7.5\text{-}8c)$$

The bracketed terms are shown in Fig. 7.5-3b as "added" stresses. These stresses are independent of x and ℓ. They are negligible for a slender beam. Even with $\ell/c = 5$, for which the beam is scarcely slender, the cubic σ_x distribution displays a maximum value that is little more than 1% of the maximum flexural stress Mc/I at the middle of the beam. Except for stresses near ends of the beam, we conclude that the elasticity solution does more to confirm the validity of elementary beam theory than to challenge it.

7.6 DISPLACEMENTS CALCULATED FROM STRESSES

If stresses are known and displacements are required, they can be calculated by integration of the strain-displacement relations, Eqs. 7.2-3. For a simple illustration of the process, consider a beam of unit thickness in pure bending, Fig. 7.6-1. The stress field

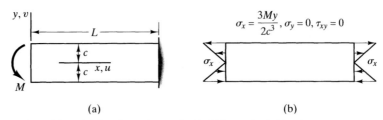

FIGURE 7.6-1 (a) Cantilever beam of unit thickness loaded by tip moment. (b) Stress field in the beam.

shown satisfies the requirements of equilibrium, compatibility, and boundary conditions on stress. From Eqs. 7.1-3 and 7.2-3,

$$\frac{\partial u}{\partial x} = \frac{3My}{2Ec^3} \quad \text{hence} \quad u = \frac{3Mxy}{2Ec^3} + f_y \tag{7.6-1}$$

where f_y is a function of y that comes from integration (at this stage we cannot say that f_y is simply a constant because we are dealing with partial derivatives). Similarly, again from Eqs. 7.1-3 and 7.2-3,

$$\frac{\partial v}{\partial y} = -\frac{3\nu My}{2Ec^3} \quad \text{hence} \quad v = -\frac{3\nu My^2}{4Ec^3} + f_x \tag{7.6-2}$$

where f_x is a function of x that comes from integration. The last strain-displacement relation involves γ_{xy}, which is zero. Substituting Eqs. 7.6-1 and 7.6-2 into Eq. 7.2-3 and grouping terms, we obtain

$$\left(\frac{3Mx}{2Ec^3} + \frac{df_x}{dx}\right) + \left(\frac{df_y}{dy}\right) = 0 \tag{7.6-3}$$

Expressions in parentheses are respectively functions of x alone and of y alone. The equation must be true for arbitrary x and y in the beam, from which we conclude that each expression in parentheses must be a constant. Otherwise, we could vary x alone (or y alone) and violate the equality. The most general expressions for which Eq. 7.6-3 is valid are

$$\frac{3Mx}{2Ec^3} + \frac{df_x}{dx} = a_1 \quad \text{and} \quad \frac{df_y}{dy} = -a_1 \tag{7.6-4}$$

where a_1 is a constant. These are ordinary differential equations. Integrations yield f_x and f_y. Hence

$$u = \frac{3Mxy}{2Ec^3} - a_1 y + a_2 \tag{7.6-5a}$$

$$v = -\frac{3\nu My^2}{4Ec^3} - \frac{3Mx^2}{4Ec^3} + a_1 x + a_3 \tag{7.6-5b}$$

where a_2 and a_3 are constants of integration. Constants a_1, a_2, and a_3 are associated with rigid body displacement of the beam. They have no effect on the stress field. To determine them we elect the following boundary conditions as appropriate to Fig. 7.6-1a: at $x = L$ and $y = 0$, both u and v are zero; at $x = L$, $\partial u/\partial y = 0$. The latter condition prevents rotation of a vertical line at $x = L$ (the same result is produced by $\partial v/\partial x = 0$ at $x = L$, since $\gamma_{xy} = 0$ in this problem). Thus

$$a_1 = \frac{3ML}{2Ec^3} \qquad a_2 = 0 \qquad a_3 = -\frac{3ML^2}{4Ec^3} \qquad (7.6\text{-}6)$$

and displacements $u = u(x,y)$ and $v = v(x,y)$ are completely determined. For example, $v = v(x,y)$ is

$$v = -\frac{3M}{4Ec^3}[(L - x)^2 + \nu y^2] \qquad (7.6\text{-}7)$$

From this equation we obtain $v = -3ML^2/4Ec^3$ at $x = y = 0$, which is complete agreement with the elementary beam formula $v = -ML^2/2EI$ for a beam of unit thickness. The Poisson ratio term in Eq. 7.6-7 shows that all points for which $y \ne 0$ in the undeformed beam move downward with respect to the beam axis. This displacement violates the fixed-end condition depicted in Fig. 7.6-1a, but according to Saint-Venant's principle the discrepancy may be ignored a short distance from the right end.

7.7 PLANE STRESS PROBLEMS IN POLAR COORDINATES

Elasticity solutions in rectangular Cartesian coordinates often closely resemble or even duplicate solutions from mechanics of materials. In polar coordinates, elasticity solutions appear that are unavailable from mechanics of materials.

There is no change in concepts previously stated, but some equations that describe the concepts have different forms. Stress-strain relations, Eqs. 7.1-3, require only mutually perpendicular directions, which now are r (radial) and θ (circumferential). Thus

$$\epsilon_r = \frac{1}{E}(\sigma_r - \nu\sigma_\theta) + \alpha T \qquad \epsilon_\theta = \frac{1}{E}(\sigma_\theta - \nu\sigma_r) + \alpha T \qquad \gamma_{r\theta} = \frac{\tau_{r\theta}}{G} \qquad (7.7\text{-}1)$$

Equilibrium Equations. Equations 7.3-2 can be transformed to polar coordinates by mathematical operations, but it is more instructive to derive them afresh. Again we consider a body of unit thickness, and essentially repeat what is done in Section 7.3, but now in polar coordinates. B_r and B_θ are body forces per unit volume. In Fig. 7.7-1, stresses σ_r, σ_θ, and $\tau_{r\theta}$ have rates of change with respect to r and θ, and therefore have different values on different edges as shown. Inner and outer edges have different lengths, $d\ell$ and $d\ell_a$, which would cause even constant σ_r (or constant $\tau_{r\theta}$) to exert different forces on these edges. The sum of forces parallel to a radial line through the center of the element, beginning with the edge of length $d\ell$ and moving counterclockwise, is

$$-\sigma_r\,d\ell - \tau_{r\theta}\,dr - \sigma_\theta\,dr\sin\frac{d\theta}{2} + \sigma_a d\ell_a + \tau_b\,dr$$
$$- \sigma_b\,dr\sin\frac{d\theta}{2} + B_r r\,dr\,d\theta = 0 \qquad (7.7\text{-}2)$$

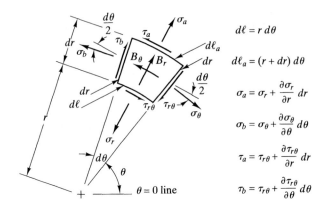

FIGURE 7.7-1 Stresses on a differential element in a plane problem, expressed in polar coordinates.

Terms that involve $\sin(d\theta/2)$ are present because forces $\sigma_\theta\, dr$ and $\sigma_b\, dr$ each have an inward radial component proportional to $\sin(d\theta/2)$. Similarly, the sum of θ-direction forces is

$$-\tau_{r\theta} d\ell - \sigma_\theta dr + \tau_{r\theta} dr \sin\frac{d\theta}{2} + \tau_a d\ell_a + \sigma_b dr$$

$$+ \tau_b dr \sin\frac{d\theta}{2} + B_\theta r\, dr\, d\theta = 0 \tag{7.7-3}$$

Terms that involve $\sin(d\theta/2)$ are present because forces $\tau_{r\theta}\, dr$ and $\tau_b dr$ each have a θ-direction component proportional to $\sin(d\theta/2)$. We simplify Eqs. 7.7-2 and 7.7-3 by gathering terms, setting $\sin(d\theta/2) = d\theta/2$, discarding as negligible those terms that contain the product of three differentials, and dividing by $r\, dr\, d\theta$. Thus we obtain the two equilibrium equations

$$r \text{ direction:} \quad \frac{\partial \sigma_r}{\partial r} + \frac{1}{r}\frac{\partial \tau_{r\theta}}{\partial \theta} + \frac{\sigma_r - \sigma_\theta}{r} + B_r = 0 \tag{7.7-4a}$$

$$\theta \text{ direction:} \quad \frac{1}{r}\frac{\partial \sigma_\theta}{\partial \theta} + \frac{\partial \tau_{r\theta}}{\partial r} + 2\frac{\tau_{r\theta}}{r} + B_\theta = 0 \tag{7.7-4b}$$

These two equations are the polar-coordinate forms of Eqs. 7.3-2.

Body forces B_r and B_θ are in general functions of r and θ. A practical application of body force loading is a disk that spins at constant angular speed ω, for which $B_r = \rho\omega^2 r$ and $B_\theta = 0$, where ρ is the mass density. This application is considered in Chapter 8.

Airy Stress Function. Let body forces be derivatives of a potential function V, specifically $B_r = -\partial V/\partial r$ and $B_\theta = -(\partial V/\partial \theta)/r$, where $V = V(r,\theta)$. Some tedious mathematics transforms Eqs. 7.4-2 to polar coordinates, with the result that stresses in terms of the Airy stress function $F = F(r,\theta)$ are

$$\sigma_r = \frac{1}{r}\frac{\partial F}{\partial r} + \frac{1}{r^2}\frac{\partial^2 F}{\partial \theta^2} + V \qquad \sigma_\theta = \frac{\partial^2 F}{\partial r^2} + V \qquad \tau_{r\theta} = -\frac{\partial}{\partial r}\left(\frac{1}{r}\frac{\partial F}{\partial \theta}\right) \qquad (7.7\text{-}5)$$

Similar transformation of ∇^2 in Eq. 7.4-4 yields the harmonic operator in polar coordinates.

$$\nabla^2 = \frac{\partial^2}{\partial r^2} + \frac{1}{r}\frac{\partial}{\partial r} + \frac{1}{r^2}\frac{\partial^2}{\partial \theta^2} \quad \text{or} \quad \nabla^2 = \frac{1}{r}\frac{\partial}{\partial r}\left(r\frac{\partial}{\partial r}\right) + \frac{1}{r^2}\frac{\partial^2}{\partial \theta^2} \qquad (7.7\text{-}6)$$

With ∇^2 thus defined, and $\nabla^4 = \nabla^2(\nabla^2)$, Eqs. 7.4-3 and 7.4-5 are again governing equations for plane stress problems. Solution methods described in Section 7.4 remain applicable.

The polar-coordinate form of the compatibility equation in terms of stress, Eq. 7.4-8, is

$$\nabla^2(\sigma_r + \sigma_\theta) + (1 + \nu)\left(\frac{\partial B_r}{\partial r} + \frac{B_r}{r} + \frac{1}{r}\frac{\partial B_\theta}{\partial \theta}\right) + E\alpha\nabla^2 T = 0 \qquad (7.7\text{-}7)$$

Plane strain-displacement relations in polar coordinates are occasionally useful in our discussion. They are as follows [6.1].

$$\epsilon_r = \frac{\partial u}{\partial r} \qquad \epsilon_\theta = \frac{u}{r} + \frac{1}{r}\frac{\partial v}{\partial \theta} \qquad \gamma_{r\theta} = \frac{1}{r}\frac{\partial u}{\partial \theta} + \frac{\partial v}{\partial r} - \frac{v}{r} \qquad (7.7\text{-}8)$$

where $u = u(r,\theta)$ and $v = v(r,\theta)$ are respectively the radial and circumferential components of displacement.

7.8 CIRCULAR HOLE

We consider a plate in uniaxial tension that contains a central hole, Fig. 7.8-1a. The following solution is exact for a plate of infinite width $2w$. When the solution is used in practice, the hole must be small; $b \ll w$ and $b \ll \ell$. However, lest the plane stress assumption become questionable, b should not be smaller than the plate thickness. The solution was published by Kirsch in 1898. We explain the solution procedure as an example of the semi-inverse method. Emphasis is given to concepts of the process rather than to details of expressions and the tedious manipulations that are unfortunately required.

Imagine first that the hole is not there. Then the stress field in Cartesian coordinates is uniaxial stress $\sigma_x = \sigma_o$ at all locations. The same stress field in polar coordinates, Fig. 7.8-1b, is obtained by use of stress transformation equations (Fig. 1.10-1) or Mohr's circle calculations. The associated Airy stress function is the last term in Eq. 7.5-1. In Cartesian coordinates and in polar coordinates, it is

$$F_1 = \frac{\sigma_o}{2} y^2 \quad \text{or} \quad F_1 = \frac{\sigma_o}{2} r^2 \sin^2\theta = \frac{\sigma_o}{4} r^2 (1 - \cos 2\theta) \qquad (7.8\text{-}1)$$

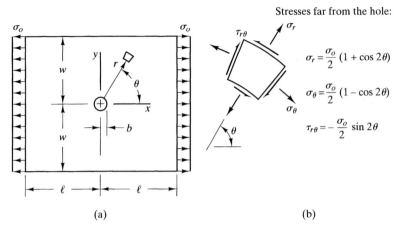

FIGURE 7.8-1 (a) Circular hole in a plate under tension. (b) "Far field" stresses (for $r \gg b$), written in polar coordinates.

This function is called F_1 because another function F_2 is also needed to account for the hole. The complete stress function for an infinite plate with a hole is

$$F = F_1 + F_2 \quad \text{where} \quad F_2 = f + g \cos 2\theta \qquad (7.8\text{-}2)$$

where $f = f(r)$ and $g = g(r)$ are independent of θ. We have *assumed* that F_2 has the form shown. We will show that f and g exist such that $\nabla^4 F = 0$ and all stress boundary conditions are satisfied. Thus we show that the assumption is correct and the solution is valid. Now $\nabla^4 F_1$ is already known to be zero, so $\nabla^4 F = 0$ yields $\nabla^4 F_2 = 0$. The biharmonic operator $\nabla^4 = \nabla^2(\nabla^2)$ can be written by applying the harmonic operator ∇^2 in Eq. 7.7-6. In applying the operator ∇^4, we note that f and g have zero derivatives with respect to θ and ordinary (not partial) derivatives with respect to r. Also, $\nabla^4 F_2 = 0$ produces an equation of the form $(\text{---}) + (\cdots)\cos 2\theta = 0$, where, if written out, expressions denoted here by (---) and $(\cdots)$ would be seen to be functions of r. The equation $(\text{---}) + (\cdots)\cos 2\theta = 0$ must be true for all values of θ; thus we obtain the *two* equations $(\text{---}) = 0$ and $(\cdots) = 0$. These two equations can be written in the following formats.

$$\frac{1}{r}\frac{d}{dr}\left\{r\frac{d}{dr}\left[\frac{1}{r}\frac{d}{dr}\left(r\frac{df}{dr}\right)\right]\right\} = 0 \qquad (7.8\text{-}3)$$

$$r\frac{d}{dr}\left(\frac{1}{r^3}\frac{d}{dr}\left\{r^3\frac{d}{dr}\left[\frac{1}{r^3}\frac{d}{dr}(r^2 g)\right]\right\}\right) = 0 \qquad (7.8\text{-}4)$$

Equation 7.8-3 comes directly from the second form of Eq. 7.7-6, independent of θ. Equation 7.8-4 can be devised after $\nabla^4 F_2 = 0$ is written out. Equations 7.8-3 and 7.8-4 are easy to integrate, working from the outside in, like peeling an onion of four layers. Equation 7.8-3 yields

$$f = a_1 + a_2 \ln r + a_3 r^2 + a_4 r^2 \ln r \qquad (7.8\text{-}5)$$

where the a_i are constants of integration. Integration of Eq. 7.8-4 yields an expression for g with four additional constants of integration. (Special formats for the differential equations are not essential. For example, in any format Eq. 7.8-4 can be solved by substituting the assumed solution $g = r^p$. One obtains a fourth order polynomial whose roots are $p = 4, 2, 0,$ and -2.)

Next we return to the stress function $F = F_1 + F_2$ and apply Eqs. 7.7-5 to obtain stresses. In doing so constant a_1 disappears, leaving a total of seven integration constants, to be determined by boundary conditions on stress. We know that $\sigma_r = 0$ and $\tau_{r\theta} = 0$ on the boundary of the hole, $r = b$. Also, we must obtain the stresses shown in Fig. 7.8-1b for $r \gg b$. Finally, stresses must remain finite as r increases without limit (this condition demands that a_4 and a constant in the expression for g both vanish). Simultaneous solution of the boundary condition equations yields the integration constants in terms of σ_o and b. Final expressions for stress in a plate of infinite width with a circular hole are

$$\sigma_r = \frac{\sigma_o}{2}\left(1 - \frac{b^2}{r^2}\right) + \frac{\sigma_o}{2}\left(1 - 4\frac{b^2}{r^2} + 3\frac{b^4}{r^4}\right)\cos 2\theta \qquad (7.8\text{-}6a)$$

$$\sigma_\theta = \frac{\sigma_o}{2}\left(1 + \frac{b^2}{r^2}\right) - \frac{\sigma_o}{2}\left(1 + 3\frac{b^4}{r^4}\right)\cos 2\theta \qquad (7.8\text{-}6b)$$

$$\tau_{r\theta} = -\frac{\sigma_o}{2}\left(1 + 2\frac{b^2}{r^2} - 3\frac{b^4}{r^4}\right)\sin 2\theta \qquad (7.8\text{-}6c)$$

The peak normal stress is $\sigma_\theta = 3\sigma_o$, on the edge of the hole at $\theta = \pm\pi/2$. For a plate of *finite* width we must use a stress concentration factor K_t instead. For example, if $w = 4b$, Eq. 2.7-2 yields $K_t = 2.42$, and $\sigma_{\text{nom}} = 4\sigma_o/3$, so that the peak normal stress is $K_t \sigma_{\text{nom}} = 3.23\sigma_o$, at $\theta = \pm\pi/2$.

Plots of σ_r and σ_θ along transverse and longitudinal centerlines of the plate for $b \ll w$ are shown in Fig. 7.8-2. The fact that stresses approach nominal values at small distances from the hole is confirmation of Saint-Venant's principle.

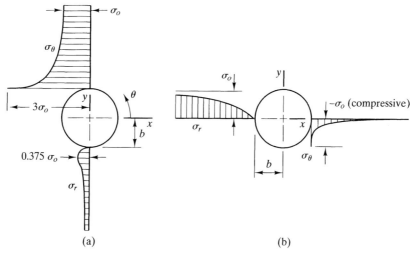

FIGURE 7.8-2 Normal stresses on transverse and longitudinal centerlines of the plate in Fig. 7.8-1a, for $w \gg b$.

210   Chapter 7   Elements of Theory of Elasticity

Solutions for a hole in a plate under other states of plane stress can be obtained by superposition. For example, if the foregoing solution is combined with another solution in which everything is rotated 90°, we obtain results for "far field" tension σ_o in all directions (see Fig. 2.7-2c). Or, if the direction of σ_o is reversed in one of these two solutions, we obtain results applicable to a far-field state of pure shear.

7.9 CONCENTRATED FORCE LOADING

In plane elasticity, a "concentrated force" P is regarded as uniformly distributed along a thickness-direction line rather than acting at a mathematical point. In the present section, P is the force on a plane body of unit thickness. For a plane body of some other thickness, P must be taken as the actual force divided by the actual thickness.

The problem of loading by a concentrated force is not an elastic contact problem. In an elastic contact problem, elastic bodies are pressed together, the contact area grows as load increases, and stresses in the contact zone are not directly proportional to load. In the concentrated force problem, force is always concentrated and stresses are directly proportional to load. Practically, force is applied over some area, however small, and local yielding may occur. The concentrated-force idealization is satisfactory unless stresses quite near the force are required.

Force on a Straight Edge. In Fig. 7.9-1a, the plane body is "semi-infinite," meaning that it extends to infinity below the straight edge to which force P is applied. Flamant published the solution to this problem in 1892. The appropriate Airy stress function is $F = a_1 r\theta \sin\theta$, where a_1 is a constant. From Eqs. 7.7-5, the resulting stresses are

$$\sigma_r = 2a_1 \frac{\cos\theta}{r} \qquad \sigma_\theta = 0 \qquad \tau_{r\theta} = 0 \tag{7.9-1}$$

This is a field of normal stress, radial from point O, as shown in Fig. 7.9-1b. Figure 7.9-1b shows the actual direction of σ_r; as usual, tensile stress is considered positive in

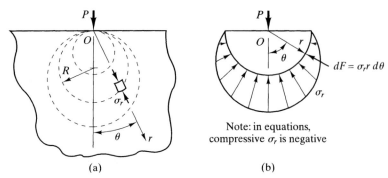

FIGURE 7.9-1 (a) Concentrated force P applied normal to the boundary of a semi-infinite plane of unit thickness. (b) The half-disc shown is loaded only by force P and radial stress σ_r.

equations. Constant a_1 is determined by stating that vertical forces sum to zero in Fig. 7.9-1b.

$$-\int_{-\pi/2}^{\pi/2} (\cos\theta)\,\sigma_r r\,d\theta - P = 0 \quad \text{yields} \quad a_1 = -\frac{P}{\pi} \quad (7.9\text{-}2)$$

Thus Eqs. 7.9-1 become, for a body of unit thickness,

$$\sigma_r = -\frac{2P\cos\theta}{\pi r} \qquad \sigma_\theta = 0 \qquad \tau_{r\theta} = 0 \quad (7.9\text{-}3)$$

This remarkably simple stress field satisfies all requirements of an elasticity solution and is therefore exact (within limits of the assumptions made). On a circle of radius R, shown dashed in Fig. 7.9-1a, $2R\cos\theta = r$; therefore, at all points on the dashed circle, normal stress directed towards point O is $\sigma_r = -P/\pi R$. In experimental stress analysis by photoelasticity, isochromatic circles are clearly visible where dashed circles appear in Fig. 7.9-1a.

Tip Force on a Wedge. Figure 7.9-2a shows a generalization of the preceding problem. Again the solution is given by Eqs. 7.9-1, and again constant a_1 is determined by stating that vertical forces sum to zero. Now the integral in Eq. 7.9-2 must have limits $-\alpha$ to α. Solving for a_1 and substituting into Eqs. 7.9-1, we obtain

$$\sigma_r = -\frac{2P\cos\theta}{(2\alpha + \sin 2\alpha)r} \qquad \sigma_\theta = 0 \qquad \tau_{r\theta} = 0 \quad (7.9\text{-}4)$$

For *transverse* tip force on a wedge, Fig. 7.9-2b, the solution is obtained by integrating from $\theta = \pi/2 - \alpha$ to $\theta = \pi/2 + \alpha$. Thus

$$\sigma_r = -\frac{2P\cos\theta}{(2\alpha - \sin 2\alpha)r} \qquad \sigma_\theta = 0 \qquad \tau_{r\theta} = 0 \quad (7.9\text{-}5)$$

If α is small, so that taper is slight, the exact σ_r from Eqs. 7.9-5 agrees well with the approximate σ_r from the flexure formula ($\sigma_r \approx My/I = Pxy/I$). The exact shear stress

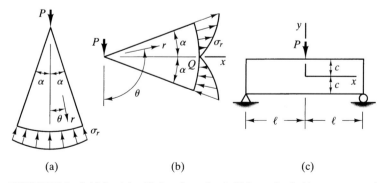

FIGURE 7.9-2 (a,b) Straight-sided wedges of unit thickness, loaded by concentrated tip forces. (c) Simply supported, centrally loaded plane beam of unit thickness.

$\tau_{r\theta}$ is zero everywhere, while shear stress $\tau = VQ/It$ of elementary beam theory is nonzero, and provides a y-direction force equal to load P. We see from Fig. 7.9-2b that the exact σ_r has y-direction components, which can be shown to yield a y-direction force equal in magnitude to load P.

Either wedge in Fig. 7.9-2 can be rotated 90° so its orientation matches that of the other, and the tip forces can be given different names. Thus, by superposition, one can obtain the solution for an in-plane tip force of any orientation, on a wedge spanning any angle $0 < \alpha < \pi$.

Remarks. In Fig. 7.9-2c, flexural stress $\sigma_x = My/I$ is locally disturbed by concentrated load P. How great is the disturbance along the vertical centerline? The beam is not semi-infinite, but Eqs. 7.9-3 can be used in approximate analyses [6.1]. The simplest such analysis is as follows. Let $r \ll c$, and integrate over $0 < \theta < \pi/2$ in Fig. 7.9-1b to obtain the horizontal force produced by σ_r over the quarter-circle. This force turns out to be P/π. Next consider the right half of the beam, loaded by rightward horizontal force P/π at $y = c$. Force P/π produces axial flexural stress and axial normal stress on the cross section at $x = 0$, both of which we calculate by elementary formulas. Superposing these stresses on flexural stress $\sigma_x = My/I$ due to the vertical load P, we obtain, at the central cross section of a beam of unit thickness,

$$\sigma_x = -\frac{3P\ell y}{4c^3} + \frac{3Py}{2\pi c^2} + \frac{P}{2\pi c} \qquad (7.9\text{-}6)$$

in which the latter two terms represent corrections to the usual flexural stress. The formula is approximate, and is not valid very close to the load point. We see that the correction terms are negligible for a slender beam, $\ell \gg c$.

Finally, consider a concentrated load in the plane of an infinite plate, Fig. 7.9-3a. At first it seems that the solution above or below the x axis is available as shown in the latter parts of Fig. 7.9-3, by using Eq. 7.9-3. But consideration of displacements shows that this result is not possible [6.1]. In Fig. 7.9-3b, points on the horizontal boundary displace away from point O; in Fig. 7.9-3c they displace toward point O. In other words, displacements are incompatible along the x axis. The correct solution shows that σ_r, σ_θ, and $\tau_{r\theta}$ are all nonzero [6.1].

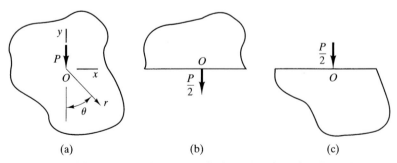

FIGURE 7.9-3 (a) Concentrated force P applied at an interior point of an infinite plane. (b, c) Regarding part (a) as the sum of two problems with semi-infinite planes (this is *not correct*).

7.10 THERMAL STRESS

Uniaxial Stress. We consider plane stress states that result from one-dimensional temperature distributions in a rectangular plate, Fig. 7.10-1a. Let T be temperature relative to an initial temperature at which the plate is undeformed and stresses are zero. Also let α, the coefficient of thermal expansion, be independent of T. Consider the case of T a function of y only and zero body forces. If also σ_y and τ_{xy} are zero and σ_x depends only on y, then $\nabla^2(\sigma_x + \sigma_y) = d^2\sigma_x/dy^2$, and Eq. 7.4-8 becomes a second-order ordinary differential equation with a simple solution, as follows.

$$\frac{d^2}{dy^2}(\sigma_x + E\alpha T) = 0 \quad \text{hence} \quad \sigma_x = -E\alpha T + a_1 y + a_2 \quad (7.10\text{-}1)$$

where a_1 and a_2 are constants of integration. This stress field is exact if it satisfies the stress-free boundary conditions. The assumption that σ_y and τ_{xy} are zero satisfies the free-edge boundary conditions on $y = \pm c$ and the zero-shear condition on $x = \pm \ell$. However, if T is an arbitrary function of y, we can satisfy the condition $\sigma_x = 0$ on $x = \pm \ell$ only approximately. This can be done by stating that σ_x produces zero net axial force F and zero net moment M; that is, for a plate of unit thickness,

$$F = 0 = \int_{-c}^{c} \sigma_x\, dy \quad \text{and} \quad M = 0 = \int_{-c}^{c} \sigma_x y\, dy \quad (7.10\text{-}2)$$

from which we solve for constants a_1 and a_2. Equation 7.10-1 then becomes

$$\sigma_x = -E\alpha T + \frac{3E\alpha y}{2c^3}\int_{-c}^{c} Ty\, dy + \frac{E\alpha}{2c}\int_{-c}^{c} T\, dy \quad (7.10\text{-}3)$$

We see that if T is either constant or linear in y, then $\sigma_x = 0$. For these special cases, the zero-stress solution is exact throughout the plate. If T is a general function of y, no values of a_1 and a_2 in Eq. 7.10-1 can produce $\sigma_x = 0$ all along edges $x = \pm \ell$, and there is stress σ_x within the plate despite the absence of external constraint. Then, according to Saint-Venant's principle, Eq. 7.10-3 is reliable in the central portion of the plate, up to a distance of about $2c$ from either end.

Let the foregoing problem be modified by adding rigid but frictionless supports, as shown in Fig. 7.10-1b. Thus $\sigma_y = 0$, $\epsilon_x = 0$, and the first of Eqs. 7.1-3 yields $0 = \sigma_x/E + \alpha T$. Combining this equation with Eq. 7.10-1, we obtain $a_1 y + a_2 = 0$, which can be true for all y only if $a_1 = 0$ and $a_2 = 0$. Thus $\sigma_x = -E\alpha T$ for any $T = T(y)$. This solution is exact, because it satisfies equilibrium, compatibility, and boundary conditions

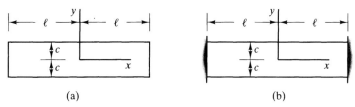

FIGURE 7.10-1 Rectangular plates of unit thickness. (a) Unsupported. (b) x-direction displacement prevented at ends.

exactly. Stress $\sigma_x = -E\alpha T$ can be appreciable; for example, it is about -240 MPa in steel for $T = 100°C$.

Disk with Axisymmetric T. Here T, σ_r, and σ_θ are functions of r only, and $\tau_{r\theta} = 0$. Equation 7.7-4b becomes $0 = 0$. Equation 7.7-4a, the equation of radial equilibrium, becomes

$$\frac{d\sigma_r}{dr} + \frac{\sigma_r - \sigma_\theta}{r} = 0 \tag{7.10-4}$$

Axisymmetric problems such as pressurized cylinders are discussed in detail in Chapter 8. The analysis there is a continuation of the development in Section 7.7. One result is a governing equation in terms of displacement, analogous to Eq. 7.1-2. This equation (Eq. 8.1-4), with thermal terms retained and body force terms discarded, becomes

$$\frac{d}{dr}\left[\frac{1}{r}\frac{d}{dr}(ru)\right] = (1 + \nu)\alpha\frac{dT}{dr} \tag{7.10-5}$$

where $u = u(r)$ is radial displacement at arbitrary radius r. Because of axial symmetry, there is no circumferential displacement. A function $u = u(r)$ that satisfies Eq. 7.10-5 satisfies compatability conditions and provides stresses that satisfy Eq. 7.10-4. Integration of Eq. 7.10-5 yields

$$u = (1 + \nu)\frac{\alpha}{r}\int_b^r T\, r_r\, dr_r + C_1 r + \frac{C_2}{r} \tag{7.10-6}$$

where b is the inner radius of the disk, Fig. 7.10-2a, and C_1 and C_2 are constants of integration. In the integral, r_r is used to denote a radius between b and the radius r at which u is calculated. Stresses are obtained by substituting Eq. 7.10-6 into Eqs. 7.7-8 and using stress-strain relations.

Long cylinders have displacement and stress formulas similar to those for a disk, but since the plane stress condition does not prevail for general axisymmetric thermal load in a cylinder, Poisson's ratio terms are different from those of a disk [1.6, 6.1]. A common application is a pipe that carries a hot fluid. Over most of the pipe length ϵ_z is

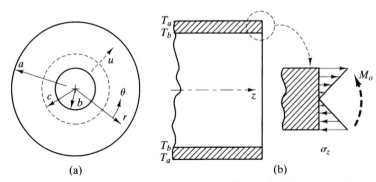

FIGURE 7.10-2 (a) Flat disk with a central hole. (b) Longitudinal cross section of a pipe carrying a hot fluid ($T_b > T_a$). Superposed moment M_o unloads the end.

constant, and axial stress σ_z develops (Fig. 7.10-2b). At a free end, σ_z must vanish. Conceptually, the free-end condition is obtained by superposing a circumferentially distributed end moment M_o on the solution for constant ϵ_z (Fig. 7.10-2b). In addition to making the net σ_z zero at the end, the effect of M_o is to flare the end outward, with an accompanying stress field that may increase the net circumferential tensile stress about 25% [6.1]. Problems of this type are considered in Section 13.8.

Hot Spot. How large are stresses in a "hot spot"? Consider a solid disk ($b=0$) with outer edge $r=a$ free. In Eq. 7.10-6 we must have $C_2=0$ so that u remains finite at $r=0$. Constant C_1 is determined from the condition that $\sigma_r=0$ at $r=a$. The resulting stress equations are

$$\sigma_r = E\alpha\left(\frac{1}{a^2}\int_0^a Tr\,dr - \frac{1}{r^2}\int_0^r Tr_r\,dr_r\right) \tag{7.10-7a}$$

$$\sigma_\theta = E\alpha\left(\frac{1}{a^2}\int_0^a Tr\,dr + \frac{1}{r^2}\int_0^r Tr_r\,dr_r - T\right) \tag{7.10-7b}$$

where, in integrals with limits 0 and r, r_r is a radius in the range $0<r_r<r$. Now let the disk be uniformly heated an amount T_o over a region of radius c, where $c<a$. For $r<c$, the respective integrals in Eq. 7.10-7a yield $T_o c^2/2$ and $T_o r^2/2$. Thus, in the central region,

$$\text{For } r<c: \quad \sigma_r = \sigma_\theta = \frac{E\alpha T_o}{2}\left(\frac{c^2}{a^2} - 1\right) \tag{7.10-8}$$

In the outer region $r>c$, $T=0$. Integrals with limits 0 and r yield $T_o c^2/2$; hence

$$\text{For } r>c: \quad \sigma_r = \frac{E\alpha T_o}{2}\left(\frac{c^2}{a^2} - \frac{c^2}{r^2}\right) \quad \text{and} \quad \sigma_\theta = \frac{E\alpha T_o}{2}\left(\frac{c^2}{a^2} + \frac{c^2}{r^2}\right) \tag{7.10-9}$$

If $c \ll a$, σ_r and σ_θ both approach $-E\alpha T_o/2$ in the region $r<c$. In the outer portion of the disk, at $r=c$, σ_r and σ_θ from Eqs. 7.10-9 combine to produce the maximum shear stress $\tau_{\max} = E\alpha T_o/2$. If we let c approach a and consider the outer rim $r=a$, we would expect on physical grounds that $\sigma_\theta = 0$, as the disk is now uniformly heated thoughout. But Eq. 7.10-9 yields $\sigma_\theta = E\alpha T_o$. The discrepancy arises because mathematically the unheated rim is still there, even though its volume has shrunk to zero. It is forced to strain circumferentially an amount αT_o by the heated portion, and it therefore carries tensile stress $E\alpha T_o$.

The foregoing solution presumes that heating is uniform in the thickness direction. If instead only a small, thin surface layer is heated, its expansion is restrained by the large volume of adjacent unheated material. Surface-normal stress is essentially zero throughout the layer. Thus, from Eqs. 7.7-1 with $\epsilon_r=0$ and $\epsilon_\theta=0$, we obtain $\sigma_r = -E\alpha T_o/(1-\nu)$. This result is a "worst case" estimate of thermal stress in a surface-layer hot spot, as surrounding elastic material loaded by the hot spot actually responds by deforming slightly.

Remarks. If $\nabla^2 T=0$, Eq. 7.4-3 suggests that temperature does not matter. More specifically, the suggestion is that if supports do nothing to inhibit thermal expansion or contraction, then a harmonic temperature field does not change the stresses. Such is the case if T is constant or linear in rectangular Cartesian coordinates, material properties are temperature-independent, and the material is homogeneous, isotropic, and linearly elastic. However, in problems of disks and cylinders with $T = T(r)$, stresses develop even when $\nabla^2 T = 0$. Disks and cylinders are *multiply connected* bodies, for which there are compatibility considerations not suggested by Eq. 7.4-3. Difficulties with compatibility are avoided if the problem is formulated in terms of displacement, as in Eq. 7.10-5 and in Chapter 8.

The foregoing discussion of thermoelasticity is brief and deals only with problems that are static and "uncoupled" in the sense that temperature does not depend on deformation. In such problems one determines the temperature field by heat transfer considerations alone, then uses the temperature field to calculate stresses and deformations, whereupon analysis is complete. This procedure is adequate for most problems. An example of *coupled* thermoelasticity, in which deformation influences temperature, appears in hypersonic flight: Aerodynamic heating may produce deformations that alter air flow enough to significantly affect local heating rates.

7.11 TORSION. WARPING FUNCTION

Shear stress τ and angle of twist θ in a shaft of *circular* cross section are given by the elementary formulas $\tau = Tr/J$ and $\theta = TL/GJ$. Plane cross sections remain plane after twisting. The largest shear stress appears at the greatest distance from the axis of the shaft. These results are familiar from elementary mechanics of materials. They are also exact according to theory of elasticity.

None of the foregoing results are correct if the shaft has a *noncircular* cross section. In such case, no simple formulas are available for τ and θ. Initially plane cross sections do not remain plane: they warp; that is, cross sections develop displacements parallel to the axis of the shaft (Fig. 7.11-1a). The largest shear stress may appear where the lateral surface is closest to the axis of the shaft, and the smallest where the lateral surface is most distant. In a square cross section, Fig. 7.11-1b, the largest shear stress appears at points A. Corners B are free of shear stress (because lateral surfaces are free surfaces).

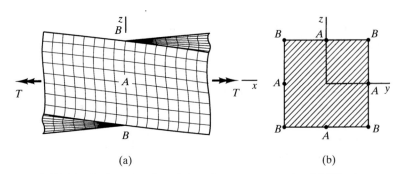

FIGURE 7.11-1 Torsion of a prismatic bar of square cross section. (a) Side view, after twisting. The grid was rectangular before twisting. (b) Cross section at an arbitrary x.

Section 7.11 Torsion. Warping Function

Basic assumptions of the torsion problem are as follows. The shaft is prismatic. Its material is homogeneous, isotropic, and linearly elastic. Displacements are small. The loading is pure torque about the longitudinal axis, applied by stress distributions on ends $x =$ constant.

In this section and the next we discuss the formulation and elasticity aspects of the torsion problem. More practical methods of solution, approximations, and application to problems of engineering interest appear in Chapter 9.

The rate of twist is prominent in torsion analysis. Its definition is

$$\text{Rate of twist:} \quad \beta = \frac{d\theta}{dx} \qquad (7.11\text{-}1)$$

where x is the axial coordinate along the shaft and θ is the angle of rotation (twist) about the x axis.

Warping Function. The formulation to be described is one of the renowned solutions of elasticity. It is due to Saint-Venant (1855). He made the insightful assumption that displacement of an arbitrary point has x, y, and z components

$$u = \beta\Psi \qquad v = -(\beta x)z \qquad w = (\beta x)y \qquad (7.11\text{-}2)$$

where $\Psi = \Psi(y,z)$ is called the *warping function*. Equations 7.11-2 are the basis of a semi-inverse solution, in which the form of the solution is assumed in part, and the remainder (the function Ψ) must be determined so that all requirements are satisfied. That this procedure is possible justifies the assumption, and provides an exact solution.

Substitution of Eqs. 7.11-2 into the strain-displacement relations, Eqs. 7.2-3 and 7.2-4, shows that four of the six strains are zero:

$$\epsilon_x = \epsilon_y = \epsilon_z = \gamma_{yz} = 0 \qquad (7.11\text{-}3)$$

Because $\epsilon_y = \epsilon_z = \gamma_{yz} = 0$ we see that *a cross section does not deform in its own plane*. In other words, when viewed along the x axis, each cross section appears to rotate as a rigid body about a point O, which is called the center of twist. In Fig. 7.11-2a, straight line OA appears to remain inextensible and straight (although it bends in the *axial* direction). Displacements v and w are y and z components of the small distance AA'.

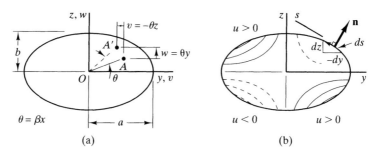

FIGURE 7.11-2 (a) Displacement components in the plane of a cross section. (b) Boundary normal **n** and length increments in the yz plane. Warping displacements (contours of constant u), for the direction of T shown in Fig. 7.11-1.

The only nonzero strains are γ_{xy} and γ_{zx}. They are obtained from Eqs. 7.2-3 and 7.2-4, and when multiplied by shear modulus G they produce the shear stresses

$$\tau_{xy} = G\beta\left(\frac{\partial \Psi}{\partial y} - z\right) \quad \text{and} \quad \tau_{zx} = G\beta\left(\frac{\partial \Psi}{\partial z} + y\right) \tag{7.11-4}$$

These are the only nonzero stresses in the torsion problem. They depend on y and z but not on axial coordinate x.

Compatibility conditions are automatically satisfied by basing the solution on a continuous and differentiable displacement field (Eqs. 7.11-2). It remains to satisfy equilibrium equations and stress boundary conditions. Equations 7.11-4 and the first equilibrium equation, Eq. 7.3-3a, yield

$$\frac{\partial^2 \Psi}{\partial y^2} + \frac{\partial^2 \Psi}{\partial z^2} = 0 \quad \text{or} \quad \nabla^2 \Psi = 0 \tag{7.11-5}$$

The remaining two equilibrium equations are identically satisfied. Boundary conditions, Eqs. 7.3-4, are applied to the lateral surface of the shaft. Surface tractions are all zero; also $\ell = 0$ because the boundary normal is in the yz plane. The latter two of Eqs. 7.3-4 are identically satisfied, and the first becomes

$$\tau_{xy}m + \tau_{zx}n = 0 \quad \text{or} \quad \tau_{xy}\frac{dz}{ds} - \tau_{zx}\frac{dy}{ds} = 0 \tag{7.11-6}$$

where m and n are direction cosines of a boundary normal (Fig. 7.11-2b). The substitutions $m = dz/ds$ and $n = -dy/ds$ come from Fig. 7.11-2b, in which dy is negative because s increases in the counterclockwise direction. Equations 7.11-4 and 7.11-6 yield

$$\frac{\partial \Psi}{\partial y}\frac{dz}{ds} - \frac{\partial \Psi}{\partial z}\frac{dy}{ds} = \frac{1}{2}\frac{d}{ds}(y^2 + z^2) \tag{7.11-7}$$

The torsion problem has been reduced to finding a function $\Psi = \Psi(y,z)$ that satisfies Eqs. 7.11-5 within the boundary and satisfies Eq. 7.11-7 on the boundary.

EXAMPLE

Consider an elliptical cross section, Fig. 7.11-2. The boundary of the cross section has the equation $(y^2/a^2) + (z^2/b^2) = 1$. Equations 7.11-5 and 7.11-7 are satisfied by the warping function

$$\Psi = \frac{b^2 - a^2}{a^2 + b^2}yz \tag{7.11-8}$$

Equations 7.11-4 and 7.11-8 yield the stresses

$$\tau_{xy} = -G\beta\frac{2a^2 z}{a^2 + b^2} \quad \text{and} \quad \tau_{zx} = G\beta\frac{2b^2 y}{a^2 + b^2} \tag{7.11-9}$$

From these equations we see that, for $a > b$, shear stress is larger at $z = \pm b$ than at $y = \pm a$. Torque associated with Eqs. 7.11-8 is determined as described in the next section (see the first form of Eqs. 7.12-5). Thus we relate T to β, and can then express shear stresses in terms of torque. These results are

$$T = \frac{\pi a^3 b^3}{(a^2 + b^2)}G\beta \quad \tau_{xy} = -\frac{2Tz}{\pi a b^3} \quad \tau_{zx} = \frac{2Ty}{\pi a^3 b} \tag{7.11-10}$$

For the special case of a circular cross section of radius c, the boundary equation has the form $y^2 + z^2 = c^2$, and Eqs. 7.11-5 and 7.11-7 are satisfied by $\Psi = 0$. Therefore there is no warping, exactly as assumed in elementary mechanics of materials. Shear stresses in Eq. 7.11-4 become $\tau_{xy} = -G\beta z$ and $\tau_{zx} = G\beta y$, in agreement with the elementary formulas $\tau = Tc/J$ and $\beta = T/GJ$.

Usually, one seeks stresses but not the warping displacement. Accordingly, in the following section we consider a stress function solution, which is more adaptable to practical stress analysis.

7.12 TORSION. PRANDTL STRESS FUNCTION

In 1903, Ludwig Prandtl suggested a way to solve the torsion problem without having to determine displacements. From the development in Section 7.11, we know that τ_{xy} and τ_{zx} are the only nonzero stresses. These stresses are now defined as derivatives of the Prandtl stress function $\phi = \phi(y,z)$.

$$\tau_{xy} = \frac{\partial \phi}{\partial z} \quad \text{and} \quad \tau_{zx} = -\frac{\partial \phi}{\partial y} \qquad (7.12\text{-}1)$$

The reason for choosing this form is that the only nontrivial equilibrium equation, Eq. 7.3-3a, is identically satisfied by any function $\phi = \phi(y,z)$ that is continuous through its second derivatives.

Compatibility conditions are introduced as follows. In Eqs. 7.11-4, we substitute for τ_{xy} and τ_{zx} from Eqs. 7.12-1, differentiate the first equation with respect to z and the second with respect to y, and subtract the second equation from the first. The result is

$$\frac{\partial^2 \phi}{\partial y^2} + \frac{\partial^2 \phi}{\partial z^2} = -2G\beta \quad \text{or} \quad \nabla^2 \phi = -2G\beta \qquad (7.12\text{-}2)$$

All that remains is to satisfy boundary conditions. Substitution of Eqs. 7.12-1 into Eq. 7.11-6 yields

$$\frac{\partial \phi}{\partial z} dz + \frac{\partial \phi}{\partial y} dy = 0 \quad \text{or} \quad d\phi = 0 \qquad (7.12\text{-}3)$$

The boundary condition $d\phi = 0$ means that ϕ is constant on the boundary. Because stresses depend on derivatives of ϕ rather on ϕ itself, without loss of generality we can take $\phi = 0$ as the boundary condition. (If there is a lengthwise hole in the shaft, ϕ must then be a nonzero constant on the boundary of the hole.) Stress analysis of a solid shaft is thus reduced to finding a function $\phi = \phi(y,z)$ such that $\nabla^2 \phi$ is independent of y and z (so that it satisfies Eqs. 7.12-2) and is zero on the boundary. Such analysis is easy to describe but not easy to do if the shape of the cross section is at all complicated.

The largest shear stress appears at a boundary of the cross section, and is equal in magnitude to the derivative of ϕ with respect to the direction normal to the boundary. This statement will be made clear by the *membrane analogy*, discussed in Section 9.2. In equation form, with n the direction normal to the boundary, Fig. 7.11-2b,

$$\text{Shear stress in direction } s \text{ normal to } n: \quad \tau_{sx} = -\frac{\partial \phi}{\partial n} \qquad (7.12\text{-}4)$$

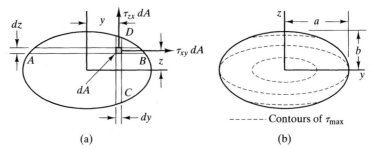

FIGURE 7.12-1 (a) Differential areas and forces used in the calculation of torque T. (b) Contours of equal maximum shear stress in an elliptical cross section.

where shear stress τ_{sx} in Fig. 7.11-2b is directed tangent to the boundary. Equation 7.12-4 is obtained by noting that Eqs. 7.12-1 are valid for arbitrarily oriented rectangular Cartesian coordinates in the plane of a cross section. Accordingly we may use orthogonal axes n and s instead of y and z. Indeed, Eq. 7.12-4 is also valid at an arbitrary point *within* the cross section, with n and s any two mutually perpendicular directions in the cross section.

Torque is developed by force increments $\tau_{zx} dA$ and $\tau_{xy} dA$ acting through the respective moment arms y and z. From Fig. 7.12-1a and Eqs. 7.12-1,

$$T = \int (\tau_{zx} y - \tau_{xy} z)\, dA = -\iint \left(\frac{\partial \phi}{\partial y} y + \frac{\partial \phi}{\partial z} z \right) dy\, dz \qquad (7.12\text{-}5)$$

Rewritten, the double integral is

$$T = -\int \left(\int_A^B \frac{\partial \phi}{\partial y} y\, dy \right) dz - \int \left(\int_C^D \frac{\partial \phi}{\partial z} z\, dz \right) dy \qquad (7.12\text{-}6)$$

where A, B, C, and D designate ends of elemental strips, Fig. 7.12-1a. Integrals over elemental strips can be integrated by parts. For the strip from A to B, $(\partial \phi / \partial y) dy$ becomes $(d\phi / dy) dy = d\phi$, so that

$$\int_A^B \frac{\partial \phi}{\partial y} y\, dy = \int_A^B y\, d\phi = [\phi y]_A^B - \int_A^B \phi\, dy = -\int_A^B \phi\, dy \qquad (7.12\text{-}7)$$

The term $[\phi y]_A^B$ vanishes because $\phi = 0$ at all points on the boundary. The integral over the strip from C to D is evaluated similarly. Equation 7.12-5 becomes

$$T = 2 \iint \phi\, dy\, dz \quad \text{or} \quad T = 2 \int \phi\, dA \qquad (7.12\text{-}8)$$

EXAMPLE

Torque T is applied to a shaft of elliptical cross section, Fig. 7.12-1b. Determine the shear stresses and the rate of twist in terms of T, G, a, and b.

Consider the Prandtl stress function

$$\phi = C \left(1 - \frac{y^2}{a^2} - \frac{z^2}{b^2} \right) \qquad (7.12\text{-}9)$$

where C is a constant. The parenthetical expression, if equated to zero, is the equation of the boundary ellipse. Substitution of ϕ into Eq. 7.12-2 yields

$$C = \frac{a^2 b^2}{a^2 + b^2} G\beta \qquad (7.12\text{-}10)$$

Function ϕ solves a torsion problem because $\nabla^2 \phi$ is constant; furthermore, ϕ pertains to an elliptical cross section because $\phi = 0$ on an elliptical boundary. Torque T in terms of β is obtained by substituting Eq. 7.12-10 into Eq. 7.12-9 and the result into Eq. 7.12-8. Integration is facilitated by noting that integrals of $y^2 dA$ and $z^2 dA$ are moments of inertia of the cross sectional area $A = \pi ab$. These quantities are tabulated [1.6, 1.7]. Hence

$$T = \frac{\pi a^3 b^3}{a^2 + b^2} G\beta \quad \text{or} \quad \beta = \frac{a^2 + b^2}{\pi a^3 b^3} \frac{T}{G} \qquad (7.12\text{-}11)$$

Now we can state ϕ in terms of T, and finally obtain stresses by applying Eqs. 7.12-1. Results are the same as in Eqs. 7.11-10. On the boundary, we see that τ_{xy} has greater magnitude than τ_{zx} if $a > b$; that is, shear stress is greatest where the boundary is closest to the axis of the shaft. If $a = b$, τ_{xy} and τ_{zx} both have largest magnitude Tr/J, where r is a radial coordinate and $J = \pi a^4/2$ is the polar moment of the cross-sectional area.

The maximum shear stress at an arbitrary point is $\tau_{\max} = dF/dA$, where dF is the resultant force in the yz plane on a differential area dA. Thus

$$dF = (\tau_{xy}^2 + \tau_{zx}^2)^{1/2} dA \quad \text{so} \quad \tau_{\max} = \frac{2T}{\pi ab} \left(\frac{y^2}{a^4} + \frac{z^2}{b^4} \right)^{1/2} \qquad (7.12\text{-}12)$$

Contours of constant $\tau_{\max}$ are shown dashed in Fig. 7.12-1b.

7.13 INELASTIC MATERIAL BEHAVIOR

Thus far in this chapter we have assumed that the material stress-strain relation is linear and time-independent. In the present section we briefly discuss plasticity and creep, for which these assumptions are not acceptable. Plasticity is considered time independent; creep is time dependent. Expositions of general theory and analysis methods for plasticity and creep may be found in many other textbooks. Simple applications of plastic analysis to specific problems appear in other chapters of this book.

In elasticity, with small displacements and a linearly elastic material, the principle of superposition is applicable, and a solution that satisfies equilibrium, compatibility, and boundary conditions is unique. In general, with plasticity and creep, separate solutions for two or more loads cannot be superposed to obtain the solution for the loads applied simultaneously, and the final state of stress and deformation may not be unique; it may depend on the order in which loads are applied. Nevertheless, many concepts and procedures used in elasticity theory are also used in the analysis of plasticity and creep.

Plasticity. General plasticity analysis requires formulation of a yield criterion, a flow rule, and a hardening rule. These concepts are not needed in theory of elasticity. A *yield criterion* relates the onset of yielding to the state of stress. The von Mises criterion is one of many (see Section 3.3). A *flow rule* relates stress increments, strain increments, and state of stress in the plastic range. If all components of the loading increase

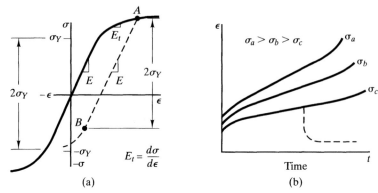

FIGURE 7.13-1 (a) Uniaxial stress-strain relations for a typical metal. (b) Qualitative creep curves at constant temperature: strain versus time at different values of uniaxial stress.

proportionally, so may all components of stress at any point. Then one may use total stresses and total strains rather than their increments, and in effect the flow rule is bypassed. A *hardening rule* describes how the yield condition changes for straining beyond initial yield.

As applied to a uniaxial stress-strain relation, Fig. 7.13-1a, the yield criterion says that plastic flow begins at stress σ_Y. The flow rule says that continued deformation proceeds along the curve whose slope is tangent modulus E_t. The "kinematic hardening" rule says that a total elastic range of $2\sigma_Y$ is preserved. Thus, if load is removed at point A, there is elastic unloading as shown by the dashed line, and yielding resumes at point B in compression. (If the material were nonlinear but elastic, rather than elastic-plastic, unloading would retrace the solid curve back to the origin.)

Creep. Creep is strain that changes with time at constant stress. Change of stress with time at constant strain is called *relaxation* and is regarded as an aspect of creep. As examples, hot turbine blades may creep, and bolt tensions in hot structures may relax. Creep and relaxation proceed faster as temperature increases. Creep may be of engineering significance at roughly 50% of the melting temperature on an absolute scale. Concrete, wood, and some plastics may creep appreciably at room temperature. Increasing the stress increases the creep rate (Fig. 7.13-1b). Increasing the temperature has much the same effect. The dashed line in Fig. 7.13-1b indicates unloading from stress σ_c: There is some immediate elastic recovery and further recovery with time, but permanent strain remains.

General analysis methods for creep build upon analysis tools developed for plasticity. Thus, many equations are similar; strain increment $d\epsilon_x$ in plastic analysis becomes strain rate $d\epsilon_x/dt$ in creep analysis.

PROBLEMS

7.1-1. Length L of a cylindrical plug is glued into a cylindrical hole in a tapered sleeve, as shown. Elastic moduli of plug and sleeve are E_p and E_s, respectively. If shear stress in the glue is to be independent of x, how should cross-sectional area A_s of the sleeve vary? Express the answer in terms of E_p, E_s, L, c, and x. Assume that uniaxial stress prevails in both plug and sleeve.

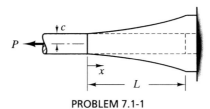

PROBLEM 7.1-1

7.1-2. In Fig. 7.1-1a let distributed load q be independent of x.
  (a) Use Eq. 7.1-1 to determine stress σ_x as a function of x.
  (b) Use Eq. 7.1-2 to determine displacement u as a function of x.
  (c) Show that the solution to part (b) yields the solution to part (a).
  (d) If the bar is tapered, so that $A = A(x)$, what equation replaces Eq. 7.1-1? Does this equation require that q be constant?

7.2-1. A plane displacement field has components $u = -a_1(y^2 + \nu x^2)$ and $v = 2a_1 xy$, where a_1 is a constant. Determine the stresses, and sketch them acting on the boundary of the region $-a \leq x \leq a, -b \leq y \leq b$.

7.2-2. Let Cartesian coordinates $x'y'$ be rotated with respect to Cartesian coordinates xy by a clockwise angle θ. Express strains $\epsilon_{x'}$ and $\gamma_{x'y'}$ in the $x'y'$ system in terms of $\epsilon_x, \epsilon_y, \gamma_{xy}, \sin \theta$, and $\cos \theta$. (Analogous stress transformation equations appear in Fig. 1.10-1.) Suggestion: Write $\epsilon_{x'} = \partial u'/\partial x'$, where $u' = u \cos \theta + v \sin \theta$, use the chain rule, and note that $\partial x/\partial x' = \cos \theta, \partial y/\partial x' = \sin \theta$, and so on.

7.2-3. A uniform cantilever beam of length L is fixed at end $x = 0$ and loaded by uniform downward load q. According to elementary beam theory, displacements are $v = -qx^2(x^2 - 4Lx + 6L^2)/24EI$ and $u = -y(dv/dx)$. Use these displacements to evaluate strains and then stresses in terms of x and y. Is Eq. 7.2-5 satisfied? Do stresses at $x = 0$ provide the correct force and moment?

7.3-1. (a) Let a nominally plane body have gradually varying thickness t, so that $t = t(x,y)$. Forces per unit length are $N_x = \sigma_x t$, $N_y = \sigma_y t$, and $N_{xy} = \tau_{xy} t$. What are the differential equations of equilibrium in terms of forces per unit length?
  (b) Let all six stresses act on the differential element of volume $dV = dx\,dy\,dz$. Derive the equilibrium equations, Eqs. 7.3-3.

7.3-2. Examine each of the following fields to see if it is a valid solution of a plane elasticity problem. If not, what condition is not met? For simplicity, let Poisson's ratio be zero. The a_i are constants.
  (a) $u = 4a_1(x^2y + y^3), v = 2a_1 y^3$
  (b) $u = a_1 xy^2, v = -a_1 x^2 y$
  (c) $\epsilon_x = a_1 x^3, \epsilon_y = a_1 x^2 y, \gamma_{xy} = a_1 xy^2$
  (d) $\sigma_x = a_1 x + a_2 y, \sigma_y = a_3 x + a_4 y, \tau_{xy} = -a_4 x - a_1 y$.

7.3-3. Axial stress in the cantilever beam of Fig. 7.3-2 is $\sigma_x = 3Pxy/2c^3$. Use this equation, a differential equation of equilibrium, and boundary conditions to determine shear stress τ_{xy}. From the other equilibrium equation, what can you say about σ_y?

7.3-4. Differential equations of equilibrium can be expressed in terms of displacements by substituting Eqs. 7.1-3 and 7.2-3 into Eqs. 7.3-2. Derive these two equations for the case of no body forces and no temperature change.

7.3-5. A plane square body occupies the region $0 \le x \le L, 0 \le y \le L$. Side $x = 0$ is bonded to a rigid support. The other three sides are free. Qualitatively sketch the distribution of surface tractions applied by the support for each of the following temperature distributions, where a_1 is a constant. (a) $T = a_1 x$. (b) $T = a_1 y$. Suggestion: Begin by sketching the deformed shape of an unsupported body, for which $\gamma_{xy} = 0$.

7.3-6. A rectangular plate has a central circular hole, as shown. Imagine that the quarter of the plate in the first quadrant is to be isolated and analyzed. What boundary conditions on stress and displacement are appropriate for such an analysis?

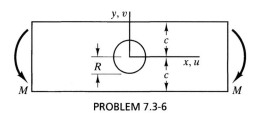

PROBLEM 7.3-6

7.4-1. (a) Derive Eq. 7.4-8. Follow instructions in the text, and note that the term $\partial^2 \tau_{xy}/\partial x \partial y$ appears, multiplied by 2. Substitute for one of these terms from Eq. 7.3-2a and for the other from Eq. 7.3-2b.

(b) Show that introduction of the Airy stress function converts Eq. 7.4-8 to Eq. 7.4-3.

7.4-2. Imagine that the Airy stress function

$$F = a_1 x^2 \quad \text{for} \quad 0 \le x \le 1, \qquad F = a_1(2x - 1) \quad \text{for} \quad 1 \le x \le 2$$

where a_1 is a constant, is proposed as the solution of a plane stress problem in the square region $0 \le x \le 2, 0 \le y \le 2$. Is such a solution acceptable? Sketch boundary tractions on the square region. On physical grounds, what can you say about compatibility?

7.4-3. (a) Determine stresses associated with the Airy stress function $F = a_1 x^2 y/2$, where a_1 is a constant. Sketch the associated boundary tractions on the rectangular region $-L \le x \le L, 0 \le y \le h$. By summing forces and moments produced by the tractions, show that the region is in static equilibrium.

(b) Repeat part (a) with reference to the Airy stress function $F = a_1(x^4 - 3x^2 y^2)$ and the square region $0 \le x \le 1, 0 \le y \le 1$.

7.4-4. Stresses $\sigma_x = a_1 y$ and $\sigma_y = a_2 x$, where a_1 and a_2 are constants, exist in the rectangular region $0 \le x \le a, 0 \le y \le b$. Shear stress is also present, but no body force. Integrate Eqs. 7.4-2 to obtain a general expression for F. What shear stress τ_{xy} is predicted?

7.4-5. Stresses σ_x and σ_y are applied as boundary tractions normal to edges of a 3 m by 3 m square region, as shown. Shear stress within the region is known to be $\tau_{xy} = 3x^2 + 7y^2 + 2x + 2.5$, in which numerical factors have units such that τ_{xy} has units Pa. Determine functions $f = f(x)$ and $g = g(y)$ required for equilibrium. Does the stress field within the region satisfy compatibility requirements?

7.4-6. Explain why a stress-free boundary condition is expressed by the equations $F = 0$ and $\partial F/\partial n = 0$, where n is a direction normal to the boundary.

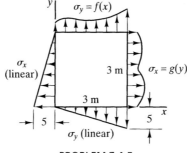

PROBLEM 7.4-5

7.5-1. (a) Show that the plane region in Fig. 7.5-1c is in static equilibrium under forces and moments produced by stresses applied to its boundary.
  (b) Repeat part (a) with reference to Fig. 7.5-2a.
  (c) In Fig. 7.5-3a, consider the portion of the beam above a horizontal line at an arbitrary value of y. Show that stresses applied to this portion produce a net y-direction force of zero.
  (d) Show that the "added σ_x" in Fig. 7.5-3b is self-equilibrating; that is, it produces no net force or moment.

7.5-2. For a uniform cantilever beam loaded by transverse tip force, show that beam theory yields the same shear stress τ_{xy} as does elasticity theory.

7.5-3. Show that the Airy stress function of Eq. 7.5-7 is biharmonic.

7.5-4. How should Eq. 7.5-7 be modified if uniformly distributed load q on the top edge of the beam in Fig. 7.5-3a is to be replaced by uniform downward traction q on the lower edge of the beam?

7.5-5. The cantilever beam shown has unit thickness. Beam theory predicts flexural stress $\sigma_x = qx^2y/2I$, where $I = 2c^3/3$. Using this σ_x, integrate Eqs. 7.3-2 to determine τ_{xy} and σ_y. Apply boundary conditions on $y = c$ to determine functions produced by integration. Is this stress field in fact possible?

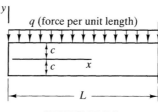

PROBLEM 7.5-5

7.5-6. Imagine that an Airy stress function F is required for the problem of a uniform cantilever beam loaded by uniformly distributed downward load q along its upper surface. Without carrying out details of the calculations, describe how F can be obtained by combining solutions to the problems of Figs. 7.5-2b and 7.5-3a.

7.6-1. Determine the displacement field associated with the stresses of Eqs. 7.4-7. As boundary conditions, impose zero displacement and zero rotation at $x = y = 0$.

7.6-2. A prismatic bar hangs under its own weight, with end $x=0$ free and end $x=L$ supported. The body force is $B_x = -\rho g$, where ρ is the mass density and g is the acceleration of gravity. Assume that σ_x is the only nonzero stress. (This assumption is exact if tractions on end $x=L$ are properly applied.) Determine σ_x in terms of ρ, g, and x. Also determine displacements u, v, and w as functions of the coordinates. As boundary conditions, set v and w to zero on the axis of the bar, and set $u=0$ at $x=L, y=z=0$.

7.6-3. Determine the displacement field associated with the stress field of Problem 7.4-4. As boundary conditions, set $u = v = \partial v/\partial x = 0$ at $x = y = 0$.

7.6-4. For the cantilever beam of Fig. 7.3-2b, determine the displacement field. As boundary conditions at $x = L, y = 0$, impose $u = v = 0$ and keep the beam axis horizontal.

7.7-1. In Eq. 7.7-2, radial force on the element edge of length ℓ_a and unit thickness is written $\sigma_a d\ell_a$, where σ_a and $d\ell_a$ are defined in Fig. 7.7-1. This force may also be written as $\{\sigma_r r + [\partial(\sigma_r r)/\partial r]\, dr\} d\theta$. Explain why. Also show that Eq. 7.7-4a is again obtained.

7.7-2. In the sketch shown, determine stresses σ_A and σ_B so as to satisfy equilibrium. Assume that $\partial \sigma_r / \partial r = \Delta \sigma_r / \Delta r$, and so on.

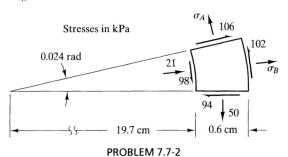

PROBLEM 7.7-2

7.7-3. (a) Show that Eqs. 7.7-4 are satisfied by Eqs. 7.7-5.

(b) Show, by transformation of ∇^2 from Cartesian coordinates, that Eq. 7.7-6 is indeed the harmonic operator in polar coordinates.

(c) Verify Eq. 7.7-7. Suggestion: Write $\nabla^4 F = \nabla^2(\nabla^2 F)$ in Eq. 7.4-3, and use Eqs. 7.7-5 and 7.7-6 to write $\nabla^2 F$ in terms of σ_r and σ_θ.

7.7-4. What problem is solved by the Airy stress function $F = a_1 r^2(1 + \cos 2\theta)$, where a_1 is a constant? Suggestion: Transform the stresses to Cartesian coordinates.

7.7-5. Investigate what problem is solved by the Airy stress function

$$F = a_1 \ln r + \frac{pb^2 r^2}{2(a^2 - b^2)}$$

where a_1 is a constant. Consider the annular region $b \leq r \leq a$. Use the boundary condition $\sigma_r = 0$ on $r = a$.

7.7-6. (a) Investigate what problem is solved by the Airy stress function $F = C\theta$, where C is a constant.

(b) What is the most general solution for F of $\nabla^4 F = 0$, if F is a function of θ but not of r?

7.7-7. Natural processes create "growth stresses" in trunks of trees, especially in hardwoods. Axial and circumferential stresses in a tree trunk of outer radius a are approximately [7.3]

$$\sigma_z = \sigma_{za}\left(1 + 2\ln\frac{r}{a}\right) \quad \text{and} \quad \sigma_\theta = \sigma_{\theta a}\left(1 + \ln\frac{r}{a}\right)$$

where σ_{za} and $\sigma_{\theta a}$ are respectively positive and negative constants.

(a) Obtain the expression for radial stress σ_r by use of Eq. 7.7-4a.

(b) Show that σ_z produces zero net axial force.

(c) Soon after growing, a new layer of wood (on the outside of the trunk) acquires stresses σ_{za} longitudinally and $\sigma_{\theta a}$ circumferentially. Thus the new layer creates increments in σ_z, σ_θ, and σ_r in wood already present. Use this information to derive the formulas the for σ_z and σ_r.

(d) A crosscut makes $\sigma_z = 0$ on the cut end. Effectively, reversed stress σ_z is superposed on growth stress σ_z, for a resultant σ_z of zero. On a wedge $0 \le \theta \le \alpha$, what moment is associated with reversed stress σ_z? (This moment tends to peel a wedge-shaped strip away from a cut log.)

(e) If growth stress σ_z protects a tree trunk from failure in high wind, what can you say about the relative tensile and compressive strengths of green wood?

7.8-1. Obtain the expression for $g = g(r)$ by integration of Eq. 7.8-4.

7.8-2. (a) Use integration to calculate the total force transferred by stress σ_θ across the x axis in Fig. 7.8-1a. Assume that $\ell \gg b$.

(b) Similarly, calculate the total force transferred by stress σ_θ across the positive y axis in Fig. 7.8-1a from $r = b$ to $r = 10b$. Assume that $w \gg b$.

7.8-3. Determine stress concentration factors produced when a circular hole is drilled in fields of (a) equal biaxial tension, and (b) pure shear.

7.8-4. (a) For what values of θ in Fig. 7.8-1 is the maximum shear stress on the boundary of a small hole the same as it is a great distance from the hole?

(b) What biaxial stress field produces exactly two stress-free points on the boundary of a small circular hole?

7.8-5. A thin-walled cylindrical tube has mean radius 80 mm and wall thickness 1.0 mm. The tube is loaded by tensile axial force $P = 10$ kN and by torque $T = 1.1$ kN·m. A small circular hole is drilled through the tube wall. What are the greatest tensile and compressive stresses at the hole?

7.9-1. (a) Consider Fig. 7.9-2a and, following the procedure described in the text, obtain constant a_1 in Eqs. 7.9-1.

(b) Consider Fig. 7.9-2b and obtain constant a_1 in Eqs. 7.9-1 by taking moments about point Q.

(c) Let $\alpha = 10°$ in Fig. 7.9-2b. Compare σ_r at $\theta = 100°$ and $\tau_{r\theta}$ at $\theta = 90°$ with stresses $\sigma = Mc/I$ and $\tau = VQ/It$ of beam theory.

7.9-2. Forces P and $2P$ act on a semi-infinite plane of unit thickness, as shown. Determine the principal stresses at point A in terms of P and a.

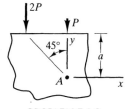

PROBLEM 7.9-2

7.9-3. Use stress σ_r from Eqs. 7.9-3 to obtain expressions for stresses σ_x and σ_y at point O in the semi-infinite plane shown. Suggestion: Consider an increment of load $q\,dx$, write the associated stress increments $d\sigma_x$ and $d\sigma_y$ in terms of $d\sigma_r$ and θ by use of stress transformation equations, and integrate.

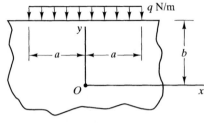

PROBLEM 7.9-3

7.9-4. Two forces P act on a semi-infinite plane of unit thickness, as shown. If we temporarily regard the Airy stress function that produces Eqs. 7.9-1 as a function of x and y rather than r and θ, the stress function for the present two-load problem can be written $F_M = F(x,y) - F(x',y')$, where $x' = x$ and $y' = y - a$. Thus $F_M = (\Delta F/\Delta y)\Delta y \approx (\partial F/\partial y)a$ becomes the Airy stress function for moment load $M = Pa$. The approximation $F_M \approx (\partial F/\partial y)a$ is satisfactory for stresses at points whose distance from point O is more than a few multiples of a. Determine F_M in terms of M and θ.

7.9-5. (a) Carry out the analysis described in the text for the beam of Fig. 7.9-2c, and thus verify Eq. 7.9-6.

(b) Evaluate Eq. 7.9-6 on the y axis at locations $y = -c$, $y = 0$, and $y = c/2$. For comparison, also evaluate σ_x according to the flexure formula $\sigma_x = My/I$ at these locations. Let $\ell = 4c$, and express the results in terms of P and c.

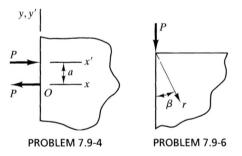

PROBLEM 7.9-4   PROBLEM 7.9-6

7.9-6. (a) A plane body of unit thickness extends to infinity in one quadrant, as shown. Use the wedges of Fig. 7.9-2 and superposition to determine the stress field in terms of P, r, and β.

(b) Show that the solution of part (a) gives zero horizontal force on a quarter-circle region from $\beta = 0$ to $\beta = \pi/2$.

7.9-7. An inclined force P acts on a semi-infinite plane of unit thickness, as shown.

(a) Determine σ_r along line OA in terms of P, β, r, and λ.

(b) How is λ related to β if β is chosen to make $\sigma_r = 0$ along line OA?

(c) How is λ related to β if, for some given λ and r, β is chosen to make σ_r a maximum?

7.9-8. Show that the Airy stress function

$$F = \frac{qr^2}{2(\tan \alpha - \alpha)}(\alpha - \theta + \sin \theta \cos \theta - \tan \alpha \cos^2 \theta)$$

is the solution to the problem of the uniformly loaded wedge shown.

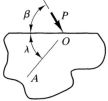
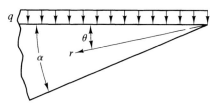

PROBLEM 7.9-7   PROBLEM 7.9-8

7.10-1. Consider the two temperature fields $T = T_o$ and $T = T_o y/c$, where T_o is a constant. For each temperature field and with $\sigma_x = 0$, determine constants a_1 and a_2 in Eq. 7.10-1.

7.10-2. For the problem of Fig. 7.10-1b, let $T = T_o y/c$, where T_o is a constant. Determine stress σ_x in the plate by beam theory. That is, remove support at one end, apply the thermal load, calculate the moment needed to restore the support condition, and use the flexure formula.

7.10-3. Consider a flat plate of arbitrary shape, as shown. The boundary condition is fixed in the xy plane ($u = v = 0$) but frictionless in the z direction (w unrestrained). If $T = az$, where a is a constant, what is the complete state of stress, strain, and displacement in the plate? Let $w = 0$ at the plate midsurface $z = 0$.

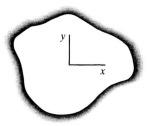

PROBLEM 7.10-3

7.10-4. A long, thin-walled pipe has temperature T_b at inner radius b and temperature T_a at outer radius a, as in Fig. 7.10-2b. Temperature may be assumed to vary linearly through the wall thickness. Away from pipe ends, what are stresses at $r = b$ and at $r = a$?

7.10-5. Verify the following equations: (a) 7.10-6, (b) 7.10-7a, and (c) 7.10-7b.

7.10-6. Imagine that electric current causes the temperature in an unsupported cylindrical bar to become $T = k(a^2 - r^2)$, where k is a constant and a is the radius of the bar. Let the bar have a zero Poisson ratio. Determine stresses σ_r and σ_θ due to T.

7.10-7. For steady-state conditions with prescribed boundary temperatures, heat flow is governed by the equation $\nabla^2 T = 0$. What then is $T = T(r)$ in Fig. 7.10-2a, in terms of r, a, b, and prescribed temperatures T_a at $r = a$ and T_b at $r = b$?

7.11-1. (a) Show that Eqs. 7.11-5 and 7.11-7 are satisfied by Eq. 7.11-8.
(b) Verify Eqs. 7.11-10.

7.11-2. For the elliptical cross section of Fig. 7.11-2, obtain an expression for axial displacement u in terms of T, G, a, b, y, and z.

7.12-1. In torsion, it is expected that shear stresses on a cross section produce zero net force in the y and z directions (why?). Prove that this is so. Suggestion: Use Fig. 7.12-1a and Eqs. 7.12-1.

7.12-2. Imagine that the elliptical cross section of Fig. 7.11-2a contains an elliptical hole whose equation is

$$\frac{y^2}{(ak)^2} + \frac{z^2}{(bk)^2} = 1$$

where k is a constant in the range $0 < k < 1$. By imagining that material is removed from a solid bar, determine β and τ_{max} in terms of T, G, a, b, and k. Let $a > b$.

7.12-3. (a) For a bar whose cross section is an equilateral triangle as shown, the Prandtl stress function has the form

$$\phi = k\left(y - \sqrt{3}z - \frac{2h}{3}\right)\left(y + \sqrt{3}z - \frac{2h}{3}\right)\left(y + \frac{h}{3}\right)$$

where k is a constant. (If equated to zero, each parenthetical expression is the equation of one side of the triangle.) Obtain k in terms of G, β, and h. Plot τ_{zx} along the y axis. Express β and τ_{max} in terms of T, G, and h. Given that $\int \phi \, dA = G\beta h^4 \sqrt{3}/90$.

(b) In part (a), stress function ϕ is based on products of expressions that define the boundary of the cross section. Will this approach work for the rectangular cross section $-a \leq y \leq a, -b \leq z \leq b$?

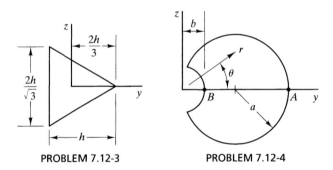

PROBLEM 7.12-3   PROBLEM 7.12-4

7.12-4. A cylindrical shaft under torsional load has a lengthwise groove of radius b, as shown in the sketch of the cross section. Consider the Prandtl stress function

$$\phi = -\frac{G\beta}{2}\left(y^2 + z^2 - 2ay + \frac{2b^2 ay}{y^2 + z^2} - b^2\right)$$

(a) Show that $\phi = 0$ on the boundary of the cross section.

(b) Show that Eq. 7.12-2 is satisfied. Suggestion: Here it is easier to work in polar coordinates $r\theta$.

(c) Determine the ratio of shear stress at B to shear stress at A. Hence, what stress concentration factor is approached if $a \gg b$?

CHAPTER 8

Pressurized Cylinders and Spinning Disks

8.1 GOVERNING EQUATIONS

This chapter deals with *axisymmetric* problems of circular disks and cylinders. Thus, nothing varies with circumferential coordinate θ. Also, although temperature terms are included in basic equations, applications in this chapter deal with the mechanical loads of pressure on cylinders and spinning of disks (Fig. 8.1-1). For *thin-walled* pressurized cylinders and *slender* spinning rings, mechanics of materials is adequate for analysis. The mechanics of materials approach fails for thick-walled cylinders and disks because the state of deformation cannot be anticipated accurately enough. For these problems we must use tools of the theory of elasticity. Accordingly, this chapter uses concepts from Chapter 7 and equations from Section 7.7.

Assumptions. We assume that the material is homogeneous, isotropic, and linearly elastic (except in Sections 8.6 and 8.7, where plastic action is considered). Displacements and strains are small. Geometry and loading are axisymmetric. States of stress and strain are assumed not to vary along the length of a cylinder or through the thickness of a disk. Cylinders are straight. Disks are thin and have uniform thickness unless stated otherwise. Spinning disks have constant angular velocity ω, where ω is measured in radians per second.

Stress and Displacement Equations. Analysis in this section includes load terms for both the pressurized cylinder and the spinning disk. Subsequently we treat the problems separately by choosing appropriate load terms for each.

Equilibrium equations in polar coordinates are derived in Section 7.7 (Eqs. 7.7-4). In an axisymmetric problem, $\tau_{r\theta}$ and B_θ are both zero, and derivatives with respect to θ are zero. Equation 7.7-4b becomes the identity $0 = 0$. In Eq. 7.7-4a, $\partial\sigma_r/\partial r$ becomes $d\sigma_r/dr$, and B_r becomes $\rho\omega^2 r$, where ρ is the mass density and ω is the constant angular velocity of spin about the z axis (Fig. 8.1-1b). For a cylinder and for a disk of uniform thickness, the equation of radial equilibrium, Eq. 7.7-4a, becomes

$$\frac{d\sigma_r}{dr} + \frac{\sigma_r - \sigma_\theta}{r} + \rho\omega^2 r = 0 \qquad (8.1\text{-}1)$$

232 Chapter 8 Pressurized Cylinders and Spinning Disks

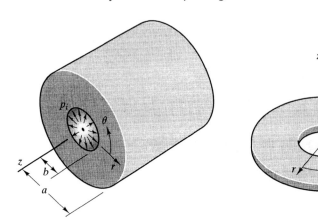

FIGURE 8.1-1 (a) Cylinder with internal pressure. (b) Spinning disk.

In strain-displacement relations, Eqs. 7.7-8, circumferential displacement v is zero, and derivatives with respect to θ are zero. Therefore $\gamma_{r\theta} = 0$. Radial and circumferential normal strains become $\epsilon_r = du/dr$ and $\epsilon_\theta = u/r$, where $u = u(r)$ is radial displacement. (The ϵ_θ expression is derived in Fig. 1.4-2b; the ϵ_r expression is the ratio of change in length to original length in the radial direction.) Strains ϵ_r and ϵ_θ can be substituted into the stress-strain relations, Eqs. 7.7-1, and these equations solved for the normal stresses. Thus

$$\sigma_r = \frac{E}{1-\nu^2}\left(\frac{du}{dr} + \nu\frac{u}{r}\right) - \frac{E\alpha T}{1-\nu} \tag{8.1-2a}$$

$$\sigma_\theta = \frac{E}{1-\nu^2}\left(\frac{u}{r} + \nu\frac{du}{dr}\right) - \frac{E\alpha T}{1-\nu} \tag{8.1-2b}$$

where $T = T(r)$ is temperature change relative to an initial temperature at which there are no thermal stresses. Equations 8.1-2 pertain to a state of *plane stress*; that is, to a state in which $\tau_{\theta z}$, τ_{zr}, and σ_z are all zero. Clearly the plane stress assumption is appropriate for a thin flat disk. It is not obvious that Eqs. 8.1-2 lead to correct results for a pressurized cylinder. The assumption that they do is justified later, in connection with Eq. 8.2-5. Temperature terms in Eqs. 8.1-2 apply only to a thin disk; they are not valid for a cylinder unless Poisson's ratio is zero.

Substitution of Eqs. 8.1-2 into Eq. 8.1-1 yields

$$\frac{d^2u}{dr^2} + \frac{1}{r}\frac{du}{dr} - \frac{u}{r^2} = (1+\nu)\alpha\frac{dT}{dr} - \frac{1-\nu^2}{E}\rho\omega^2 r \tag{8.1-3}$$

Terms on the right-hand side of Eq. 8.1-3 apply only to a thin disk; they do not apply to a cylinder unless Poisson's ratio is zero. With an alternative form for the left-hand side, Eq. 8.1-3 is

$$\frac{d}{dr}\left[\frac{1}{r}\frac{d}{dr}(ru)\right] = (1+\nu)\alpha\frac{dT}{dr} - \frac{1-\nu^2}{E}\rho\omega^2 r \tag{8.1-4}$$

Integration of Eq. 8.1-4 yields

$$u = C_1 r + \frac{C_2}{r} + (1+\nu)\frac{\alpha}{r}\int_b^r T r_r \, dr_r - \frac{1-\nu^2}{E}\rho\omega^2\frac{r^3}{8} \qquad (8.1\text{-}5)$$

where C_1 and C_2 are constants of integration, r_r in the integral is a radius in the range $b < r_r < r$, and b is the radius of a central hole. If there is no hole, $b = 0$.

In the remainder of this chapter we consider only mechanical loads. (Thermal stresses in a disk are discussed in Section 7.10.) Accordingly, we now discard temperature terms. Thus, substitution of Eq. 8.1-5 into Eqs. 8.1-2 yields

$$\sigma_r = \frac{E}{1-\nu^2}\left[(1+\nu)C_1 - \frac{1-\nu}{r^2}C_2\right] - \frac{3+\nu}{8}\rho\omega^2 r^2 \qquad (8.1\text{-}6a)$$

$$\sigma_\theta = \frac{E}{1-\nu^2}\left[(1+\nu)C_1 + \frac{1-\nu}{r^2}C_2\right] - \frac{1+3\nu}{8}\rho\omega^2 r^2 \qquad (8.1\text{-}6b)$$

in which terms that contain ω are applicable to a thin disk but not to a cylinder unless Poisson's ratio is zero.

In obtaining Eqs. 8.1-6 we have satisfied the differential equations of equilibrium and have satisfied compatibility by basing the formulation on a continuous displacement field $u = u(r)$. All that remains in order to produce a valid elasticity solution is to select C_1 and C_2 so that boundary conditions are satisfied. For each problem, pressurized cylinder and spinning disk, there is a different set of constants C_1 and C_2. In what follows we elect to treat the problems separately. If pressure and spinning exist simultaneously, perhaps with temperature change as well, separate solutions for each may be superposed if conditions are linearly elastic.

8.2 PRESSURIZED CYLINDERS

Stress Formulas. Consider a cylinder of outer radius a and inner radius b (Fig. 8.1-1a). Formulas for stress due to internal pressure p_i and external pressure p_o come from Eqs. 8.1-6. To evaluate constants C_1 and C_2 we set $\omega = 0$ and apply the two boundary conditions

$$\begin{aligned}\sigma_r &= -p_i \quad \text{at} \quad r = b \\ \sigma_r &= -p_o \quad \text{at} \quad r = a\end{aligned} \qquad (8.2\text{-}1)$$

Negative signs appear because positive pressure is a compressive stress. After obtaining expressions for C_1 and C_2, we substitute them into Eqs. 8.1-6 and obtain the following formulas for stresses in the pressurized cylinder.

$$\sigma_r = p_i\frac{b^2}{a^2-b^2}\left(1 - \frac{a^2}{r^2}\right) - p_o\frac{a^2}{a^2-b^2}\left(1 - \frac{b^2}{r^2}\right) \qquad (8.2\text{-}2a)$$

$$\sigma_\theta = p_i\frac{b^2}{a^2-b^2}\left(1 + \frac{a^2}{r^2}\right) - p_o\frac{a^2}{a^2-b^2}\left(1 + \frac{b^2}{r^2}\right) \qquad (8.2\text{-}2b)$$

These pressure vessel formulas, known as the Lamé solution, date from 1833. (Analogous pressure vessel formulas for spheres are also available [1.6, 1.7, 6.1], and are also due to Lamé.) Stress fields of Eqs. 8.2-2 are shown in Fig. 8.2-1. We see that stresses vary strongly with r, in contrast to the uniform distribution of σ_θ assumed in the elementary solution for a *thin*-walled cylinder.

A nonzero axial stress σ_z is present in a cylinder with end caps. Internal and external pressures create axial forces by acting on the respective projected areas πb^2 and πa^2. The net axial force is resisted by stress σ_z in the cylinder, which acts on cross-sectional area $\pi(a^2 - b^2)$. Therefore

$$\sigma_z = \frac{p_i b^2 - p_o a^2}{a^2 - b^2} \qquad (8.2\text{-}3)$$

Equations 8.2-2 and 8.2-3 are not reliable close to the juncture between a cylinder and an end cap because of "discontinuity stresses." These stresses tend towards zero over a short distance from the juncture, as predicted by Saint-Venant's principle. At $r = b$, where stresses are usually largest in magnitude, $0 < \sigma_z < \sigma_\theta$ for loading by p_i and $\sigma_\theta < \sigma_z < 0$ for loading by p_o. If end caps are absent, and pressure p_i is produced by frictionless pistons that compress fluid in the cylinder, then no axial force is carried by the cylinder, and $\sigma_z = 0$. Equations 8.2-2 remain valid whether or not there are end caps.

The maximum shear stress is $\tau_{max} = (\sigma_1 - \sigma_3)/2$, where σ_1 and σ_3 are the maximum and minimum principal stresses. If loading is by internal pressure alone or by external pressure alone, the largest τ_{max} in the cylinder appears at $r = b$, and is

$$\text{At } r = b: \quad \tau_{max} = \frac{pa^2}{a^2 - b^2} \qquad (p = p_i \text{ or } p = p_o) \qquad (8.2\text{-}4)$$

This stress is useful in the maximum shear stress failure criterion.

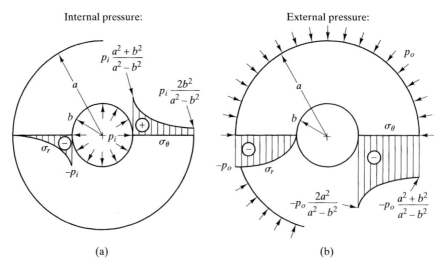

FIGURE 8.2-1 Stresses σ_r and σ_θ in a thick-walled cylinder under internal and external pressure. Stresses are shown approximately to scale for $p_i = p_o$ and $a = 3b$.

Remarks. Equations 8.1-2 pertain to a plane stress condition, yet are used to develop Eqs. 8.2-2 for the pressurized cylinder. Such use must be justified, as it is not obvious that σ_z and τ_{zr} are zero in a long cylinder without end caps. Substitution of Eqs. 8.2-2 into the equation for axial strain ϵ_z yields

$$\epsilon_z = \frac{1}{E}[\sigma_z - \nu(\sigma_r + \sigma_\theta)] = \frac{\sigma_z}{E} - \frac{2\nu}{E}\frac{p_i b^2 - p_o a^2}{a^2 - b^2} \quad (8.2\text{-}5)$$

We see that if σ_z constant (as in Eq. 8.2-3), then ϵ_z is independent of r, which enables the following argument. Imagine that before loading the cylinder is sliced like a sausage into disks of uniform thickness dz. Pressure p_i and/or p_o is then applied to each disk, with $\sigma_z = 0$. Equations 8.2-2 clearly apply to each disk because each is in a state of plane stress. And, according to Eq. 8.2-5, each disk changes thickness uniformly; that is, the flat surfaces remain exactly flat. Therefore the disks can be stacked to form the cylinder without having to apply z-direction loads to close gaps. Accordingly Eqs. 8.2-2 apply to cylinders as well as disks. If uniform σ_z from Eq. 8.2-3 is superposed, it produces uniform radial and circumferential strains, each $-\nu\sigma_z/E$, but no radial or circumferential stress.

Numerators and denominators in stress equations can be divided by b^2, so as to express stresses in terms of the ratio a/b. For example, from Eqs. 8.8-2b and 8.2-4, with $p_o = 0$,

$$\sigma_\theta = p_i \frac{1}{(a/b)^2 - 1}\left(1 + \frac{a^2}{r^2}\right) \qquad \tau_{\max} = \frac{p_i(a/b)^2}{(a/b)^2 - 1} \quad (8.2\text{-}6)$$

Thus we see that stresses in a pressurized cylinder depend on proportions of the cylinder rather than on its size. From Eq. 8.2-6 we can also conclude that if $\tau_{\max}$ is not to exceed an allowable value, a small increase in p_i may require a large increase in a/b. For example, imagine that with $a/b = 3$, p_i produces the allowable $\tau_{\max}$. To maintain this $\tau_{\max}$ under pressure $1.1p_i$, a/b must be increased to 6.7. The proportion $a/b = 3$ could be maintained by using a stronger material, if possible, so as to increase the allowable $\tau_{\max}$ to $1.1\,\tau_{\max}$.

In some structures, such as a bar under axial load, stress can be made arbitrarily small by making the cross section arbitrarily large. Such is not the case for a pressurized cylinder. From Eq. 8.2-2b or Fig. 8.2-1,

$$\text{At } r = b, \text{ for } a \gg b: \quad \sigma_\theta \approx p_i - 2p_o \quad (8.2\text{-}7)$$

From Eqs. 8.2-6 and 8.2-7 we see that when p_i is applied to a cylinder without residual stresses, neither σ_θ nor $\tau_{\max}$ can be less than p_i, even if a/b approaches infinity. (Equation 8.2-7 also shows that a small circular hole in a field of uniform stress p_o creates a stress concentration factor of 2.)

For large a/b, the outer portion of a cylinder is little stressed by p_i. For example, Fig. 8.2-1a yields $\sigma_{\theta a}/\sigma_{\theta b} = 0.077$ for $a/b = 5$. This result suggests that stresses near the bore (at $r = b$) will not change much if the shape of the outer boundary is irregular rather than circular. As a simple rule, one might say that if the distance from a hole to the boundary or to another hole is $5b$ or more, then σ_θ and $\tau_{\max}$ at a pressurized circular hole in a plane body of any shape are both approximately p_i.

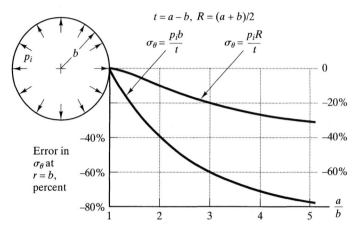

FIGURE 8.2-2 Circumferential stress at the bore ($r = b$) under internal pressure: percentage error of formulas for thin-walled cylinders (Eqs. 8.2-8) versus a/b.

How reliable are elementary formulas for circumferential stress in a thin-walled cylinder under internal pressure? In the present notation, the elementary formula $\sigma = pr_i/t$ becomes $\sigma_\theta = p_i b/(a - b)$. If a is not much larger than b, and with R the mean radius $R = (a + b)/2$, Eq. 8.2-2b yields an approximate mean-radius formula that is more accurate at $r = b$ than the elementary formula. Thus, formulas for thin-walled cylinders are

$$\text{Elementary:} \quad \sigma_\theta = \frac{p_i b}{t} \qquad \text{Mean radius:} \quad \sigma_\theta = \frac{p_i R}{t} \qquad (8.2\text{-}8)$$

Percentage errors of these formulas, relative to the exact σ_θ at $r = b$ from Eq. 8.2-2b, are plotted in Fig. 8.2-2. Data show that if the error in σ_θ at $r = b$ is not to exceed 5% in magnitude, the elementary formula is usable for a/b up to 1.1 and the mean radius formula is usable for a/b up to 1.6.

EXAMPLE

A cylinder with end caps is to have inner radius $b = 10$ mm. It must carry internal pressure 140 MPa with safety factor $SF = 1.7$. Failure is defined as the onset of yielding. The material yields at uniaxial tensile stress $\sigma_Y = 530$ MPa. According to the maximum shear stress failure criterion, what must be outer radius a? Under the working pressure of 140 MPa, what is the radial expansion at $r = b$ and at $r = a$? Use $E = 204$ GPa and $\nu = 0.29$.

Radius a can be determined from Eq. 8.2-4, using $\tau_{max} = \sigma_Y/2 = 265$ MPa, design pressure $p_i = (SF)140 = 238$ MPa, and $b = 10$ mm. We obtain

$$a = 31.3 \text{ mm} \qquad (8.2\text{-}9)$$

At the working pressure $p_i = 140$ MPa, from Eqs. 8.2-2 and 8.2-3, at $r = b = 10$ mm,

$$\sigma_r = -140 \text{ MPa} \qquad \sigma_\theta = 172 \text{ MPa} \qquad \sigma_z = 16 \text{ MPa} \qquad (8.2\text{-}10)$$

Circumferential strain is $\epsilon_\theta = [\sigma_\theta - \nu(\sigma_z + \sigma_r)]/E$. Evaluating this expression at $r = b$, we obtain

$$\epsilon_{\theta b} = \frac{1}{204{,}000}[172 - 0.29(16 - 140)] = 0.00102 \qquad (8.2\text{-}11)$$

Using now the strain-displacement relation from Eq. 7.7-8, we obtain radial displacement at $r = b$.

$$\epsilon_\theta = \frac{u}{r} \qquad \text{hence} \qquad u_b = b\epsilon_{\theta b} = 0.0102 \text{ mm} \qquad (8.2\text{-}12)$$

Radial displacement at $r = a = 31.3$ mm is calculated in similar fashion. With $\sigma_{\theta a} = 31.8$ MPa,

$$u_a = a\epsilon_{\theta a} = \frac{31.3}{204{,}000}[31.8 - 0.29(16 + 0)] = 0.0042 \text{ mm} \qquad (8.2\text{-}13)$$

8.3 SHRINK FITS. COMPOUND CYLINDERS

Introduction. As argued in connection with Eq. 8.2-7, a cylinder without prestress cannot carry extreme internal pressure even if its wall is very thick. Also, except at the bore $r = b$, most material is not used effectively because it is understressed. The allowable pressure can be raised, and material used more effectively, by prestressing before internal pressure is applied. The goal of prestressing is to create residual compressive σ_θ near the bore when $p_i = 0$, so that when $p_i > 0$ the net σ_θ is smaller than it would otherwise be.

The present section discusses prestressing by shrinking an outer cylinder (the jacket) onto an inner cylinder. The process could be generalized, resulting in a construction of many jackets [8.1]. However, each jacket is less effective than the one before in increasing strength, and jackets increase manufacturing costs because precise dimensions are important. The effect of several jackets is mimicked by wrapping an inner cylinder with several layers of wire under tension [3.2, 8.1]. As compared with shrink-fit construction, a wrapped cylinder is more easily damaged, harder to repair, and more flexible if bent as a beam. Another way to achieve prestressing is to produce plastic flow by high internal pressure. Release of the high pressure leaves the cylinder with a favorable pattern of residual stresses. This method is discussed in Section 8.6.

In what follows we consider two cylinders of the same material, the larger one having an inside diameter slightly less than required to fit over the smaller one. The larger cylinder is heated to make it expand, slipped over the smaller cylinder, and allowed to cool. Thus the cylinders exert contact pressure p_c on one another and are stressed because of it. Internal pressure p_i is subsequently applied. If p_i and inner radius b are prescribed, important quantities that must be determined are outer radii of the inner cylinder and the jacket cylinder, and the "interference," which is the small difference between the outer diameter of the inner cylinder and the inner diameter of the jacket.

Shrink-Fit Theory. In what follows we develop equations that describe a construction in which cylinder and jacket have equal strength, which means that cylinder and jacket reach the same allowable stress under the combined action of prestress pressure p_c and internal pressure p_i. We take the allowable stress to be a prescribed magnitude of the maximum shear stress, $\tau_{max} = (\sigma_1 - \sigma_3)/2$, where σ_1 and σ_3 are the extreme principal stresses. The largest shear stresses appear at $r = b$ in the inner cylinder and at $r = c$ in the outer cylinder (Fig. 8.3-1). At these locations σ_1 and σ_3 are respectively circumferential and radial; axial stress σ_z is the intermediate principal stress σ_2.

Stresses due to shrink-fit pressure p_c can be superposed on stresses due to internal pressure p_i. Cylinders are treated separately to obtain stresses due to p_c; the compound cylinder can be treated as a single unit to obtain stresses due to p_i. Normal stresses at critical locations are shown in Fig. 8.3-1. In the notation of Fig. 8.3-1, when p_i and p_c are both present, maximum shear stresses at $r = b$ in the inner cylinder and at $r = c$ in the jacket cylinder are

$$(\tau_{max})_b = \frac{(\sigma_A + \sigma_E) - (\sigma_B + \sigma_F)}{2} \qquad (\tau_{max})_c = \frac{(\sigma_C + \sigma_G) - (\sigma_D + \sigma_H)}{2} \qquad (8.3\text{-}1)$$

For a material that yields at stress σ_Y in uniaxial tension, $\tau_{max} = \sigma_Y/2$. Accordingly, we wish to impose the conditions

$$(\tau_{max})_b = \frac{1}{2}\sigma_Y \qquad (\tau_{max})_c = \frac{1}{2}\sigma_Y \qquad (8.3\text{-}2)$$

By equating shear stresses in Eqs. 8.3-1 we obtain

$$p_c \left(\frac{c^2}{c^2 - b^2} + \frac{a^2}{a^2 - c^2} \right) = p_i \frac{a^2(c^2 - b^2)}{c^2(a^2 - b^2)} \qquad (8.3\text{-}3)$$

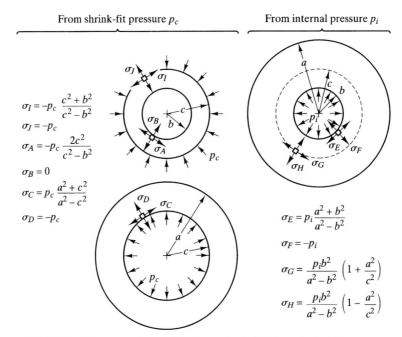

FIGURE 8.3-1 Stress notation used in the text for shrink-fit analysis. All stresses are considered positive if tensile.

(Note that p_c is the shrink-fit pressure alone; it is not the magnitude of σ_r at $r=c$ due to the combination of p_i and the shrink fit.) If p_c is eliminated between Eq. 8.3-3 and either of Eqs. 8.3-1, with $\tau_{max} = \sigma_Y/2$, there results

$$p_i = \frac{\sigma_Y}{2}\left[2 - \left(\frac{b^2}{c^2} + \frac{c^2}{a^2}\right)\right] \tag{8.3-4}$$

For design, one might prescribe p_i, σ_Y, and b, and seek values of a and c. Equation 8.3-4 is satisfied by infinitely many combinations of a and c. There is an optimal choice, discussed in the following subsection. For now we continue with shrink-fit theory.

At this stage, we presume that p_i, σ_Y, and b have been prescribed, radii a and c have been selected, and p_c has been calculated from Eq. 8.3-3. It remains to calculate the required interference and temperature change. Before cylinders are joined, the inner diameter of the jacket must be too small by an amount we will call 2Δ. (It is equally correct to say that the outer diameter of the inner cylinder must be 2Δ units too large.) After cylinders are joined, pressure p_c makes the diameters equal. Therefore

$$\Delta = u_{co} - u_{ci} \tag{8.3-5}$$

where u_{co} and u_{ci} are respectively the radial displacements of outer and inner cylinders at $r=c$ due to p_c alone (note that p_i is absent from these calculations). From Eqs. 7.7-1 and 7.7-8, $\epsilon_\theta = (\sigma_\theta - \nu\sigma_r)/E$ and $u = r\epsilon_\theta$. Axial stress σ_z is omitted from the calculation because it is uniform and does not contribute to Δ. Therefore $u = r(\sigma_\theta - \nu\sigma_r)/E$ at an arbitrary radius r. Accordingly, at $r=c$, in the notation of Fig. 8.3-1,

$$u_{co} = \frac{c}{E}(\sigma_C - \nu\sigma_D) \quad \text{and} \quad u_{ci} = \frac{c}{E}(\sigma_I - \nu\sigma_J) \tag{8.3-6}$$

Writing stresses in their expanded form, we obtain from Eqs. 8.3-5 and 8.3-6

$$\Delta = \frac{2c^3 p_c}{E}\left[\frac{a^2 - b^2}{(a^2 - c^2)(c^2 - b^2)}\right] \tag{8.3-7}$$

The cylinders cannot be put together with p_c already applied, so the outer cylinder is first heated enough to create the required Δ. Since $\Delta = c\epsilon_\theta = c\alpha T$, the required uniform temperature change T is $T = \Delta/c\alpha$. As a practical matter, some clearance is necessary, so a larger T must be used. In choosing this temperature difference, consideration must be given to difficulties of assembly and the effect of radiative heat transfer.

In cooling, the jacket shrinks axially as well as radially, thus developing axial friction forces between cylinder and jacket. Friction forces modify σ_r and σ_θ and create additional stresses, including shear stress τ_{zr}. These effects are neglected in the foregoing analysis.

Optimal Proportions. Equation 8.3-4 states the pressure p_i that can be sustained when τ_{max} is the same in cylinder and jacket. The value of c that maximizes p_i is obtained by differentiating p_i with respect to c and setting the derivative to zero. Thus

$$c = \sqrt{ab} \tag{8.3-8}$$

For this optimal value of c, Eqs. 8.3-3, 8.3-4, and 8.3-7 yield

$$a = \frac{b}{1 - (p_i/\sigma_Y)} \qquad \Delta = \frac{p_i c}{E} \qquad p_c = \frac{p_i(a - b)}{2(a + b)} \qquad (8.3\text{-}9)$$

If p_i, σ_Y, and b are prescribed, an optimal design is completed by using the first of Eqs. 8.3-9 to obtain a, Eq. 8.3-8 to obtain c, and the second of Eqs. 8.3-9 to obtain Δ. This Δ can be achieved by heating the jacket uniformly an amount $T = \Delta/c\alpha$.

With the maximum shear stress limited to $\sigma_Y/2$ and with $c = \sqrt{ab}$, and for given values of a and b, one can obtain a simple formula for the ratio of the p_i allowed by a compound cylinder to the p_i allowed by a single cylinder. Similarly, for given values of p_i and σ_Y, one can obtain a formula for the ratio of the weight of a compound cylinder to the weight of a single cylinder of the same material. It is left as an exercise to show that these ratios are

$$p_i \text{ ratio} = \frac{2a}{a + b} \qquad \text{weight ratio} = \frac{(2\sigma_Y - p_i)(\sigma_Y - 2p_i)}{2(\sigma_Y - p_i)^2} \qquad (8.3\text{-}10)$$

We see that the p_i ratio is scarcely more than unity if $a \approx b$, but approaches 2 if $a \gg b$. In the weight ratio expression, $\sigma_Y > 2p_i$ due to limitations of a single cylinder (see Eq. 8.2-6).

EXAMPLE 1

A solid disk and an annular disk, of the same thickness but different materials, fit together exactly (Fig. 8.3-2a). What state of stress is produced by uniformly reducing the temperature of the annular disk an amount T_o? (The same state of stress would be produced by uniformly heating the solid disk an amount T_o if both disks have the same coefficient of thermal expansion.)

This problem is easily solved by the same considerations as used to obtain Eqs. 8.3-6 and 8.3-7. The compatibility condition is

$$\text{At } r = c: \quad (u \text{ due to } p_c \text{ and } T_o)_{\text{outer}} = (u \text{ due to } p_c)_{\text{inner}} \qquad (8.3\text{-}11)$$

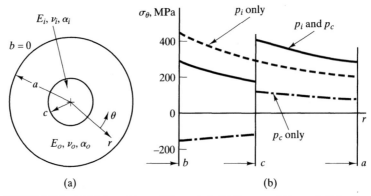

FIGURE 8.3-2 (a) Disk problem of Example 1. (b) Stresses of Example 2, for the design pressure $p_i = 1.7(140) = 238$ MPa.

Stresses in the solid disk are $\sigma_r = -p_c$ and $\sigma_\theta = -p_c$ throughout. Equation 8.3-11 becomes

$$\frac{c}{E_o}\left[p_c\frac{a^2+c^2}{a^2-c^2} - \nu_o(-p_c)\right] - c\alpha_o T_o = \frac{c}{E_i}[-p_c - \nu_i(-p_c)] \qquad (8.3\text{-}12)$$

The temperature term is negative because temperature decrease makes the annular disk contract. If both disks are of the same material, Eq. 8.3-12 yields

$$\text{Same material:} \quad p_c = \frac{E\alpha T_o(a^2-c^2)}{2a^2} \qquad (8.3\text{-}13)$$

Stresses in the solid disk are $\sigma_r = \sigma_\theta = -p_c$. Stresses in the annular disk, from Eqs. 8.2-2 and 8.3-13, are

$$\text{For } r > c: \quad \sigma_r = \frac{E\alpha T_o}{2}\left(\frac{c^2}{a^2} - \frac{c^2}{r^2}\right), \quad \sigma_\theta = \frac{E\alpha T_o}{2}\left(\frac{c^2}{a^2} + \frac{c^2}{r^2}\right) \qquad (8.3\text{-}14)$$

These equations are the same as obtained previously in a different way (Eqs. 7.10-9).

Equations of the present example are approximate for the shrink fit of a thin disk on a solid shaft (Fig. 2.7-3c). The shaft is stiffer than a disk in its response to p_c because it is not in a state of plane stress. Also, there are stress concentration effects.

EXAMPLE 2

Reconsider the example problem of Section 8.2: For the same data and with $\alpha = 12(10^{-6})/°C$, design a compound cylinder. That is, determine the required dimensions, shrink-fit pressure p_c, the radius difference Δ, and the required uniform temperature change. Also determine stresses at $r = b$ and at $r = c$ under the design pressure. How much weight is saved as compared with a single cylinder?

From Eqs. 8.3-8 and 8.3-9, under design pressure $p_i = SF(140) = 238$ MPa,

$$a = 18.15 \text{ mm} \qquad c = 13.5 \text{ mm} \qquad p_c = 34.5 \text{ MPa} \qquad (8.3\text{-}15)$$

From Eq. 8.3-9, $\Delta = 0.0157$ mm. This small Δ reminds us of the need for accurate machining. A temperature change $\Delta/c\alpha = 97°C$ is required to create Δ, and of course more would be needed to make sure the cylinders do not seize before they are fully in place.

From Eq. 8.2-2b with $p_i = 238$ MPa and $p_c = 34.5$ MPa, circumferential stresses at $r = b$ in the inner cylinder and at $r = c$ in the outer cylinder are

$$\sigma_{\theta b} = -p_c\frac{2c^2}{c^2-b^2} + p_i\frac{a^2+b^2}{a^2-b^2} = 292 \text{ MPa} \qquad (8.3\text{-}16a)$$

$$\sigma_{\theta c} = p_c\frac{a^2+c^2}{a^2-c^2} + p_i\frac{b^2}{a^2-b^2}\left(1+\frac{a^2}{c^2}\right) = 411 \text{ MPa} \qquad (8.3\text{-}16b)$$

From Eq. 8.2-2a, the corresponding radial stresses are $\sigma_{rb} = -238$ MPa and $\sigma_{rc} = -119$ MPa. As expected, stresses σ_r and σ_θ combine to produce $\tau_{max} = \sigma_Y/2$ at $r = b$ and at $r = c$. From Eq. 8.2-3, $\sigma_z = 104$ MPa, which is clearly the intermediate principal stress at both locations and so does not enter the calculation of τ_{max}. (If stresses at the working pressure are desired, Eqs. 8.3-16 can be applied with $p_i = 140$ MPa and $p_c = 34.5$ MPa.)

Distributions of circumferential stress are shown in Fig. 8.3-2b. The "p_i and p_c" curves display the stresses of Eqs. 8.3-16 at $r = b$ and $r = c$. The "p_i only" curve would occur without the prestressing of a shrink fit. The "p_c only" curves are initial stresses produced by the shrink fit. We see that shrink-fit stresses by themselves produce no yielding because their τ_{max} values are less than the allowable value $\sigma_Y/2$.

Weight of the compound cylinder, relative to weight of the single cylinder discussed in the example problem of Section 8.2, is as follows. For a unit length and with γ the weight density,

$$\frac{\text{(weight, compound)}}{\text{(weight, single)}} = \frac{\gamma\pi(18.15^2 - 10^2)}{\gamma\pi(31.3^2 - 10^2)} = 0.26 \qquad (8.3\text{-}17)$$

This result can also be obtained from Eq. 8.3-10.

8.4 SPINNING DISKS, PART I

Spinning disks serve as support for blades in a steam or gas turbine. In aircraft engines, low weight is important. To reduce weight and rotational inertia, a turbine disk is usually thickest near the center rather than being flat (that is, of uniform thickness). The simpler case of a flat disk is discussed in the present section. The disk of variable thickness is discussed in Section 8.5.

Flywheel. A flywheel of constant cross section, Fig. 8.4-1a, is a simple special case of a spinning disk. Let ω be the angular velocity of the flywheel, measured in radians per second. Let R be the distance from the axis of revolution to the centroid of the cross-sectional area A of the rim. A particle mass m at radius R would exert outward inertia force $mR\omega^2$. Similarly, if $R \gg t$, an incremental slice of the rim exerts the inertia force shown in Fig. 8.4-1b, where ρ is the mass density. Circumferential stress σ_θ in the rim is assumed uniform because the rim is thin. Each force $\sigma_\theta A$ in Fig. 8.4-1b has a component parallel to the inertia force proportional to $\sin(d\theta/2)$, or to $d\theta/2$ since $d\theta$ is infinitesimal. Thus from equilibrium considerations, *if spokes exert no radial forces on the rim*,

$$(\rho A R\, d\theta)\, R\omega^2 - 2\left(\sigma_\theta A \sin \frac{d\theta}{2}\right) = 0 \qquad \text{hence} \qquad \sigma_\theta = \rho\omega^2 R^2 \qquad (8.4\text{-}1)$$

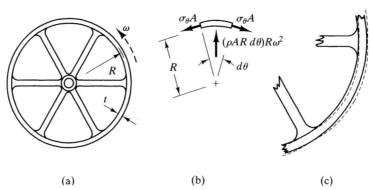

(a) (b) (c)

FIGURE 8.4-1 (a) Rotating flywheel. (b) Differential element of the rim, showing inertia force and force due to σ_θ. (c) Flexural deformation of the rim between spokes.

Radial expansion of the rim is then $R\epsilon_\theta = \rho\omega^2 R^3/E$. In general, spokes expand less, so they pull inward on the rim and reduce σ_θ. Stresses that result from flexure of the rim, Fig. 8.4-1c, can be approximated as follows if spokes are numerous. Regard a rim segment between spokes as a straight beam fixed at both ends and uniformly loaded. If spokes expand 3/4 as much as the rim, we would then superpose 1/4 of the flexural stresses on 3/4 of σ_θ from Eq. 8.4-1. Here we have not discussed how the factor of 3/4 is obtained, and we have ignored the effect of additional circumferential stress associated with the radial deflection shown in Fig. 8.4-1c. Also, starting and stopping will flex the spokes. A computational solution is well suited to such problems.

Flat Disks. Solutions for spinning disks of uniform thickness are easy to obtain from equations in Section 8.1. In a disk with no central hole, stresses and radial displacement must remain finite as r approaches zero. Accordingly, from either Eq. 8.1-5 or Eqs. 8.1-6, we conclude that $C_2 = 0$. To obtain C_1, we use Eq. 8.1-6a and set $\sigma_r = 0$ at the outer radius $r = a$. Upon substituting C_1, Eqs. 8.1-6 become, for a *flat solid disk*,

$$\sigma_r = \frac{3+\nu}{8}\rho\omega^2 a^2\left(1 - \frac{r^2}{a^2}\right) \tag{8.4-2a}$$

$$\sigma_\theta = \frac{3+\nu}{8}\rho\omega^2 a^2\left(1 - \frac{1+3\nu}{3+\nu}\frac{r^2}{a^2}\right) \tag{8.4-2b}$$

In a disk with a central hole of radius b, radial stress must vanish at both boundaries, $r = a$ and $r = b$. Thus Eq. 8.1-6a yields expressions for C_1 and C_2. Upon substituting these expressions Eqs. 8.1-6 become, for a *flat disk with a central hole*,

$$\sigma_r = \frac{3+\nu}{8}\rho\omega^2 a^2\left(1 + \frac{b^2}{a^2} - \frac{b^2}{r^2} - \frac{r^2}{a^2}\right) \tag{8.4-3a}$$

$$\sigma_\theta = \frac{3+\nu}{8}\rho\omega^2 a^2\left(1 + \frac{b^2}{a^2} + \frac{b^2}{r^2} - \frac{1+3\nu}{3+\nu}\frac{r^2}{a^2}\right) \tag{8.4-3b}$$

From Eq. 8.4-3a, the maximum σ_r appears at $r = \sqrt{ab}$, where it has the value

$$(\sigma_r)_{max} = \frac{3+\nu}{8}\rho\omega^2 a^2\left(1 - \frac{b}{a}\right)^2 \tag{8.4-4}$$

Nowhere does σ_r exceed the maximum σ_θ, whether or not the disk has a central hole.

If a radial stress σ_o were applied at $r = a$, as representative of turbine blades attached to the rim, stresses could be calculated by superposing the stresses of Eqs. 8.4-3 on the pressure vessel stresses of Eqs. 8.2-2, with $p_i = 0$ and $p_o = -\sigma_o$.

Equations 8.4-2, 8.4-3, and 8.4-4 pertain to a disk thin enough that plane stress conditions prevail. We cannot make the argument that spinning disk equations pertain also to a spinning cylinder, because ϵ_z depends on r in a spinning disk, but ϵ_z must be independent of r in a long cylinder (except near its ends).

The stresses of Eqs. 8.4-2 and 8.4-3 are plotted in Fig. 8.4-2b. Clearly, circumferential stress σ_θ is significantly increased by a central hole. Indeed, for $a \gg b$, σ_θ at $r = b$ is almost doubled by the hole. In other words, a central pinhole has a stress concentration factor of 2.

Chapter 8   Pressurized Cylinders and Spinning Disks

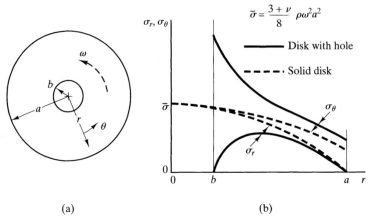

(a)   (b)

FIGURE 8.4-2 (a) Disk with central hole of radius b. (b) Radial and circumferential stresses in spinning disks of uniform thickness.

The peripheral speed of the disk is $v_a = a\omega$. If Eqs. 8.4-2 and 8.4-3 are changed in form by the substitution $\omega = v_a/a$, results show that size does not matter, in the sense that two disks having the same ν, ρ, and v_a (and the same a/b ratio, if there is a central hole) have the same stresses at a given value of r/a.

EXAMPLE

A steel disk, 0.03 m thick and 0.76 m in diameter, is attached to a solid steel shaft, 0.1600 m in diameter, by means of a shrink-fit. The hole in the disk is 0.1598 m in diameter before it is attached to the shaft. What is the largest normal stress at standstill? At what speed will the shrink-fit loosen, and what are stresses at this speed? Let $E = 200$ GPa, $\nu = 0.27$, and $\rho = 7860$ kg/m^3.

For shrink-fit analysis at standstill, we will assume that the shaft can be treated as a solid disk of the same thickness as the annular disk (see remarks that follow Eq. 8.3-14). Due to contact pressure p_c, stresses in the shaft are $\sigma_r = -p_c$ and $\sigma_\theta = -p_c$. And, from Eqs. 8.2-2, stresses at $r = b$ in the annular disk are $\sigma_r = -p_c$ and $\sigma_\theta = 1.093 p_c$. Circumferential strains must create a radial gap of 0.0001 m at $r = 0.08$ m. Hence, using Eq. 8.3-5 and applying the radial displacement equation $u = r(\sigma_\theta - \nu\sigma_r)/E$ to each part, we write

$$0.0001 = \frac{0.08}{E}[1.093 p_c - \nu(-p_c)] - \frac{0.08}{E}[-p_c - \nu(-p_c)] \qquad (8.4\text{-}5)$$

from which $p_c = 119$ MPa. The largest normal stress at standstill is therefore $\sigma_\theta = 1.093 p_c = 131$ MPa, at $r = 0.08$ m in the annular disk.

Until the shrink-fit loosens, stresses due to spinning can be calculated as if the disk-plus-shaft assembly were a single solid disk. These stresses are superposed on stresses caused by p_c. The loosening speed ω_L is defined as the speed at which net radial stress at $r = 0.08$ m becomes zero. Thus, using Eq. 8.4-2a,

$$0 = -119(10^6) + \frac{3.27}{8} 7860 \, \omega_L^2 (0.38)^2 \left[1 - \left(\frac{0.08}{0.38}\right)^2\right] \qquad (8.4\text{-}6)$$

from which $\omega_L = 518$ rad/s, which is 4950 rpm. Stresses at this speed are due entirely to spinning, and can be obtained from Eqs. 8.4-3. For example, at $r = 0.08$ m in the annular disk, Eq. 8.4-3b yields $\sigma_\theta = 252$ MPa. (As an alternative method of solution, we could calculate this σ_θ from the equation $0.0001 = b\sigma_\theta/E$, then use Eq. 8.4-3b at $r = b$ to obtain $\omega_L = 518$ rad/s.)

8.5 SPINNING DISKS, PART II

Variable Thickness. Turbine disks are usually thicker near the center than at the rim. The motivation for taper is clear from Fig. 8.4-2b: highest stresses appear near the center, and can be reduced by having less material near the rim and more near the center. For a given allowable stress, a tapered disk has less weight and less rotational inertia than a flat disk.

Equations applicable to a tapered disk are as follows. It is assumed that taper is not so pronounced that that the plane-stress assumption must be abandoned. That is, it is assumed that σ_r and σ_θ are the only nonzero stresses and that they are constant through the thickness. With thickness h now a function of r, the equation of radial equilibrium involves the forces shown in Fig. 8.5-1b. After adding force components parallel to inertia force F_r, discarding terms that contain the product of three or more differentials, replacing $\sin(d\theta/2)$ by $d\theta/2$, and combining terms, we obtain

$$\frac{1}{rh}\frac{d}{dr}(\sigma_r rh) - \frac{\sigma_\theta}{r} + \rho\omega^2 r = 0 \qquad (8.5\text{-}1)$$

Substitution of σ_r and σ_θ from Eqs. 8.1-2, with $T = 0$, converts Eq. 8.5-1 to

$$\frac{d^2u}{dr^2} + \frac{1}{r}\frac{du}{dr} - \frac{u}{r^2} + \frac{1}{h}\left(\frac{du}{dr} + \nu\frac{u}{r}\right)\frac{dh}{dr} = -\frac{1-\nu^2}{E}\rho\omega^2 r \qquad (8.5\text{-}2)$$

If thickness h is constant, Eq. 8.5-1 reduces to Eq. 8.1-1, and Eq. 8.5-2 reduces to Eq. 8.1-3 with $T = 0$.

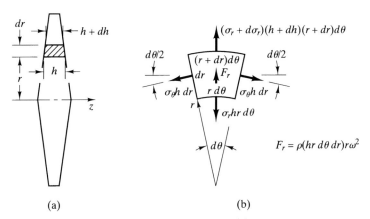

FIGURE 8.5-1 (a) Cross section of a tapered disk. (b) Forces that act on a differential element.

The only known solution of Eq. 8.5-2 applicable to a disk with a central hole comes from adopting a *hyperbolic profile*, $h = H/r^s$, where H is a constant and s is a positive number. A disk of hyperbolic profile is not optimal, but it is better than a flat disk. A disk of arbitrary profile can be analyzed computationally, and software may also have the ability to optimize the profile.

Disk of Uniform Strength. Rather than analyze, and seek $u = u(r)$ when $h = h(r)$ is prescribed, we can design, and use Eq. 8.5-2 to obtain $h = h(r)$ when $u = u(r)$ is prescribed. This approach, devised in Sweden about 1900, determines the shape of a solid disk such that the maximum shear stress is the same at every point. A representative shape for such a disk is shown in Fig. 8.5-2a.

The state of stress at all points of the disk is assumed to be $\sigma_z = 0$ and $\sigma_r = \sigma_\theta = \sigma_o$, where σ_o is a stress chosen by the designer, perhaps by applying a safety factor to the yield strength of the material. Thus $\tau_{max} = \sigma_o/2$ everywhere in the disk. Equations 8.1-2 yield

$$\frac{du}{dr} = \frac{u}{r} \quad \text{and} \quad u = \frac{1-\nu}{E} r \sigma_o \tag{8.5-3}$$

This u function satisfies compatibility and the stress-strain relation. The equilibrium equation is Eq. 8.5-2, which becomes, after substitution of Eq. 8.5-3 and rearrangement of terms,

$$\frac{dh}{h} = -\frac{\rho \omega^2}{\sigma_o} r \, dr \quad \text{hence} \quad h = h_o \exp\left(-\frac{\rho \omega^2}{2\sigma_o} r^2\right) \tag{8.5-4}$$

The latter equation is obtained by integration of the former, with the constant of integration obtained from the boundary condition $h = h_o$ at $r = 0$.

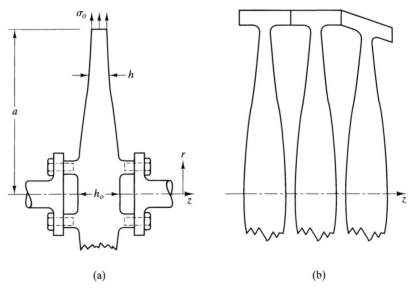

FIGURE 8.5-2 (a) Cross section of a disk of uniform strength. (b) Cross sections of disks of uniform strength, joined without a central shaft.

Equation 8.5-4 requires that the rim $r = a$ carry radial stress σ_o. This stress is supplied by turbine blades, which load the disk but do not contribute to its strength. In design, known values of ρ, ω, σ_o, and h at $r = a$ can be used to determine h_o, whereupon Eq. 8.5-4 supplies $h = h(r)$. Since σ_o provided by turbine blades is directly proportional to ω^2, the ratio ω^2/σ_o in Eq. 8.5-4 is constant, and the disk remains a disk of uniform strength for any ω.

Shaft Attachment. If an annular disk is attached to a continuous shaft of radius b, there are high stresses in the disk at $r = b$, caused both by shrink-fit attachment and by spinning (Figs. 8.2-1a and 8.4-2b). The condition $\sigma_r = \sigma_o$ cannot be provided at the edge of a hole; therefore, a disk designed according to Eq. 8.5-4 cannot have a central hole. One way to omit the hole is shown in Fig. 8.5-2a, where shaft segments terminate at the disk rather than penetrating it. With this arrangement there are only minor stress concentrations, but there may be alignment difficulties. Another arrangement is shown in Fig. 8.5-2b, where the central shaft is omitted altogether. Adjacent disks are connected at edges of flared rims that also serve for the attachment of turbine blades. This arrangement also has much greater bending stiffness than a central shaft with disks attached.

8.6 PLASTIC ACTION IN THICK-WALLED CYLINDERS

As shown early in this chapter, a high-pressure cylinder may be massive. Weight can be saved by installing a favorable pattern of stresses before internal pressure is applied, by means of shrink-fit construction as discussed in Section 8.3. It can also be done by *autofrettage* ("self-binding"), in which part or all of the material is made plastic by applied pressure, which is then released, leaving behind favorable residual stresses. If autofrettage causes all the material to yield and unloading is elastic, the residual stress pattern is that of a cylinder with infinitely many jackets of infinitesimal thickness, each optimally shrink-fitted. Subsequent loading up to the autofrettage pressure is carried without additional yielding.

In preceding sections of this chapter, failure is defined as the onset of yielding. When plastic action is allowed, failure may be defined as the fully plastic condition. Clearly, for a given safety factor, the latter definition allows a greater working pressure.

Yield Criterion. In developing equations for plastic action under internal pressure, we will assume that $\sigma_\theta > \sigma_z > \sigma_r$, use the maximum shear stress yield criterion, and assume that the material does not strain harden. Reasons for these choices are as follow.

Arguments of plasticity theory show that, in plastic material of the cylinder, the increment of axial strain is

$$d\epsilon_z = \frac{d\epsilon_p}{\sigma_e}\left[\sigma_z - \frac{\sigma_\theta + \sigma_r}{2}\right] \tag{8.6-1}$$

where $d\epsilon_p$ is the "effective" plastic strain increment, and σ_e is defined by Eq. 2.6-12. If plane strain conditions prevail in the cylinder, then $\epsilon_z = 0$, and Eq. 8.6-1 yields $\sigma_z = (\sigma_\theta + \sigma_r)/2$. On the other hand, if the plane stress condition $\sigma_z = 0$ prevails, previous *elastic* equations for the internally pressurized cylinder show that $\sigma_\theta > \sigma_z > \sigma_r$. We will see that the same is true under plastic conditions if the fraction of the cylinder that yields is not greater than would be of practical use. Accordingly, we can say with reasonable confidence that σ_θ and σ_r are the extreme principal stresses in the actual cylinder. Therefore $\tau_{max} = (\sigma_\theta - \sigma_r)/2$. If σ_Y is the yield stress in a tensile test, then $\tau_{max} = \sigma_Y/2$ when the material undergoes plastic flow. Thus, for plastic material without strain hardening, Fig. 2.7-1b, according to the maximum shear stress yield criterion,

$$\sigma_\theta - \sigma_r = \sigma_Y \tag{8.6-2}$$

Alternatively, the von Mises yield criterion might be chosen. With plane strain conditions, $\sigma_z = (\sigma_\theta + \sigma_r)/2$, and Eqs. 2.6-12 and 3.3-2 predict that plastic flow occurs when $\sigma_\theta - \sigma_r = 2\sigma_Y/\sqrt{3}$, where $2/\sqrt{3} = 1.15$. Therefore, by using Eq. 8.6-2, and by neglecting strain hardening, we slightly underpredict the internal pressure required to make a cylinder fully plastic. Experiments show that residual stresses produced by unloading from the plastic state are in good agreement with an "average" yield criterion; that is, based on $\sigma_\theta - \sigma_r = 1.078\sigma_Y$ [8.2].

The discussion of the present section pertains to loading by internal pressure, for which $\sigma_\theta > \sigma_r$ at all points. For loading by external pressure, $\sigma_r > \sigma_\theta$, and Eq. 8.6-2 must be replaced by $\sigma_r - \sigma_\theta = \sigma_Y$.

Fully Plastic Condition. The equilibrium equation, Eq. 8.1-1, is independent of material properties and is therefore valid for both elastic and plastic conditions. However, Eqs. 8.1-2 apply only to linearly elastic conditions. Also, when all material has yielded and there is no strain hardening, radial displacement u cannot be determined. Instead of using Eqs. 8.1-2, we combine Eq. 8.1-1 with Eq. 8.6-2, then integrate. Thus, with $\omega = 0$,

$$\frac{d\sigma_r}{dr} - \frac{\sigma_Y}{r} = 0 \quad \text{from which} \quad \sigma_r = \sigma_Y \ln r + C \tag{8.6-3}$$

where C is a constant of integration. From the boundary condition $\sigma_r = -p_i$ at inner radius $r = b$, we determine that $C = -p_i - \sigma_Y \ln b$. Equations 8.6-2 and 8.6-3 then yield

Stresses at radius r in plastic region
$$\begin{cases} \sigma_r = -p_i + \sigma_Y \ln \dfrac{r}{b} & (8.6\text{-}4a) \\ \sigma_\theta = -p_i + \sigma_Y \left(1 + \ln \dfrac{r}{b}\right) & (8.6\text{-}4b) \end{cases}$$

Pressure p_i that makes the entire cylinder yield is called p_{fp}, the fully plastic pressure. It is obtained from Eq. 8.6-4a by setting $\sigma_r = 0$ at outer radius $r = a$.

$$\text{Fully plastic pressure:} \quad p_{fp} = \sigma_Y \ln \frac{a}{b} \tag{8.6-5}$$

Stresses given by Eqs. 8.6-4 with $p_i = p_{fp}$ are shown in Fig. 8.6-1a. The σ_θ and σ_r curves are vertically equidistant, as demanded by Eq. 8.6-2. For $a/b > 2.72$, σ_θ is negative at $r = b$ when $p = p_{fp}$, and, depending on σ_z, Eq. 8.6-2 may not be a valid assumption.

Residual stresses, which are stresses that remain when internal pressure has been released, are calculated by superposing the plastic stresses of Eqs. 8.6-4 on

Section 8.6   Plastic Action in Thick-Walled Cylinders   249

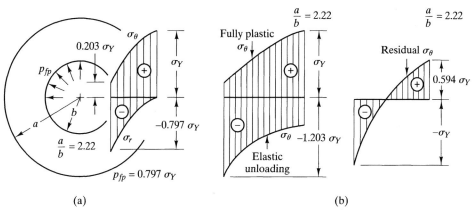

FIGURE 8.6-1 (a) Stresses in a cylinder when fully plastic pressure p_{fp} is applied, for $a/b = 2.22$. (b) Development of the residual circumferential stress pattern, for $a/b = 2.22$.

elastic stresses from Eqs. 8.2-2, with $p_o = 0$ and the *negative* internal pressure $p_i = -p_{fp}$ (if unloading from the fully plastic state). The residual radial stress is zero at $r = b$ and at $r = a$. Figure 8.6-1b shows the residual circumferential stress for the case $a/b = 2.22$. This a/b is the largest for which the residual σ_θ does not exceed σ_Y in compression. In other words, if the Bauschinger effect is ignored and the bore is not to yield in compression when p_{fp} is removed, a/b must not exceed 2.22. If no "reversed yielding" has occurred and p_i is reapplied, elastic conditions prevail until p_i reaches p_{fp}.

If material is to be brought to the verge of yielding by internal pressure $p_i = p_Y$, what subsequent increase in p_i is allowed by autofrettage to pressure p_{fp}? For a single cylinder, we obtain p_Y from Eq. 8.2-4 with $\tau_{max} = \sigma_Y/2$. For a compound cylinder, we obtain p_Y from the first of Eqs. 8.3-9. The ratio of p_{fp} to p_Y, for the particular case $a/b = 2.22$, is 2.00 for a single cylinder and 1.45 for a compound cylinder of optimal proportions.

When $a/b = 2.22$ and $p_i = p_{fp}$, axial stress σ_z of Eq. 8.2-3 is equal to σ_θ at $r = b$. Elsewhere σ_z is intermediate to σ_θ and σ_r. Thus we confirm that stresses in a fully plastic cylinder do not violate the assumptions contained in Eq. 8.6-2. One can show that the same is true of stresses in a partially yielded cylinder.

Equations that approximate the *burst pressure* are available. For this calculation one acknowledges that, after considerable strain at yield stress σ_Y, the stress-strain curve rises due to strain hardening. With σ_{ult} the ultimate strength of the material, two equations proposed for burst pressure are

$$p_b \approx \sigma_{ult} \ln \frac{a}{b} \quad \text{and} \quad p_b \approx \frac{2\sigma_Y}{\sqrt{3}}\left(2 - \frac{\sigma_Y}{\sigma_{ult}}\right) \ln \frac{a}{b} \qquad (8.6\text{-}6)$$

The first equation for p_b comes from Eq. 8.6-5, and incorporates the assumption that the stress-strain diagram is approximately flat-topped in the vicinity of σ_{ult}. The second equation seems to agree better with experimental evidence [8.3].

Partial Plasticity. The plastic region may extend only to radius c, leaving the outer portion of the cylinder elastic (Fig. 8.6-2a). Stresses in the elastic portion $c < r \leq a$ can be calculated from Eqs. 8.2-2. With p_c the magnitude of σ_r at $r = c$, these stresses are

$$\sigma_r = p_c \frac{c^2}{a^2 - c^2}\left(1 - \frac{a^2}{r^2}\right) \quad \text{and} \quad \sigma_\theta = p_c \frac{c^2}{a^2 - c^2}\left(1 + \frac{a^2}{r^2}\right) \quad (8.6\text{-}7)$$

To obtain p_c, we note that yielding impends in the elastic material at $r = c$. Accordingly, from Eqs. 8.6-2 and 8.6-7,

$$p_c = \sigma_Y \frac{a^2 - c^2}{2a^2} \quad (8.6\text{-}8)$$

With this value of p_c, Eqs. 8.6-7 become

Elastic region,
$r > c$, with
yielding at $r = c$
$$\begin{cases} \sigma_r = \sigma_Y \dfrac{c^2}{2a^2}\left(1 - \dfrac{a^2}{r^2}\right) & (8.6\text{-}9\text{a}) \\[6pt] \sigma_\theta = \sigma_Y \dfrac{c^2}{2a^2}\left(1 + \dfrac{a^2}{r^2}\right) & (8.6\text{-}9\text{b}) \end{cases}$$

The internal pressure that produces yielding out to any radius c in the range $b \leq c \leq a$ is obtained by equating radial stresses at $r = c$ from Eqs. 8.6-4a and 8.6-9a and solving for p_i. Assigning the name p_p to this value of p_i, we obtain

$$\text{Yielding in range } b \leq r < c: \quad p_p = \sigma_Y\left[\ln\frac{c}{b} + \frac{a^2 - c^2}{2a^2}\right] \quad (8.6\text{-}10)$$

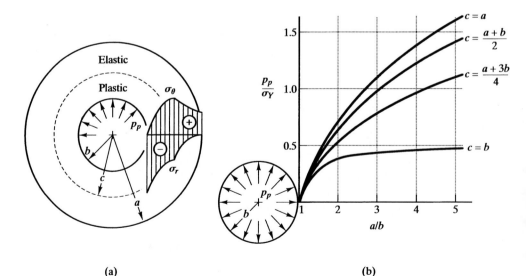

(a)   (b)

FIGURE 8.6-2 (a) Stresses in a partially yielded cylinder. (b) Relations between internal pressure, cylinder proportions, and depth of yield ($c = b$ when yielding begins; $c = a$ when fully plastic).

Figure 8.6-2b shows the ratio p_p/σ_Y versus a/b for depths of yield equal to 0, 1/4, 1/2, and 1 times the wall thickness. We see that if even a small depth of yield is allowed, a substantial increase in internal pressure is made possible.

For given values of a, b, and σ_Y, what is the largest internal pressure p_p for which no reversed yielding will occur when p_p is released? Neglecting the Bauschinger effect, we ask that σ_θ at $r=b$ not exceed σ_Y in compression when stresses from Eqs. 8.6-4b and 8.2-2b (with $p_i = -p_p$) are superposed. Thus

$$(-p_p + \sigma_Y) - p_p \frac{b^2}{a^2 - b^2}\left(1 + \frac{a^2}{b^2}\right) = -\sigma_Y \quad \text{yields} \quad p_p = \sigma_Y \frac{a^2 - b^2}{a^2} \quad (8.6\text{-}11)$$

The latter equation can also be used to obtain radius a when b, σ_Y, and p_p are prescribed. Although unyielded material provides reserve strength, depth of yielding beyond $r=c$ is avoided so that there will not be cyclic yielding as internal pressure is repeatedly applied and released.

It is reasonable to ask if beneficial residual stresses can be induced by the service load rather than as an extra step in manufacture. Autofrettage by service load is acceptable if dimensional changes associated with yielding need not be adjusted by machining before the cylinder is put to use.

EXAMPLE

A cylinder for which $b = 10$ mm, $\sigma_Y = 530$ MPa, and $\sigma_{ult} = 880$ MPa must sustain internal pressure $p_p = 490$ MPa. What external radius a is required? What is the radius c of the elastic-plastic boundary, and what are σ_θ and σ_r at that radius? What is the safety factor, based on the burst pressure?

We will assume that it is acceptable to bring the bore to the verge of reversed yielding when p_p is released. Accordingly, Eq. 8.6-11 is applicable; it provides $a = 36.4$ mm. Radius c (not needed for design) is then obtained from Eq. 8.6-10 as $c = 17.1$ mm. The value of σ_θ at $r=c$ can be obtained from either Eq. 8.6-4b or Eq. 8.6-9b. Both equations yield $\sigma_{\theta c} = 323$ MPa. Then, from Eq. 8.6-2, $\sigma_{rc} = \sigma_{\theta c} - \sigma_Y = -207$ MPa. Axial stress σ_z, Eq. 8.2-3, is 40 MPa, making it the intermediate principal stress; therefore assumptions contained in Eq. 8.6-2 are not violated. From the first of Eqs. 8.6-6, the burst pressure is about 1140 MPa, which implies a safety factor of $1140/490 = 2.3$.

The cylinder of the foregoing example can be compared with single and compound elastic cylinders. According to Eq. 8.2-4, a single cylinder with $\sigma_Y = 530$ MPa cannot contain 490 MPa without yielding, even for infinite a/b. A compound cylinder need not yield, but according to the first of Eqs. 8.3-9 its external radius must be 133 mm, so that its weight is 14 times that of the autofrettaged cylinder.

8.7 PLASTIC ACTION IN SPINNING DISKS

If yielding is allowed in a spinning disk, what rotational speed defines failure? A reasonable answer is ω_{fp}, the speed that makes the disk fully plastic. In calculating ω_{fp} we will use the maximum shear stress yield criterion and ignore strain hardening. Thus, for reasons like those explained in connection with Eq. 8.6-2, we are almost sure to underestimate the bursting speed. Nevertheless, ω_{fp} is the speed at which large dimensional changes commence, which may be an important consideration.

Analysis. From Fig. 8.4-2b we see that σ_r does not exceed σ_θ in a spinning *elastic* disk, whether or not there is a central hole. With some confidence we can say that the same is true if the disk is fully plastic. If also $\sigma_z = 0$, as previously assumed, then the extreme principal stresses are σ_θ and zero. Thus the maximum shear stress yield criterion becomes

$$\tau_{max} = \frac{\sigma_\theta - 0}{2} \quad \text{and} \quad \tau_{max} = \frac{\sigma_Y}{2} \quad \text{hence} \quad \sigma_\theta = \sigma_Y \tag{8.7-1}$$

Because σ_θ is independent of r in the fully plastic condition, ω_{fp} can be calculated simply, without recourse to a differential equation. We cut the disk in half, as shown in Fig. 8.7-1a, and use inertia forces F_a and F_b in an equilibrium equation. F_a accounts for the entire half disk of radius a, and F_b accounts for the smaller disk of radius b cut from it. Centroidal distances are $r_a = 4a/3\pi$ for F_a and $r_b = 4b/3\pi$ for F_b. Stress σ_o represents the average radial stress at $r = a$ due to turbine blades attached to the rim, if present. With disk thickness h taken as constant, equilibrium of forces in Fig. 8.7-1 requires

$$2ah\sigma_o + F_a - F_b - (2a - 2b)h\sigma_Y = 0 \tag{8.7-2}$$

With F_a and F_b as stated in Fig. 8.7-1, Eq. 8.7-2 provides

$$\omega_{fp}^2 = \frac{3}{\rho(a^3 - b^3)}[(a - b)\sigma_Y - a\sigma_o] \tag{8.7-3}$$

Let us compare ω_{fp} with the speed ω_Y at which yielding begins, in a flat disk with $\sigma_o = 0$ and no central hole. Yielding begins at the center of the disk. From Eq. 8.4-2b, with $r = 0$ and $\sigma_\theta = \sigma_Y$,

$$\omega_Y^2 = \frac{8\sigma_Y}{(3 + \nu)\rho a^2} \tag{8.7-4}$$

For $\nu = 0.27$, Eqs. 8.7-3 and 8.7-4 provide $\omega_{fp}/\omega_Y = 1.11$. In other words, only an 11% increase in speed will cause plastic action to spread from the center of the disk to its rim. If

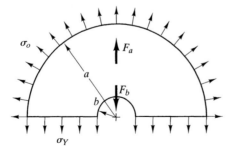

F_a associated with $0 < r < a$:
$$F_a = m_a r_a \omega_{fp}^2 = \rho \frac{\pi a^2}{2} h \left(\frac{4a}{3\pi}\right) \omega_{fp}^2$$

F_b associated with $0 < r < b$:
$$F_b = m_b r_b \omega_{fp}^2 = \rho \frac{\pi b^2}{2} h \left(\frac{4b}{3\pi}\right) \omega_{fp}^2$$

h = thickness, ρ = mass density

FIGURE 8.7-1 Half of a disk with a central hole: quantities used in the calculation of fully plastic speed ω_{fp}. F_a and F_b are inertia forces.

there is a small central hole ($b \ll a$) then ω_{fp} is practically unchanged, but since the elastic σ_θ is doubled by a small central hole, ω_Y^2 is halved if $\sigma_\theta = \sigma_Y$ is to be maintained near the center of the disk. Therefore $\omega_{fp}/\omega_Y = 1.57$ for a flat disk with a small central hole and $\nu = 0.27$.

Remarks. Stress $\sigma_r = \sigma_r(r)$ in the plastic condition can be determined by integration of the differential equation. The process is similar to what is done with Eq. 8.6-3, and is left as an exercise. If stopping the disk after fully plastic spinning produces no reversed yielding, residual stresses at standstill can be obtained by superposing plastic stresses $\sigma_\theta = \sigma_Y$ and $\sigma_r = \sigma_r(r)$ and the *negative* of elastic stresses from Eqs. 8.4-2 or 8.4-3, with $\omega = \omega_{fp}$ in the elastic equations. If fully plastic loading includes radial stress σ_o at $r = a$, unloading also contributes $\sigma_r = \sigma_\theta = -\sigma_o$ throughout a solid disk, or stresses given by Eqs. 8.2-2 for a disk with a central hole.

As argued following Eq. 8.7-4, for a flat disk the ratio ω_{fp}/ω_Y is not substantially greater than unity. The ratio is exactly unity for a disk of uniform strength (Eq. 8.5-4) because yielding begins at all points simultaneously. Accordingly, residual stresses produced by plastic action are not as beneficial to a spinning disk as they are to a pressurized cylinder.

The bursting speed can be estimated from Eq. 8.7-3 by replacing σ_Y by the tensile strength σ_{ult}; however, for some materials this estimate is almost 30% high [1.6]. The maximum speed may be governed by failure of attached turbine blades. A fragmenting disk can do great damage, so safety precautions for testing a disk at its design speed are mandatory. The dangers of bursting a pressurized cylinder are much less, provided that the working fluid is liquid rather than gas and care is taken to eliminate any gas trapped in the cylinder or entrained in the fluid. One should also restrain the vessel against the impulse of liquid that may be ejected, propelled by release of elastic energy stored in vessel walls.

PROBLEMS

8.2-1. Show that if the vessel wall is thin, Eq. 8.2-2b reduces to $\sigma_\theta = p_i R/t$, where $R = (a + b)/2$ is the mean radius and $t = a - b$ is the wall thickness.

8.2-2. Imagine that, due to an oversight, the term $C_1 r$ is omitted from Eq. 8.1-5. Constant C_2 can be determined by use of the known σ_r at $r = b$, or by slicing the cylinder in half lengthwise and considering the equilibrium of an entire half-cylinder loaded by p_i and σ_θ. Calculate C_2 both ways, with $T = 0$, $\omega = 0$, and $p_o = 0$. Why do the resulting expressions for C_2 not agree? At $r = b$ and for $a/b = 3$, compare the two approximations for σ_θ with the exact σ_θ.

8.2-3. (a) A cylinder with end caps, for which $a = 40$ mm and $b = 22$ mm, is made of a material that yields in tension at 400 MPa. Based on the maximum shear stress failure criterion, what is the allowable internal pressure if the safety factor is to be 1.6?

(b) Repeat part (a), but use the von Mises failure criterion.

8.2-4. A cylinder with ends caps has the following dimensions and properties: $a = 20$ mm, $b = 10$ mm, $E = 200$ MPa, and $\nu = 0.28$. Measurements show that a circumferential line on the outside changes in length from 10.0000 mm to 10.0040 mm when internal pressure p_i is applied. What is p_i?

8.2-5. Consider a cylinder of length L under internal pressure, sliced in half lengthwise by a plane that contains the axis. Calculate the force produced by stress σ_θ on the cross section between $r=b$ and $r=c$, where $b<c<a$. Then, for $a=3b$ and $c=2b$, express this force as a fraction of the pressure force $2p_i bL$. For these dimensions, what fraction of the total weight of the cylinder lies between $r=b$ and $r=c$?

8.2-6. Let $\sigma_z = 0$ and $p_o = 0$, and assume that inner radius b and design pressure p_i are prescribed. In the following situations, obtain a formula for the required outer radius a of the cylinder.
  (a) The material is brittle and fails at normal stress σ_{ult} in uniaxial tension.
  (b) The material is ductile and yields at normal stress σ_Y in uniaxial tension. Use the maximum shear stress failure criterion.
  (c) Repeat part (b), but use the von Mises failure criterion.
  (d) Let $p_i = \sigma_{ult}/4$ in part (a) and $p_i = \sigma_Y/4$ in parts (b) and (c). In each part, determine a in terms of b.

8.2-7. Internal pressure p_i and external pressure p_o both act on a cylinder with end caps.
  (a) If $\sigma_\theta = 0$ at $r=b$, what are p_o and σ_θ at $r=a$, in terms of p_i, a, and b?
  (b) For $a=2b$ and $\sigma_\theta = 0$ at $r=b$, does the maximum shear stress appear at $r=a$ or at $r=b$? What is this shear stress, in terms of p_i?

8.2-8. Internal pressure p_i and external pressure p_o both act on a cylinder. For what range of the ratio p_i/p_o is the magnitude of σ_θ greater at $r=a$ than at $r=b$? Express the answer in terms of a/b.

8.2-9. A thick-walled *sphere* is loaded by internal pressure p_i.
  (a) Using the sketch shown, determine the differential equation of equilibrium that relates radial stress σ_r and tangential stress σ_t.
  (b) In terms of radial displacement u, strains are $\epsilon_r = du/dr$ and $\epsilon_t = u/r$. Use this information and the stress-strain relation to obtain an equation analogous to Eq. 8.1-3 (without T and ω terms).
  (c) Solve the equation of part (b); that is, obtain a result analogous to Eq. 8.1-5.

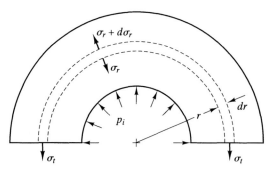

PROBLEM 8.2-9

8.3-1. A shrink fit is used to attach a solid disk within an annular disk of outer radius a. In the manner of Fig. 8.2-1, qualitatively sketch the distributions of σ_r and σ_θ for $0 < r < a$.

8.3-2. A compound cylinder has dimensions $b=8$ mm, $c=11$ mm, and $a=15$ mm. When loaded by internal pressure p_i and by external pressure $p_o = 6$ MPa, circumferential stresses at $r=b$ and at $r=c$ in the inner cylinder are equal. Determine the relation between p_i and the shrink-fit pressure p_c.

8.3-3. Two thin-walled cylinders of different material are bonded together and also bonded to rigid walls, as shown. Assume that walls serve only to prevent axial displacement, and make approximations consistent with $t_i \ll R$ and $t_o \ll R$. Set up, but do not solve, sufficient equations to determine circumferential and longitudinal stresses produced by (a) uniform heating, and (b) internal pressure p_i.

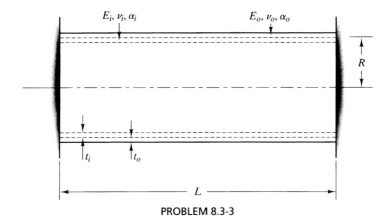

PROBLEM 8.3-3

8.3-4. The shrink-fit pressure between a jacket and a solid cylinder ($b=0$) is p_c.
   (a) In terms of a, c, p_c, and E, what was the radius difference Δ, Eq. 8.3-5, before doing the shrink fit?
   (b) If cylinder and jacket were both exactly the same length L before the shrink fit, what is the difference in their lengths afterwards? Assume that $\sigma_z = 0$.

8.3-5. A compound cylinder has dimensions $b=1$ mm, $c=2$ mm, and $a=3$ mm. If dimension b does not change from its pre-shrink-fit value when the cylinder is loaded by both internal pressure p_i and shrink-fit pressure p_c, what is p_c as a fraction of p_i? Let $\nu=0.25$.

8.3-6. Consider a compound cylinder for which $a=3b$ and $c=2b$. Under the combination of internal pressure p_i and shrink-fit pressure p_c, both cylinder and jacket have maximum circumferential stress $\bar{\sigma}_\theta$. Determine p_i and p_c in terms of $\bar{\sigma}_\theta$.

8.3-7. A compound cylinder, for which $E=72$ GPa, $b=40$ mm, $c=60$ mm, and $a=120$ mm, is to be designed for internal pressure 110 MPa. What must be the interference 2Δ in diameter if the net σ_θ at $r=c$ in the outer cylinder is to be 90 MPa?

8.3-8. A flat plate of thickness t is locally and suddenly heated, such that the temperature is raised an amount T throughout a small circular region of radius c and thickness t. Show that the maximum shear stress in the heated zone is $\alpha E T/4$.

8.3-9. In a compound cylinder for which $a=3b$ and $c=2b$, elastic moduli of inner and outer cylinders have the relation $E_i = kE_o$, where k is a constant. Assume that $\nu_i = \nu_o$ and that the cylinders fit exactly ($p_c = 0$). Loading is by external pressure p_o only. Obtain an expression for σ_θ at $r=b$ in terms of p_o, k, and Poisson's ratio. Check the answer against Eq. 8.2-2b for the special case $k=1$.

8.3-10. Dimensions of a compound cylinder are $a=75$ mm, $b=25$ mm, and $c=50$ mm. Let $E=200$ GPa. The shrink-fit interference in diameter is $2\Delta=0.02$ mm.
   (a) What is the shrink-fit pressure p_c?
   (b) What internal pressure p_i will produce the same τ_{max} at $r=b$ as at $r=c$ in the outer cylinder?

8.3-11. Show that for optimal design of a compound cylinder, Eqs. 8.3-8 and 8.3-9, radial stress at $r=c$ is $-p_i/2$ when internal pressure p_i is applied.

8.3-12. An optimally proportioned compound cylinder is loaded by the design value of internal pressure. Show that σ_θ at $r=c$ in the jacket is always greater than σ_θ at $r=b$ for all values of a/b.

8.3-13. Derive the ratios stated in Eqs. 8.3-10.

8.3-14. (a) Use the dimensions and shrink-fit pressure p_c of Example 2, Section 8.3, and calculate the radial stress at $r=c$ when $p_i = 150$ MPa (rather than the design pressure of 238 MPa).

(b) Consider the jacket alone, loaded at $r=c$ by the magnitude of σ_r calculated in part (a), and determine σ_θ at $r=a$.

(c) Show that the same σ_θ is obtained by superposing the σ_θ due to p_i on the entire compound cylinder and the σ_θ due to p_c on the jacket.

8.3-15. (a) A compound cylinder is to have a bore of 5 mm radius. The design pressure, at which yielding begins, is to be $p_i = 180$ MPa. Using $\sigma_Y = 400$ MPa and the maximum shear stress failure criterion, determine radii a and c of an optimal design.

(b) At the design pressure, what are the effective stresses σ_e (Eq. 2.6-12) at the bore and at $r=c$ in the jacket? Assume that $\sigma_z = 0$.

(c) Repeat part (b), but let the cylinder have end caps ($\sigma_z \neq 0$).

8.3-16. A compound cylinder for which $a=30$ mm and $b=10$ mm is optimally designed ($c = \sqrt{ab}$), but due to a manufacturing error, Δ is made 50% larger than Eq. 8.3-9 requires. Consider net circumferential stresses at $r=a$ and at $r=c$ in the jacket when the design value of p_i is applied. By what factors have these stresses been increased by the manufacturing error?

8.3-17. If a cylinder is to contain very high internal pressure without yielding, the layer $b < r < c$ may be composed of wedges, as shown. A cylindrical gasket at $r=b$ prevents p_i from penetrating into the flat contact surfaces between wedges. There are many wedges, so that $\sigma_\theta \approx 0$ for $b < r < c$. What is τ_{max} at $r=c$ in the outer cylinder, in terms of p_i, a, b, and c? Show that this τ_{max} is minimum for $a/c = \sqrt{3}$. For $a/c = \sqrt{3}$, determine p_i in terms of a, b, and σ_Y.

8.4-1. (a) Cut the flywheel of Fig. 8.4-1 in half along a diameter and ignore the spokes. Obtain the formula for σ_θ by considering the force exerted by σ_θ and the inertia force of the half-flywheel.

(b) If straight and uniform spokes are to exert no radial forces on a flywheel, what must be the relation between ρ and E of the flywheel and the corresponding properties of the spokes?

8.4-2. Show that Eq. 8.4-3b indicates a stress concentration factor of 2 as b approaches zero.

8.4-3. Imagine that a solid flat disk is cut in half along a diameter. Show that the force exerted by σ_θ on the cross section thus exposed balances the inertia force of the half-disk.

8.4-4. Let circumferential stress in a flat disk with a central hole be approximated by use of Eq. 8.4-1. If $\nu = 0.27$ and the error in peak stress is not to exceed 5% in magnitude, what is the maximum permissible a/b ratio?

8.4-5. The stress $\sigma_\theta = 252$ MPa is calculated in text following Eq. 8.4-6 by considering the rotation of an annular disk. Show that the same stress is obtained by superposing σ_θ in a spinning solid disk on σ_θ from the shrink fit.

8.4-6. In the example problem of Section 8.4, what are σ_r and σ_θ at $r=0.08$ m in the annular disk if $\omega = \omega_L/2$?

8.4-7. In the example problem of Section 8.4, imagine that the disk is held motionless while torque is applied to one end of the shaft. If the coefficient of static friction is 0.4, what torque is needed to make the shrink fit slip? What then is the shear stress in the shaft?

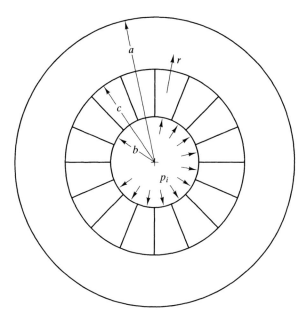

PROBLEM 8.3-17

8.4-8. A flat steel disk has outer radius 100 mm, Poisson's ratio 0.27, and mass density 7860 kg/m³. If shear stress in the disk is not to exceed 70 MPa, what speed ω is permissible if
   (a) the disk has no hole?
   (b) the disk has a central hole of radius 40 mm?

8.4-9. A flat steel turbine disk of 0.76 m outside diameter is to rotate at 3000 rpm. At this speed, turbine blades and shrouding produce a radial rim loading of 14.0 MPa. If τ_{max} is not to exceed 60 MPa, is a small central hole permissible?

8.4-10. The disk shown has uniform thickness. The wedge-shaped slots are bounded by radial lines and remove half the material in the range $a < r < c$.
   (a) Obtain a formula for the average radial stress σ_o at $r = a$ due to spinning.
   (b) Let $a = 41$ cm, $b = 10$ cm, $c = 66$ cm, and $\rho = 7860$ kg/m³. What is σ_θ at $r = b$ due to $\omega = 1800$ rpm?

8.4-11. A flat disk of uniform thickness h is shrunk onto a shaft of the same material, causing contact pressure p_c when $\omega = 0$. Assume that the shaft behaves as a solid disk of thickness h. The coefficient of static friction between disk and shaft is μ.
   (a) At what speed ω_o does the contact pressure fall to zero?
   (b) At what speed is it possible to transmit the greatest power between disk and shaft? Express the answer in terms of ω_o.
   (c) Obtain an expression for the maximum power.

8.4-12. A flat aluminum disk of 26 cm outer radius is shrunk onto a solid steel disk of the same thickness and 4 cm radius. Assume that the modulus and density of aluminum are each one-third of the corresponding values for steel. Let $E_{st} = 200$ GPa and assume that Poisson's ratio is 0.3 for both materials. For the shrink fit, Δ of Eq. 8.3-5 is 0.004 cm.
   (a) What is the shrink-fit pressure p_c?
   (b) At what speed ω will the shrink fit loosen?

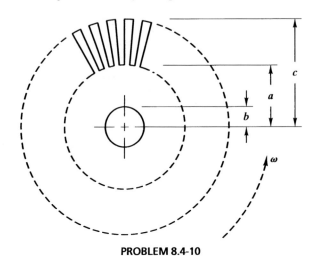

PROBLEM 8.4-10

8.4-13. A stepped disk is made by welding together two disks, both thin, flat, and of the same material, but differing in thickness, as shown. The stepped disk is free of stress at standstill. There is no central hole.

(a) At speed ω, a radial stress σ_c appears at $r = c$ in the outer disk. Set up equations from which σ_c can be obtained in terms of quantities shown in the sketch. Ignore stress concentrations at the step.

(b) Outline a calculation strategy that will lead to a good choice of a/c.

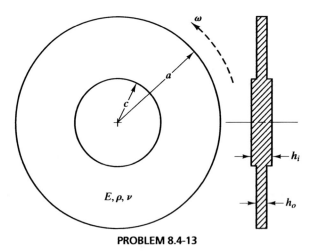

PROBLEM 8.4-13

8.5-1. A steel disk of outer radius 0.60 m is designed as a disk of uniform strength for rotational speed 3600 rpm. If, at radius $r = 0.40$ m, $\sigma_\theta = 80$ MPa and $h = 0.020$ m, what is the center thickness h_o? Let $\rho = 7860$ kg/m^3.

8.5-2. Disks of uniform strength are to be designed. In each case $\rho = 7860$ kg/m^3, the outer radius is 0.60 m, and the rim thickness is 0.02 m. Calculate h_o at $r = 0$, h at $r = 0.20$ m, and h at $r = 0.40$ m in each of the following cases.

(a) $\sigma_o = 200$ MPa at $\omega = 1800$ rpm.
(b) $\sigma_o = 200$ MPa at $\omega = 3600$ rpm.
(c) $\sigma_o = 140$ MPa at $\omega = 3600$ rpm.
(d) $\sigma_o = 140$ MPa at $\omega = 4200$ rpm.

8.5-3. In a simple approximate analysis of a spinning disk with a central hole and having variable thickness [8.4], σ_r is ignored and σ_θ is taken as $\sigma_\theta = k/r$, where k is a constant. Constant k is determined by considering radial forces on a slice that spans angle $d\theta$ circumferentially and extends from the inner radius to the outer radius.

(a) Establish the equation that yields k if $h = h(r)$ is prescribed.

(b) If the approximate analysis is applied to a *flat* disk with $a = 4b$, determine k and the percentage error of σ_θ at $r = b$. Let $\nu = 0.27$.

8.6-1. (a) Write expressions for σ_r and σ_θ as functions of r for the fully plastic case $p_i = p_{fp}$.

(b) What is the fully plastic value of *external* pressure p_o?

(c) Repeat part (a) for the case $p_o = p_{fp}$.

8.6-2. A cylinder is to be made of a material whose tensile yield strength is $\sigma_Y = 280$ MPa. The inside radius, working pressure p_i in service, and safety factor are to be 120 mm, 60 MPa, and 2.0 respectively. In parts (a), (b), and (c), determine outer radius a. Also, in parts (b) and (c), determine the weight of the cylinder as a fraction of the weight of the cylinder in part (a).

(a) Use a single cylinder, with failure defined as the onset of yielding according to the maximum shear stress yield criterion.

(b) Use a compound cylinder, with failure defined as the onset of yielding according to the maximum shear stress yield criterion.

(c) Failure is defined as reaching the fully plastic condition.

8.6-3. Consider a compound cylinder for which $b = 30$ mm, $c = 45$ mm, $a = 75$ mm, and two different materials are used. For inner and outer cylinders respectively, tensile yield strengths are 300 MPa and 450 MPa. Use the maximum shear stress yield criterion to determine the fully plastic internal pressure.

8.6-4. For $a/b = 2.22$, Eq. 8.6-5 yields $p_{fp} = 0.797\sigma_Y$. Determine, if possible, the a/b ratios of single and compound cylinders of the same material, if $p_i = 0.797 \sigma_Y$ and τ_{max} must be limited to $\sigma_Y/2$ in both cases. Also, in comparison with the autofrettaged cylinder, how much does the compound cylinder weigh?

8.6-5. Obtain a formula for the fully plastic internal pressure of a thick-walled *sphere*. Suggestion: Consider the answer to Problem 8.2-9(a).

8.6-6. Consider the cylinder designed in the example problem of Section 8.2. What internal pressure p_{fp} makes the cylinder fully plastic? Qualitatively sketch residual stresses σ_r and σ_θ that remain when p_{fp} is released. If internal pressure p_i is reapplied, what can you say about the value of p_i that produces renewed yielding?

8.6-7. Repeat Problem 8.6-6, but now consider the compound cylinder designed in Example 2 of Section 8.3.

8.6-8. The fully plastic internal pressure is applied to a cylinder for which $a < 2.22b$, then released. Show by integration that the residual σ_θ distribution is self-equilibrating.

8.6-9. A cylinder has inside and outside radii $b = 1.0$ cm and $a = 4.0$ cm. The material yields in tension at stress σ_Y. Yielding to an intermediate radius c is produced by internal pressure p_i. After p_i is released, material at $r = b$ is on the verge of reversed yielding. Determine p_i and c. Ignore possible strain hardening and the Bauschinger effect.

8.6-10. Imagine that after applying the fully plastic internal pressure to a cylinder and then releasing it, σ_θ at the bore $r=b$ must be $\sigma_\theta = -\beta\sigma_Y$, where $0 < \beta \le 1$.
  (a) Determine a formula for the ratio a/b in terms of β.
  (b) What is a/b if $\beta = 0.5$?
  (c) What is the largest a/b for which the formula is valid? Physically, what happens if a/b exceeds this value?

8.7-1. Obtain ω_{fp}, Eq. 8.7-3, by working with the differential equation. That is, use Eq. 7.7-4a, set $\sigma_\theta = \sigma_Y$, integrate to obtain an expression for σ_r, evaluate the constant of integration, and solve for ω_{fp}.

8.7-2. Consider a disk of hyperbolic profile with a central hole. Let $h = H/r$, where H is a constant.
  (a) Obtain a formula for the fully plastic speed ω_{fp}, based on the maximum shear stress yield criterion.
  (b) For the case $a = 80$ mm and $b = 10$ mm, by what factor is ω_{fp} increased over ω_{fp} for a disk of uniform thickness?

8.7-3. Obtain formulas for σ_r and σ_θ as functions of r in the following disks of uniform thickness spinning at fully plastic speed, with $\sigma_o = 0$.
  (a) A solid disk.
  (b) A disk with a central hole, for which $a = 4b$.

8.7-4. (a) Determine the fully plastic speed ω_{fp} in terms of σ_Y and ρ for the slotted disk of Problem 8.4-10. Use the maximum shear stress yield criterion. Let $a = 41$ cm, $b = 10$ cm, and $c = 66$ cm.
  (b) By what factor has ω_{fp} been reduced by the slotted material? That is, compare ω_{fp} of part (a) with ω_{fp} for a disk with no material hung on the rim at $a = 41$ cm.
  (c) Does the slotted material yield at speed ω_{fp} of part (a)?

8.7-5. A flat disk with a central hole is spun at speed ω_{fp}, then stopped. Let $\sigma_o = 0$ and assume that unloading is elastic. Show by integration that the residual σ_θ distribution is self-equilibrating.

8.7-6. A flat disk with a central hole is spun to its fully plastic speed. What is the smallest a/b for which there will be no reversed yielding when spinning is stopped? Let $\sigma_o = 0$ and $\nu = 0.3$.

CHAPTER 9

Torsion

9.1 INTRODUCTION. SAINT-VENANT TORSION THEORY

Assumptions. In this chapter the material is regarded as homogeneous, isotropic, and linearly elastic. Displacements are small, so that displacements and stresses are directly proportional to load. Torsion members are prismatic. Exceptions to some of these assumptions are made for brief discussions of pretwisted strips (Section 9.14) and plastic torsion (Section 9.15).

The foregoing assumptions are made in classical torsion theory, which was devised by Saint-Venant and is discussed in Sections 7.11 and 7.12. Important results of the theory are summarized in the following subsection for use in the present chapter. These results are valid for a prismatic member loaded only by torque about its longitudinal axis, with all cross sections free to warp (see Fig. 7.11-1a). Torque must be constant along the length of the member. Torque is applied to both ends of the member, by shear stresses that have the same distribution over end cross sections as over every other cross section.

Summary of Results. Equations that follow come from Sections 7.11 and 7.12, and are used early in the present chapter. For this use it is not essential that their derivation be mastered.

Cross sections warp axially, but do not distort in their own planes. Thus, in Fig. 7.11-1, the cross section shown has x-direction displacements but no y- or z-direction displacements. With x the axial direction, all stresses are independent of x. Normal stresses σ_x, σ_y, σ_z, and shear stress τ_{yz} are all zero. The only nonzero stresses are shear stresses τ_{xy} and τ_{zx}, shown in Fig. 9.1-1. (Because of boundary conditions, τ_{xy} and τ_{zx} are zero at corner B, so corner B is entirely free of stress.) By definition,

$$\text{Rate of twist:} \quad \beta = \frac{d\theta}{dx} \tag{9.1-1}$$

where θ is the angle of twist. Rate of twist β is independent of axial coordinate x in pure torsion. In terms of the Prandtl stress function $\phi = \phi(y,z)$, with G the shear modulus and A the cross-sectional area,

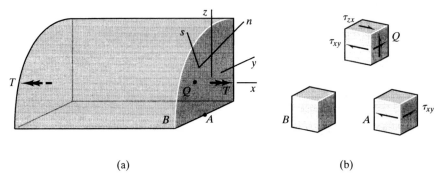

FIGURE 9.1-1 (a) Torsion of a prismatic bar. Coordinate x is axial. (b) Stresses at selected points in a typical cross section, in the xyz coordinate system. Stresses are shown in directions defined as positive.

$$\text{Governing equation:} \quad \frac{\partial^2 \phi}{\partial y^2} + \frac{\partial^2 \phi}{\partial z^2} = -2G\beta \quad (9.1\text{-}2)$$

$$\text{Boundary condition:} \quad \phi \text{ is constant on a boundary} \quad (9.1\text{-}3)$$

$$\text{Torque:} \quad T = 2 \int \phi \, dA \quad (9.1\text{-}4)$$

$$\text{Shear stress:} \quad \tau_{sx} = -\frac{\partial \phi}{\partial n} \quad (9.1\text{-}5)$$

Orthogonal axes n and s originate at an arbitrary point and are arbitrarily inclined, but lie in the plane of a cross section (Fig. 9.1-1a). Equation 9.1-5 says that shear stress in direction s is equal in magnitude to the slope of ϕ in direction n. At a given point, the largest shear stress τ_{sx} appears when n is directed normal to a contour $\phi =$ constant.

A torsion problem is solved by determining a function $\phi = \phi(y,z)$ that satisfies Eqs. 9.1-2 and 9.1-3. The associated torque and stress are then obtained from Eqs. 9.1-4 and 9.1-5. Compact formulas for rate of twist β and torsional stiffness T/β are

$$\beta = \frac{T}{GK} \qquad \frac{T}{\beta} = GK \quad (9.1\text{-}6)$$

Factor K is determined by solving the torsion problem. In general, K cannot exceed J, the centroidal polar moment of inertia of the cross-sectional area. Usually K is much smaller than J. Only for a circular cross section is K equal to J.

Outline of the Chapter. The membrane analogy is discussed first. The analogy makes it easier to deal with the Prandtl stress function ϕ. Particular attention is given to thin-walled open sections. Next, closed sections of one or more cells are discussed. The effects of partial or complete restraint of warping in thin-walled open sections are then considered. Restraint of warping is found to produce significant normal stresses and additional shear stresses. Final sections of the chapter deal with bimoment, large angles of rotation, pretwisted bars, and plastic action.

Repeated reference is made to "open" and "closed" sections. Open sections include standard I and channel sections. Closed sections include those of pipes. In mathematical terms, open sections are simply connected and closed sections are multiply connected.

9.2 MEMBRANE ANALOGY

Equation 9.1-2 has the same form as the equation that describes small lateral deflections of a stretched membrane, initially flat, that is then loaded by uniform pressure on one side. Lateral deflection of the membrane corresponds to the Prandtl stress function ϕ. This correspondence is called the *membrane analogy* or the *soap film analogy*. It was suggested by Prandtl in 1903 and has been used for the experimental solution of torsion problems. The main value of the analogy is that it enables us to draw qualitative conclusions about stress, torque, and rate of twist. This assessment is done by imagining a membrane stretched over an opening of the same shape as the cross section to be studied, and visualizing the inflated shape of the membrane. In some cases this thought process leads directly to useful quantitative results.

The governing equation of the membrane is obtained as follows. Let tension S in the membrane be the same at all points and in all directions, and remain essentially unchanged when lateral pressure is applied. These properties are supplied by a soap film, or by a membrane whose lateral deflection ζ is small. Dimensions of S are [force/length]. For small ζ, small-angle approximations such as $\sin(\partial\zeta/\partial y) = \partial\zeta/\partial y$ can be used. In Fig. 9.2-1a, force increments $S\,dz$ on differential element $dy\,dz$ of the membrane have the net upward component

$$-S\,dz\,\sin\frac{\partial\zeta}{\partial y} + S\,dz\,\sin\left(\frac{\partial\zeta}{\partial y} + \frac{\partial^2\zeta}{\partial y^2}\,dy\right) = S\frac{\partial^2\zeta}{\partial y^2}\,dy\,dz \qquad (9.2\text{-}1)$$

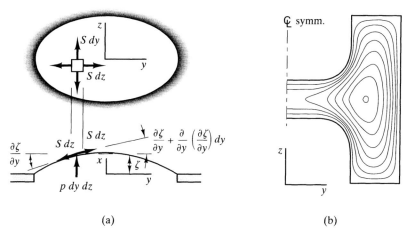

(a) (b)

FIGURE 9.2-1 (a) Equilibrium of a differential element of an inflated membrane. (b) Membrane for the right half of a stocky I section [9.1]. Contours (lines of constant elevation ζ) are separated by equal elevation changes $\Delta\zeta$.

A similar result is obtained from force increments $S\,dy$. Pressure p exerts an upward force $p\,dy\,dz$. Equating to zero the sum of all three upward force components, and dividing by $S\,dy\,dz$, we obtain

$$\frac{\partial^2 \zeta}{\partial y^2} + \frac{\partial^2 \zeta}{\partial z^2} = -\frac{p}{S} \qquad (9.2\text{-}2)$$

This equation says that the sum of curvatures in mutually perpendicular directions has the constant value $-p/S$ all over the membrane. Upon comparing Eqs. 9.1-2 and 9.2-2, we see that ζ corresponds to ϕ, and p/S corresponds to $2G\beta$. Shear stresses correspond to slopes of the membrane. Torque corresponds to twice the volume between the deflected membrane and the surface $\zeta = 0$.

Figure 9.2-1b shows experimentally determined contours of a membrane over an I-shaped hole in a flat plate [9.1]. All contours are separated by the same elevation increment $\Delta\zeta$. Contour lines are most closely spaced near the fillets. Therefore these locations have the greatest membrane slope and the greatest shear stress in a twisted bar of this cross section. The reader should try to visualize the deflected membrane that provides these contours.

What if the cross section contains a hole? Where there is no material, there can be no shear stress, so the membrane must have zero slope. In the membrane analogy one imagines that a hole is represented by a rigid but weightless plate, Fig. 9.2-2a, attached to the membrane on its boundary and allowed only to translate in the x direction. The only x-direction forces on the "floating plate" come from pressure p and plate-normal components of membrane tension S [9.2]. Torque on the shaft is related to volume under the entire deflected surface, including volume between the floating plate and the surface $\zeta = 0$.

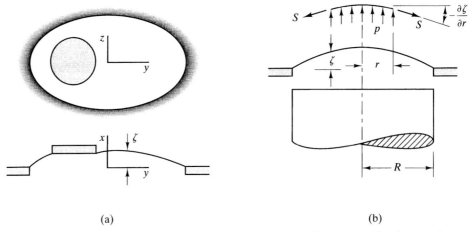

FIGURE 9.2-2 (a) Plan and section views of the membrane analogy for a cross section that contains a hole. (b) Membrane analogy for a bar of solid circular cross section.

EXAMPLE

To establish confidence in the membrane analogy, we apply it to a solid shaft of circular cross section, for which results are already known from elementary theory.

The inflated membrane over a circular hole in a flat plate is axisymmetric (Fig. 9.2-2b). Summation of vertical forces on the portion of radius r yields a differential equation, which can then be integrated to provide elevation $\zeta = \zeta(r)$.

$$p\pi r^2 + S\frac{d\zeta}{dr}2\pi r = 0 \quad \text{hence} \quad \zeta = \frac{p}{4S}(R^2 - r^2) \tag{9.2-3}$$

The boundary condition $\zeta = 0$ at $r = R$ is used in obtaining the latter equation. Having established the shape of the membrane, we can convert to stress-function notation by substituting ϕ for ζ and $2G\beta$ for p/S. Torque is then obtained by use of Eq. 9.1-4.

$$\phi = \frac{G\beta}{2}(R^2 - r^2) \quad \text{hence} \quad T = 2\int_0^R \phi(2\pi r\, dr) = G\beta J \tag{9.2-4}$$

where $J = \pi R^4/2$. Thus $\beta = T/GJ$, which is the correct result. From Eqs. 9.1-5 and 9.2-4, with $n = r$ and $G\beta = T/J$, we obtain $\tau = Tr/J$, which also is correct.

9.3 MEMBRANE ANALOGY FOR THIN-WALLED OPEN SECTIONS

In this section the membrane analogy is used to obtain simple formulas that relate torque, rate of twist, and shear stress in thin-walled open sections.

Narrow Rectangular Cross Section. Consider the cross section in Fig. 9.3-1a, with $b \gg t$. By visualizing the inflated membrane, we conclude that its shape is practically the same over most of the length $-b/2 < y < b/2$. Temporarily we neglect what happens near ends $y = \pm b/2$. The differential equation that describes elevation ζ of the membrane is obtained by summing x-direction forces on a portion of the shaded strip

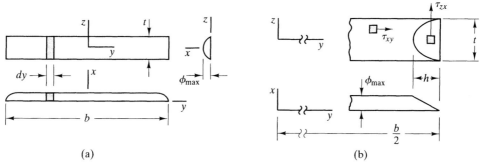

FIGURE 9.3-1 (a) Membrane analogy for a narrow rectangular cross section of dimensions b by t. Plan and elevation views are shown. (b) Construction for approximate calculation of torque directly from shear stresses.

in Fig. 9.3-1a, between $+z$ and $-z$ (see Fig. 9.2-2b and Eq. 9.2-3 for a similar calculation). This equation and its integrated form are

$$p(2z)\,dy + 2\left(S\frac{d\zeta}{dz}\,dy\right) = 0 \quad \text{hence} \quad \zeta = -\frac{p}{2S}z^2 + C \qquad (9.3\text{-}1)$$

Since $\zeta = 0$ at $z = \pm t/2$, the integration constant is $C = pt^2/8S$. We convert to stress-function notation by the substitutions $\zeta = \phi$ and $p/S = 2G\beta$. Thus

$$\zeta = \frac{p}{2S}\left(\frac{t^2}{4} - z^2\right) \quad \text{and} \quad \phi = G\beta\left(\frac{t^2}{4} - z^2\right) \qquad (9.3\text{-}2)$$

Torque T is obtained by use of Eq. 9.1-4. Integration is facilitated by a handbook formula for area under a parabola (area $= 2t\phi_{\max}/3$ in the present notation). Thus, with $\phi_{\max} = G\beta t^2/4$,

$$T = 2\int \phi\,dA = 2b\left(\frac{2t\phi_{\max}}{3}\right) = G\beta\frac{bt^3}{3} \qquad (9.3\text{-}3)$$

The largest shear stress appears on the long edges $z = \pm t/2$. This stress can be expressed in terms of β, or in terms of T by substitution from Eq. 9.3-3:

$$\tau = \left|\frac{d\phi}{dz}\right|_{t/2} = G\beta t \quad \text{or} \quad \tau = \frac{Tt}{bt^3/3} \qquad (9.3\text{-}4)$$

Equations 9.3-3 and 9.3-4 can be written in the following forms.

$$\beta = \frac{T}{GK} \quad \tau = \frac{Tt}{K} \qquad (9.3\text{-}5)$$

where

$$\text{For } b \gg t: \quad K = \frac{bt^3}{3} \qquad (9.3\text{-}6)$$

Factor K is analogous to polar moment J of the cross-sectional area, which is used for analysis of circular cross sections. However, J is *emphatically NOT* the same as K. For a thin-walled open section, K is a small fraction of J.

End Effects. The foregoing analysis implies that shear stress is infinite at $y = \pm b/2$ because the stress function $\phi = \phi(y,z)$ terminates in vertical cliffs. Actually there are small zones of smooth transition to zero (Fig. 9.3-1a). Let us idealize a transition zone as shown in Fig. 9.3-1b, where h is a small distance comparable to t. We now ask whether it is possible to obtain torque T by using shear stresses directly, instead of using Eq. 9.1-4.

A y-direction force increment $\tau_{xy}\,dA$ has moment arm z, and produces torque increment $dT = -\tau_{xy} z\,dA$, negative because T is considered positive when its vector is in the positive x direction. Thus, for $b \gg h$, torque due to τ_{xy} is approximately

$$\int (-\tau_{xy} z)\,dA = -\int_{-t/2}^{t/2} \frac{d\phi}{dz} zb\,dz = 2G\beta b \int_{-t/2}^{t/2} z^2\,dz = \frac{1}{2}G\beta\frac{bt^3}{3} \qquad (9.3\text{-}7)$$

Section 9.3 Membrane Analogy for Thin-Walled Open Sections

In Fig. 9.3-1b, for $b \gg h$, a z-direction force increment $\tau_{zx} \, dA$ has moment arm $b/2$ and produces torque increment $dT = \tau_{zx}(b/2) \, dA$. Each end of the cross section makes this contribution. The stress is $\tau_{zx} = \phi_{max}/h$. Thus, using the formula for integration also used in Eq. 9.3-3, torque due to τ_{zx} is approximately

$$2\left[\tau_{zx}\frac{b}{2}\int dA\right] = 2\frac{\phi_{max}}{h}\frac{b}{2}\frac{2th}{3} = 2\frac{G\beta t^2}{4h}\frac{bth}{3} = \frac{1}{2}G\beta\frac{bt^3}{3} \qquad (9.3\text{-}8)$$

We see that the torques of Eqs. 9.3-7 and 9.3-8 are each half of the correct total stated in Eq. 9.3-3. It would be easy to overlook torque produced by stress τ_{zx} because τ_{zx} acts on much less area than τ_{xy} and is in fact smaller in magnitude than τ_{xy}. Yet τ_{zx} contributes substantially to torque because its moment arm is large. To reduce chances for mistakes in torque calculation, it is better to use Eq. 9.1-4 than the method of Eqs. 9.3-7 and 9.3-8.

More General Shapes. Equations 9.3-5 can be applied to thin-walled open sections of arbitrary shape. The appropriate K in each case can be established by the membrane analogy, as follows.

Let the narrow rectangular cross section in Fig. 9.3-1 be bent in the yz plane into another shape, such as a circle (Fig. 9.3-2a). The inflated membrane continues to form a parabolic hill over the cross section (again with small transitions to zero at ends). If thickness t is uniform, Eq. 9.3-6 is changed only in that dimension b becomes the midline length $b = 2\pi R$, where R is the mean radius. The angle section in Fig. 9.3-2c is more complicated because there are two thicknesses, and the inflated membrane exhibits two different parabolic hills. However, p/S is the same for both parts of the membrane and $G\beta$ is the same for both parts of the stress function. According to Eq. 9.3-3, the two parts carry torques $T_1 = G\beta b_1 t_1^3/3$ and $T_2 = G\beta b_2 t_2^3/3$. The total torque is $T = T_1 + T_2 = GK$, where $K = (b_1 t_1^3 + b_2 t_2^3)/3$. In general, for a thin-walled cross section built of n parts, each having midline length b_i and uniform thickness t_i,

$$\text{For } b_i \gg t_i: \quad K = \frac{1}{3}\sum_{i=1}^{n} b_i t_i^3 \qquad (9.3\text{-}9)$$

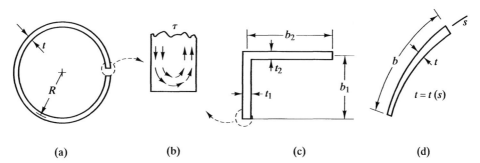

FIGURE 9.3-2 Thin-walled open cross sections. (a) Tube of circular cross section, slit open lengthwise. (b) Shear stress distribution. (c) Angle section. (d) Arbitrary section.

If thickness t changes continuously, as in Fig. 9.3-2d, one can imagine dividing the total midline length into many increments, each having thickness t_i and length ds_i. Thus Eq. 9.3-9 becomes an integral:

$$\text{For} \quad t = t(s) \quad \text{and} \quad b \gg t: \quad K = \frac{1}{3}\int_0^b t^3\, ds \tag{9.3-10}$$

Equation 9.3-4 or 9.3-5 shows that when t is not uniform the largest shear stress on the boundary of a cross section appears where t is largest. However, we have thus far neglected the effect of stress raisers such as re-entrant corners (see Section 9.4).

9.4 REMARKS AND FORMULAS

Qualitative Insights. For a prismatic member in a state of pure twist, useful conclusions about shear stress and torsional stiffness can be reached by visualizing the inflated membrane and making simple arguments.

In Fig. 9.4-1a, with $t_2 > t_1$, the membrane has greater elevation across section C than across section D. This conclusion is obvious when one considers that the membrane profile at C contains the membrane profile at D as its upper portion. The greater the thickness t, the greater is the membrane slope at the boundary of the cross section. Accordingly, if local stress raisers are temporarily ignored, one expects to find the largest shear stress where the largest inscribed circle touches the boundary. There the nominal shear stress, exclusive of any stress concentration factor, is given by Eq. 9.3-5, with t taken as the diameter of the inscribed circle. Thus, in Fig. 9.4-1a we predict that $\tau_B > \tau_C > \tau_D$.

In Fig. 9.4-1b we might expect to find the largest shear stress at points G. However, stresses at corners E and F are likely to be larger because re-entrant corners are local stress raisers. Point E is a sharp re-entrant corner, where theory of elasticity

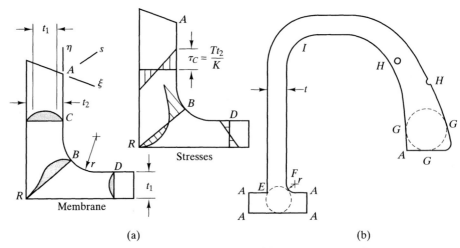

FIGURE 9.4-1 (a) Angle cross section. (b) Arbitrary cross section.

predicts infinite shear stress [6.1]. (Similar local stresses appear at corners of a keyway in a shaft, where fatigue failure often begins.) Stress is finite at fillet F and can be reduced by increasing the small fillet radius r. In Fig. 9.4-1a, if the section is thin-walled and $t_1 = t_2 = r$, the stress concentration factor at B is 1.55 [9.3]. Further increases in r increase the stress at B because the inscribed circle becomes large. Stress at B can be reduced by removing material from right-angle corner R, which makes the shape more like region I in Fig. 9.4-1b, thus reducing the diameter of the inscribed circle.

Points H in Fig. 9.4-1b identify a circular hole and a semicircular notch. Visualization of the membrane shows that they are stress raisers. If the radius of the notch or hole is much less than the thickness of the cross section and its radius of curvature, it can be shown that the local shear stress is doubled [6.1].

In Fig. 9.4-1a, point A is a sharp external corner, axes ξ and η are tangent to adjacent sides at the corner, and s is an arbitrary direction. Because the membrane (or the stress function ϕ) has zero elevation along boundaries, $\partial\phi/\partial\xi = 0$ and $\partial\phi/\partial\eta = 0$ at A. Therefore, according to the chain rule,

$$\frac{\partial\phi}{\partial s} = \frac{\partial\phi}{\partial\xi}\frac{\partial\xi}{\partial s} + \frac{\partial\phi}{\partial\eta}\frac{\partial\eta}{\partial s} = 0 \tag{9.4-1}$$

That $\partial\phi/\partial s = 0$ means there is no shear stress in any direction at a sharp external corner. There are five such corners in Fig. 9.4-1b, all labeled A. The smaller the interior angle at a sharp external corner, the less torsionally active is that corner.

In Fig. 9.4-2a, imagine that the square boundary is achieved by gradually distorting the inscribed circle, with the inflated membrane maintained at the same p/S. It is easy to see that in doing so the volume under the membrane and its maximum slope both increase. Hence we conclude that, for a given $G\beta$, a square cross section of side length $2a$ carries greater torque and has greater stress than a circular cross section of diameter $2a$.

Finally let us compare square and rectangular cross sections of the same area, Fig. 9.4-2, again for the same p/S. Clearly the square membrane encloses greater volume and has greater maximum slope. Generalizing, we conclude that when two cross sections have the same area and the same G, the more compact cross section is stiffer, and for the same β is more highly stressed (unless the less compact section contains holes or notches).

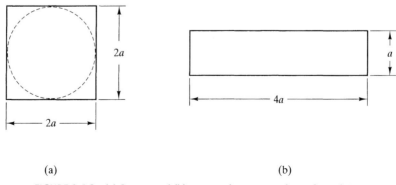

(a)   (b)

FIGURE 9.4-2 (a) Square and (b) rectangular cross sections of equal area.

Tabulations and Formulas. Table 9.4-1 provides analytically obtained data for selected aspect ratios of a rectangular cross section [9.1]. We see that the largest shear stress always appears at the middle of the longer sides. For large a/b the tabulation agrees with results in Section 9.3. When a/b is not large, Eq. 9.3-6 predicts too large a factor K because the zero elevation of the membrane along the shorter sides is not taken into account. An approximate stress formula for the rectangle, having an error less than 5% in magnitude for all $a/b \geq 1$, is [9.4]

$$\tau_A \approx \frac{15a + 9b}{5a^2b^2} T \tag{9.4-2}$$

For a cross section built of two or more narrow rectangles, as for I and T sections, K is usually greater than predicted by Eq. 9.3-9. One reason is that the membrane rises higher where parts intersect. This effect is seen in Fig. 9.2-1b (where, however, the section shown is not thin-walled). Another reason is that re-entrant corners are usually provided with fillets rather than being sharp. For these reasons, Eq. 9.3-9 should be replaced by

$$K = \frac{\alpha}{3} \sum_{i=1}^{n} b_i t_i^3 \tag{9.4-3}$$

As approximations for standard rolled sections, $\alpha = 1$ for narrow rectangles and angle sections, $\alpha = 1.15$ for channel, cross, T, and Z sections, and $\alpha = 1.30$ for I and wide-flange sections [9.5, 9.6]. More elaborate formulas are available for these and other shapes of cross section [1.6, 1.7].

Saint-Venant obtained the following approximate formula for stiffness in pure torsion.

$$\frac{T}{\beta} \approx \frac{GA^4}{40J} \tag{9.4-4}$$

where A is the area of the cross section and $J = I_y + I_z$ is the polar moment of inertia of A about its centroid. Equation 9.4-4 should not be used for a closed section, or for an

TABLE 9.4-1 Formulas and data for pure twisting of a solid rectangular cross section.

$$\tau_A = \frac{T}{C_\tau ab^2}$$
$$\beta = \frac{T}{C_\beta Gab^3}$$
$$\tau_A = C_{AB} G\beta b$$
$$(C_{AB} = C_\beta/C_\tau)$$

a/b	C_τ	τ_A/τ_B	C_β	C_{AB}
1.0	0.208	1.00	0.141	0.675
1.2	0.219	1.07	0.166	0.759
1.5	0.231	1.16	0.196	0.848
2.0	0.246	1.26	0.229	0.930
2.5	0.258	1.31	0.249	0.968
3.0	0.267	1.33	0.263	0.985
4.0	0.282	1.34	0.281	0.997
6.0	0.298	1.35	0.298	1.000
10.0	0.312	1.35	0.312	1.000
∞	0.333	1.35	0.333	1.000

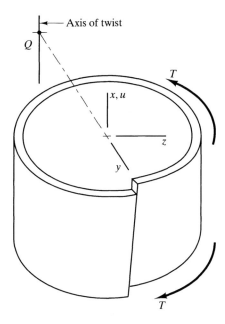

FIGURE 9.4-3 Warping displacement of a slit tube.

open section that has projecting arms and re-entrant corners. Otherwise, it is applicable to any shape of cross section (rectangular, elliptical, triangular, etc.), with an error of no more than about ±12%.

Warping and Center of Twist. In general, torsion causes cross sections to warp. Thus, with x the axial direction and u the axial displacement, $u = u(y,z)$. Figure 9.4-3a shows the warping displacement of a tube of circular cross section that is slit open lengthwise. This deformation is easily demonstrated by twisting a rolled-up sheet of paper. It is possible to identify an *axis of twist*, about which all cross sections of a prismatic member rotate when torque is constant along the member. Point Q, called the *center of twist*, is the intersection of the axis of twist with the plane of a cross section. It is usually agreed that the center of twist is coincident with the *shear center*, which is summarized in Section 9.9 and fully discussed in Sections 10.5 to 10.8. However, some investigators have disputed the coincidence [9.7]. If a cross section has two or more axes of symmetry, the center of twist coincides with the centroid of the cross section.

9.5 PURE TWIST OF THIN-WALLED TUBES OF ONE CELL

Consider a thin-walled prismatic member of arbitrarily shaped closed cross section, Fig. 9.5-1a. Thickness t may vary, although an abrupt change in t produces a stress concentration. Torque is constant along the member; therefore there are no normal stresses. Only the shear stress shown in Fig. 9.5-1b is significant. It is reasonable to assume that this stress is practically constant through the wall thickness (see remarks that follow Eq. 9.5-9). The outer surface of coupon $ABCD$, and its counterpart on the inside, are free surfaces and therefore carry no shear stress.

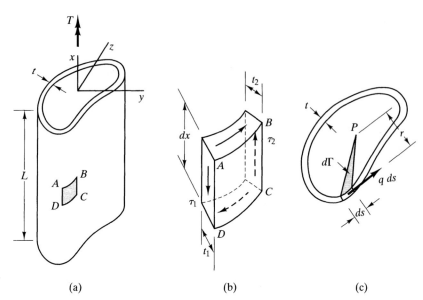

FIGURE 9.5-1 (a) Thin-walled tube under torsion. (b) Coupon cut from the tube wall. (c) The increment of torque is $dT = r(q\,ds)$.

Consider the equilibrium of x-direction forces on the arbitrary coupon $ABCD$. Thickness and shear stress do not vary in the x direction. Therefore

$$\tau_1 t_1\, dx = \tau_2 t_2\, dx \qquad (9.5\text{-}1)$$

We see that the product τt *is constant around the tube*. This quantity is given the symbol q and is called *shear flow*. Dimensions of q are [force/length]. If τ is known, q follows, and vice versa.

$$q = \tau t \quad\text{or}\quad \tau = \frac{q}{t} \qquad (9.5\text{-}2)$$

Stress $\tau = q/t$ in a closed section does not change direction across thickness t, in contrast to τ in an open section, which changes direction as shown in Fig. 9.3-2b. In an open section, $q = 0$ unless warping is partially or completely restrained.

In what follows, the formula for shear flow is obtained by a method that is related to methods used later in this chapter to analyze restraint of warping. The formula may also be obtained by the membrane analogy.

Formulas. In Fig. 9.5-1c, let P be an arbitrarily located point and s a coordinate along the midline of the cross section. Force increment $q\,ds$ is directed tangent to the midline, and r is the moment arm of this force. The increment of torque is $dT = r(q\,ds)$. The shaded triangular area is $d\Gamma = r\,ds/2$. Therefore, $dT = 2q\,d\Gamma$. Integrating all around the midline, and recalling that q is independent of s, we obtain

$$T = 2q\Gamma \quad\text{or}\quad q = \frac{T}{2\Gamma} \quad\text{and}\quad \tau = \frac{T}{2\Gamma t} \qquad (9.5\text{-}3)$$

Area Γ is *area enclosed by the midline*; it is *NOT* cross-sectional area A of the tube, which is much less. The symbol Γ is adopted so that area enclosed will not be confused with A.

An expression for rate of twist β can be obtained by the conservation of energy argument. Imagine that a tube of length L is twisted by a torque that gradually increases from zero to T. Simultaneously, the angle of twist between ends increases from zero to βL, and T does work $T(\beta L)/2$. This work is stored as strain energy. Strain energy per unit volume is $\tau^2/2G$, where τ is the shear stress when T is fully applied, and G is the shear modulus. The increment of volume is $Lt\,ds$. Therefore

$$\frac{1}{2}T(\beta L) = \oint \frac{\tau^2}{2G} Lt\,ds \qquad (9.5\text{-}4)$$

We can substitute for τ from Eq. 9.5-2, to obtain an expression for β in terms of q. Additionally we can substitute for q from Eq. 9.5-3 to obtain an expression for β in terms of T. Thus

$$\beta = \frac{q}{2G\Gamma}\oint \frac{ds}{t} \quad \text{or} \quad \beta = \frac{T}{4G\Gamma^2}\oint \frac{ds}{t} \qquad (9.5\text{-}5)$$

Equations 9.5-3 and 9.5-5 date from 1896 and are known as Bredt's formulas. They show that stress and rate of twist are reduced when Γ is increased. Therefore, for a given perimeter and thickness, a tube of circular cross section is best because it encloses greatest area.

Stress concentrations exist if there are internal corners. Data are available for a thin-walled closed tube of uniform thickness with a right-angle corner, like region B in Fig. 9.4-1a with $t_1 = t_2$ [6.1, 9.3]. An approximate fit to available data in the range $0 < t/r < 4$ yields $K_t \approx 1 + t/3r$ as the stress concentration factor for a closed tube, where the nominal stress is given by Eq. 9.5-3.

If G is not constant, Eqs. 9.5-3 are unchanged because they are based entirely on statics. In calculating β, a variable G can be modeled by using a constant G but a variable thickness. For example, if G is doubled over a lengthwise strip of the tube, in Eqs. 9.5-5 one can retain the original G but double the thickness where coordinate s spans the strip.

EXAMPLE

Consider a closed tube of circular cross section, Fig. 9.5-2a, with mean radius R and uniform thickness t, and loaded by torque T_c. From Eqs. 9.5-3 and 9.5-5, the resulting shear stress τ_c and rate of twist β_c are

$$\tau_c = \frac{T_c}{2\pi R^2 t} \qquad \beta_c = \frac{T_c}{4G(\pi R^2)^2}\left(\frac{2\pi R}{t}\right) = \frac{T_c}{2\pi G R^3 t} \qquad (9.5\text{-}6)$$

How accurate are these formulas? For a cross section of outer radius a and inner radius b, the exact stress at the outer surface is $\tau_e = T_c a/J$ and the exact rate of twist is $\beta_e = T_c/GJ$, where $J = \pi(a^4 - b^4)/2$. For $b = 0.8a$, for which $R/t = 4.5$ (a rather thick wall), $\tau_c/\tau_e = 0.911$ and $\beta_c/\beta_e = 1.012$. The ratio τ_c/τ_e is unity for $a \gg t$ and for $b = 0$. The minimum value of τ_c/τ_e is 0.828, and occurs when $b = 0.414a$, for which $R/t = 1.207$.

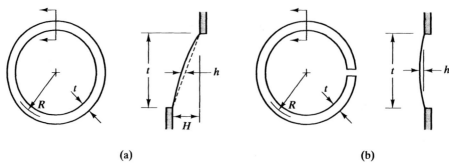

FIGURE 9.5-2 (a) Axial view of a closed tube of circular cross section, with enlarged sectional view of the membrane analogy. (b) Similar sketches for a tube slit open lengthwise.

Consider the effect of slitting the tube open, as in Fig. 9.5-2b. Clearly this results in a smaller membrane slope and a much smaller volume enclosed. If torque applied to the open section is called T_o, Eqs. 9.3-5 and 9.3-6 yield, in the open section,

$$\tau_o = \frac{T_o t}{2\pi R t^3/3} \quad \text{and} \quad \beta_o = \frac{T_o}{G(2\pi R t^3/3)} \tag{9.5-7}$$

If $T_c = T_o$, Eqs. 9.5-6 and 9.5-7 yield

$$\frac{\tau_o}{\tau_c} = 3\frac{R}{t} \quad \text{and} \quad \frac{\beta_o}{\beta_c} = 3\left(\frac{R}{t}\right)^2 \tag{9.5-8}$$

Thus, if $R/t = 10$, opening the tube increases stress by a factor of 30 and rate of twist by a factor of 300 for the same torque. Or, for the same rate of twist, $\beta_c = \beta_o$ in Eqs. 9.5-6 and 9.5-7, the ratio of torques is

$$\frac{T_c}{T_o} = 3\left(\frac{R}{t}\right)^2 \tag{9.5-9}$$

Remarks about Approximations. In deriving Eqs. 9.5-3 and 9.5-5, we assume that shear stress does not vary through the wall thickness. The membrane analogy, Fig. 9.5-2a, shows that this assumption corresponds to a membrane of constant slope H/t. An added bulge of height h is actually present but is ignored. Volume enclosed by this bulge is equal to the *entire* volume under the membrane that represents an open section of the same dimensions and the same $G\beta$. Thus $T_c + T_o$ is a better estimate than T_c of torque in the closed tube. If $t \ll R$, then $T_o \ll T_c$, and Eqs. 9.5-6 can be used with confidence.

These observations may be extended to a tube with fins, Fig. 9.5-3. For a single-cell tube with n attached fins,

$$T = G\beta\left[\frac{4\Gamma^2}{\oint_{\text{tube}}(ds/t)} + \frac{1}{3}\oint_{\text{tube}} t^3\, ds + \sum_{i=1}^{n}\left(\frac{1}{3}\int_{\text{fin}} t_i^3\, ds_i\right)\right] \tag{9.5-10}$$

The first term in parentheses comes from Eq. 9.5-5 and, in most cases, is far larger than the remaining two terms, which come from Eqs. 9.3-5 and 9.3-10. Equation 9.5-10

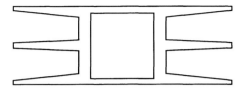

FIGURE 9.5-3 Cross section of a tube with fins.

shows that the error of neglecting fins may be less than the error of neglecting the adjustment discussed in the preceding paragraph. Unless fins are large and numerous, Eq. 9.5-3 provides a satisfactory estimate of stress; except, of course, for stress concentration effects.

9.6 PURE TWIST OF THIN-WALLED MULTICELL TUBES

Closed sections having more than one cell occur in aircraft and in ships. A thin-walled multicell tube can be analyzed by a simple extension of the single-cell analysis described in Section 9.5. In what follows, a two-cell tube is considered, and then results are generalized to allow a tube of n cells. It is assumed that the thickness of each cell wall is small in comparison with other dimensions of the cross section.

Shear flow q in a cell can be treated like loop current in an electric circuit. Thus, in Fig. 9.6-1a, q_1 and q_2 are regarded as flowing completely around their respective cells. Along a wall shared by two cells, shear flow is the algebraic sum of shear flows in adjacent cells: $q_w = q_1 - q_2$ in Fig. 9.6-1. To demonstrate this we consider junction A, Fig. 9.6-1b, and state that axially directed forces sum to zero. For a tube of axial length L,

$$-q_1 L + q_2 L + q_w L = 0 \quad \text{hence} \quad q_w = q_1 - q_2 \qquad (9.6\text{-}1)$$

Thus, in contrast to a single-cell tube, shear flow is not constant around a cell of a multicell tube.

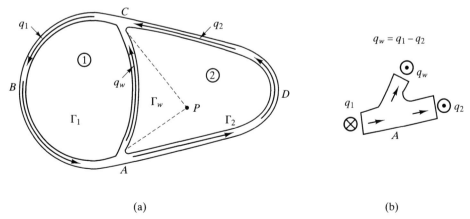

FIGURE 9.6-1 (a) Cross section of a thin-walled tube of two cells. Γ_w is the area of the three-sided subregion PCA. (b) Detail of the juncture at A. Symbols $\otimes$ and $\odot$ are arrows directed into and out of the paper respectively.

The argument used to obtain torque carried by a single-cell tube, Fig. 9.5-1c, also serves to obtain torque in a multicell tube. The location of point P in Fig. 9.6-1 is arbitrary. Force increment $q\,ds$ in a cell wall has moment arm r about point P. Summing contributions to torque from shear flows in all cell walls in Fig. 9.6-1a, we obtain

$$T = \int_{CBA} q_1 r\,ds + \int_{ADC} q_2 r\,ds - \int_{AC} q_w r\,ds \qquad (9.6\text{-}2a)$$

$$T = 2(\Gamma_1 + \Gamma_w)q_1 + 2(\Gamma_2 - \Gamma_w)q_2 - 2\Gamma_w(q_1 - q_2) \qquad (9.6\text{-}2b)$$

$$T = 2\Gamma_1 q_1 + 2\Gamma_2 q_2 \qquad (9.6\text{-}2c)$$

where Γ_1 and Γ_2 are cell areas, enclosed by midlines of cell walls.

Generalizing to the case of n cells, with i a typical cell, we have

$$T = 2\sum_{i=1}^{n} \Gamma_i q_i \qquad \beta = \frac{1}{2G\Gamma_i}\oint \frac{q_i}{t_i}ds_i \qquad (9.6\text{-}3)$$

where the latter equation comes from Eq. 9.5-5 and applies to any cell. Here we have recognized that β is the same for all cells. However, shear flow q_i is placed inside the integral because it is not constant all around a cell. If T is given, unknowns are β and the n shear flows q_i. The solution of such a problem is illustrated as follows.

EXAMPLE

Torque $T = 2(10^6)$ N·mm is applied to a three-cell tube whose cross section is shown in Fig. 9.6-2. Let $G = 70$ GPa. Determine the rate of twist, shear flows, and shear stresses. If corners resemble region B in Fig. 9.4-1a with $r = 1.2$ mm, what is the maximum shear stress?

We assign a positive direction, say counterclockwise, to shear flows q_1, q_2, and q_3 in cells 1, 2, and 3, then apply the second of Eqs. 9.6-3 to each cell. Thus

$$2G\beta = \frac{1}{40(80)}\left[\frac{40+80+40}{4}q_1 + \frac{35}{3}(q_1 - q_2) + \frac{45}{3}(q_1 - q_3)\right]$$

$$2G\beta = \frac{1}{40(35)}\left[\left(\frac{40}{4} + \frac{35}{5}\right)q_2 + \frac{40}{3}(q_2 - q_3) + \frac{35}{3}(q_2 - q_1)\right] \qquad (9.6\text{-}4)$$

$$2G\beta = \frac{1}{40(45)}\left[\left(\frac{45}{5} + \frac{40}{4}\right)q_3 + \frac{45}{3}(q_3 - q_1) + \frac{40}{3}(q_3 - q_2)\right]$$

Simultaneous solution of these three equations yields

$$q_1 = 82.50(2G\beta) \qquad q_2 = 84.15(2G\beta) \qquad q_3 = 87.88(2G\beta) \qquad (9.6\text{-}5)$$

The first of Eqs. 9.6-3 then yields

$$2(10^6) = 2(2G\beta)[40(80)(82.50) + 40(35)(84.15) + 40(45)(87.88)] \qquad (9.6\text{-}6)$$

from which $\beta = 13.2(10^{-6})$ rad/mm. Substitution of this β and $G = 70{,}000$ MPa into Eqs. 9.6-5 yields numerical values of the shear flows. These shear flows, and shear stresses $\tau_i = q_i/t_i$, are shown in Fig. 9.6-2b.

The maximum shear stress can be approximated by use of K_t stated in the paragraph that follows Eq. 9.5-5. In order to overestimate rather than underestimate τ_{max}, we use the largest

Section 9.6 Pure Twist of Thin-Walled Multicell Tubes

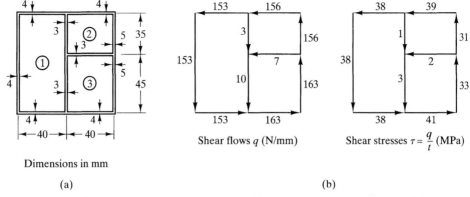

Dimensions in mm

(a)

Shear flows q (N/mm)

Shear stresses $\tau = \dfrac{q}{t}$ (MPa)

(b)

FIGURE 9.6-2 (a) Dimensions of a tube of three cells. (b) Calculated results for load $T = 2(10^6)$ N·mm.

nominal shear stress in Fig. 9.6-2, and base K_t on the larger thickness at the lower right corner. Thus

$$\tau_{\max} = K_t \tau_{\text{nom}} \approx \left[1 + \frac{5}{3(1.2)}\right] 41 = 98 \text{ MPa} \qquad (9.6\text{-}7)$$

Remarks. Symmetry of a cross section with respect to one or more axes simplifies calculations. In Fig. 9.6-3a, each of the four cells has the same q; therefore internal walls are unstressed (except for the small stresses described at the end of Section 9.5). In Fig. 9.6-3b, each of the four triangular cells has the same q; therefore there are only two different shear flows to calculate. Figure 9.6-3c is similar, and the radial walls are unstressed. The latter two cases differ from the first in that they have cells not bounded by a portion of the outer wall. These cells may carry significant shear flows. Visualization of the membrane analogy supports statements made in this paragraph.

If it can be confidently predicted that internal walls carry little shear flow, one can estimate stress and rate of twist in a multicell tube by neglecting internal walls and

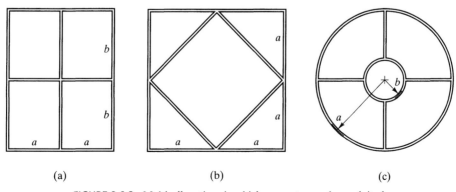

(a)   (b)   (c)

FIGURE 9.6-3 Multicell sections in which symmetry can be exploited.

treating the tube as if it had one cell. (Internal walls may be present to carry loads other than torsion, to stiffen thin external walls against buckling, or to form storage compartments.)

9.7 TORSION WITH RESTRAINT OF WARPING: I SECTIONS

Introduction. Saint-Venant torsion theory, discussed in preceding sections of this chapter, describes pure torsion of a prismatic member, where cross sections are free to warp and torque is constant along the length. In Fig. 9.7-1, these conditions are not realized. The tendency to warp is different in segments AB, BC, CD, and DE, but compatibility between segments requires that warping displacements match on sections B, C, and D. Thus some cross sections may warp more, others less, than predicted by Saint-Venant torsion theory. These disturbances alter the torsional stiffness and create axial normal stresses and additional shear stresses.

In Figs. 9.7-1 and 9.7-2, torques are applied by couple-forces F. With this manner of loading the Saint-Venant shear stress distribution, as shown in Fig. 9.3-2b, does not exist quite near the load points. There is also local distortion, as described in the last paragraph of Section 9.10. These effects have much less influence on overall behavior than warping restraint, and are ignored in what follows.

Our discussion is limited to members having thin-walled open cross sections. Such members warp much more than solid bars or closed tubes and are much more affected by restraint of warping. We deal with warping of the *midline* of a cross section out of its original plane. Warping of edges of a cross section relative to the midline is considered negligibly small if the cross section is thin-walled.

Analysis of torsion of I sections with restraint of warping began with Timoshenko in 1905. The work was extended to general sections by Vlasov [9.5]. It happens that an I or wide-flange section, with two axes of symmetry, can be treated by a combination of Saint-Venant torsion theory and elementary beam theory. This approach is difficult for general shapes of cross section. Analysis of open cross sections of general shape is discussed in Section 9.12.

Qualitative Remarks. Consider the beam in Fig. 9.7-2a. Because of symmetry, the cross section at $x = 0$ is prevented from warping, as if it were welded to a rigid wall. If length L is large, this restraint is scarcely felt near $x = L$, and there torque $T = Fh$ generates Saint-Venant shear stress $\tau = Tt/K$ (Eq. 9.3-5). Warping must be unrestrained if

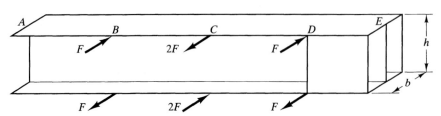

FIGURE 9.7-1 An I or wide-flange beam, with three torsional loads and right end reinforced by cover plates.

Section 9.7 Torsion with Restraint of Warping: I Sections

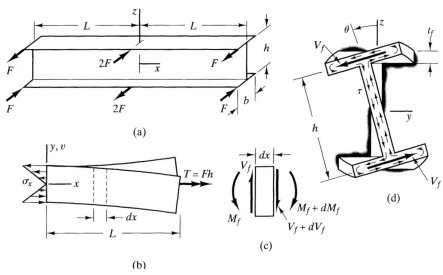

FIGURE 9.7-2 (a) An I or wide-flange beam, loaded such that the cross section at $x=0$ does not warp. (b) Top view of the upper flange, shown deflected. (c) Differential element of the upper flange. (d) Axial view, showing a rotated cross section at arbitrary x.

stress $\tau = Tt/K$ is to exist; therefore, the state of stress near $x=0$ must be different. Indeed, because warping is zero at $x=0$, the Saint-Venant shear stress is zero at $x=0$. The qualitative nature of flange stresses near $x=0$ is most easily seen by imagining that the web is discarded, so that the two flanges behave as independent cantilever beams. Thus we expect to see flexural stress σ_x due to M_f and transverse shear stress τ_{xy} due to V_f. (Note that Saint-Venant shear stress changes direction across the midline of the cross section, while shear stress due to V_f is constant through the flange thickness.) In the transition region $0 < x < L$, over which warping changes from zero toward unrestrained, all of the aforementioned types of stress coexist (Fig. 9.7-2d), in magnitudes that depend on x.

It happens that warping restraint propagates a surprisingly large distance along a thin-walled beam, and produces significant stresses. The analysis problem is to account for warping restraint and obtain expressions for stresses and angle of twist.

Analysis. The approach taken is to obtain an expression for $\beta = \beta(x)$, from which all other quantities of interest can be obtained. The following analysis uses elementary beam theory and is well suited to an I or wide-flange section. For other shapes, the analysis method discussed in Section 9.12 is more appropriate.

Consider an incremental length dx of the upper flange somewhere between $x=0$ and $x=L$, Fig. 9.7-2c. Its moment-curvature relation, from elementary beam theory, is

$$EI_f \frac{d^2v}{dx^2} = -M_f \quad \text{where} \quad I_f = \frac{t_f b^3}{12} \tag{9.7-1}$$

Here I_f is the area moment of inertia of one flange alone about the z axis, v is the y-direction deflection of the flange, and M_f is considered positive in the direction shown in Fig. 9.7-2c. When a cross section rotates through a small positive angle θ about the x axis, the upper flange has displacement $v = -(h/2)\theta$, where h is the distance between flange midlines. Equation 9.7-1 becomes

$$M_f = \frac{EI_f h}{2} \frac{d^2\theta}{dx^2} = \frac{EI_f h}{2} \frac{d\beta}{dx} \qquad (9.7\text{-}2)$$

where β is the rate of twist. Moment equilibrium in Fig. 9.7-2c requires that $V_f = -dM_f/dx$, where V_f is the transverse shear force in a flange. Therefore

$$V_f = -\frac{dM_f}{dx} = -\frac{EI_f h}{2} \frac{d^2\beta}{dx^2} \qquad (9.7\text{-}3)$$

Forces V_f in upper and lower flanges produce torque hV_f about the x axis. An additional torque $GK\beta$ is contributed by shear stresses associated with warping (Eq. 9.3-5). The two torques combine to resist applied torque T.

$$T = hV_f + GK\beta \quad \text{or} \quad T = -\frac{EI_f h^2}{2} \frac{d^2\beta}{dx^2} + GK\beta \qquad (9.7\text{-}4)$$

So that results of the present analysis will have the same form as general expressions in Section 9.12, we adopt the notation

$$\text{I section:} \quad J_\omega = I_f \frac{h^2}{2} = \frac{t_f b^3}{12} \frac{h^2}{2} = \frac{b^3 h^2 t_f}{24} \qquad (9.7\text{-}5)$$

(More exactly, I_f should be taken as half the moment of inertia of the entire cross section about the z axis.) With Eq. 9.7-5, Eq. 9.7-4 can be written in the form

$$\frac{d^2\beta}{dx^2} - k^2\beta = -k^2 \frac{T}{GK} \quad \text{where} \quad k^2 = \frac{GK}{EJ_\omega} \qquad (9.7\text{-}6)$$

EJ_ω is called the *warping rigidity* of the cross section. It is associated with *restraint* of warping rather than freedom to warp. J_ω is tabulated for various cross sections in [1.6, 1.7], where it is denoted by other symbols. The solution of Eq. 9.7-6, for T independent of x, is

$$\beta = C_1 \sinh kx + C_2 \cosh kx + \frac{T}{GK} \qquad (9.7\text{-}7)$$

where C_1 and C_2 are constants of integration, to be determined from boundary conditions. For the problem of Fig. 9.7-2a, boundary conditions are

$$\beta = 0 \quad \text{at} \quad x = 0 \quad \text{and} \quad \frac{d\beta}{dx} = 0 \quad \text{at} \quad x = L \qquad (9.7\text{-}8)$$

The former condition states that there is no warping at $x = 0$. The latter condition is obtained by noting that $\sigma_x = 0$ at a free (unrestrained) end; therefore $M_f = 0$ and Eq. 9.7-2 yields $d\beta/dx = 0$. Hence

$$C_1 = \frac{T}{GK} \tanh kL \quad \text{and} \quad C_2 = -\frac{T}{GK} \qquad (9.7\text{-}9)$$

Section 9.7   Torsion with Restraint of Warping: I Sections   281

(Additional remarks about boundary conditions appear following Eq. 9.13-2.) Equations 9.7-7 and 9.7-9 yield

$$\beta = \frac{T}{GK}(\tanh kL \sinh kx - \cosh kx + 1) \qquad (9.7\text{-}10)$$

The angle of twist at end $x = L$ is

$$\theta_L = \int_0^L \beta\, dx = \frac{TL}{GK}\left(1 - \frac{\tanh kL}{kL}\right) \qquad (9.7\text{-}11)$$

If warping were unrestrained, we would have $\theta_L = TL/GK$. The parenthetical expression in Eq. 9.7-11 is a reduction factor associated with restraint of warping. If $kL > 2$, then $\tanh kL \approx 1$, so that Eq. 9.7-11 can be simplified for $kL > 2$. For $kL = 2$, θ_L is reduced about 50% by restraint of warping imposed at $x = 0$.

By substitution of Eq. 9.7-10 into Eqs. 9.7-2 and 9.7-3, we obtain the following expressions for bending moment and shear force in a flange.

$$M_f = \frac{EI_f h}{2}\left(\frac{Tk}{GK}\right)(\tanh kL \cosh kx - \sinh kx) \qquad (9.7\text{-}12)$$

$$V_f = -\frac{EI_f h}{2}\left(\frac{Tk^2}{GK}\right)(\tanh kL \sinh kx - \cosh kx) \qquad (9.7\text{-}13)$$

Shear stress associated with warping is calculated from Eq. 9.3-4, $\tau = G\beta t$, using the β and t appropriate to the location of interest. Stresses associated with M_f and V_f are calculated from beam formulas, as illustrated by the following example.

Limitations.   The foregoing development is considered reliable if the member is not too short. "Too short" means that transverse shear deformation significantly reduces the effective torsional stiffness [1.6]. This result may appear in Fig. 9.7-2 if b becomes comparable to L. For the problem of Fig. 9.7-2a, the effective torsional stiffness over length L is T/θ_L. Equation 9.7-11 provides a formula for T/θ_L. An upper bound on T/θ_L is GJ/L, where $J = I_x + I_y$ is the polar moment of inertia of the cross-sectional area. GJ/L is an upper bound because use of J implies that there is no warping of any cross section and hence none of the reduction of stiffness that accompanies warping. Therefore, if T/θ_L from Eq. 9.7-11 exceeds GJ/L, we are warned that the member is too short for the foregoing theory to be reliable.

EXAMPLE

An I beam is fixed at end $x = 0$ and unrestrained at end $x = L$. The unrestrained end is loaded by torque T. Numerical data are shown in Fig. 9.7-3. Determine significant stresses, and the angle of twist at end $x = L$.

Equations 9.7-12 and 9.7-13 yield the bending moment and transverse shear force in a flange at the fixed end.

$$M_{x=0} = \frac{EI_f h}{2}\left(\frac{Tk}{GK}\right)\tanh kL = 268{,}000 \text{ N}\cdot\text{mm} \qquad (9.7\text{-}14)$$

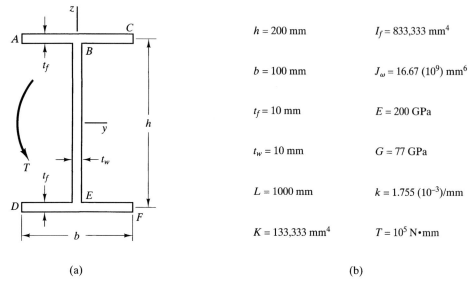

FIGURE 9.7-3 (a) An I section loaded by torque. (b) Data used in the example problem.

$$V_{x=0} = \frac{EI_f h}{2}\left(\frac{Tk^2}{GK}\right) = \frac{EJ_\omega}{h}\left(\frac{T}{EJ_\omega}\right) = \frac{T}{h} = 500 \text{ N} \qquad (9.7\text{-}15)$$

Flexural stresses of largest magnitude at $x = 0$ are

$$(\sigma_x)_{x=0} = \pm \frac{M_{x=0}(b/2)}{I_f} = \pm 16.1 \text{ MPa} \qquad (9.7\text{-}16)$$

For the direction of torque shown in Fig. 9.7-3a, σ_x is tensile at C and D, compressive at A and F.

Since a flange has a rectangular cross section, the maximum transverse shear stress is 1.5 times the average value, and appears at B and E in Fig. 9.7-3a. At the fixed end,

$$(\tau_q)_{x=0} = 1.5\frac{V_{x=0}}{bt_f} = 0.75 \text{ MPa} \qquad (9.7\text{-}17)$$

Stress τ_q is given the subscript q to distinguish it from the Saint-Venant shear stress $\tau = G\beta t$, and because in Section 9.12 it will be convenient to regard it as associated with a shear flow.

From Eq. 9.7-10, the rate of twist at $x = L$ is

$$\beta_L = \frac{T}{GK}(\tanh kL \sinh kL - \cosh kL + 1) = 6.47(10^{-6}) \text{ rad/mm} \qquad (9.7\text{-}18)$$

The term T/GK is $9.74(10^{-6})$ rad/mm, which is 50% larger than β_L. Thus we see that even though the length of the beam is five times its depth, restraint at $x = 0$ has significant influence at $x = L$. In other words, Saint-Venant's principle is not generally applicable to a thin-walled cross section. From Eq. 9.3-4, Saint-Venant shear stress associated with β_L is

$$\tau_L = G\beta_L t_f = 4.98 \text{ MPa} \qquad (9.7\text{-}19)$$

Although the load is torsional, in this example the largest stress is flexural and is more than twice the shear stress that would exist if there were no restraint of warping.

From Eq. 9.7-11, the angle of twist at $x = L$ is

$$\theta_L = \frac{TL}{GK}(1 - 0.537) = 4.51(10^{-3}) \text{ rad} \tag{9.7-20}$$

This calculation shows that θ_L is less than half what it would be if there were no restraint of warping at $x = 0$.

9.8 SECTORIAL AREA

Torsional analysis of thin-walled members is facilitated by making use of *sectorial area*. Sectorial area, and properties calculated from it, depend on the shape and size of the cross section, and can be discussed independently of the loading on the member. In this sense sectorial properties are analogous to I, the moment of inertia of an area, whose calculation can be discussed independently of its use in the flexure formula $\sigma = Mc/I$.

Sectorial area is discussed in the present section. Properties calculated from it are discussed in the following section. A reader who consults other references will find that authors disagree about terminology, symbols, and sign convention [4.2, 6.4, 9.5, 9.8].

Consider an arbitrary thin-walled cross section, Fig. 9.8-1a. For the calculation of sectorial area we use only the midline; thickness t is ignored. In Fig. 9.8-1b, a "pole" P is located arbitrarily. A straight line segment extends from P to an arbitrary point Q on the midline. An area is "swept" as point Q moves along the midline, starting at I and going a distance s. By definition, *sectorial area is twice the area swept*. An increment of the area swept is shown in Fig. 9.8-1c. Distance r is measured normal to length increment ds. Sectorial area $\omega = \omega(s)$ can be stated as

$$\text{Sectorial area:} \quad \omega = \int_0^s r \, ds \tag{9.8-1}$$

Clearly the value of ω at an arbitrary location Q depends on the location of pole P, the location of initial point I (where $\omega = 0$), and the geometry of the cross section. The

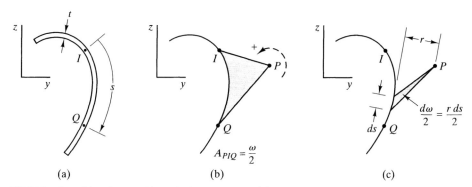

FIGURE 9.8-1 (a) Arbitrary thin-walled cross section. (b) Midline of the cross section, showing area A_{PIQ} swept by a line that rotates about P. (c) Increment of area swept.

284 Chapter 9 Torsion

most useful locations for P and I are discussed in the following section. If I is relocated on the midline, but P is not moved, ω changes by a constant at all points. We adopt the sign convention that $d\omega = r\,ds$ is positive when the sweeping line segment rotates counterclockwise about P.

EXAMPLE 1

For the cross-section midline shown in Fig. 9.8-2, construct two sectorial area diagrams: Place initial point I at G, then at C. (A sectorial area diagram shows the value of ω at every point on the midline.)

In Fig. 9.8-2b, $\omega = 0$ at I and no area is swept as an imaginary point Q moves along IP, so $\omega = 0$ along IP. The area swept from I to B is $a^2/2$, and the sweeping line rotates clockwise, so $\omega = -a^2$ at B. From B to A the triangular area swept has base $2a$ and altitude a, and the change in ω from B to A is $2a^2$. This change is added to the value of ω at B to yield a^2 as the value of ω at A. Similar remarks apply to the lower portion of the diagram.

In Fig. 9.8-2c, $\omega = 0$ at I. As an imaginary point Q moves along IGB the sweeping line rotates clockwise, and we obtain $\omega = -a^2$ at G and $\omega = -2a^2$ at B. Returning to the branch point at G and moving along GP, no area is swept, so ω remains $-a^2$ along GP. This diagram differs from that of Fig. 9.8-2b by $-a^2$ at all points.

Transfer Theorem. A transfer theorem can be used to calculate sectorial area about a second pole when the sectorial area about a first pole is known. The theorem, Eq. 9.8-6, is derived as follows.

In terms of lettered triangles in Fig. 9.8-3a, the shaded area A_{CAI} is

$$A_{CAI} = A_{CBI} + A_{BAI} - A_{CBA} = \frac{1}{2}[z\,dy + dy\,dz - (y + dy)\,dz] \quad (9.8\text{-}2)$$

This area is an increment of sectorial area about pole $P1$; that is

$$d\omega_{P1} = -2A_{CAI} = y\,dz - z\,dy \quad (9.8\text{-}3)$$

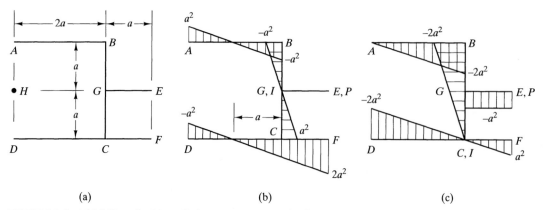

(a)   (b)   (c)

FIGURE 9.8-2 (a) Midline of a thin-walled open cross section. (b, c) Sectorial area diagrams for the same pole P but different initial points I.

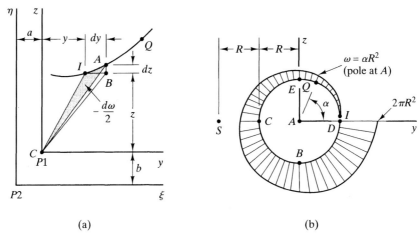

FIGURE 9.8-3 (a) Calculation of sectorial area with parallel axes. (b) Sectorial area diagram for an open circle with pole at A and initial point at I, where $\omega = 0$.

The negative sign is required because the direction of sweep is clockwise in going from I to A. At arbitrary point Q, sectorial area about pole $P1$ is

$$\omega_{P1} = \int_I^Q (y\,dz - z\,dy) \tag{9.8-4}$$

Sectorial area about pole $P2$ in Fig. 9.8-3a is obtained from Eq. 9.8-4 merely by replacing y by ξ and z by η, where $\xi = y + a$ and $\eta = z + b$.

$$\omega_{P2} = \int_I^Q [\xi\,d\eta - \eta\,d\xi] = \int_I^Q [(y + a)\,dz - (z + b)\,dy] \tag{9.8-5}$$

The integral of $y\,dz - z\,dy$ is recognized from Eq. 9.8-4 as ω_{P1}. Therefore ω_{P2}, the sectorial area about pole $P2$, is

$$\omega_{P2} = \omega_{P1} + a(z_Q - z_I) - b(y_Q - y_I) \tag{9.8-6}$$

EXAMPLE 2

In Fig. 9.8-3b, initial point I is at $\alpha = 0$. Determine sectorial area expressions associated with poles at A, E, and S.

With pole A, the area swept is a sector of a circle. Hence with I at $\alpha = 0$, sectorial area for pole A is $\omega_A = \alpha R^2$. Sectorial area ω_E for the same initial point and pole E is obtained from Eq. 9.8-6 by setting $\omega_{P1} = \omega_A$ and $\omega_{P2} = \omega_E$, with $a = 0$ and $b = -R$ (b is negative because $P1$ is below $P2$). Thus

$$\omega_E = \alpha R^2 - (-R)(R\cos\alpha - R) = R^2(\alpha + \cos\alpha - 1) \tag{9.8-7}$$

Similarly, for the same initial point but pole at S, sectorial area is obtained from Eq. 9.8-6 by setting $\omega_{P1} = \omega_A$ and $\omega_{P2} = \omega_S$ with $a = 2R$ and $b = 0$. Thus

$$\omega_S = \alpha R^2 + 2R(R\sin\alpha - 0) = R^2(\alpha + 2\sin\alpha) \tag{9.8-8}$$

9.9 SECTORIAL PROPERTIES

The following four sectorial properties are useful.

$$\text{Sectorial static moment:} \quad S_\omega = \int_A \omega \, dA \quad (9.9\text{-}1a)$$

$$\text{Sectorial linear moment:} \quad S_{y\omega} = \int_A y \omega \, dA \quad (9.9\text{-}1b)$$

$$\text{Sectorial linear moment:} \quad S_{z\omega} = \int_A z \omega \, dA \quad (9.9\text{-}1c)$$

$$\text{Sectorial moment of inertia:} \quad J_\omega = \int_A \omega^2 \, dA \quad (9.9\text{-}1d)$$

where y and z are axes in the plane of the cross section, ω is sectorial area, and dA is an increment of area A of the cross section. Integration is over the entire cross-sectional area A. If millimeters are used for length measurement, units of the respective quantities are mm^4, mm^5, mm^5, and mm^6. If thickness t of the cross section is uniform, then $dA = t \, ds$, $S_\omega = t \int \omega \, ds$, and so on. In a cross section with branches, ds is taken as positive in each branch, as $dA = t \, ds$ is always positive. Values of sectorial properties depend on the geometry of the cross section and the locations of yz axes, initial point I, and pole P. For analyses that involve warping of cross sections, one usually makes the restrictions that y and z are *centroidal* axes of the cross section and that ω is the *principal* sectorial area (described in what follows). With these restrictions, J_ω may be called the "warping constant." Sectorial properties for various sections are tabulated in [1.7].

Principal Sectorial Area. A *principal* sectorial area diagram is one constructed with initial point I located such that $S_\omega = 0$, and pole P located at the shear center of the cross section. The following description of shear center is adequate for the present discussion. Imagine a thin-walled prismatic cantilever beam, loaded by a transverse force F at the unsupported end. Let centroidal axes yz lie in the plane of the cross section (as in Fig. 9.9-1a, for example). If F passes through a certain point in the yz plane, the beam bends without twisting, regardless of how F is oriented in the yz plane. This point is the shear center. It is also the center of twist; that is, the point that has no translation when the member is loaded by torque without any restraint of warping. In general, the shear center does not lie on the cross section itself. The shear center coincides with the centroid if the cross section has two axes of symmetry. The shear center can be located by calculations described in Sections 10.5 to 10.8.

To determine principal sectorial area, the shear center must be located, but we need not actually locate an initial point I such that $S_\omega = 0$. Instead we recall that relocation of I changes ω by a constant, and we seek the appropriate constant. Accordingly, we begin with the principal pole (that is, P at the shear center) and an arbitrary location of initial point I on the midline, then determine the resulting ω. The principal sectorial area ω_p differs from ω by a constant ω_o. Also, we require that $S_\omega = 0$. That is,

$$\omega_p = \omega - \omega_o \quad \text{and} \quad \int_A \omega_p \, dA = 0 \quad (9.9\text{-}2)$$

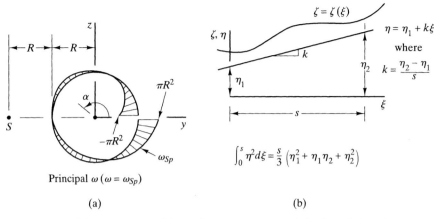

FIGURE 9.9-1 (a) Principal sectorial area diagram for an open circle. The pole is at S. (b) Useful relations for integration.

from which

$$\omega_o = \frac{1}{A} \int_A \omega \, dA \qquad (9.9\text{-}3)$$

where A is the entire area of the cross section. If sectorial properties are calculated using the principal sectorial area ($\omega = \omega_p$ in Eqs. 9.9-1), and if y and z are centroidal axes of the cross section, then $S_{y\omega}$ and $S_{z\omega}$ are also zero. This result is demonstrated in Section 10.7.

When the principal sectorial area is used, calculations of warping displacement are not contaminated by translation along the x axis or by rigid body rotations about y and z axes.

EXAMPLE 1

Determine the principal sectorial area and associated sectorial properties for an open circular cross section of mean radius R and uniform thickness t, with $R \gg t$.

Part of the solution is already available from calculations associated with Fig. 9.8-3b: Point S happens to be the shear center, and ω with respect to it is stated in Eq. 9.8-8, where it is called ω_S. Applying Eq. 9.9-3 to ω_S, we obtain

$$\omega_o = \frac{1}{2\pi Rt} \int_0^{2\pi} \omega_S t R \, d\alpha = \pi R^2 \qquad (9.9\text{-}4)$$

Hence, Eq. 9.9-2 yields the principal sectorial area. Here we call it ω_{Sp}, merely to distinguish it from ω_A, ω_S, and so on.

$$\omega_{Sp} = \omega_S - \omega_o = R^2(\alpha + 2\sin\alpha - \pi) \qquad (9.9\text{-}5)$$

This result is illustrated in Fig. 9.9-1a. Zeros of ω_{Sp} appear at $\alpha = 1.246$, $\alpha = \pi$, and $\alpha = 5.037$ radians.

In Fig. 9.9-1a, positive and negative regions of ω_{Sp} are equal, so it is obvious by inspection that $S_\omega = 0$. It is also obvious by inspection that $S_{y\omega} = 0$, since each positive contribution $y\omega \, dA$ is matched by an equal negative contribution. It is not obvious by inspection that $S_{z\omega}=0$. All these results may be verified from Eqs. 9.9-1 by straightforward integration, with $y = R \cos \alpha$ and $z = R \sin \alpha$. Similarly, with $\omega = \omega_{Sp}$, Eq. 9.9-1d yields

$$J_\omega = \frac{2\pi}{3}(\pi^2 - 6)R^5 t = 8.104 R^5 t \qquad (9.9\text{-}6)$$

Calculation Aids. Calculations for a cross section midline composed of straight line segments are simplified by the following formula. It provides the integral of the product of two functions, one of which must be linear. Let the product be $\zeta\eta$, where $\zeta = \zeta(\xi)$ is an arbitrary function of ξ, and $\eta = \eta_1 + k\xi$ is a linear function of ξ (Fig. 9.9-1b). Then

$$\int_0^s \zeta\eta \, d\xi = \int_0^s \zeta(\eta_1 + k\xi) \, d\xi = A(\eta_1 + k\xi_g) \qquad (9.9\text{-}7)$$

The integral of $\zeta \, d\xi$ is A, the area under the curve $\zeta = \zeta(\xi)$. The integral of $\xi(\zeta \, d\xi)$ is $A\xi_g$, the first moment of A about the ζ axis, where ξ_g is the ξ coordinate of the centroid of A. The final result says that the integral is equal to the area under the arbitrary function times the ordinate of the linear function directly under the centroid of the arbitrary function.

Another integration formula, shown in Fig. 9.9-1b, is useful in calculating J_ω for a cross section whose midline is composed of straight line segments. Ordinates η_1 and η_2 can be positive or negative. For complicated shapes, available computer software can calculate various properties of a cross section, including sectorial properties.

EXAMPLE 2

Determine the principal sectorial area and associated sectorial properties for the thin-walled I section shown in Fig. 9.9-2a.

The cross section is symmetric with respect to both y and z axes. Therefore the centroid and the shear center coincide, at $y = z = 0$. With this point both pole and initial point, we obtain the

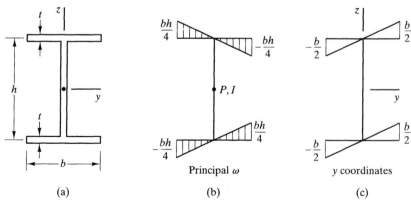

FIGURE 9.9-2 (a) I section of uniform thickness, with centroidal coordinates yz. (b) Principal sectorial area diagram. (c) Plot of y coordinates along the midline.

sectorial area ω shown in Fig. 9.9-2b. By inspection, $S_\omega = 0$; therefore ω is the principal sectorial area. To evaluate $S_{y\omega}$, we can proceed quadrant by quadrant. In the first quadrant, with ω the arbitrary function and y the linear function, Eq. 9.9-7 yields

$$\text{First quadrant:} \quad t \int y\omega \, ds = t\left[\frac{1}{2}\left(-\frac{bh}{4}\right)\left(\frac{b}{2}\right)\right]\left[\frac{2}{3}\left(\frac{b}{2}\right)\right] = -t\frac{b^3 h}{48} \quad (9.9\text{-}8)$$

A result of the same magnitude is obtained in each quadrant, but signs alternate from quadrant to quadrant. Summing over all four quadrants, we obtain $S_{y\omega} = 0$. In similar fashion, one concludes that $S_{z\omega} = 0$. These results are of course anticipated, since we have used the principal ω, and $S_{y\omega}$ and $S_{z\omega}$ are known to be zero when ω is principal. Also, in this particular example, the results $S_{y\omega} = 0$ and $S_{z\omega} = 0$ can be seen by inspection.

As for J_ω, we note that the diagram is composed of straight lines, and there are four parts that contribute equally to J_ω. Therefore, we can use the integration formula in Fig. 9.9-1b with $\eta_1 = 0$ and $\eta_2 = bh/4$ to obtain the sectorial moment of inertia or "warping constant" J_ω as follows:

$$J_\omega = t \int \omega^2 \, ds = 4t\left[\frac{b/2}{3}\left(\frac{bh}{4}\right)^2\right] = \frac{b^3 h^2 t}{24} \quad (9.9\text{-}9)$$

9.10 WARPING OF THIN-WALLED CROSS SECTIONS IN PURE TORSION

We consider thin-walled prismatic members, with an x axis that passes through centroids of cross sections. In the present section, torque is constant along the member and warping is unrestrained. Therefore warping is the x-direction distortion $u = u(y,z)$. Thus, u is independent of x (Fig. 9.10-1a), and cross sections have no distortion other than warping. The following expressions for warping are useful in developing expressions for stresses associated with restraint of warping (Section 9.11).

It is convenient to begin with a closed tube. As shown in Section 9.5, torque T produces shear flow $q = T/2\Gamma$ around the cross section, where Γ is the area enclosed by the midline. Shear strain γ is associated with q. If the tube is then slit open lengthwise, $q = 0$; therefore $\gamma = 0$. In this way, expressions for warping of an open cross section can be obtained.

Contributions to shear strain in a closed section are as follows. In Fig. 9.10-1c, let point S be the shear center (briefly described in Section 9.9). With θ the angle of twist,

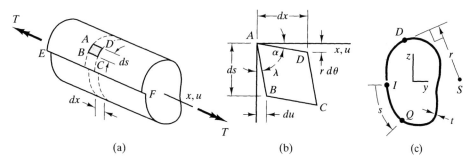

FIGURE 9.10-1 (a) Thin-walled tube, showing warping displacement if cut open along EF. (b) If not cut open, there exists a nonzero shear strain $\gamma = \alpha + \lambda$ on the midsurface. (c) Axial view of the tube.

the s-direction displacement of D relative to A is $r\,d\theta$. Warping displacement du of B relative to A also contributes to shear strain. From Fig. 9.10-1b, shear strain γ on the midline is

$$\gamma = \alpha + \lambda \quad \text{or} \quad \gamma = \frac{r\,d\theta}{dx} + \frac{du}{ds} \tag{9.10-1}$$

Now $\gamma = \tau_q/G = q/Gt$, where G is the shear modulus. Thus, with $\beta = d\theta/dx$ the rate of twist, Eq. 9.10-1 yields

$$\lambda = -\alpha + \gamma \quad \text{or} \quad \frac{du}{ds} = -\beta r + \frac{q}{Gt} \tag{9.10-2}$$

Open Sections. For an open section without warping restraint, such as that of Fig. 9.10-1a, $q = 0$, and therefore $\gamma = 0$. Then $\lambda = -\alpha$ in Fig. 9.10-1b, and $du/ds = -\beta r$. Hence, by integration,

$$u = -\beta \int_0^s r\,ds \quad \text{or} \quad u = -\beta\omega \tag{9.10-3}$$

We see that warping distortion along the midline of a thin-walled open member in pure torsion follows the distribution of sectorial area ω and is directly proportional to rate of twist β.

Equation 9.10-3 does not define u uniquely because ω depends on the locations assigned to initial point I and pole P. If I is relocated, ω changes by a constant; if P is relocated, ω changes in proportion to a and b in Eq. 9.8-6. These changes correspond to the respective rigid body motions of x-direction translation and rotations about y and z axes. These motions are absent if ω is the principal sectorial area discussed in Section 9.9 [9.5]. Such is the case in Fig. 9.9-1a and in Fig. 9.9-2b. With $u = -\beta\omega$, Fig. 9.9-2b predicts the warping displacement depicted in Fig. 9.7-2b. If warping were prevented, ω in Figs. 9.9-1a and 9.9-2b would be proportional to the distribution of the resulting axial stress σ_x (see Eq. 9.11-1).

Equation 9.10-3 shows that an open section does not warp if ω is everywhere zero. Example cross sections of this type are shown in Fig. 9.10-2. Each consists of slender straight arms that emanate from a common point S. This point is also the shear center and the center of twist.

Closed Sections. Consider a single-cell closed section; that is, omit longitudinal cut EF in Fig. 9.10-1a. In Eq. 9.10-2, we substitute for q and β from Eqs. 9.5-3 and 9.5-5, and integrate. Thus

$$u = -\frac{T}{2G\Gamma}\left[\frac{\omega}{2\Gamma}\oint\frac{ds}{t} - \int_0^s \frac{ds}{t}\right] \tag{9.10-4}$$

in which Γ is the area enclosed by the midline, and u and ω are both functions of s. If s makes a complete circuit of the midline, then $\omega = 2\Gamma$, and $u = 0$. This result is expected, as in a closed tube there is no jump in u when s crosses the initial point where $s = 0$.

Section 9.10 Warping of Thin-Walled Cross Sections in Pure Torsion

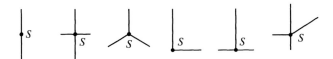

FIGURE 9.10-2 Examples of thin-walled open cross sections whose midlines do not warp when the member is twisted. Points S are shear centers.

If a tube has more than one cell, Eq. 9.10-4 can be applied to each cell. Then q is not constant around a cell, and must be kept inside the integral. Rigid body motions in each cell are such that adjacent cells have the same u along a common wall.

As an example, consider a single-cell closed tube of rectangular cross section, Fig. 9.10-3a. With G constant and $\Gamma = bh$, Eq. 9.10-4 becomes

$$u = -\frac{T}{2Gbh}\left[\frac{\omega}{bh}\left(\frac{h}{t_h} + \frac{b}{t_b}\right) - \int_0^s \frac{ds}{t}\right] \quad (9.10\text{-}5)$$

Warping displacements at corners are conveniently evaluated by placing the pole for ω at the shear center, $y = z = 0$, and the initial point at a midside such as point I. Thus at corner A we obtain $bh/4$ for ω and $h/2t_h$ for the integral in Eq. 9.10-5. At corner B, these quantities become $3bh/4$ and $h/2t_h + b/t_b$ respectively. Warping displacements at the four corners are

$$u_A = -\frac{T}{8Gbh}\left(\frac{b}{t_b} - \frac{h}{t_h}\right) \qquad u_B = u_D = -u_A \qquad u_C = u_A \quad (9.10\text{-}6)$$

Warping displacements are shown in Fig. 9.10-3a. These displacements are proportional to the axial stress σ_x that would arise if warping were prevented. Equation 9.10-6 shows that there is no tendency to warp if $b/t_b = h/t_h$.

If the tube were slit open lengthwise, warping displacements would increase greatly and would have a different distribution. Whether a member is open or closed, torque applied as couple-forces produces local distortion (Fig. 9.10-3b), which can be

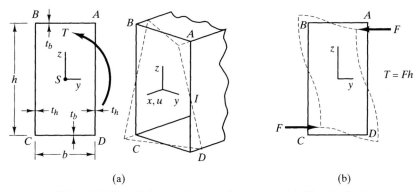

FIGURE 9.10-3 (a) Thin-walled tube of rectangular cross section. Warping displacement for $t_b = t_h$ and $h = 2b$ is shown by dashed lines. (b) Distortion of a loaded cross section unless braced in the yz plane.

9.11 STRESSES IN OPEN SECTIONS WHEN β IS NOT CONSTANT

If warping is partially or completely restrained at one or more cross sections, rate of twist β is not constant along the member. Normal and shear stresses arise in consequence. This behavior is illustrated for the special case of an I beam in Section 9.7. Similar analysis of a thin-walled member of arbitrary cross section is addressed in Section 9.12.

Axial Normal Stress. For a thin-walled open section, axial displacement $u = u(y,z)$ is given by Eq. 9.10-3. Axial strain is $\epsilon_x = \partial u/\partial x$ (Eq. 7.2-1). Axial normal stress σ_x is the only normal stress of consequence, so $\sigma_x = E\epsilon_x$. Putting all this together, we obtain

$$\sigma_x = -E\omega\frac{d\beta}{dx} \quad \text{or} \quad \sigma_x = -E\omega\frac{d^2\theta}{dx^2} \qquad (9.11\text{-}1)$$

Thus σ_x arises in an open section when β is not constant, and varies along the midline in the same way as sectorial area ω. If the member is loaded by torque alone, then σ_x must generate no axial force and no moment about axes in the plane of a cross section; that is

$$\int_A \sigma_x \, dA = 0 \qquad \int_A z\sigma_x \, dA = 0 \qquad \int_A y\sigma_x \, dA = 0 \qquad (9.11\text{-}2)$$

where integration extends over the entire cross sectional area A. By substituting Eq. 9.11-1 into Eqs. 9.11-2 and taking note of Eqs. 9.9-1, we see that Eqs. 9.11-2 demand that S_ω, $S_{y\omega}$, and $S_{z\omega}$ all vanish. In other words, when the rate of twist is not constant, we are obliged to use the principal sectorial area for stress calculation.

Additional Shear Stress. The Saint-Venant shear stress $\tau = Tt/K$, discussed in Section 9.3, is present when β is nonzero. This shear stress changes direction across the thickness of an open section and is zero at the midline. An *additional* shear stress is present when $d^2\beta/dx^2$ is nonzero. The latter shear stress is associated with a shear flow q; that is, the additional shear stress is $\tau_q = q/t$. This shear stress is essentially constant across the thickness of a thin-walled section.

Consider a differential element, Fig. 9.11-1. For equilibrium of x-direction forces, we must have

$$\left(\frac{\partial \sigma_x}{\partial x}dx\right)t\,ds + \left(\frac{\partial q}{\partial s}ds\right)dx = 0 \quad \text{or} \quad \frac{\partial q}{\partial s} = -\frac{\partial \sigma_x}{\partial x}t \qquad (9.11\text{-}3)$$

Substituting σ_x from Eq. 9.11-1, setting $t\,ds = dA$, and integrating, we obtain

$$\frac{\partial q}{\partial s} = E\frac{d^2\beta}{dx^2}\omega t \quad \text{hence} \quad q = E\frac{d^2\beta}{dx^2}\int \omega\,dA \qquad (9.11\text{-}4)$$

The integral, and hence q, are functions of distance traversed along the midline. A constant of integration has been omitted from Eq. 9.11-4, consistent with the under-

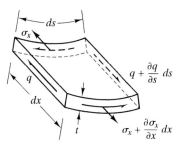

FIGURE 9.11-1 Differential element of a member such as that in Fig. 9.10-1.

standing that integration will always begin at an end of the midline, where $q = 0$ (for example, $q = 0$ along EF in Fig. 9.10-1a). So that q will again be zero when the entire midline has been traversed, ω is taken as the *principal* sectorial area.

The foregoing argument contains a contradiction. Equation 9.11-1 is based on Eq. 9.10-3, which in turn is based on the assumption that there is no shear stress on the midline. Equation 9.11-1 is used to obtain Eq. 9.11-4, according to which there *is* shear stress on the midline. Nevertheless, Eq. 9.11-4 has been found to be sufficiently accurate. The situation is analogous to that in elementary beam theory, where the assumption that plane sections remain plane leads to the flexure formula $\sigma = Mc/I$, which is then used to obtain the shear stress formula $\tau = VQ/It$, according to which plane sections do *not* remain plane.

9.12 TORSION WITH RESTRAINT OF WARPING: GENERAL SECTIONS

The present section is a generalization of Section 9.7, whose introductory paragraphs the reader may wish to review. We consider a prismatic member of arbitrary thin-walled open cross section loaded by torque T, as in Fig. 9.12-1 (although support conditions other than one end fixed are possible). Shear center S is chosen as the moment center for torque. This choice is convenient when a member is loaded by transverse force, because a transverse force creates torque proportional to the distance between its line of action and an x-parallel line through S.

Cross sections warp, but not freely because of restraint at $x = 0$. This restraint also reduces β. Total applied torque T is supported by two contributions:

$$T = T_q + T_{SV} \quad \text{or} \quad T = T_q + GK\beta \tag{9.12-1}$$

where T_{SV} is the "Saint-Venant torque" created by shear stresses described in Section 9.3. The remaining "Vlasov" torque T_q is associated with restraint of warping. Torque T_q is supplied by shear flow q, which produces shear stress $\tau_q = q/t$. (Since there is zero transverse force on the member, q must provide zero net force in y and z directions; accordingly, the q arrows must change directions as shown in Fig. 9.12-1b.) From Fig. 9.12-1b, torque T_q about S, positive in the direction shown for T, is

$$T_q = \int_I^F r(q\,ds) \quad \text{or} \quad T_q = \int_I^F q\,d\omega \tag{9.12-2}$$

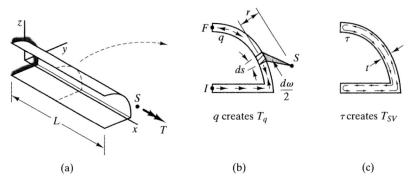

FIGURE 9.12-1 (a) Torque applied to a thin-walled open member fixed at one end. (b) Restraint of warping produces shear flow q. (c) Shear stress associated with freedom of cross sections to warp.

Integration by parts yields

$$T_q = [q\omega]_I^F - \int_I^F \omega\left(\frac{\partial q}{\partial s} ds\right) = -\int_I^F \frac{\partial q}{\partial s}\omega\, ds \qquad (9.12\text{-}3)$$

where the bracketed expression is zero because $q=0$ at free edges I and F. Substituting from Eq. 9.11-4, with $dA = t\, ds$, we obtain

$$T_q = -E\frac{d^2\beta}{dx^2}\int_A \omega^2\, dA \quad \text{or} \quad T_q = -EJ_\omega\frac{d^2\beta}{dx^2} \qquad (9.12\text{-}4)$$

where J_ω is calculated from the principal sectorial area. Equation 9.12-1 becomes

$$\frac{d^2\beta}{dx^2} - k^2\beta = -k^2\frac{T}{GK} \quad \text{where} \quad k^2 = \frac{GK}{EJ_\omega} \qquad (9.12\text{-}5)$$

This equation is the same as Eq. 9.7-6, so its general solution for constant torque is Eq. 9.7-7, namely

$$\beta = C_1 \sinh kx + C_2 \cosh kx + \frac{T}{GK} \qquad (9.12\text{-}6)$$

Boundary conditions are the same in Figs. 9.7-2b and 9.12-1a; accordingly, Eqs. 9.7-8 to 9.7-11 apply also to Fig. 9.12-1a. (Additional remarks about boundary conditions appear in text following Eq. 9.13-2.) Short members should be treated with caution; see remarks that follow Eq. 9.7-13.

EXAMPLE

Using sectorial properties, determine stresses in the example problem of Section 9.7. Data appear in Fig. 9.7-3.

The Saint-Venant shear stress is maximum at $x = L$. It is calculated without reference to sectorial properties, and is given by Eq. 9.7-19 as $\tau_L = 4.98$ MPa. All that remains is to evaluate

σ_x and τ_q at $x = 0$, where these stresses are largest. For this we need $d\beta/dx$ and $d^2\beta/dx^2$ at $x = 0$. From Eq. 9.7-10,

$$\text{At } x = 0: \quad \frac{d\beta}{dx} = \frac{Tk}{GK}\tanh kL \quad \text{and} \quad \frac{d^2\beta}{dx^2} = -\frac{Tk^2}{GK} \quad (9.12\text{-}7)$$

Data in Fig. 9.7-3 provide numerical values of these quantities.

$$\text{At } x = 0: \quad \frac{d\beta}{dx} = 16.1(10^{-9})/\text{mm}^2 \quad \text{and} \quad \frac{d^2\beta}{dx^2} = -30.0(10^{-12})/\text{mm}^3 \quad (9.12\text{-}8)$$

Equation 9.11-1 and Fig. 9.9-2b show that σ_x has largest magnitude at corners of the flanges, where ω is largest. At corners in the first and third quadrants, $\omega = -bh/4 = -100(200)/4 = -5000 \text{ mm}^2$. At these locations, from Eq. 9.11-1,

$$(\sigma_x)_{x=0} = -E\omega\frac{d\beta}{dx} = -200{,}000(-5000)16.1(10^{-9}) = 16.1 \text{ MPa} \quad (9.12\text{-}9)$$

which agrees with the result in Eq. 9.7-16. Equation 9.11-4 and Fig. 9.9-2b show that q has largest magnitude at the midwidth of a flange, where the integral of $\omega \, dA$ is largest. Integrating across the top flange from the left edge to the middle, we obtain

$$\text{At } y = 0, \quad z = \frac{h}{2}: \quad \int \omega \, dA = \frac{1}{2}\left(\frac{b}{2}\right)\left(\frac{bh}{4}\right)t = \frac{b^2 ht}{16} = 1.25(10^6) \text{ mm}^4 \quad (9.12\text{-}10)$$

At the midwidth of a flange, from Eq. 9.11-4, the magnitude of shear flow at the fixed end is

$$q_{x=0} = E\frac{d^2\beta}{dx^2}\int \omega \, dA = 200{,}000(30.0)10^{-12}(1.25)10^6 = 7.5 \text{ kN/mm} \quad (9.12\text{-}11)$$

Hence at this location the shear stress $\tau_q = q/t$ is

$$(\tau_q)_{x=0} = \frac{q_{x=0}}{t} = \frac{7.5}{10} = 0.75 \text{ MPa} \quad (9.12\text{-}12)$$

which agrees with the result in Eq. 9.7-17.

9.13 BIMOMENT

In elementary mechanics of materials, one learns to analyze a prismatic member loaded by an axial force that does not pass through centroids of cross sections. The procedure is to translate the force to the centroid, thus introducing bending moments, and superpose direct axial stress and flexural stresses. If this procedure is applied to the member in Fig. 9.13-1a, one obtains axial force N and bending moments M_y and M_z, and stresses that result. At the end of the beam in Fig. 9.13-1b, N, M_y, and M_z are each represented by a statically equivalent set of four forces F applied at flange tips. We see that the three sets of forces do not add up to the single force $4F$ shown in Fig. 9.13-1a. But force $4F$ results if we also add a set of four forces F that amounts to opposing couples, so that the set is statically equivalent to a null force system. This force system is a *bimoment*. If flanges were not connected by the web, bimoment would deflect them as independent cantilever beams, somewhat like the beam flanges in Fig. 9.7-2.

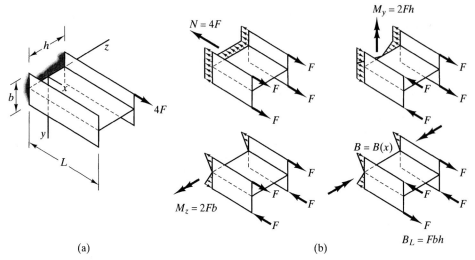

FIGURE 9.13-1 (a) Axial load $4F$ at the edge of one flange of an I beam. (b) The same loading, in terms of four component loadings. Stress distributions are shown.

One can now see that in the twisted I beam of Fig. 9.7-2, a bimoment exists at all cross sections where M_f is nonzero.

Bimoment is associated with warping of cross sections and twist of the member. In a stocky member, these effects are localized and usually unimportant. In a thin-walled member, distortions may be significant and may propagate a considerable distance along the member. The theory of bimoments is due mainly to Vlasov [9.5]. The discussion here is brief and limited to prismatic members of thin-walled open cross section.

Bimoment B is defined as the integral of $\sigma_x \omega$ over the area of the cross section.

$$B = \int_A \sigma_x \omega \, dA \quad \text{or} \quad B = \sum F_i \omega_i \tag{9.13-1}$$

B has dimensions [force·length^2] and may be positive or negative. An increment $dB = \sigma_x \omega \, dA$ is positive if the product $\sigma_x \omega$ is positive. In general, $B = B(x)$. The integral form pertains to an arbitrary cross section. The summation form can be used at a loaded end where there are concentrated axial forces F_i (large σ_x on a small ΔA). Each F_i contributes $F_i \omega_i$ to B, where ω_i is the value of ω where F_i is applied. Thus, from Figs. 9.9-2b and 9.13-1, we obtain $B_L = Fbh$ at $x = L$.

Bimoment is associated with a rate of twist β in a beam. When an expression for $\beta = \beta(x)$ has been obtained, angles of rotation can be determined by integration. In what follows, displacements are emphasized. Axial stress σ_x can be obtained from Eq. 9.11-1, in which $d\beta/dx$ is related to B by Eq. 9.13-5.

Bimoment Analysis. Consider a member loaded only by bimoment, as in Fig. 9.13-2a. Then $T = 0$ in Eq. 9.12-5. Hence the equation, and its solution $\beta = \beta(x)$, are

$$\frac{d^2\beta}{dx^2} - k^2\beta = 0 \quad \text{hence} \quad \beta = C_1 \sinh kx + C_2 \cosh kx \tag{9.13-2}$$

where $k^2 = GK/EJ_\omega$.

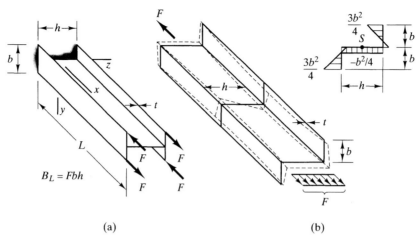

FIGURE 9.13-2 (a) Bimoment $B_L = Fbh$ applied to end $x = L$ of an I beam. (b) Z section of uniform thickness twisted by bimoment, and its principal sectorial area diagram.

Boundary conditions for a member loaded by bimoment B and/or torque T are as follows. At a fixed end, $\theta = 0$ and $\beta = 0$. At an unloaded free end, $T = 0$ and $B = 0$. One may also identify a simply supported end as an end where there is no rotation and no axial stress; thus $\theta = 0$ and $B = 0$.

For the problem of Fig. 9.13-2a, end $x = 0$ is fixed. Therefore $\beta = 0$ at $x = 0$, making $C_2 = 0$. At $x = L$, $B = B_L$. Equation 9.13-2 and the first of Eqs. 9.13-5 yield

$$C_1 = -\frac{B_L}{EJ_\omega k \cosh kL} \quad \text{and} \quad \beta = -\frac{B_L \sinh kx}{EJ_\omega k \cosh kL} \quad (9.13\text{-}3)$$

Since $\theta = 0$ at $x = 0$, and $EJ_\omega k^2 = GK$, the angle of rotation at arbitrary x is

$$\theta = \int_0^x \beta \, dx = \frac{B_L}{GK}\left(\frac{1 - \cosh kx}{\cosh kL}\right) \quad (9.13\text{-}4)$$

The positive sense of θ is counterclockwise when looking in the negative x direction. We see that end $x = L$ in Fig. 9.13-2a rotates clockwise, an amount B_L/GK if kL is large. Note that this rotation is produced by a loading that exerts no torque about the x axis.

Bimoment at arbitrary x, from Eqs. 9.9-1d, 9.11-1, and 9.13-1, is

$$\text{In general:} \quad B = -EJ_\omega \frac{d\beta}{dx} \qquad \text{Here:} \quad B = B_L \frac{\cosh kx}{\cosh kL} \quad (9.13\text{-}5)$$

Let s be distance from the loaded end, $s = L - x$. Equation 9.13-5 yields the conclusion that if the beam is long, $B \approx B_L e^{-ks}$. Thus B, and the σ_x associated with it, decline to about 5% of B_L when $s = 3/k$. For the case of an I section of uniform thickness t,

$$\text{For} \quad s = \frac{3}{k}: \quad s_{0.05} \approx \frac{3}{2t}\sqrt{\frac{(1+\nu)b^3h^2}{2b+h}} \quad (9.13\text{-}6)$$

For example, let $\nu = 0.3$ and $h = 2b = 20t$. Then Eq. 9.13-6 yields $s = 171t$, or $s = 8.6h$, which is a considerable distance. This example shows that Saint-Venant's principle is not generally applicable to thin-walled members loaded by bimoment (or by torque with restraint of warping, where bimoment develops).

In Fig. 9.13-2b, axial force F is applied to the web of a Z section. Because the principal sectorial area is a nonzero constant along the web, this loading applies bimoment as well as axial force to each end, and the section twists as shown. If instead axial load at each end were applied by concentrated forces $F/2$ at points where $\omega = 0$, there would be no bimoment and no twisting.

Physically, the deformation shown in Fig. 9.13-2b can be anticipated as follows. At ends, the loading shown amounts to a stress $\sigma_x = F/ht$ applied to the web. Regard this loading as the sum of two loadings: (a) $\sigma_x/2$ applied to the web and flanges, and (b) $\sigma_x/2$ applied to the web and $-\sigma_x/2$ applied to the flanges. Load (a) produces only extension. Load (b) is associated with bending of the flanges in opposite directions, as a sketch of this loading will clearly suggest.

9.14 LARGE ROTATION, AXIAL LOAD, AND PRETWISTING

A prismatic bar of circular cross section lengthens slightly when twisted in the elastic range. This behavior has been known since about 1910 and is called the *Poynting effect*. Clearly the effect must be proportional to an even power of rate of twist β, since the sign of β can make no difference. It happens that lengthening is a second-order effect; that is, it is proportional to β^2. It is so small as to be of no importance for shafts of circular cross section because β is small in the elastic range.

If the cross section is not circular but (say) a narrow rectangle, β^2 in the elastic range may not be negligible, the member tends to shorten rather than lengthen, and an appreciable axial stress σ_x may develop. Fibers become inclined to the axis of the member; thus σ_x develops force components that contribute to torque. If the member is pretwisted before loading, there are first-order relations between torque load and the amount of length change and between axial tension load and the amount of untwisting. These considerations find application in propellers and turbine blades.

In the present section, the following restrictions apply. The member is linearly elastic and straight. The cross section has symmetry about two or more axes, is thin-walled and open, with straight radial arms, and centroid at the intersection of the arms. Examples of such cross sections include the first three in Fig. 9.10-2. (Thus, midlines of the arms experience no warping when the member is twisted, and axial stresses generate no bimoment and no bending moments about centroidal axes of a cross section.) If there is pretwist, centroids of all cross sections lie on the same straight line. Rates of twist and pretwist are small. Analyses that allow for more general cross sections, a helical axis, warping of cross sections, and bending loads, are available [9.9–9.11].

Axial Stress Due to Twisting. We begin with a narrow rectangular cross section, without pretwist. Other shapes are considered subsequently.

Imagine that supports at ends of the member prevent it from changing length. What axial force F_x do supports apply? Consider a representative length dx, Fig. 9.14-1a. From Fig. 9.14-1c, axial strain ϵ_x of a fiber a distance r from the axis of the member is

Section 9.14 Large Rotation, Axial Load, and Pretwisting

$$\epsilon_x = \frac{ds - dx}{dx} = \frac{ds}{dx} - 1 = [1 + (r\beta)^2]^{1/2} - 1 \approx \frac{(r\beta)^2}{2} \quad (9.14\text{-}1)$$

where β is the rate of twist. The approximation is acceptable if $r\beta$ is small in comparison with unity. Strain ϵ_x is accompanied by stress $\sigma_x = E\epsilon_x$, which varies quadratically from zero on the x axis to a maximum on the narrow edges of the cross section. The accompanying axial tensile force F_x is

$$F_x = \int_{-b/2}^{b/2} E\epsilon_x(t\,dr) = \frac{E\beta^2}{2}\int_{-b/2}^{b/2} r^2 t\,dr = \frac{Eb^3 t}{24}\beta^2 \quad (9.14\text{-}2)$$

If in fact the member lacks axial restraint, the σ_x distribution can be determined by superposing on $E\epsilon_x$ the σ_x that results from an axial compressive force F_x. We take the latter stress to be its average value, $\sigma_x = -F_x/bt$. Thus, in a member loaded by torque and without axial restraint,

$$\sigma_x = \frac{E\beta^2}{2}\left(r^2 - \frac{b^2}{12}\right) \quad \text{or} \quad \sigma_x = \frac{E\tau^2}{2G^2 t^2}\left(r^2 - \frac{b^2}{12}\right) \quad (9.14\text{-}3)$$

where the latter form comes from the substitution $\beta = \tau/Gt$ (Eq. 9.3-4). The σ_x distribution is shown in Fig. 9.14-1d. This σ_x distribution of course cannot exist at a free end, but should be well established a distance b from a free end, in accord with Saint-Venant's principle.

Torque associated with σ_x is determined as follows. Force on an area increment $t\,dr$ of the cross section is $dF_x = \sigma_x t\,dr$. This force increment is inclined at a small angle λ to the x axis (Fig. 9.14-1c), and therefore has component $\lambda\,dF_x$ in a direction normal to the long side of the cross section. The moment arm of $\lambda\,dF_x$ about the x axis is r. Hence, torque associated with σ_x is

$$\int r\lambda\,dF_x = \int_{-b/2}^{b/2} r(\beta r)\frac{E\beta^2}{2}\left(r^2 - \frac{b^2}{12}\right)t\,dr = \frac{Eb^5 t}{360}\beta^3 \quad (9.14\text{-}4)$$

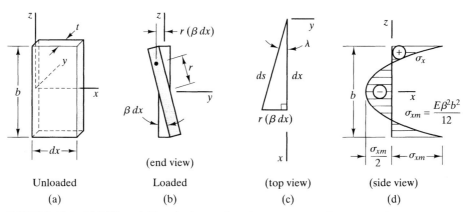

FIGURE 9.14-1 (a) Differential length of a member of narrow rectangular cross section. (b, c) Geometric relations after twisting, at radius r. (d) Axial stress σ_x if there is zero net axial force.

This torque adds to torque associated with Saint-Venant shearing stresses, Eq. 9.3-5. The total torque is

$$T = G\frac{bt^3}{3}\beta + \frac{Eb^5t}{360}\beta^3 = G\frac{bt^3}{3}\beta\left[1 + \frac{Eb^4}{120Gt^2}\beta^2\right] \quad (9.14\text{-}5)$$

Depending on the magnitudes of β and b/t, the latter term may be significant.

EXAMPLE

A long prismatic member has a rectangular cross section of dimensions 1.25 mm by 100 mm, with $E = 200$ GPa and $G = 77$ GPa. Torque $T = 5600$ N·mm is applied. What are the rate of twist and the stresses, according to linear theory? What are these quantities if the nonlinear terms of Eq. 9.14-5 are included?

Linear theory, Eqs. 9.3-4 and 9.3-5, yields

$$\beta = \frac{T}{G(bt^3/3)} = 1.12(10^{-3}) \text{ rad/mm} \qquad \tau = G\beta t = 108 \text{ MPa} \quad (9.14\text{-}6)$$

Nonlinear theory, Eq. 9.14-5, yields

$$5600 = 5.013(10^6)\beta\left[1 + \frac{200{,}000(100^4)}{120(77{,}000)(1.25^2)}\beta^2\right] \quad (9.14\text{-}7)$$

from which $\beta = 0.68(10^{-3})$ rad/mm. The associated maximum σ_x is then obtained from Eq. 9.14-3 at $r = b/2$ or from Fig. 9.14-1d, and shear stress from $\tau = G\beta t$.

$$\sigma_{xm} = 77 \text{ MPa} \qquad \tau = 66 \text{ MPa} \quad (9.14\text{-}8)$$

We see that the maximum σ_x can be comparable to τ and can significantly reduce the rate of twist.

Initial Axial Stress. Let the member of Fig. 9.14-1 carry an initial axial tensile force F_o, so that there exists a uniform axial tensile stress $\sigma_o = F_o/bt$ before twisting. The contribution of σ_o to torque is obtained by the same calculation procedure as used in Eq. 9.14-4.

$$\int r\lambda\, dF_o = \int_{-b/2}^{b/2} r(\beta r)\sigma_o t\, dr = \frac{\sigma_o b^3 t}{12}\beta \quad (9.14\text{-}9)$$

The total torque needed to produce rate of twist β is obtained from Eqs. 9.14-5 and 9.14-9.

$$T = G\frac{bt^3}{3}\beta\left[1 + \frac{Eb^4}{120Gt^2}\beta^2 + \frac{\sigma_o b^2}{4Gt^2}\right] \quad (9.14\text{-}10)$$

We see that σ_o can have an appreciable effect. It is a first-order effect, as the σ_o term in brackets does not contain β. In the preceding example problem, an added σ_o of about 100 MPa is sufficient to reduce β by half.

Other Cross Sections. A cross section that is not a narrow rectangle, yet meets restrictions stated at the outset of the present section, is easily treated. In Fig. 9.14-1 we

Section 9.14 Large Rotation, Axial Load, and Pretwisting

need only regard $t\,dr$ as area element dA and r as its distance from the centroid of the cross section. Then we identify the integral of $r^2 dA$ in Eq. 9.14-2 as J, the centroidal polar moment of inertia of cross-sectional area A. Equation 9.14-3 becomes

$$\sigma_x = \frac{E\beta^2}{2}\left(r^2 - \frac{J}{A}\right) \qquad (9.14\text{-}11)$$

As for torque, if we add initial stress σ_o, apply the first integral in Eq. 9.14-4 to the net axial stress $\sigma_x + \sigma_o$, and add torque $GB K$ from Eq. 9.3-5, we obtain

$$T = GK\beta\left[1 + \frac{E\beta^2}{2GK}\left(\int_A r^4\,dA - \frac{J^2}{A}\right) + \frac{\sigma_o J}{GK}\right] \qquad (9.14\text{-}12)$$

For the special case of a rectangular cross section with $b \gg t$, we obtain $K = bt^3/3$, $\int r^4\,dA = b^5 t/80$, $J = b^3 t/12$, $A = bt$, and verify that Eq. 9.14-12 reduces to Eq. 9.14-10.

Effect of Pretwist. We consider members manufactured with a uniform rate of twist α and having a type of cross section permitted by restrictions stated at the outset of the present section. In what follows we make approximations that require α to be small, so that the fiber most distant from the member axis is inclined no more than about 10 degrees to the member axis. We also require that β, the additional rate of twist due to load, be small in comparison with α [9.12].

The analysis method is similar to that used previously. Since β is now only a fraction of the total rate of twist, Fig. 9.14-1c and Eq. 9.14-1 must be revised as shown in Fig. 9.14-2. For zero change in length of the member, axial strain is now $\epsilon_x = r^2\alpha\beta$; previously, in Eq. 9.14-1, it was $\epsilon_x = r^2\beta^2/2$. The strain expressions differ in that $\beta/2$ is replaced by α. Accordingly, since neither α nor β is a function of r, we need only replace $\beta/2$ by α in preceding integrals. Thus, in the absence of axial force loading, axial stress is

$$\sigma_x = E\alpha\beta\left(r^2 - \frac{J}{A}\right) \qquad (9.14\text{-}13)$$

Let us now superpose uniform axial tensile stress σ_o. Torque associated with the net axial stress is calculated as in Eq. 9.14-4, with λ replaced by $r\alpha$ (since $\beta \ll \alpha$); that is

$$\int_A r(r\alpha)(\sigma_x + \sigma_o)\,dA = E\alpha^2\beta\left(\int_A r^4\,dA - \frac{J^2}{A}\right) + \alpha\sigma_o J \qquad (9.14\text{-}14)$$

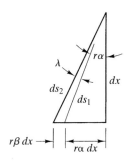

FIGURE 9.14-2 Revision of Fig. 9.14-1c to allow for pretwist rate α. Rate of twist β is produced by applied torque. As in Fig. 9.14-1c, member length is held constant for calculation of ϵ_x.

For α small and $\beta \ll \alpha$:
$$ds_1 \approx \left[1 + \tfrac{1}{2}(r\alpha)^2\right]dx$$
$$ds_2 \approx \left[1 + \tfrac{1}{2}(r\alpha + r\beta)^2\right]dx$$
$$\epsilon_x = \frac{ds_2 - ds_1}{ds_1} \approx r^2\alpha\beta$$

The total torque needed to produce rate of twist β is obtained from Eqs. 9.3-5 and 9.14-14.

$$T = GK\beta\left[1 + \frac{E\alpha^2}{GK}\left(\int_A r^4\,dA - \frac{J^2}{A}\right)\right] + \alpha\sigma_o J \qquad (9.14\text{-}15)$$

For a rectangular cross section with $\sigma_o = 0$ and $b/t = 20$, torsional stiffness T/β is approximately tripled if edges $r = \pm b/2$ of the cross section are inclined at 10 degrees to the member axis.

If $T = 0$, Eq. 9.14-15 shows that positive σ_o creates negative β. In other words, a pretwisted member partially untwists when pulled. A propeller blade behaves in this manner, although its cross section is closed, its properties are functions of x, and σ_o varies with x.

A pretwisted member changes length when twisted. This change is a first-order effect, not second order as is the case when there is no pretwist. The amount of length change due to torque T, here called Δ_T, can be calculated by use of Maxwell's reciprocal theorem. First, one uses Eq. 9.14-15 to calculate β due to an axial force $\sigma_o A$ alone; that is, with $T = 0$. Then, for this β

$$T(\beta L) = (\sigma_o A)\Delta_T \qquad \text{hence} \qquad \Delta_T = \frac{T\beta L}{\sigma_o A} \qquad (9.14\text{-}16)$$

where Δ_T is the change in length L due to an applied torque T. Since β is directly proportional to σ_o when $T = 0$, the value of σ_o used to calculate β does not matter. When T is in the same direction as the pretwist, Δ_T is negative.

An axial force $\sigma_o A$ causes partial untwisting. To determine the resulting change in length, one can first use Eq. 9.14-15 with $\beta = 0$. Thus $T = \alpha\sigma_o J$ is the torque that prevents twisting. The torque is removed by superposing an equal torque in the opposite sense. Thus the total change in length is given by Eq. 9.14-16 with $T = -\alpha\sigma_o J$ and β calculated as described above Eq. 9.14-16, plus $\sigma_o L/E$ to account for lengthening due to direct axial stress.

9.15 TORSION WITH PLASTIC ACTION

We consider prismatic members whose material is linearly elastic up to yield strength τ_Y in shear, with a flat-topped stress-strain diagram thereafter. Thus, strain hardening is neglected. We seek the fully plastic torque T_{fp} in terms of τ_Y and dimensions of the cross section. It happens that for simple shapes of cross section, T_{fp} is easy to calculate by a simple extension of the membrane analogy.

Fully Plastic Condition. Consider a prismatic bar of arbitrary cross section, with x the axial coordinate. Let the bar be twisted in pure torsion until all material yields. With τ_{xy} and τ_{zx} the only nonzero stresses, the associated principal stresses can be calculated from Eq. 2.3-1. The result is

$$\sigma_1 = -\sigma_3 = [\tau_{xy}^2 + \tau_{zx}^2]^{1/2} \qquad \sigma_2 = 0 \qquad (9.15\text{-}1)$$

According to the maximum shear stress yield criterion,

$$\tau_Y = \frac{\sigma_1 - \sigma_3}{2} \qquad \text{hence} \qquad \tau_Y^2 = \tau_{xy}^2 + \tau_{zx}^2 \qquad (9.15\text{-}2)$$

As in linearly elastic analysis, shear stresses can be defined in terms of the Prandtl stress function ϕ (Eqs. 7.12-1). Thus, differential equations of equilibrium are satisfied. If compatibility conditions are to be expressed in terms of stress, conditions must be linearly elastic, which is not the case here. Instead of a compatibility equation, we use Eq. 9.15-2, which becomes

$$\left(\frac{\partial \phi}{\partial y}\right)^2 + \left(\frac{\partial \phi}{\partial z}\right)^2 = \tau_Y^2 \qquad (9.15\text{-}3)$$

In linear elasticity, ϕ is constant on a boundary of the cross section, and torque T is twice the volume under the ϕ surface. These conclusions are reached by equilibrium considerations, without reference to material properties. Therefore, they remain valid for plastic conditions. The rate of twist cannot be determined for plastic conditions. One can only say that twist is sufficient to cause plastic deformation in all of the material (or, in practice, almost all of the material).

The left-hand side of Eq. 9.15-3 is the square of the gradient of ϕ; that is, it is the square of the maximum slope of the ϕ surface. Therefore, Eq. 9.15-3 says that the maximum slope is constant and equal to τ_Y. To Nadai [3.1], writing in 1923, this constant slope suggested the *sand hill analogy*. A hill of sand, with its entire surface at the angle of repose, has the same maximum slope everywhere and is thus a model of the Prandtl stress function for the fully plastic condition.

A sand hill provides qualitative information and, for simple shapes of cross section, a quantitative relation between T_{fp}, τ_Y, and dimensions of the cross section. As with the membrane analogy, a sand hill experiment is likely to be conducted in imagination rather than in the laboratory. Thus, in imagination we take a horizontal flat plate, whose shape is that of the cross section to be analyzed, and pour sand on it until it refuses to carry more. At this time the constant-maximum-slope condition is satisfied, and the boundary condition is $\phi = 0$ on the edge of the plate. The height of the sand hill is τ_Y times the shortest horizontal distance from the top of the hill to the nearest $\phi = 0$ boundary. The volume under the hill can then be determined. Twice this volume is equal to T_{fp}. In the rectangular cross section of Fig. 9.15-1a, for example, the volume is composed of a pyramid and a prism, both of height $(b/2)\tau_Y$. Thus

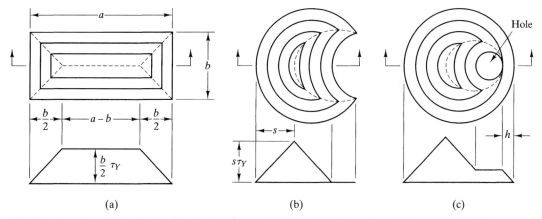

FIGURE 9.15-1 Sand hill analogies, showing top views and cross-sectional views through the centers. Solid lines are contours $\phi =$ constant. Ridge lines are shown dashed.

$$T_{fp} = 2\int \phi\, dA = 2\left[\frac{b^2}{3}\left(\frac{b}{2}\tau_Y\right) + \frac{(a-b)b}{2}\left(\frac{b}{2}\tau_Y\right)\right] \qquad (9.15\text{-}4)$$

For the special case of a solid square cross section of side length a, $T_{fp} = \tau_Y a^3/3$. From Table 9.4-1, the torque T_Y that initiates yielding of a square cross section is $T_Y = 0.208\,\tau_Y a^3$, so $T_{fp}/T_Y = 1.60$. (For a solid circular cross section, $T_{fp}/T_Y = 4/3$, which the reader may wish to verify by making use of the sand hill analogy.)

Top views in Fig. 9.15-1ab show contours (lines of equal elevation) of sand hills on solid cross sections, with a uniform elevation increment between contours. Thus, when viewed from above, contours are equidistant; indeed, they can be constructed in this manner, working from the boundary inward. The sand hill has ridges if there are external corners. Viewed from above, ridge lines bisect external corners. Shear stresses are directed tangent to contours.

The sand hill analogy can also be applied to a cross section that has one or more holes. Consider Fig. 9.15-1c. We imagine that a tube, in this case a circular one like a stovepipe, is pushed up through a hole in the flat plate. Sand is poured on the plate and around the tube until no more can be carried. Then the tube is lowered, allowing sand to spill into the tube and away, just until the entire outer surface of the tube is obscured by sand. In Fig. 9.15-1c, the entire upper end of the tube is now a distance $\tau_Y h$ above the horizontal plate. Thus we achieve the boundary conditions $\phi = 0$ on the edge of the plate and $\phi =$ constant on the edge of the hole, and the proper ϕ surface is obtained. For calculation of torque, volume of the sand hill is regarded as including volume between the $\phi = 0$ plate elevation and the $\phi =$ constant elevation over each hole. In Fig. 9.15-1c, volume calculation is not easy. In other cases it may be simple: For example, if a square cross section has a central square hole with sides parallel to the outer boundary.

Partial Plasticity. Consider the application of a torque less than T_{fp} but sufficient to cause yielding. The following qualitative analysis, based on a combination of the membrane and sand hill analogies, can indicate where yielding begins and how it spreads as torque increases [3.1].

Where elastic and plastic material meet, both have shear stress τ_Y. Since there is no discontinuity of stress across the elastic-plastic interface, there is no abrupt change in elevation or slope of the ϕ surface at the interface. Therefore the curved ϕ surface that represents the elastic portion joins smoothly with the constant slope ϕ surface that represents the plastic portion. In the analogy, one imagines a transparent roof shaped like the surface of the sand hill, with an initially flat membrane at its base, like the floor of an attic. Lateral pressure from below inflates the membrane and provides the (elastic) membrane analogy. Initiation of yielding corresponds to the first contact of the membrane with the roof. Further inflation produces areas of contact, which correspond to yielded regions (Fig. 9.15-2). Full contact corresponds to the fully plastic condition. Full contact is approached only with extreme lateral pressure, which corresponds to an extreme rate of twist. By this time the material will have strain hardened, members having a noncircular cross section will no longer be prismatic, and secondary stresses will have appeared. Clearly torque T_{fp} is an approximation, and the actual ultimate torque is likely to be greater than T_{fp}.

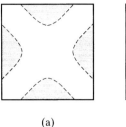
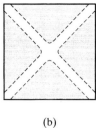

FIGURE 9.15-2 Partially plastic square cross sections. Plastic regions are shaded. (a) Moderate amount of yielding. (b) Almost complete yielding.

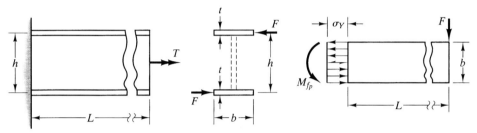

FIGURE 9.15-3 Approximate treatment of fully plastic torque associated with restraint of warping in an I beam. Side, end, and top views are shown.

Restraint of Warping. The fully plastic torque that can be carried by a thin-walled member is increased by restraint of warping. A simple approximate method is available for such problems [9.13]. Consider the problem depicted in Fig. 9.15-3. The net fully plastic torque is taken as the torque T_{fp} predicted by the sand hill analogy plus a contribution that comes from fully plastic bending in the flanges. The latter contribution is formulated with the aid of Fig. 9.15-3. Flanges are regarded as independent beams, each loaded by one of the couple-forces F, where $F = T/h$. When a flange is fully plastic, $FL = M_{fp}$, where $M_{fp} = \sigma_Y b^2 t/4$. Hence $F = M_{fp}/L$ and $T = (M_{fp}/L)h$. The net fully plastic torque is

$$(T_{fp})_{\text{net}} \approx T_{fp} + \frac{h}{L} M_{fp} \tag{9.15-5}$$

PROBLEMS

9.2-1. For a circular cross section with a central circular hole, use the membrane analogy to develop formulas for $\tau_{\max}$ and β in terms of torque T and dimensions of the cross section.

9.3-1. A prescribed rate of twist β is imposed on a prismatic member whose cross section is rectangular, with $b = 8t$. If mistakenly analyzed by the elementary formula $\tau = Tc/J$, what is the percentage error in the calculated maximum shear stress?

9.3-2. (a) A prismatic member has the thin-walled open cross section shown, where $R \gg t$. Express $\tau_{\max}$ and β in terms of $R, t, G,$ and torque T.

(b) If $R = 10t$, what are the ratios of the correct $\tau_{\max}$ and β to the values predicted by elementary formulas ($\tau = TR/J$ and $\beta = T/GJ$)?

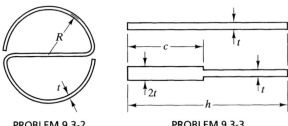

PROBLEM 9.3-2   PROBLEM 9.3-3

9.3-3. Members having the cross sections shown are each twisted by torque T. Assume that $h \gg t$ and $c \gg 2t$, and neglect stress concentrations. What should be the ratio c/h so that τ_{max} in the stepped section is half that in the uniform section? How then do rates of twist compare? And, in the stepped section, what fraction of its torque is supplied by the thicker portion?

9.3-4. For the cross section shown, determine the torsional stiffness in terms of G and the maximum shear stress in terms of T. Use numerical integration. Dimensions are in millimeters.

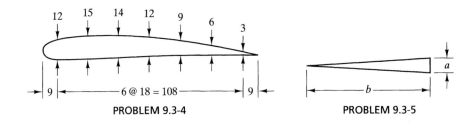

PROBLEM 9.3-4   PROBLEM 9.3-5

9.3-5. The cross section shown is an isoceles triangle for which $b \gg a$.
   (a) Use the membrane analogy to determine τ_{max} and β in terms of T, G, a, and b.
   (b) What are the percentage errors of τ_{max} and β if Eq. 9.3-6 is used, with the average thickness $t = a/2$?

9.3-6. The sketch represents the axial view of a cantilever beam of length L. Each arm of the cross section has length a and thickness t, where $a \gg t$. Lateral load P acts at end $x = L$. At the fixed end $x = 0$, determine the largest flexural and shear stresses, in terms of dimensions and load P. Neglect stress concentration effects. On a sketch of the cross section, indicate exactly where these stresses appear.

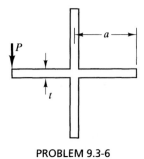

PROBLEM 9.3-6

9.4-1. Consider an elliptical cross section, Fig. 7.11-2. Let a ≫ b.
   (a) Use equations in Section 9.3 to determine τ_{max} in terms of T, a, and b. (It is helpful to note that $I_y = \pi a b^3/4$.)
   (b) What is the torsional stiffness, according to both the membrane analogy and Eq. 9.4-4?

9.4-2. Consider the square and rectangular cross sections of equal area in Fig. 9.4-2. What is the ratio of maximum shear stresses, square to rectangular, for (a) the same β, and (b) the same T? (c) What is the torsional stiffness of each section, according to both Table 9.4-1 and Eq. 9.4-4?

9.4-3. Two torsion members carry the same torque T and have the same material properties and cross-sectional area A. One cross section is square, the other circular. Determine the rate of twist and maximum shear stress in each, in terms of T, A, and G.

9.4-4. A bar of square cross section, 20 mm on a side, is used as a shaft. The material yields in tension at 300 MPa. Determine the allowable torque, based on the von Mises criterion and a safety factor of 2 against the onset of yielding. In a 400 mm length, what angle of twist is produced by this torque? Let $G = 79$ GPa.

9.5-1. Use the membrane analogy to derive Eqs. 9.5-3 and 9.5-5.

9.5-2. Let b and a represent inner and outer radii of a closed tube of circular cross section.
   (a) Determine the ratios of τ_c and β_c from Eq. 9.5-6 to the exact values of these quantities, in terms of b/a.
   (b) Plot these ratios versus b/a, for $0 < b/a < 1$.
   (c) Verify the results described following Eq. 9.5-6 (a minimum of $\tau_c/\tau_e = 0.828$, when $b = 0.414a$).

9.5-3. Consider two tubes of circular cross section, of wall thickness $t = 1$ mm, and the same material. One tube is closed and of mean radius $R = 10$ mm. If the other is slit open lengthwise, what must be its radius if it is to have the same torsional stiffness as the closed tube? Do you think the open tube might buckle?

9.5-4. By using integration, in a manner analogous to Eq. 9.3-7, relate torque to maximum shear stress in the slit tube of Fig. 9.5-2b. The resulting formula should agree with Eq. 9.3-4.

9.5-5. (a) A thin-walled closed tube has uniform wall thickness t and a rectangular cross section of dimensions a by b. Express τ and β in terms of t, G, torque T, aspect ratio $r = a/b$, and perimeter $p = 2(a + b)$. Show that among all rectangular cross sections of the same perimeter, a square cross section is stressed and twisted the least.
   (b) Torque T is applied to two thin-walled closed tubes of the same perimeter and wall thickness. One cross section is circular, the other square. Determine the ratios of τ and β in the square cross section to τ and β in the circular cross section. Neglect stress concentrations.

9.5-6. A thin-walled closed tube has the cross section shown. The midline is a circle of radius R. Thickness t varies linearly with $|\alpha|$, for $0 < |\alpha| < \pi$.
   (a) What is the torsional stiffness? What, and precisely where, is τ_{max}? Express results in terms of T, G, R, and t_o.
   (b) Determine the quantities requested in part (a) for a tube of radius R and uniform wall thickness $1.5t_o$.
   (c) The tube of part (a) is now slit open lengthwise at $\alpha = \pi$. What, and precisely where, is τ_{max}?

9.5-7. A pipe of mean radius R is constructed by bending sheet metal until edges overlap, then riveting lengthwise along the overlap, as shown in end view. If s is the lengthwise spacing of rivets, what is the average shear force per rivet when a torque T is applied?

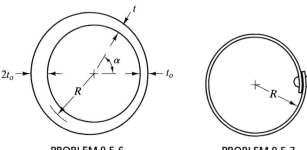

PROBLEM 9.5-6   PROBLEM 9.5-7

9.5-8. A thin-walled closed member of circular cross section has length L, uniform wall thickness, and a gentle linear taper of its mean radius from r_0 at one end to r_L at the other. Derive an expression for the relative angle of twist between ends.

9.6-1. Use the membrane analogy to derive Eqs. 9.6-3, for the specific case of a thin-walled tube of two cells.

9.6-2. Consider a thin-walled multicell tube without any isolated internal cells (thus the latter two cross sections in Fig. 9.6-3 are excluded). Assume that all external cell walls are of uniform thickness. What geometric condition must be satisfied if internal cell walls are to have no shear flow? Suggestion: Consider the membrane analogy.

9.6-3. Consider an I section of web depth $h = 20t$ and flange width $b = 10t$, where t is the uniform thickness of web and flanges. Cover plates of thickness t are now welded between the flanges (as in portion DE in Fig. 9.7-1). By what factor has torsional stiffness been increased by the cover plates?

9.6-4. How beneficial are the internal walls in Fig. 9.6-2? Find out by calculating percentage changes in β and the largest q if the internal walls are removed.

9.6-5. (a) Let all walls of the cross section shown in Fig. 9.6-3b have the same thickness t, where $t \ll a$. Determine expressions for the torsional stiffness of the section and the shear flow in each wall in terms of torque T and dimensions.

    (b) Repeat part (a) for the cross section shown in Fig. 9.6-3c.

9.6-6. All wall thicknesses are 0.5 mm in the two-cell closed tube shown. Dimensions are in millimeters. The maximum shear stress (neglecting stress concentrations) is 60 MPa. Determine the applied torque and the rate of twist in terms of G.

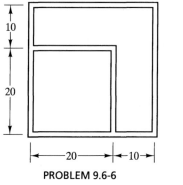

PROBLEM 9.6-6

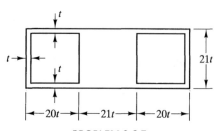

PROBLEM 9.6-7

9.6-7. The section shown consists of a solid central rectangle with identical square tubes attached. For a given torque, by what factors would the maximum shear stress and rate of twist increase if the square tubes were omitted? Use convenient but reasonable approximations. Neglect stress concentrations.

9.7-1. A prismatic member of length L is made of material whose modulus E is low. Imagine that over a length $L/2$, E is changed from low to practically infinite. If torque remains the same and the cross section is circular, by what factor is the angle of twist over length L reduced? Qualitatively, how does the answer change if the cross section is rectangular? Why?

9.7-2. (a) A member is loaded by a torque uniformly distributed along the length, of intensity T_q (dimensions [force·length/length]). What equation replaces Eq. 9.7-6? Suggestion: Substitute $\beta = d\theta/dx$ and differentiate.

(b) What are boundary conditions in terms of θ at a free end, at a fixed end, and at a simply supported end (where only θ is restrained)?

9.7-3. Consider the I section described in Fig. 9.7-3. However, let $L = 2000$ mm, and let each end of the member be welded to a massive block that prevents warping. Torque T is applied to each block, so as to twist the member. Determine the maximum axial stress σ_x and the relative rotation between blocks as follows.

(a) Solve the problem simply, by making use of results already determined in the example problem of Section 9.7.

(b) Solve the problem formally, making use of Eq. 9.7-7 and appropriate boundary conditions.

9.7-4. (a) The free end of a cantilever beam of length L is loaded by lateral force P, as shown. Use numerical data provided in Fig. 9.7-3, and determine maximum and minimum values of axial stress σ_x at the fixed end.

(b) Determine the y and z components of deflection of the point to which force P is applied.

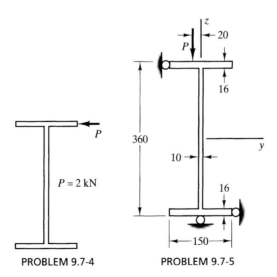

PROBLEM 9.7-4   PROBLEM 9.7-5

9.7-5. The sketch represents the end view of an I beam, 6000 mm long. Dimensions shown are in millimeters. Simple supports shown are located at each end. They prevent lateral deflection and rotation about the x axis but do not restrain warping or rotations about y and z axes. At midspan, where lateral load P acts, determine the largest axial stress σ_x. At an end determine the largest τ_{zx} in the web and the largest τ_{xy} in a flange. Express stresses in terms of P. Indicate where these stresses appear on the respective cross sections. Let $E = 200$ GPa and $G = 77$ GPa.

9.8-1. Plot sectorial area diagrams for the following choices of pole P and initial point I in Fig. 9.8-2a.

    (a) P at C, I at D.   **(b)** P at G, I at C.
    (c) P at A, I at B.   **(d)** P at H, I at B.

9.8-2. Use Eq. 9.8-6 to acomplish the following changes of pole. Sketch the resulting sectorial area diagrams.

    (a) In Fig. 9.8-2c, move the pole from E to G.
    (b) In Problem 9.8-1b, move the pole from G to B.
    (c) In Problem 9.8-1c, move the pole from A to C.

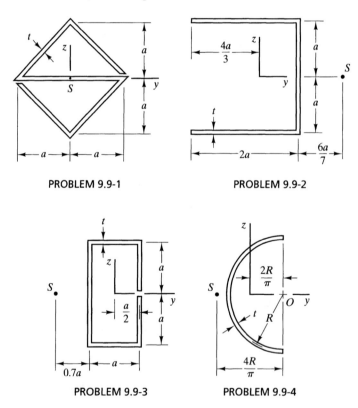

PROBLEM 9.9-1

PROBLEM 9.9-2

PROBLEM 9.9-3

PROBLEM 9.9-4

9.9-1. **(a)** For the thin-walled cross section shown, determine and sketch the principal sectorial area diagram. Thickness t is uniform. Point S is the shear center. Axes yz are centroidal.

    (b) Determine J_ω based on the principal sectorial area diagram.

9.9-2. Repeat the instructions of Problem 9.9-1 for the channel section shown.

9.9-3. Repeat the instructions of Problem 9.9-1 for the open section shown.

9.9-4. Repeat the instructions of Problem 9.9-1 for the half-circle section shown.

9.9-5. Evaluate the warping constant J_ω for the Z section shown in Fig. 9.13-2b.

9.10-1. In Fig. 9.10-3a, relocate initial point I to corner A, and use Eq. 9.10-4 to recalculate the axial displacements of corners A, B, C, and D. What explains the difference between these displacements and those described by Eq. 9.10-6?

9.10-2. If a closed section has no warping, $\lambda = 0$ in Eq. 9.10-1. Without recourse to sectorial area, show that this implies $b/t_b = h/t_h$ in Fig. 9.10-3.

9.10-3. At the loaded end of the I beam in Problem 9.7-4, determine the x-direction displacement at the tips of each flange.

9.10-4. (a) Imagine a closed tube whose cross section is an equilateral triangle. Each side has length a and thickness t, where $a \gg t$. Use Eq. 9.10-4 to show that there is no warping when the tube is twisted.

(b) Now let each of the three sides have a different thickness. By argument rather than calculation, show that there is still no warping.

9.11-1. For the open circular cross section of Fig. 9.9-1a, Eq. 9.11-4 yields $q = EtR^3(d^2\beta/dx^2)f$, where $f = f(\alpha)$. Determine f. Obtain the numerical value of f for these values of α: 0, 1.246, π, 5.037, and 2π. Sketch the distribution of q along the midline of the cross section.

9.11-2. For the channel section of Problem 9.9-2, use Eq. 9.11-4 to plot q around the section (in terms of E, $d^2\beta/dx^2$, a, and t). Indicate the direction of q. Locate points where $q = 0$ and points where q is maximum or minimum.

9.12-1. For the open circular cross section in Fig. 9.9-1a, let $R = 25$ mm and $t = 2$ mm. Also let $E = 100$ GPa and $G = 40$ GPa. A member of length $L = 0.5$ m is fixed at end $x = 0$. At end $x = L$, force P in the z direction is applied at the centroid of the cross section. At end $x = 0$ determine σ_x in terms of P, at $\alpha = 0$ (above the slit) and at $\alpha = \pi/2$. At end $x = L$, determine the Saint-Venant shear stress and the angle of twist in terms of P.

9.12-2. For the cross section of Problem 9.9-1, let $a = 40$ mm and $t = 2$ mm [hence $J_\omega = 689(10^6)$ mm^6]. Also let $E = 200$ GPa and $G = 77$ GPa. In a member of length $L = 1.5$ m, warping is prevented at each end while torque compels ends to rotate 0.050 radians with respect to one another. What torque is applied? What are significant normal and shear stresses, at ends and at the middle? Is the member long enough for transverse shear deformation to be ignored?

9.12-3. For the cross section of Problem 9.9-2, let $a = 50$ mm and $t = 2$ mm [hence $J_\omega = 1.191(10^9)$ mm^6]. Also let $E = 200$ GPa and $G = 77$ GPa. A member of length $L = 0.9$ m is fixed at end $x = 0$ and unrestrained at end $x = L$. Torque $T = 60$ N·m is applied at end $x = L$. What are significant normal and shear stresses at the ends? By what factor is θ at $x = L$ reduced by restraint of warping at $x = 0$? Is the member long enough for transverse shear deformation to be ignored?

9.12-4. For the cross section of Problem 9.9-3, let $a = 30$ mm and $t = 1$ mm [hence $J_\omega = 89.9(10^6)$ mm^6]. Also let $E = 200$ GPa and $G = 77$ GPa. A member of length $L = 0.6$ m is fixed at end $x = 0$. At end $x = L$, downward force P is applied to the upper left corner; that is, P in the $-z$ direction acts along the midline of the web. Determine the axial stress of greatest magnitude at end $x = 0$, in terms of P.

9.12-5. A cantilever beam of length L has the right-angle cross section shown and is tip-loaded by transverse force P at the centroid of the cross section. For what ratio L/a is the maximum flexural stress equal to twice the maximum shear stress due to twisting?

9.12-6. A prismatic member is loaded by torques T_1 and T_2 as shown. When Eq. 9.12-5 is applied to this problem, how many constants of integration appear? What conditions serve to determine them?

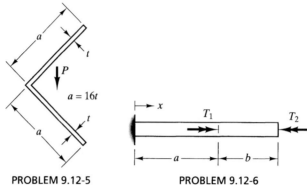

PROBLEM 9.12-5   PROBLEM 9.12-6

9.13-1. (a) For the I beam in Fig. 9.13-2a, let web and flanges have the same thickness t. Also let $E = 2.5G$, $b = 10t$, $h = 20t$, and $L = 240t$. Bimoment $B_L = Fbh$ is applied at $x = L$. Determine θ at $x = L$ and the maximum axial stress at $x = 0$, in terms of B_L, G, and t.

(b) Alter the support conditions in part (a): Let end $x = 0$ be free and end $x = L$ be simply supported. Again B_L is applied at end $x = L$. Determine θ at $x = 0$.

9.13-2. For the member depicted in Fig. 9.13-2a, use numerical data in Fig. 9.7-3. Obtain expressions for the largest magnitudes of τ from Eq. 9.3-4 and $\tau_q = q/t$ from Eq. 9.11-4. Determine the ratios of τ and τ_q to the largest σ_x, assuming that σ_x varies linearly across each flange.

9.13-3. (a) Let a member have simply supported ends. Loading consists of bimoments B_0 at end $x = 0$ and B_L at end $x = L$. Explain how rotation θ can be determined as a function of x.

(b) Let the member in Fig. 9.13-2b have simply supported ends. Imagine that a torque T is to be applied at midspan to prevent rotation there. Explain how T can be calculated in terms of force F.

9.13-4. Show that Maxwell's reciprocal theorem applies to the bimoment problem. Specifically, show that a torque T times end rotation θ_L in Fig. 9.13-2a is equal to bimoment B_L acting through the warping displacements at $x = L$ due to T in Fig. 9.7-2.

9.14-1. In the example problem of Section 9.14, imagine that with T maintained at 5600 N·mm, the rate of twist is to be halved by the addition of uniform axial stress. What uniform axial stress is required, and what then are the maximum axial stress and the shear stress due to twisting?

9.14-2. Axial force and torque are applied to a member of the cross section shown. The three arms are each 1 mm thick and 15 mm long. Angles between arms are each 120°. If the largest σ_x is 60 MPa, what is the torque, and what is the maximum shear stress due to twisting? Let $E = 2.6G$ and $G = 77$ GPa.

PROBLEM 9.14-2

9.14-3. The cross section of a torsion member is a narrow rectangle, 1 mm by 22 mm. Poisson's ratio is 0.30. The pretwist rate is $\alpha = 0.016$ rad/mm.
  (a) By what factor is torsional stiffness increased by pretwist?
  (b) In terms of torque T, what is the maximum shear stress due to twisting and what is the maximum axial normal stress?
  (c) What is the change in length, in terms of T, G, and L?
  (d) If T is replaced by axial tensile force F, what is the change in length, in terms of F, G, and L?

9.14-4. Consider a pretwisted member for which $E = 2.6G$. The cross section is a narrow rectangle for which $b = 20$ mm and $t = 1$ mm. Pretwist inclines outer edges of the cross section at 10° to the axis of the member.
  (a) By what factor is torsional stiffness increased by pretwist? Let $\sigma_o = 0$.
  (b) When $\beta = 0.07°$/mm, what uniform axial stress must be applied so that torque will be the same as torque in a member with the same β but no pretwist? Let $E = 200$ GPa.

9.15-1. Show that the left-hand side of Eq. 9.15-3 can be regarded as the square of the maximum slope of the ϕ surface.

9.15-2. Sketch sand hill analogies, in the manner of Fig. 9.15-1, for the cross sections shown. Cross section (c) contains a central square hole.

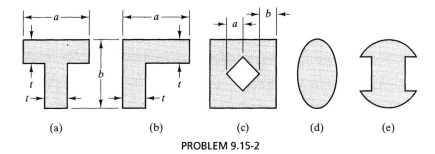

PROBLEM 9.15-2

9.15-3. For each of the following cross sections, determine the fully plastic torque in terms of τ_Y and the dimensions.
  (a) Problem 9.15-2a.
  (b) Problem 9.15-2b.
  (c) Problem 9.15-2c, with $a = b$.

9.15-4. A solid bar of circular cross section and outer radius R is twisted until the elastic core has radius $R/2$. Use the membrane and sand hill analogies to obtain an expression for applied torque in terms of τ_Y and R. Check the expression by using equations of statics.

9.15-5. For the I beam in Fig. 9.15-3, let $h = 2b$ and $t = b/10$. Let the web thickness be $b/20$. If $\tau_Y = \sigma_Y/2$, for what value of L/b do both terms in Eq. 9.15-5 contribute equally to the net fully plastic torque?

CHAPTER 10

Unsymmetric Bending and Shear Center

10.1 UNSYMMETRIC BENDING OF STRAIGHT BEAMS

This chapter is concerned with stresses and deflections produced by transverse load on a member that is straight and prismatic, and whose material is homogeneous, isotropic, and linearly elastic. It is assumed that the member bends without twisting, which means that the load must act through the shear center (whose location is discussed in the latter portion of the chapter). Loads that twist the beam, or have an axial component, create additional stresses and deflections that can be calculated separately and superposed on stresses and deflections discussed in this chapter.

Symmetric bending is the subject of elementary bending theory. There it is assumed that the member has a longitudinal plane of symmetry and that transverse loads act in that plane. Then, as in Fig. 10.1-1, moment vector **M** is parallel to the neutral axis. Flexural stress in Fig. 10.1-1 is $\sigma_x = M_y z/I_y$, where $M = M_y = Ps$. Lateral deflection due to bending is directed normal to the neutral axis. Note that **M** is perpendicular to the plane of loads. This circumstance is dictated by statics and has nothing to do with the shape of the cross section; therefore it is also true of unsymmetric bending.

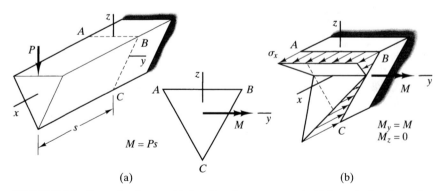

FIGURE 10.1-1 Example of symmetric bending. Axes yz are centroidal axes in the plane of a typical cross section.

Section 10.1 Unsymmetric Bending of Straight Beams   315

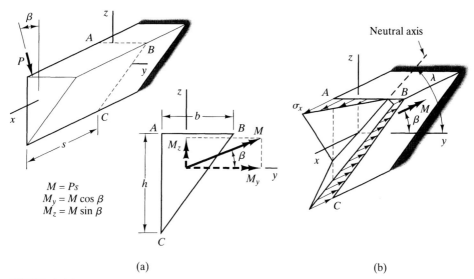

FIGURE 10.1-2 Example of unsymmetric bending. Axes yz are centroidal axes in the plane of a typical cross section.

Unsymmetric bending means that transverse loads do not act in a longitudinal plane of symmetry, or the member has no such plane. The latter case is depicted in Fig. 10.1-2. In general, moment vector **M** is *not* parallel to the neutral axis. Lateral deflection due to bending is directed normal to the neutral axis, but in general is not parallel to the plane of loads.

If the load in Fig. 10.1-2 were vertical, lateral deflection would have a horizontal component. In practice, there may exist constraints such that only vertical deflection is allowed. In that case, constraints apply horizontal load components, such that the neutral axis is horizontal under the resultant of horizontal and vertical load components.

In what follows, y and z are centroidal axes in the plane of a cross section. They need not be principal axes. As in Fig. 10.1-2, their orientation can be suited to the geometry of the cross section, for convenience in the calculation of properties of the cross section. **M** is regarded as the moment vector applied to the portion of the beam behind the yz plane in which stresses are sought. The front, or positive x portion of the beam, has been cut away to expose **M**. The direction of **M** is determined by the right-hand rule. Components of **M** have magnitudes M_y and M_z, which are considered positive when directed in positive y and z directions, as shown in Fig. 10.1-2. (Subsequent formulas are of course not restricted to triangular cross sections.)

Analysis. We assume that axial stress σ_x is the only nonzero normal stress. Therefore $\sigma_x = E\epsilon_x$. As in elementary beam theory, we also assume that plane sections remain plane when the beam is bent. This assumption dictates that axial strain ϵ_x varies linearly with distance from the neutral axis. Since the position of the neutral axis is yet to be established, at this stage we can say only that ϵ_x is a linear function of y and z. Therefore σ_x is also a linear function of y and z; that is

$$\left.\begin{array}{l}\sigma_x = E\epsilon_x \\ \epsilon_x = a'_o + a'_y y + a'_z z\end{array}\right\} \quad \text{hence} \quad \sigma_x = a_o + a_y y + a_z z \quad (10.1\text{-}1)$$

where $a_o = Ea'_o$ and so on. Constants a_o, a_y, and a_z are determined from the following equilibrium equations.

$$0 = \int_A \sigma_x \, dA \qquad M_y = \int_A z\sigma_x \, dA \qquad M_z = -\int_A y\sigma_x \, dA \quad (10.1\text{-}2)$$

In these expressions, $\sigma_x \, dA$ is a force increment dF produced by σ_x acting on a differential element dA of cross-sectional area A. The first equation says that since there is no net axial force on the beam, axial force increments dF sum to zero. The second equation says that moment increments $z \, dF$ sum to M_y. The third equation is similar, and requires a negative sign because when y and dF are both positive, moment increment $y \, dF$ is opposite in direction to M_z.

Because y and z are centroidal axes, Eq. 10.1-1 and the first of Eqs. 10.1-2 yields $0 = a_o A$; therefore $a_o = 0$. This means that the neutral axis of bending passes through the centroid of the cross section. Equation 10.1-1 and the latter two of Eqs. 10.1-2 involve integrals of $y^2 \, dA$, $z^2 \, dA$, and $yz \, dA$. Respectively, these three integrals are area moments of inertia I_z and I_y and product of inertia I_{yz} (discussed in Section 1.3). Thus

$$M_y = I_{yz} a_y + I_y a_z \quad \text{and} \quad M_z = -I_z a_y - I_{yz} a_z \quad (10.1\text{-}3)$$

Solution of these two equations for a_y and a_z yields

$$a_y = \frac{-M_y I_{yz} - M_z I_y}{I_y I_z - I_{yz}^2} \quad \text{and} \quad a_z = \frac{M_y I_z + M_z I_{yz}}{I_y I_z - I_{yz}^2} \quad (10.1\text{-}4)$$

Hence Eq. 10.1-1 becomes

$$\sigma_x = \frac{(-M_y I_{yz} - M_z I_y)y + (M_y I_z + M_z I_{yz})z}{I_y I_z - I_{yz}^2} \quad (10.1\text{-}5)$$

Angle λ between the y axis and the neutral axis is determined by stating that the neutral axis is the locus of points at which $\sigma_x = 0$, and noting that $\tan \lambda = z/y$. Hence, from Eq. 10.1-1 with $a_o = 0$, and from Eq. 10.1-4,

$$\tan \lambda = -\frac{a_y}{a_z} \quad \text{or} \quad \tan \lambda = \frac{M_y I_{yz} + M_z I_y}{M_y I_z + M_z I_{yz}} \quad (10.1\text{-}6)$$

Again, orthogonal axes y and z need have no particular orientation in the plane of the cross section but *must be centroidal*.

Remarks. If net axial force N is also applied, the foregoing analysis yields $a_o = N/A$ rather than $a_o = 0$. Uniform axial stress N/A is superposed on σ_x of Eq. 10.1-5. The presence of N translates the neutral axis away from the centroid but does not change its orientation.

An alternative expression for σ_x is obtained by eliminating M_z between Eqs. 10.1-5 and 10.1-6. The result is

$$\sigma_x = \frac{z - y \tan \lambda}{I_y - I_{yz} \tan \lambda} M_y \qquad (10.1\text{-}7)$$

Although M_z does not appear explicitly in Eq. 10.1-7, it influences σ_x via λ, Eq. 10.1-6. As $\tan \lambda$ becomes large in magnitude, Eq. 10.1-7 becomes numerically sensitive, and fails entirely for $\lambda = \pm\pi/2$. This difficulty is removed by electing axes y and z that are differently oriented in the plane of the cross section.

If y and z are principal centroidal axes (discussed in Section 1.3), then $I_{yz} = 0$, and Eqs. 10.1-5 and 10.1-7 both reduce to

$$\text{Principal axis form } (I_{yz} = 0): \quad \sigma_x = -\frac{M_z y}{I_z} + \frac{M_y z}{I_y} \qquad (10.1\text{-}8)$$

Thus, σ_x is obtained by applying the elementary flexure formula about each principal axis, and superposing results. This method is preferred when the orientation of principal axes is known by inspection, as for a cross section having an axis of symmetry. It may also be a good method for a standard rolled section such as an angle with unequal legs, for which principal axis data are available in tabulations.

10.2 NUMERICAL EXAMPLES

The simply supported beam shown in Fig. 10.2-1 has an angle cross section. Dimensions and properties of the cross section are shown in Fig. 10.2-1b. On a cross section 1.0 m from the left end, the largest tensile and compressive flexural stresses are required.

Locations of highly stressed points must be determined. Flexural stress varies linearly with distance from the neutral axis; therefore we seek stresses at points most distant from the neutral axis. Since the load is parallel to the z axis, $M_z = 0$, which simplifies the problem slightly.

Method 1. With $M_z = 0$, Eq. 10.1-6 yields the neutral axis orientation as

$$\tan \lambda = \frac{I_{yz}}{I_z} = \frac{1.200(10^6)}{1.627(10^6)} \quad \text{hence} \quad \lambda = 36.4° \qquad (10.2\text{-}1)$$

A sketch to scale, Fig. 10.2-1a, immediately shows that points A and B are most distant from the neutral axis, and therefore carry the extreme flexural stresses. Point A has coordinates $y = 25$ mm and $z = -35$ mm. Bending moments are $M_y = -6(10^6)$ N·mm and $M_z = 0$. Also, $I_y I_z - I_{yz}^2 = 3.290(10^{12})$ mm^8. Equation 10.1-5 yields

$$\sigma_{xA} = \frac{[-1.200(25) + 1.627(-35)]10^6}{3.290(10^{12})}(-6)10^6 = 159 \text{ MPa} \qquad (10.2\text{-}2)$$

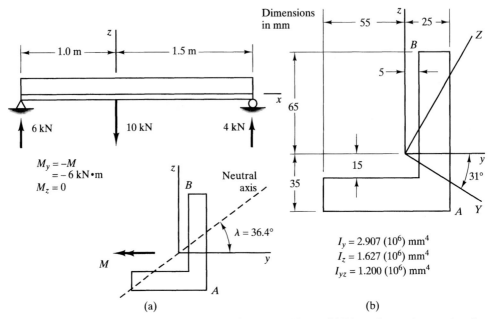

FIGURE 10.2-1 (a) An angle section used as a simply supported beam. (b) Dimensions and properties of the cross section. The cross section is viewed looking leftward along the beam.

Similarly, at point B, where $y = 5$ mm and $z = 65$ mm,

$$\sigma_{xB} = \frac{[-1.200(5) + 1.627(65)]10^6}{3.290(10^{12})}(-6)10^6 = -182 \text{ MPa} \qquad (10.2\text{-}3)$$

Signs of σ_{xA} and σ_{xB} are as expected on physical grounds. Equations 10.2-2 and 10.2-3 provide correct algebraic signs because we have observed sign conventions used in the derivation of Eq. 10.1-5.

Clearly one cannot naively assume that y is the neutral axis and use the expression $M_y z/I_y$. Stresses would be incorrect and the region of compressive stress in the horizontal leg would be missed.

In Fig. 10.2-1 we chose to view the cross section by looking leftward along the beam. Had we chosen instead to look rightward, the cross section would appear rotated 180° about the z axis from the position shown in Fig. 10.2-1b, and I_{yz} would have opposite sign. However, calculated stresses would be the same.

Method 2. From Eqs. 10.1-4, with $M_y = -6(10^6)$ N·mm and $M_z = 0$,

$$a_y = \frac{-(-6)10^6(1.200)10^6}{3.290(10^{12})} = 2.19 \text{ MPa/mm}$$

$$a_z = \frac{(-6)10^6(1.627)10^6}{3.290(10^{12})} = -2.97 \text{ MPa/mm} \qquad (10.2\text{-}4)$$

The first of Eqs. 10.1-6 then yields $\lambda = \arctan(2.19/2.97) = 36.4°$, as shown in Fig. 10.2-1a. Again points A and B are identified as the important locations. Equations 10.1-1 and 10.2-4 yield

$$\sigma_{xA} = 2.19(25) - 2.97(-35) = 159 \text{ MPa}$$
$$\sigma_{xB} = 2.19(5) - 2.97(65) = -182 \text{ MPa} \quad (10.2\text{-}5)$$

These results are of course the same as in Eqs. 10.2-2 and 10.2-3. Method 2 may be perceived as slightly simpler than Method 1 because its algebraic expressions are more compact.

Method 3. Equation 10.1-8 can be used if y and z are principal axes of the cross section. In the present example, we name the principal axes Y and Z (Fig. 10.2-1b). The 31° angle, as well as principal moments of inertia $I_Y = 3.627(10^6)$ mm^4 and $I_Z = 0.907(10^6)$ mm^4, are calculated in Eqs. 1.3-11 and 1.3-12. Y and Z coordinates of points can be determined from transformation equations stated in the paragraph preceding Eq. 1.3-4 (in different notation). For point A,

$$Y_A = (-35)\sin(-31°) + 25\cos(-31°) = 39.5 \text{ mm}$$
$$Z_A = (-35)\cos(-31°) - 25\sin(-31°) = -17.1 \text{ mm} \quad (10.2\text{-}6)$$

Also, $M_Y = M_y \cos 31° = -5.14$ kN·m and $M_Z = M_y \sin 31° = -3.09$ kN·m. Equation 10.1-8 becomes

$$\sigma_{xA} = -\frac{-3.09(10^6)39.5}{0.907(10^6)} + \frac{-5.14(10^6)(-17.1)}{3.627(10^6)} = 159 \text{ MPa} \quad (10.2\text{-}7)$$

Stress σ_{xB} is calculated similarly. Method 3 seems appropriate only if the orientation of principal axes, and properties with respect to them, are known from previous calculation or from a table, as they are for standard rolled sections. Otherwise, error is invited by the length of calculations and the need to be very careful with signs.

10.3 BEAM DEFLECTIONS IN UNSYMMETRIC BENDING

In elementary beam theory, load is applied in a plane of symmetry of the member. Therefore load acts along a principal axis of the cross section, and lateral deflection is parallel to that axis. Thus, in Fig. 10.1-1, deflection analysis begins with the small-deflection moment-curvature relation

$$\text{For } I_{yz} = 0: \quad \frac{d^2w}{dx^2} = -\frac{M_y}{EI_y} \quad (10.3\text{-}1)$$

where w is the z-direction deflection. The negative sign in Eq. 10.3-1 is needed because positive M_y is associated with negative curvature. For a cantilever beam of length L, loaded by tip force P in the negative z direction, integration leads to the familiar formula $w_L = -PL^3/3EI_y$ for tip deflection, parallel to the z axis. There is no y component of deflection.

In unsymmetric bending, a load not directed along a principal axis of the cross section produces a deflection component normal to the direction of load as well as a deflection component parallel to it. One way to calculate the resultant lateral deflection is similar to Method 3 of the preceding section: Resolve the load into perpendicular components, one parallel to each principal axis, calculate deflection due to each load component, and combine deflection components vectorially to obtain the resultant lateral deflection. This method has the disadvantages noted for Method 3. It is usually easier to adopt the method described in what follows.

Analysis. In Fig. 10.3-1a, Δ is the resultant deflection of a point on the neutral axis of the cross section. It is important to note that *the direction of Δ is perpendicular to the neutral axis*. Deflection Δ has components v and w, positive in y and z directions respectively. Therefore, from Fig. 10.3-1a,

$$v = -w \tan \lambda \quad \text{and} \quad \Delta = \frac{w}{\cos \lambda} = \sqrt{v^2 + w^2} \qquad (10.3\text{-}2)$$

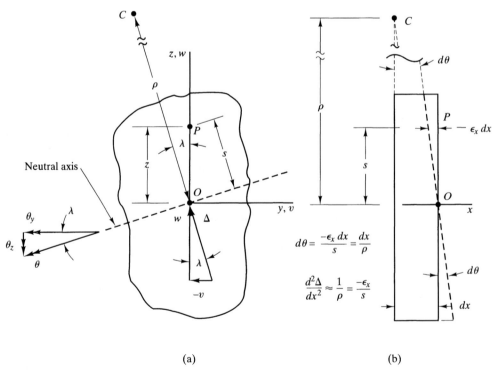

(a) \hspace{4cm} (b)

FIGURE 10.3-1 (a) Deflection Δ and rotation θ of an arbitrary cross section. C is the center of curvature and ρ is the radius of curvature. (b) Geometric relations in the plane of curvature.

Section 10.3 Beam Deflections in Unsymmetric Bending

Angle λ can be obtained from Eq. 10.1-6, using known moments of inertia and known components of the bending moment vector. All that remains is to obtain an expression for w.

Figure 10.3-1b is a view normal to the plane that contains Δ and the axis of the beam. Geometric relations shown are also used in elementary beam theory (see Fig. 1.6-1 and Eq. 1.6-1). Here these relations are written in the notation and sign convention of the present analysis. For an arbitrary point P on the z axis, $s = z \cos \lambda$. Also, $\Delta = w/\cos \lambda$. Therefore

$$\frac{d^2 \Delta}{dx^2} = \frac{-\epsilon_x}{s} \quad \text{yields} \quad \frac{d^2 w}{dx^2} = -\frac{\epsilon_x}{z} \tag{10.3-3}$$

Next we write $\epsilon_x = \sigma_x/E$, where σ_x comes from Eq. 10.1-7, with $y=0$ to place point P of Fig. 10.3-1 on the z axis. Thus

$$\frac{d^2 w}{dx^2} = -\frac{M_y}{E(I_y - I_{yz} \tan \lambda)} \tag{10.3-4}$$

where, as stated by Eq. 10.1-6,

$$\tan \lambda = \frac{M_y I_{yz} + M_z I_y}{M_y I_z + M_z I_{yz}} \tag{10.3-5}$$

Equations 10.3-1 and 10.3-4 differ in their moment of inertia terms, but otherwise have the same form. Accordingly, the usual procedures of elementary beam theory can be used to calculate w, then Eqs. 10.3-2 used to provide v and Δ. Rotation of the cross section—in other words, slope of the beam—can be determined in similar fashion. The following example illustrates procedures.

Calculations are not difficult if the cross section is constant and the neutral axis has the same orientation over the entire length of the beam. Otherwise hand calculations may be tedious, whether or not the cross section has an axis of symmetry. Bending of a pretwisted bar is discussed in [9.11].

EXAMPLE

Determine the deflection and slope of the tip of the cantilever beam in Fig. 10.3-2a, in terms of q, Q, L, and E. Properties of the cross section are stated in Fig. 10.3-2a.

Loads q and Q must be treated separately because the neutral axis changes inclination along the beam when q and Q act simultaneously. We consider load q first. With load Q temporarily ignored, M_z is zero, and

$$\tan \lambda = \frac{I_{yz}}{I_z} = \frac{2.646(10^6)}{3.000(10^6)} \quad I_y - I_{yz} \tan \lambda = 6.666(10^6) \text{ mm}^4 \tag{10.3-6}$$

Since q acts downward, deflection w must be downward (in the negative z direction). The handbook formula for tip deflection is $qL^4/8EI$. Therefore

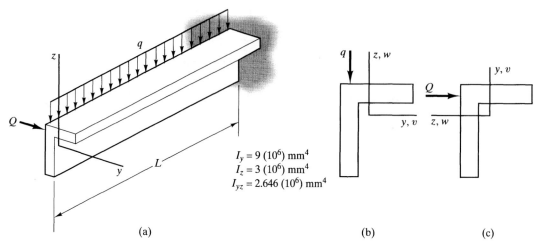

FIGURE 10.3-2 (a) Cantilever beam of angle cross section. (b) Axes for calculating deflection caused by load q. (c) Axes for calculating deflection caused by load Q.

$$w = -\frac{qL^4}{8E(6.666)10^6} = -18.75(10^{-9})\frac{qL^4}{E} \quad \text{(down)}$$

$$v = -w \tan \lambda = 16.54(10^{-9})\frac{qL^4}{E} \quad \text{(right)}$$

(10.3-7)

where labels "down" and "right" mean, respectively, directions $-z$ and $+y$ in Fig. 10.3-2a. As for slope at the tip, the handbook formula is $qL^3/6EI$. Therefore

$$\theta_y = -25.0(10^{-9})\frac{qL^3}{E} \qquad \theta_z = \theta_y \tan \lambda = -22.1(10^{-9})\frac{qL^3}{E} \qquad (10.3\text{-}8)$$

where negative signs indicate rotations directed opposite to those shown in Fig. 10.3-1a.

For load Q, Eq. 10.3-4 fails because $M_y = 0$. We therefore adopt new axes, as shown in Fig. 10.3-2c. For this choice of axes, $I_y = 3(10^6)$ mm^4, $I_z = 9(10^6)$ mm^4, $I_{yz} = -2.646(10^6)$ mm^4, and

$$\tan \lambda = \frac{I_{yz}}{I_z} = \frac{-2.646(10^6)}{9.000(10^6)} \qquad I_y - I_{yz} \tan \lambda = 2.222(10^6) \text{ mm}^4 \qquad (10.3\text{-}9)$$

The handbook formula for tip deflection is $QL^3/3EI$. Therefore, with algebraic signs pertaining to Fig. 10.3-2c and labels "right" and "down" pertaining to Fig. 10.3-2a,

$$w = -\frac{QL^3}{3E(2.222)10^6} = -150.0(10^{-9})\frac{QL^3}{E} \quad \text{(right)}$$

$$v = -w \tan \lambda = -44.10(10^{-9})\frac{QL^3}{E} \quad \text{(down)}$$

(10.3-10)

The handbook formula for slope at the tip is $QL^2/2EI$. Therefore, in the coordinate system of Fig. 10.3-2c,

Section 10.4   Transverse Shear Stress and Shear Flow   323

$$\theta_y = -225(10^{-9})\frac{QL^2}{E} \qquad \theta_z = \theta_y \tan \lambda = 66.2(10^{-9})\frac{QL^2}{E} \qquad (10.3\text{-}11)$$

With positive directions up and to the right in Fig. 10.3-2a, combined deflection components are

$$\begin{aligned} -18.75(10^{-9})\frac{qL^4}{E} - 44.10(10^{-9})\frac{QL^3}{E} & \quad \text{(vertical)} \\ 16.54(10^{-9})\frac{qL^4}{E} + 150.0(10^{-9})\frac{QL^3}{E} & \quad \text{(horizontal)} \end{aligned} \qquad (10.3\text{-}12)$$

Components of slope may be combined in similar fashion.

10.4   TRANSVERSE SHEAR STRESS AND SHEAR FLOW

Shear flow is force per unit length. It is produced by twisting, as described in Section 9.5, and also by transverse loading. Shear flow acts on both longitudinal and transverse sections, and is closely related to transverse shear stress. Expressions for shear flow and shear stress due to transverse loading are developed in the present section. Their uses include determining whether stresses are within allowable limits and locating the shear center of a cross section.

Members are assumed to be homogeneous, isotropic, linearly elastic, and prismatic (straight and without taper). If such a beam is thin-walled, a cross section displays only one shear stress of appreciable magnitude, directed tangent to the midline of the cross section. Shear stress directed normal to the midline must vanish on immediately adjacent free surfaces and therefore cannot be large.

Formulas for shear stress and shear flow are based on the flexure formula and therefore contain the same restrictions, namely those cited in the preceding paragraph. Another assumption in the flexure formula is that plane sections remain plane, which is not quite true if transverse shear force is present. However, if transverse shear force is constant along a member, all cross sections warp identically, so that axial strain ϵ_x, which is due to relative rotation of adjacent cross sections, is the same whether transverse shear force is constant or zero. When transverse shear force varies along a member, analyses by theory of elasticity show that shear stress formulas remain very nearly correct if the beam is slender [6.1]. The formulas are less accurate for short beams and beams with flanges so wide that "shear lag" is significant (Section 14.6).

Symmetric Cross Sections.   We first examine the direction of shear stress at a given location in a cross section. Consider, for example, the T section beam in Fig. 10.4-1. On the small coupon $ABCDEFGH$, axial force N is produced by flexural stress σ_x on face $ABCD$. On parallel face $EFGH$, axial force is $N + dN$; it is larger than N because bending moment, and consequently σ_x, is larger there. The only other axial force on the coupon is dF, which is produced by shear stress. Force dF must act in the direction shown on face $ACEG$ if axial forces are to sum to zero on the coupon. Shear force on face $ABCD$ must therefore have the direction shown by the arrow because shear arrows must both point either toward or away from line AC. This kind of argument,

324 Chapter 10 Unsymmetric Bending and Shear Center

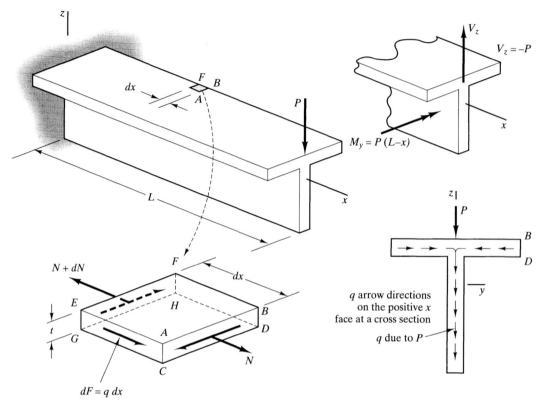

FIGURE 10.4-1 Diagrams used to establish the direction of shear flow q and obtain an expression for its magnitude. Axes yz are centroidal.

applied to various coupons cut from a beam, serves to establish the direction of shear stress at all locations on a cross section.

Next we review the elementary formula for transverse shear stress, usually written as $\tau = VQ/It$ in elementary textbooks. Again the T section beam in Fig. 10.4-1 is used as a vehicle for the discussion, although the formula $\tau = VQ/It$ is not restricted to T sections. The formula is obtained from the argument of the preceding paragraph, by expressing it quantitatively. In Fig. 10.4-1, the average shear stress across thickness t is $\tau_{ave} = dF/t\,dx$. Shear flow is $q = dF/dx$. In what follows we obtain a formula for q, and from it obtain $\tau_{ave} = q/t$.

In Fig. 10.4-1, axes yz are centroidal, and the neutral axis of bending coincides with the y axis. With area increment $dA = dy\,dz$, axial forces due to flexural stresses on coupon $ABCDEFGH$ are

$$N = \int \sigma_x \, dA \quad \text{and} \quad N + dN = \int (\sigma_x + d\sigma_x) \, dA \qquad (10.4\text{-}1)$$

Shear flow q is force per unit length, directed along the midline of a cross section. It acts on both longitudinal and transverse cross sections. Equilibrium of axial forces on coupon $ABCDEFGH$ requires that

$$q\,dx + N - (N + dN) = 0 \quad \text{hence} \quad q = \frac{dN}{dx} = \int \frac{d\sigma_x}{dx}\,dA \quad (10.4\text{-}2)$$

Stress σ_x is given by the flexure formula.

$$\sigma_x = \frac{M_y z}{I_y} \quad \text{hence} \quad q = \int \frac{dM_y}{dx}\frac{z}{I_y}\,dA = \frac{dM_y}{dx}\frac{1}{I_y}\int z\,dA \quad (10.4\text{-}3)$$

Equilibrium of a differential length of the beam requires that $dM_y/dx = V_z$, where positive directions of M_y and V_z are those shown in Fig. 10.4-1. (The relation $dM_y/dx = V_z$ is available from Fig. 1.7-1a, with the substitutions $M = -M_y$ and $V = -V_z$.) Equation 10.4-3 yields q, and the average shear stress across thickness t follows:

$$q = \frac{V_z A_s \bar{z}}{I_y} \qquad \tau_{\text{ave}} = \frac{dF}{t\,dx} = \frac{q}{t} \quad (10.4\text{-}4)$$

where A_s is the area of the face $ABCD$ and $\bar{z}$ is the z distance between the centroid of A_s and the centroid of the entire cross section. In generally, A_s is the entire cross-sectional area lying to one side of the thickness-direction line along which shear flow and shear stress are sought. $A_s\bar{z}$ is the first moment of A_s about the neutral axis. The notation $A_s\bar{z} = Q$ is common in elementary texts. In Fig. 10.4-1, shear flow is largest at the neutral axis and is zero at free edges; that is, zero at both ends of the cross and at the bottom of the stem. An example application of Eq. 10.4-4 appears in Section 10.5.

The formula for τ_{ave}, Eq. 10.4-4, is known as *Jourawski's formula*, after the Russian engineer who in 1844 was the first to provide practical formulas for transverse shear stress in beams. Shear stress varies from point to point along dimension t, and τ_{ave} is its average value. In a thin-walled cross section, thickness t is measured normal to the midline, and τ_{ave} is almost equal to the maximum τ along dimension t. In a stocky cross section, τ_{ave} may seriously underestimate τ_{max}. For example, let Eq. 10.4-4 be applied to a solid rectangular cross section of width b and depth h, with V_z parallel to dimension h and $t = b$. Equation 10.4-4 yields $\tau_{\text{ave}} = 3V_z/2bh$ at mid-depth (see also Fig. 4.3-2). As compared with τ_{max} at mid-depth, τ_{ave} errs by being about 3% low for $b/h = 0.5$, 11% low for $b/h = 1$, 28% low for $b/h = 2$, and 75% low for $b/h = 10$ [6.1].

General Cross Sections. Let the cross section be of arbitrary shape. Transverse shear force has components V_y and V_z (Fig. 10.4-2a). Let V_y and V_z be directed through the shear center, so that shear stress contains no contribution from twisting of the member about its axis or from restraint of warping. Positive directions for shear forces V_y and V_z and bending moments M_y and M_z are shown in Fig. 10.4-2a.

The arguments of Eqs. 10.4-1 to 10.4-4 need to be altered in only one respect. Now σ_x is not simply $M_y z/I_y$; rather, σ_x is given by Eq. 10.1-5. With this substitution, and the relations

$$\frac{dM_y}{dx} = V_z \qquad \frac{dM_z}{dx} = -V_y \qquad \int z\,dA = A_s\bar{z} \qquad \int y\,dA = A_s\bar{y} \quad (10.4\text{-}5)$$

the formula for shear flow becomes

326 Chapter 10 Unsymmetric Bending and Shear Center

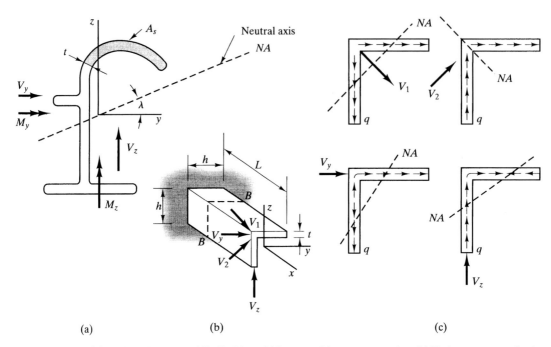

FIGURE 10.4-2 (a) Positive directions of V_y, V_z, M_y, and M_z on an arbitrary cross section. (b) Various transverse loads on a cantilever beam. (c) Shear flow on the negative x face at cross section BB, produced by various loads.

$$q = \frac{(-V_z I_{yz} + V_y I_y)\bar{y} + (V_z I_z - V_y I_{yz})\bar{z}}{I_y I_z - I_{yz}^2} A_s \quad (10.4\text{-}6)$$

An alternative form of this relation is obtained by taking σ_x from Eq. 10.1-7.

$$q = \frac{V_z(\bar{z} - \bar{y} \tan \lambda)}{I_y - I_{yz} \tan \lambda} A_s \quad (10.4\text{-}7)$$

A positive q points in the positive x direction on the "non A_s" surface of a longitudinal cut. That is, for positive q in Fig. 10.4-2a, shaded portion A_s exerts on the unshaded portion a longitudinal force directed out of the paper. In Fig. 10.4-1, we see from coupon $ABCDEFGH$ that q is *negative* when load P is applied in the direction shown.

If y and z are principal centroidal axes of the cross section, then $I_{yz} = 0$, and Eq. 10.4-6 becomes

$$\text{Principal axis form } (I_{yz} = 0): \quad q = \frac{V_y A_s \bar{y}}{I_z} + \frac{V_z A_s \bar{z}}{I_y} \quad (10.4\text{-}8)$$

which can be regarded as the application of Eq. 10.4-4 in each of two mutually perpendicular directions, followed by superposition of results.

Remarks. The foregoing equations are sufficient to determine shear flow and average shear stress in an *open* cross section—that is, in a cross section without isolated holes. Things are not as simple if the cross section is closed, because the cut that isolates A_s may cut across more than one thickness of material. For example, in a closed tube of circular cross section with varying wall thickness and z-direction load, consider shear flows where the wall is cut by a diameter along the y axis. Then q from Eq. 10.4-6 is the *sum* of the two shear flows at ends of the diameter. More information is needed to obtain the individual shear flows (see Section 10.8). If wall thickness is uniform, the two shear flows are equal; each is half the total.

A circular cross section of uniform wall thickness is an instance where symmetry provides useful information. In general, shear flow is zero on an axis of symmetry of the cross section if transverse shear force is directed along this axis. Then each half of the cross section carries half the total transverse shear force, and Eq. 10.4-4 can be applied to either half.

Examples in Fig. 10.4-2c are instructive. Neutral axes of bending for separate load cases are labeled *NA*. Loads V_1 and V_2 are each parallel to a principal axis of the cross section. In addition, V_1 is directed along an axis of symmetry, so $q = 0$ at the symmetry axis when V_1 is applied. Principal axes are easily located for this cross section. They could be used in place of the yz axes indicated in Fig. 10.4-2b; then Eq. 10.4-8 is applicable. Shear flow due to any transverse load on a symmetric or unsymmetric cross section can be obtained from Eq. 10.4-8 if the load is resolved into components parallel to principal axes.

Under load V_y in Fig. 10.4-2c, shear flow changes direction in the vertical leg. This result should be expected, because z-direction forces must sum to zero. Specifically, the integral of $q\,dz$ along the z-parallel leg must vanish. Similar remarks apply to loading by force V_z. A change in direction of q occurs where $q = 0$. Where $q \neq 0$, it is helpful to check the direction of q by applying the qualitative analysis described early in the present section.

Shear flows shown in Figs. 10.4-1 and 10.4-2c are statically equivalent to the transverse shear force in the plane of a cross section. That is, in each case the resultant force produced by q is the same magnitude and direction as the transverse shear force, and exerts the same moment about the x axis because the transverse loads shown are directed through shear centers of the cross sections.

10.5 SHEAR CENTER: INTRODUCTION

When transverse load is applied to a beam, will the beam twist as well as bend? It will unless the load is directed through the *shear center* (also called the *flexural center*). Consider the beam in Fig. 10.5-1. The cross section is symmetric about the y axis. Its shear center S is located on the y axis left of the vertical web, as shown. Horizontal load V_y is the resultant of shear flows distributed over a cross section. Only when V_y acts through S does the beam bend without twisting. A vertical load P along the web, Fig. 10.5-1b, creates torque of magnitude Pe_y, and the beam twists as well as bends. The twisting component of deflection consists of rotation of cross sections about the *elastic*

328 Chapter 10 Unsymmetric Bending and Shear Center

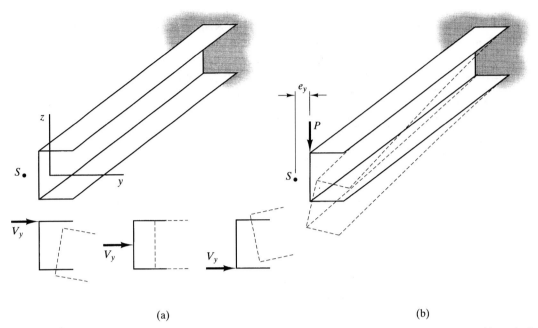

FIGURE 10.5-1 Cantilever beam. Point S, shown in the plane of the end cross section, is the shear center. (a) Dashed lines indicate displaced positions of the end due to horizontal load. (b) Deflection produced by vertical tip load P.

axis, which is the longitudinal axis that passes through shear centers of all cross sections.

If a beam bends without twisting, shear stresses associated with torsion are absent. Then the only transverse forces on a cross section are those associated with shear flows discussed in Section 10.4. Until about 1920* no one asked whether the resultant force produced by shear flow passes through the centroid of the cross section. It does not, unless the cross section has an axis of symmetry and transverse load on the beam acts along that axis. The shear center can be defined as the point through which the resultant of transverse shear forces always passes, regardless of the orientation of transverse load on the beam. If transverse load on the beam acts though the shear center, internal and external transverse force resultants appear collinear when viewed along the longitudinal axis.

Other definitions of shear center are possible [9.7]. It might be defined as the load point for which the centroid of the cross section does not rotate, or for which the average rotation of the cross section is zero. From Fig. 10.5-1a we see that a load not through the shear center has more displacement and so does more work; accordingly shear center might be defined as the load point for which strain energy in the beam is minimized. With any plausible definition, the shear center of a compact or thick-walled

---

*Early studies related to shear centers were performed between 1913 and 1921 by Timoshenko, by Griffith and Taylor, and by Maillart. The term "shear center" was apparently introduced by Maillart, the Swiss designer of graceful structures in reinforced concrete.

cross section is usually close to its centroid. The precise location usually does not matter because a small torsional load causes little twist in such a member. This tolerance of imprecision is fortunate, because the mathematics of locating the shear center of a compact or thick-walled cross section may be considerable; numerical methods may be preferred [10.1]. In a thin-walled section, the shear center may be far from the centroid, and torsion may be important. Fortunately, in a cross section of a thin-walled member the only shear stress of consequence (directed tangent to the midline of the cross section) is practically constant along a thickness-direction line when twist is zero. Consequently the shear center can be located with comparative ease.

Calculation methods discussed in this and the following two sections are applicable to members that are thin-walled, straight, prismatic, and have small deflections. Also, the material is assumed to be homogeneous, isotropic, and linearly elastic. These assumptions enter via previous equations in this chapter, which are used in shear center calculations. Nonlinearity is not considered, so an arbitrary magnitude of transverse load can be applied in calculating the location of the shear center.

EXAMPLE

The thin-walled rectangular tube in Fig. 10.5-2a is to be cut lengthwise along edge CD, and the shear center of the resulting open section is to be located. The following method of solution [10.2] is simple and provides insight into the shear center problem, but it is not a general method. It requires that the open section be obtainable simply by making a lengthwise cut in a closed section. It also requires that shear flow in the closed section can be easily calculated, which usually means that the closed section must have an axis of symmetry parallel to the direction of transverse load. More general methods of locating the shear center are discussed in subsequent sections.

Let the closed section in Fig. 10.5-2a have uniform thickness t. It therefore has two axes of symmetry, and its shear center is at its centroid. Along edge CD, shear flow q_1 produced by central load P can be calculated from Eq. 10.4-4. With $A_s = bt$, the area of the upper flange, q from Eq. 10.4-4 is the sum of equal shear flows at both ends of the flange. That is, $q = 2q_1$. Therefore

$$q_1 = \frac{1}{2} \frac{V_z A_s \bar{z}}{I_y} = \frac{1}{2} \frac{P(bt)(h/2)}{2\frac{th^3}{12} + 2bt\left(\frac{h}{2}\right)^2} = \frac{3Pb}{2h(h+3b)} \tag{10.5-1}$$

in which the moment of inertia of a flange about its own centroid, $bt^3/12$, has been omitted from I_y as negligible.

If load P is eccentric an amount e_y, Fig. 10.5-2b, torque $T = Pe_y$ appears. From Eq. 9.5-3, the additional shear flow due to T, constant around the closed tube, is

$$q_2 = \frac{T}{2\Gamma} = \frac{Pe_y}{2bh} \tag{10.5-2}$$

Since the tube is thin-walled, the rate of twist associated with q_2 is negligible in comparison with what it would be if the tube were slit open (see the argument associated with Eq. 9.5-8).

Shear flows q_1 and q_2 are oppositely directed. If they are equal in magnitude, net shear flow along CD becomes zero. The value of e_y for which $q_1 = q_2$ is given by

330 Chapter 10 Unsymmetric Bending and Shear Center

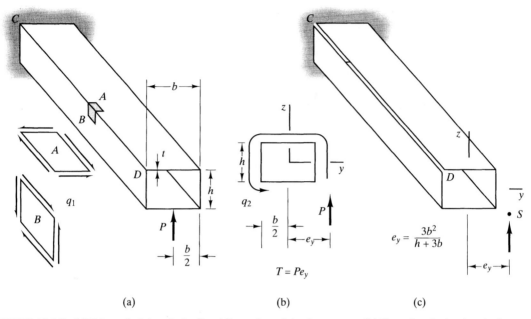

FIGURE 10.5-2 (a) Thin-walled closed tube. Load P acts through its shear center. (b) Shear flow in the closed tube due to torque $T = Pe_y$. (c) The member slit open lengthwise along CD, loaded through its shear center.

$$\frac{3Pb}{2h(h + 3b)} = \frac{Pe_y}{2bh} \quad \text{hence} \quad e_y = \frac{3b^2}{h + 3b} \tag{10.5-3}$$

But if net shear flow is zero along CD, the tube can be slit open along CD with no effect on stresses or deformations. These deformations include bending but, as noted in text following Eq. 10.5-2, practically zero twist. Accordingly, e_y is the y coordinate of shear center S of the open tube in Fig. 10.5-2c. The z coordinate of S is negative, with magnitude obtainable from Eq. 10.5-3 by interchange of b and h. That is,

$$e_z = -\frac{3h^2}{b + 3h} \tag{10.5-4}$$

Shear center S in Fig. 10.5-2c lies outside the area enclosed by the midline of the cross section unless $h > 3b$ or $b > 3h$.

10.6 SHEAR CENTER: SPECIAL OPEN SECTIONS

One Axis of Symmetry. If a cross section has an axis of symmetry, shear center S is known to lie on that axis. Therefore only one coordinate of S need be calculated. In what follows, the procedure for calculating this coordinate in a thin-walled open cross section is explained by example, using the channel section shown in Fig. 10.6-1. We

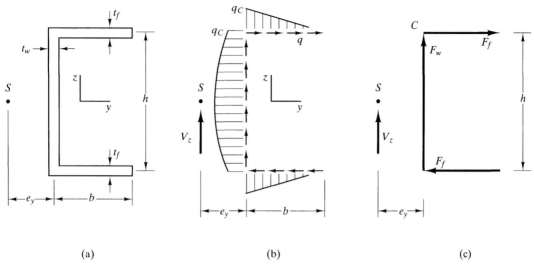

FIGURE 10.6-1 Thin-walled, symmetric channel cross section. (a) Dimensions. (b) Shear flow q. (c) Force system F_f and F_w produced by q, statically equivalent to V_z.

elect to locate S relative to the web midline rather than locate the y coordinate of S. Accordingly there is no need to locate the centroid of the cross section, other than to realize that it lies on the axis of symmetry.

Figure 10.6-1b can be regarded as analogous to the last part of Fig. 10.4-1, which shows shear flow q produced by tip load on a cantilever beam. By definition, the resultant force produced by q passes through the shear center. In Fig. 10.6-1, if the resultant force and transverse load V_z are statically equivalent, they have the same moment about an arbitrarily chosen longitudinal axis. This condition is sufficient to locate the line of action of V_z; that is, to locate the shear center. Implicit in this argument is the assumption that there is no twist; therefore there are no torsional shear stresses to contribute to the moment equation. Calculation details are as follows.

Moment of inertia I_y, neglecting the small contribution $bt_f^3/12$ of each flange, is

$$I_y = \frac{1}{12}t_w h^3 + 2\left[bt_f\left(\frac{h}{2}\right)^2\right] \tag{10.6-1}$$

From Eq. 10.4-4, we conclude that q varies linearly along each flange, quadratically along the web, and in magnitude is symmetric about the y axis. At the corners, where $q = q_C$,

$$q_C = \frac{V_z A_s \bar{z}}{I_y} = \frac{V_z(bt_f)(h/2)}{I_y} \tag{10.6-2}$$

Forces produced by q are F_f in each flange and F_w in the web. These forces can be calculated by integration: of $q\,dy$ over distance b to calculate F_f, and of $q\,dz$ over distance h

to calculate F_w. We elect corner C as the moment center. Therefore F_w is not needed. Since q varies linearly along each flange, F_f can be obtained without integration.

$$F_f = \frac{q_C}{2}b = \frac{V_z b^2 h t_f}{4I_y} \tag{10.6-3}$$

If V_z and the resultant of forces F_f and F_w are statically equivalent, they have the same moment about an arbitrary point. The moment equation about corner C, and the resulting e_y, are

$$F_f h = V_z e_y \quad \text{hence} \quad e_y = \frac{3b^2 t_f}{h t_w + 6 b t_f} \tag{10.6-4}$$

If the section is very deep or has very narrow flanges, then $b \ll h$; hence $e_y \approx 0$, as is reasonable. At the other extreme, $b \gg h$, for which $e_y \approx b/2$.

An alternative physical viewpoint, which leads to the same value of e_y, is as follows. Consider a slice of the beam, isolated by parallel cutting planes normal to the beam axis and distance dx apart. To one cut face, apply force V_z. To the other, apply resisting forces due to q (arrows F_f and F_w reversed in Fig. 10.6-1c). Equilibrium of moments about a longitudinal axis through corner C requires that $F_f h - V_z e_y = 0$, from which Eq. 10.6-4 is again obtained.

S Located by Inspection. If a cross section has a *center of symmetry* for both geometry and material properties, this center is also the shear center. The Z section of Fig. 10.6-2a is a case in point. The centroid is a center of symmetry because the geometry is repeated after an in-plane rotation about the centroid, in this case a rotation of 180°. For any orientation of transverse force V, arguments of Section 10.4 show that symmetrically located areas dA carry equal shear flows q and hence equal force increments $dF = q \, dA$. Forces dF exert equal but oppositely directed moment increments

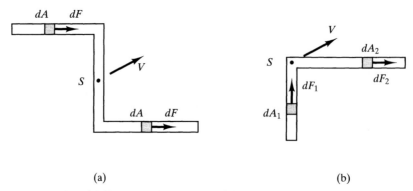

FIGURE 10.6-2 (a) The center of symmetry and the shear center coincide, at point S. (b) The shear center of a thin-walled angle section is at the corner.

about the centroid. The sum of these moment increments over the cross section is zero, which means that the resultant of transverse shear force passes through the centroid. Therefore the centroid is also shear center S.

For all locations on legs of a thin-walled angle section, Fig. 10.6-2b, shear force increments are directed through the corner, where legs intersect. Therefore the resultant of transverse shear forces passes through the corner, and the corner is the shear center.

The argument applied to the angle section can be applied to any cross section that consists of any number of narrow rectangles that all emanate from a single point. This point is the shear center. Examples of such a cross section appear in Fig. 9.10-2.

10.7 SHEAR CENTER: GENERAL OPEN SECTIONS

Consider a thin-walled open cross section that has neither axis of symmetry nor center of symmetry. Its shear center may be located using the method explained in Section 10.6, but calculations become more tedious. To obtain both coordinates of the shear center by that method, calculations must be done twice. Shear flow q is given by Eq. 10.4-6 (or Eq. 10.4-7) rather than Eq. 10.4-4. Forces produced by q can be obtained easily in some cases, as in Eq. 10.6-3, or can be obtained by integration along a midline.

Formulas that use sectorial properties (Eqs. 10.7-6) may be more convenient, especially for cross sections composed of narrow rectangles that intersect at right angles. The formulas are derived as follows. To minimize confusion with signs, we adopt the alternative physical viewpoint described in the paragraph that follows Eq. 10.6-4. Thus, Fig. 10.7-1 shows *resisting* shear flow rather than statically equivalent shear flow.

In Fig. 10.7-1, axes y and z are centroidal but need not be principal. Coordinates e_y and e_z of shear center S are to be determined. They are measured in y and z directions relative to an arbitrarily located point P, which is used as moment center. Shear

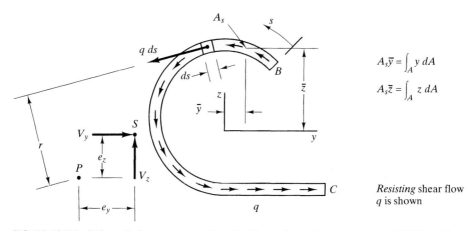

FIGURE 10.7-1 Thin-walled open cross section of arbitrary shape. Axes yz are centroidal. Shear flow q resists the resultant of V_y and V_z.

flow q resists transverse forces V_y and V_z. If there is no twisting, so that the resultant of q passes through S, then the net moment about arbitrary point P must vanish; that is

$$-V_y e_z + V_z e_y + \int_B^C qr\, ds = 0 \qquad (10.7\text{-}1)$$

where integration covers the entire midline, from B to C. Equation 10.4-6 provides an expression for q. Thus Eq. 10.7-1 becomes

$$-V_y e_z + V_z e_y + \frac{-V_z I_{yz} + V_y I_y}{I_y I_z - I_{yz}^2} \int_B^C A_s \bar{y} r\, ds + \frac{V_z I_z - V_y I_{yz}}{I_y I_z - I_{yz}^2} \int_B^C A_s \bar{z} r\, ds = 0 \quad (10.7\text{-}2)$$

From Section 9.8, we recognize that $r\, ds = d\omega$, where ω is sectorial area. Integrals can be evaluated by parts. With $r\, ds = d\omega$, the first integral becomes

$$\int_B^C A_s \bar{y}\, d\omega = A_s \bar{y} \omega \Big|_B^C - \int_B^C \omega\, d(A_s \bar{y}) = -\int_B^C \omega\, d(A_s \bar{y}) \qquad (10.7\text{-}3)$$

Term $A_s \bar{y} \omega$ vanishes at B and C, since $A_s = 0$ at B, and $A_s \bar{y} = 0$ at C because it is the first moment of the entire cross-sectional area about the centroidal z axis. From definitions of centroidal distances, written in Fig. 10.7-1, we obtain $d(A_s \bar{y}) = y\, dA$. Thus we make a final transformation of the integral. The second integral in Eq. 10.7-2 is treated similarly. These results are

$$\int_B^C A_s \bar{y} r\, ds = -\int_A \omega y\, dA \quad \text{and} \quad \int_B^C A_s \bar{z} r\, ds = -\int_A \omega z\, dA \qquad (10.7\text{-}4)$$

where area integrals cover the entire area A of the cross section. Integrals that contain ω are recognized from Eqs. 9.9-1 as sectorial linear moments, where they are given the symbols $S_{y\omega}$ and $S_{z\omega}$. Equation 10.7-2 can now be written as

$$V_y\left(-e_z - \frac{I_y S_{y\omega} - I_{yz} S_{z\omega}}{I_y I_z - I_{yz}^2}\right) + V_z\left(e_y + \frac{I_{yz} S_{y\omega} - I_z S_{z\omega}}{I_y I_z - I_{yz}^2}\right) = 0 \qquad (10.7\text{-}5)$$

This equation can be true for arbitrary values of V_y and V_z only if the two expressions in parentheses vanish separately. Thus we obtain the following equations, which locate the shear center relative to point P.

$$e_y = \frac{I_z S_{z\omega} - I_{yz} S_{y\omega}}{I_y I_z - I_{yz}^2} \qquad e_z = \frac{I_{yz} S_{z\omega} - I_y S_{y\omega}}{I_y I_z - I_{yz}^2} \qquad (10.7\text{-}6)$$

Point P can be placed at any convenient position. It serves as a pole for calculation of sectorial area ω. The choice of initial point does not matter. Relocating it changes ω by a constant but does not change $S_{y\omega}$ or $S_{z\omega}$.

A statement that follows Eq. 9.9-3 can now be verified. If $e_y = 0$ and $e_z = 0$, then $S_{y\omega} = 0$ and $S_{z\omega} = 0$. That is, we conclude that if axes y and z are centroidal and ω is constructed with pole P at the shear center, then sectorial linear moments are zero.

EXAMPLE

Locate the shear center of the cross section in Fig. 10.7-2a. Axes y and z are centroidal, and $I_y = 20.8(10^6)$ mm^4, $I_z = 10.5(10^6)$ mm^4, and $I_{yz} = -6.0(10^6)$ mm^4.

To determine ω, we elect point P as both pole and initial point. The resulting sectorial area diagram appears in Fig. 10.7-2b. To calculate $S_{y\omega}$ and $S_{z\omega}$, we elect the method of Eq. 9.9-7, with $dA = t\,dy$. The centroid of the ω diagram is at $y = (-100/6)$ mm and $z = -120$ mm. Also, $t = 6$ mm. Thus

$$S_{y\omega} = \int_A y\omega\, dA = \frac{-20{,}000(100)}{2}\left(-\frac{100}{6}\right)6 = 1.00(10^8)\ \text{mm}^5 \qquad (10.7\text{-}7a)$$

$$S_{z\omega} = \int_A z\omega\, dA = \frac{-20{,}000(100)}{2}(-120)6 = 7.20(10^8)\ \text{mm}^5 \qquad (10.7\text{-}7b)$$

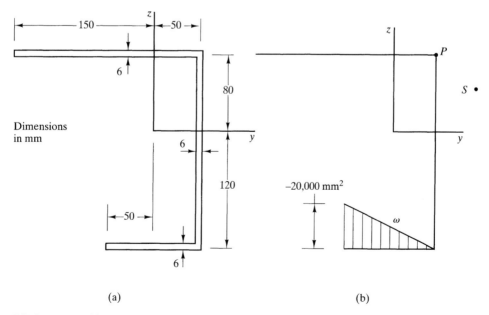

FIGURE 10.7-2 (a) Dimensions of an unsymmetric cross section of uniform thickness. (b) Midline of the cross section, with sectorial area diagram for pole and initial point at point P.

Substitution of Eqs. 10.7-7 into Eqs. 10.7-6 yields the location of the shear center relative to point P:

$$e_y = 44.7 \text{ mm} \quad \text{and} \quad e_z = -35.1 \text{ mm} \quad (10.7\text{-}8)$$

In the yz system, the shear center has coordinates $y = 50.0 + 44.7 = 94.7$ mm and $z = 80.0 - 35.1 = 44.9$ mm.

10.8 SHEAR CENTER: REMARKS

Closed Sections. Compared with a closed section, shear flow in an open section is easy to determine because area A_s (Fig. 10.4-2a) is known to have zero shear flow on its outer edge. Cutting an area A_s from a closed section exposes shear flows on two or more edges. Equations in Section 10.4 provide the algebraic sum of these shear flows but not their individual values.

A calculation procedure for a thin-walled closed tube of one or more cells is as follows. Each cell is cut open, and the section is analyzed by methods previously described for open sections. Then cells are closed again by superposing a shear flow q_i around each cell i. Net shear flow in each cell must make all cells have the same rate of twist (see Eq. 9.6-3). If the rate of twist is set to zero, unknowns to be determined are the q_i and a coordinate of the shear center. If transverse load is at a prescribed location, unknowns are the q_i and the rate of twist. Details of the procedure are described in the first edition of this book and in [10.3].

In comparison with a thin-walled open section, a closed section twists much less due to torque, and its shear center is usually much closer to its centroid.

Shear Center and Center of Twist. The following argument indicates that the two centers coincide. Consider Fig. 10.8-1. The cross section shown may be regarded as the tip of a cantilever beam, to which transverse load P and torque T are applied. According to Maxwell's reciprocal theorem, work done by P in moving through displacement created by T is equal to work done by T in rotating through twist created by P, that is

$$Pw_T = T\theta_P \quad \text{hence} \quad w_T = \frac{T}{P}\theta_P \quad (10.8\text{-}1)$$

If $d = 0$ in Fig. 10.8-1, so that P is directed through shear center S, then $\theta_P = 0$; therefore $w_T = 0$, which means that torque causes the section to rotate about S.

A transverse load not directed through the shear center creates both bending moment and torque, the torque being equal to the load times its perpendicular distance from the shear center. If the section is thin-walled and open, applicable analysis tools for determining stress and deflection can be found in Sections 9.11, 9.12, 10.1, and 10.3. Torque causes cross sections to rotate about the shear center, except as modified by complications discussed in what follows.

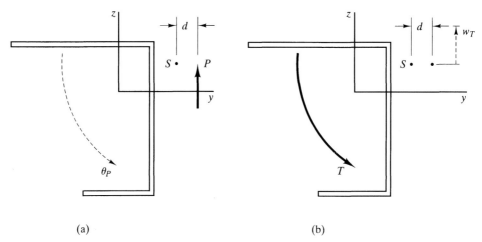

FIGURE 10.8-1 Arbitrary cross section, showing (a) rotation θ_P due to transverse load P, and (b) lateral displacement w_T of the load point due to torque T.

Complications. Numerical Methods. Analysis tools described in this chapter are based on simplifying assumptions stated at the outset. But a practical member may have taper, varying cross section, cutouts, and internal stiffeners, all of which modify the shear flow distribution. Shear flow is also modified by restraint associated with supports. Accordingly the actual shear center location may vary along a beam, and preceding equations may not locate it accurately [10.3]. Thin-walled structures may be also sensitive to small imperfections associated with manufacture. Two nominally identical members having different imperfections may respond to load rather differently.

Numerical methods are often appropriate. Open or closed sections that satisfy assumptions made in this chapter can be treated by special programs [10.4]. A member that does not satisfy the assumptions can be analyzed by finite element methods. The most direct way to locate the shear center of a finite element model appears to be identifying the point about which the loaded cross section rotates when torque is applied. Alternatively, one can note that angle of twist is directly proportional to torque produced by a transverse load. Thus, in Fig. 10.8-1a, $\theta_P = (a_1 + a_2 y)P$, where a_1 and a_2 are constants and y is the horizontal coordinate of load P. For positions y_1 and y_2 of load P, one can compute θ_{P1} and θ_{P2} numerically, then solve for a_1 and a_2. Twist is zero when $\theta_P = 0$; thus the y coordinate of the shear center is $y = -a_1/a_2$.

PROBLEMS

10.1-1. (a) In Fig. 10.1-2, show that if load P acts parallel to AC, then the neutral axis passes through B.

(b) In Fig. 10.1-2, if load P acts parallel to BC, for what value of h/b is the neutral axis perpendicular to BC?

10.1-2. Show that for a solid circular cross section of radius R, Eq. 10.1-8 reduces to $\sigma_x = MR/I$, where $M^2 = M_y^2 + M_z^2$ and I is taken about a diameter.

10.1-3. In Fig. 9.13-1, axial displacement u can be described by $u = u_o - \theta_z y + \theta_y z - \beta \omega$, where components of motion of a cross section are u_o in axial translation, θ_y and θ_z in rigid body rotation about y and z axes, and $\beta \omega$ in warping. Here β is rate of twist and ω is sectorial area. Quantities $u_o, \theta_y, \theta_z,$ and β are functions of x. Axial stress is $\sigma_x = E(\partial u/\partial x)$. Evaluate integrals of $\sigma_x \, dA$, $\sigma_x y \, dA$, $\sigma_x z \, dA$, and $\sigma_x \omega \, dA$ over cross-sectional area A. Recognize these integrals as force and moments, in terms of which σ_x can be expressed. Obtain this expression for σ_x. (Its bending moment terms should agree with Eq. 10.1-5).

10.2-1. (a) Let a cantilever beam have a rectangular cross section of dimensions b by h. Show that if transverse tip load is directed along a diagonal of the cross section, the neutral axis coincides with the other diagonal.

(b) By what factor would the maximum flexural stress be reduced if P were reoriented so as to act parallel to dimension h? Obtain numerical values of this factor for $h = b$ and for $h = 10b$.

10.2-2. Let a column have a rectangular cross section of dimensions b by h and be loaded by an axial compressive force. If the force acts within a certain central area of the cross section, there is no tensile stress in the column. Define this area, known as the *kern*, in terms of b and h and show it on a sketch of the cross section.

10.2-3. In the beam of Fig. 10.2-1, view the cross section from the left end, as suggested in the last paragraph of Method 1. Again calculate flexural stresses at locations A and B.

10.2-4. In Fig. 10.2-1, how must the 10 kN load be oriented with respect to the z axis if lateral deflection of the beam is to have no y component? For this orientation, by what factors are σ_{xA} and σ_{xB} changed from values calculated in Section 10.2?

10.2-5. Bending moment M acts on the triangular cross section shown.

(a) For $\beta = 30°$ and $M = 6$ kN·m, determine flexural stresses at A, B, and C, and the orientation of the neutral axis.

(b) For what angle β is the neutral axis parallel to AC?

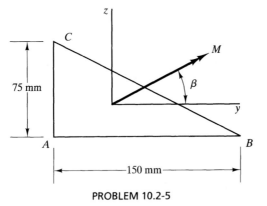

PROBLEM 10.2-5

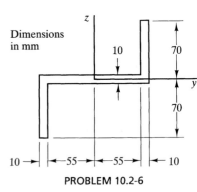

PROBLEM 10.2-6

10.2-6. (a) A cantilever beam 1.6 m long has the Z section shown. Load $P = 2500$ N in the $-z$ direction is applied at the tip. Determine the flexural stress of greatest magnitude. Where on the cross section does this stress appear?

(b) Repeat part (a) with load P reoriented so it acts in the $-y$ direction.

(c) Repeat part (a) with load P applied at the point $y = 60$ mm, $z = 0$, and directed axially along the beam, away from the fixed end.

10.2-7. A beam carries bending moment M and has the cross section shown. It is known that points A on the cross section are free of stress and that flexural stresses at points B have magnitude 200 MPa. What is the magnitude and orientation of M?

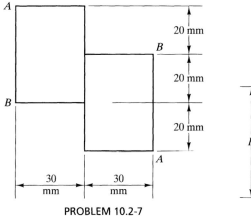

PROBLEM 10.2-7

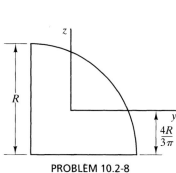

PROBLEM 10.2-8

10.2-8. A simply supported beam, 2.0 m long, carries a uniformly distributed load of 6500 N/m in the $-z$ direction. The cross section is a solid quarter circle, as shown, for which $I_y = I_z = 0.05488R^4$ and $I_{yz} = -0.01647R^4$. Axes yz are centroidal. Determine R such that the maximum tensile flexural stress is 24 MPa.

10.2-9. Determine the locations and magnitudes of greatest tensile and compressive flexural stress in the cantilever beam shown. Thickness t is small and uniform. Axes y and z are centroidal. Ignore possible twisting of the beam about its axis.

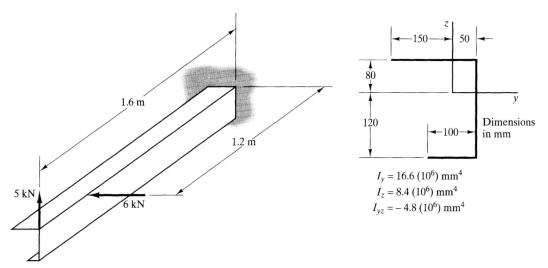

$I_y = 16.6\,(10^6)\text{ mm}^4$
$I_z = 8.4\,(10^6)\text{ mm}^4$
$I_{yz} = -4.8\,(10^6)\text{ mm}^4$

PROBLEM 10.2-9

10.3-1. The cross section of a cantilever beam of length L is a solid equilateral triangle, b units on a side. Transverse tip load P is applied. If P can have any orientation in the plane of the tip cross section, what are the largest and smallest possible magnitudes of (a) flexural stress, and (b) tip deflection? Express answers in terms of P, L, E, and b.

10.3-2. A cantilever beam has a narrow rectangular cross section, 40 mm by 500 mm. The 500 mm dimension is inclined at 6° to the vertical. Before being allowed to deflect under its own weight, the beam axis is horizontal. Determine the angle with respect to the vertical of the tip deflection.

10.3-3. The cross section of a cantilever beam is a right triangle, as shown. A transverse load P produces resultant deflection Δ parallel to the hypotenuse of the cross section. Determine the y and z components of P.

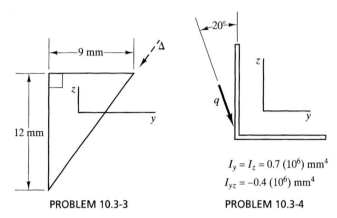

$I_y = I_z = 0.7\,(10^6)\ \text{mm}^4$
$I_{yz} = -0.4\,(10^6)\ \text{mm}^4$

PROBLEM 10.3-3   PROBLEM 10.3-4

10.3-4. The cross section shown is that of a simply supported beam, 2 m long, loaded by a uniformly distributed lateral load q whose plane is 20° from the vertical. Determine the magnitude and direction of lateral deflection at midspan, in terms of q and elastic modulus E.

(a) Use the method associated with Eqs. 10.3-2 and 10.3-4.

(b) Use deflection components parallel to principal axes of the cross section.

10.3-5. In Problem 10.3-4, add a roller support at midspan. This support contacts the central cross section adjacent to its corner on the z-parallel side. Thus, motion is prevented in the y direction but is unrestrained in the z direction. Determine (a) the force exerted by the support, and (b) the midspan deflection, in terms of q and elastic modulus E.

10.3-6. In problem 10.2-9, determine the magnitude and direction of tip deflection. Let $E = 70$ GPa and consider only flexural deformations.

10.3-7. A long beam of unsymmetric cross section, such as that in Fig. 10.2-1, is placed on an elastic foundation and loaded by a vertical force. Assume that normal and tangential forces exerted by the foundation are directly proportional to its displacements, with constant k_n normal to the foundation and constant k_t tangent to it. Describe steps of a calculation procedure by which stresses and deflections due to flexure can be obtained.

10.3-8. The cantilever beam shown has a rectangular cross section and is manufactured with a 90° twist over length L. Determine the magnitude and direction of tip deflection due to tip load P, in terms of P, L, E, and I, where $I = bh^3/12$. Assume that Poisson's ratio is zero and that $b \gg h$.

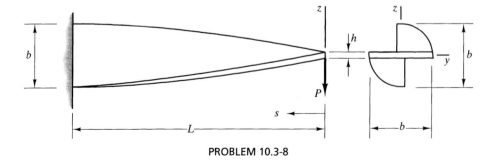

PROBLEM 10.3-8

10.4-1. In Eq. 10.4-4, show that $A_s \bar{z}$ may refer to area on either side of the location of interest without changing the calculated magnitude of q at that location.

10.4-2. Two narrow cross sections are shown, (a) rectangular and (b) triangular. Transverse shear force V_z acts on each. Show by integration that q, as calculated from Eq. 10.4-4, produces force V_z.

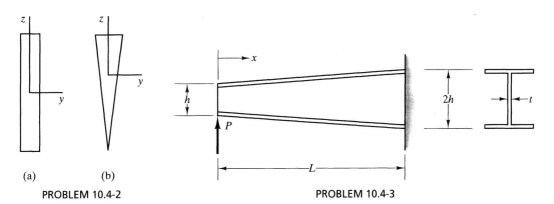

PROBLEM 10.4-2

PROBLEM 10.4-3

10.4-3. Average transverse shear stress throughout the web of an I beam, τ_{ave}, is calculated by dividing transverse shear force by cross-sectional area of the web. A tapered I beam is shown in side view and in cross section. What are the values of τ_{ave} in this beam at $x = 0$, at $x = L/2$, and at $x = L$, in terms of P, h, and t?

10.4-4. The cross section shown consists of two identical right triangular cross sections, glued together. Transverse shear force V acts at angle α. What angle α maximizes shear stress in the glue layer, and what is this shear stress in terms of V?

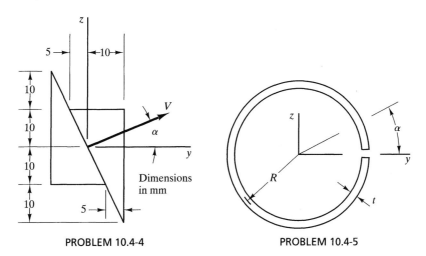

PROBLEM 10.4-4

PROBLEM 10.4-5

10.4-5. Transverse shear force V_z in the z direction is applied to the thin-walled, open circular cross section shown.
  (a) Express shear flow q in terms of V_z, R, and angle α.
  (b) At what α is q maximum, and what is q_{max}?

10.4-6. For the thin-walled angle section in Fig. 10.4-2b, $I_y = I_z = 5th^3/24$ and $I_{yz} = th^3/8$, where axes y and z are centroidal. Horizontal tip load V_y is applied to the beam, as shown in Fig. 10.4-2c.
  (a) Determine the value of z for which $q = 0$.
  (b) Show by integration that q in the vertical leg produces zero net vertical force.
  (c) Determine the location and magnitude of the largest shear flow.
  (d) Consider the corner, where legs meet. Shear flow there can be calculated from Eq. 10.4-6, or from Eq. 10.4-8 (with axes reoriented so that y and z become principal axes, and with appropriate components of the original load). Show that the two calculations yield the same q.

10.4-7. Consider the Z section in Problem 10.2-6. Demonstrate that if the neutral axis of bending passes through the middle of each flange, then shear flow q is zero along the y axis. When then is the maximum q, in terms of the transverse shear force? Where on the cross section does q_{max} appear?

10.4-8. (a) Let a y-parallel shear force V_y act on the Z section of Problem 10.2-6. In the z-parallel leg $0 < z < 70$ mm, obtain an expression for q in terms of V_y.
  (b) Repeat part (a), but consider the y-parallel web $0 < y < 60$ mm.
  (c) Use the results of parts (a) and (b) to sketch the distribution of q over the cross section. At what locations does q vanish?
  (d) One might ignore asymmetry and make the estimate $q_{max} \approx (V_y/A_{web})t$, as is commonly done for an I section. What is the percentage error of this estimate?

10.5-1. Use the method described in Section 10.5 to locate shear centers of the open cross sections shown. Each member is thin walled, of uniform thickness, and contains a lengthwise cut.
  (a) Circle.   (b) Equilateral triangle.   (c) Square (use the axes shown).

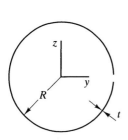 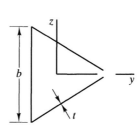 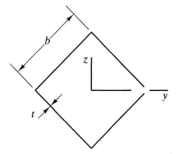

PROBLEM 10.5-1

10.5-2. In Problem 10.5-1a, let twist of the closed section be taken into account. Thus the result of Problem 10.5-1a is modified by a small term that depends on R and t. Obtain this modified result.

10.6-1. Locate shear centers of the cross sections shown, in terms of the given dimensions. Each section is thin-walled and open, of uniform thickness t, and symmetric with respect to its horizontal centerline. Assume frictionless contact along an overlap [parts (c) and (h)]. Where possible, check answers by examining limiting cases (as by letting b approach zero, for example).

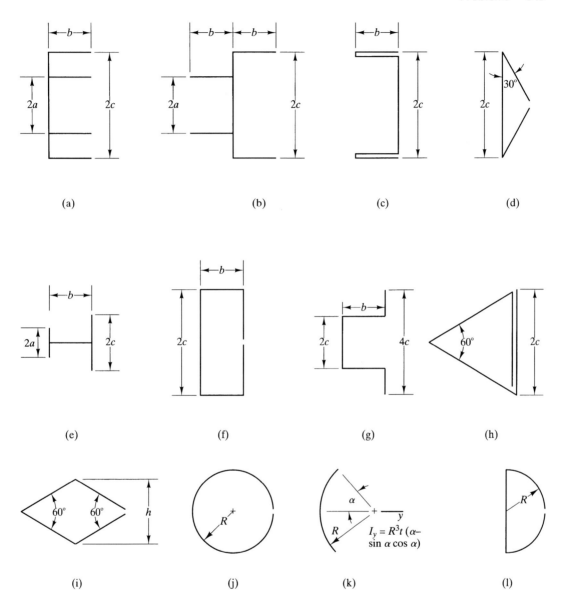

PROBLEM 10.6-1

10.6-2. The cross section shown consists of two flanges, each of cross-sectional area A, joined by a thin semicircular web. Assume that flanges carry all flexural stress and the web carries all transverse shear stress. Locate the shear center in terms of R.

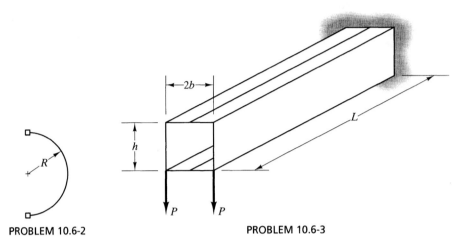

PROBLEM 10.6-2

PROBLEM 10.6-3

10.6-3. For the channel section in Fig. 10.6-1, let $h = 2b$ and $t_f = t_w$. Two such channels are welded together lengthwise as shown, to form a closed square section. In terms of P, what force is transmitted across each weld? Is this force uniformly distributed lengthwise? Explain.

10.6-4. (a) Several cantilever beams of arbitrary cross section have the same length. Before loading, their elastic axes are parallel and lie in common plane AB, as shown. Imagine that beam tips are pinned to a rigid bar, which lies along AB. Show that the rigid bar remains parallel to AB when load V_z is applied if e_y has the value shown.

(b) Use this formula for e_y to solve Problem 10.6-1e.

(c) Use this formula for e_y to solve Problem 10.6-1h.

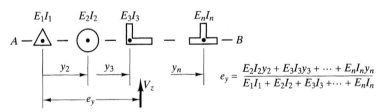

PROBLEM 10.6-4

10.6-5. For the cantilever beam of Fig. 10.4-2b, let $L = 1000$ mm, $h = 120$ mm, and $t = 8$ mm. Load V_z is applied. Consider the absolute maximum shear stress in the beam, on whatever plane it acts. By what factor is this stress increased if V_z is directed through the centroid of the cross section, rather than through the shear center as shown? At what locations on the cross section do you expect to find these shear stresses?

10.7-1. Consider the cross section in Fig. 10.7-2a. Locate the shear center by placing the pole for ω at the lower right corner of the cross section.

10.7-2. Use Eqs. 10.7-6 to locate shear centers of the following cross sections.

(a) Problem 10.6-1c.   (b) Problem 10.6-1e.   (c) Problem 10.6-1f.
(d) Problem 10.6-1h.   (e) Fig. 10.5-2c.

10.7-3. Locate the shear center of the unsymmetric Z section shown. Axes yz are centroidal. Assume that $a \gg t$.

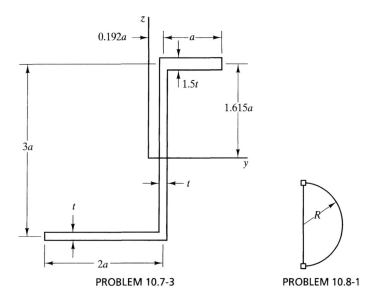

PROBLEM 10.7-3 PROBLEM 10.8-1

10.8-1. The half-circle cross section of Problem 10.6-2 is augmented by a thin flat web between flanges, to form the closed D-section shown. Both webs have the same thickness.
  (a) Locate the shear center.
  (b) Consider a cantilever beam having this cross section. Apply transverse tip load P, parallel to the flat web and a distance R to the right of it. Obtain an expression for the rate of twist.

CHAPTER 11

Plasticity in Structural Members and Collapse Analysis

11.1 INTRODUCTION

Elastic structural design under static load is guided by the proposition that when the design load is reached, yielding impends. By "design load," we mean the working load multiplied by a safety factor. In reality, yielding appears before the calculated elastic design load is reached, because of stress concentrations at connections and load points, unfavorable residual stresses, and construction errors. But the load-carrying capacity of a structure is not exhausted when yielding impends. Also, elastic analysis may predict unrealistic stresses because of factors difficult to anticipate and therefore ignored: misalignment of components, support settlement, slip at connections, and so on.

Limit design allows yielding and is guided by the proposition that when the design load is reached, the structure verges on collapse. *Plastic collapse* is said to occur when yielding is so extensive that displacements increase drastically with little change in load. The plastic collapse load exceeds the elastic limit load if the structure is statically indeterminate. For a given safety factor, limit design allows a greater working load than elastic design. Thus, less material is needed to carry a given working load. Plastic collapse is not affected by minor defects of construction; that is, if supports and connections are capable of exerting full resistance, modest support settlement or connection slip neither increases nor reduces the plastic collapse load, although larger deflections may be needed in order to reach that load.

In subsequent sections we assume that material does not strain harden or strain soften, and that deflections are small enough that the deformed geometry is essentially the same as the original geometry. Strain hardening strengthens a structure; it is a conservative approximation to neglect it. Strain softening causes the resistance of a member to drop after reaching a maximum. For example, the static load that can be supported by a reinforced concrete beam declines when the concrete crumbles. Static load capacity may also drop if deflections are large, for example if sidesway in a building frame causes vertical loads to contribute significantly to bending moments in columns. Plastic collapse analysis is *not* appropriate if failure is due to buckling before the plastic collapse load is reached, or if failure is due to fatigue, fracture, or excessive deflection.

Fortunately, for most beams and frames of structural steel, the assumptions of zero strain hardening and small deflections lead to good estimates of load-carrying capacity. Deflections associated with a collapse load are usually small enough that the analysis remains valid, and deflections associated with the *working* load are usually acceptably small. Of course the analyst cannot assume that such will always be the case.

Typically a beam or frame is subjected to two or more loads, simultaneously applied. Here another assumption is invoked for collapse analysis: that loads are proportional, in the sense that they have prescribed ratios to one another when the collapse load is reached. So long as other ratios do not cause premature collapse, the prescribed ratios need not be enforced until the collapse load is reached.

In this chapter we consider plastic collapse of beams and frames, for which the collapse condition is reached by development of "plastic hinges" while most material of the structure remains elastic. Loads are assumed to be static; cycles of loading are not discussed. Typically we seek the load-carrying capacity of a given structure. We will see that plastic collapse analysis is usually simpler than elastic analysis because it does not require that simultaneous equations be written and solved. Deflections are discussed briefly. Theorems of plastic collapse analysis are discussed in Section 11.7. General plasticity theory is briefly summarized in Section 7.13. Plasticity and its structural applications are discussed in many books, including [3.1, 11.1–11.7].

A plastic collapse load may be called a fully plastic load when geometry and loading are such that all material of a member or a structure becomes plastic. Fully plastic loads for the following problems are discussed in previous sections: torsion of a circular cross section (Section 1.9), pressurized cylinder (Section 8.6), spinning disk (Section 8.7), and torsion of a general cross section (Section 9.15). In the present chapter, M_{fp} denotes fully plastic bending moment capacity at a cross section, and P_c or q_c denotes the collapse value of an externally applied load on a beam or a frame. Plastic collapse conditions for plates and columns are discussed in Sections 12.9 and 14.5.

11.2 STRESS AND DEFLECTION IN LOADING AND UNLOADING

In this section we review idealized elastic-plastic material behavior and illustrate its application in plastic analysis by means of a simple structure.

Material. Figure 11.2-1a illustrates idealized material behavior. In uniaxial tension or compression there is a definite yield stress σ_Y. Further straining occurs at the same stress. That is, the material becomes "perfectly plastic," meaning that it does not strain harden. (Strain hardening behavior can be included in plastic analysis, but calculations become more difficult. Strain hardening is not considered in this chapter.) The elastic portion of a stress-strain relation is not needed in order to calculate a plastic collapse load. Accordingly, if deflections are not to be calculated, we may assume that the material is rigid until it yields, as shown in Fig. 11.2-1b.

In the perfectly plastic regime, work done by applied loads is not stored as recoverable strain energy. Plastic components of strain do not disappear upon release of load. This "permanent set" is also displayed by mechanical analogues shown in Fig. 11.2-1. Plastic deformation is analogous to sliding of weight W with coefficient of friction μ. During sliding, energy is dissipated by friction, and when load is removed the block

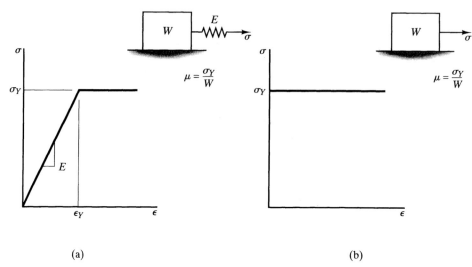

FIGURE 11.2-1 Idealized stress-strain relations and their mechanical analogues. (a) Elastic, then perfectly plastic. (b) Rigid, then perfectly plastic.

does not move back, although the spring in Fig. 11.2-1a recovers its original length. The "sliding block" analogue may be helpful in visualization and in considering work absorbed by plastic deformation.

EXAMPLE

Consider the structure in Fig. 11.2-2a. All vertical members have the same cross-sectional area A and are made of the material depicted in Fig. 11.2-1a. The horizontal member is rigid. We ask for the relation between load P and its displacement v, as P increases from zero to its maximum possible value.

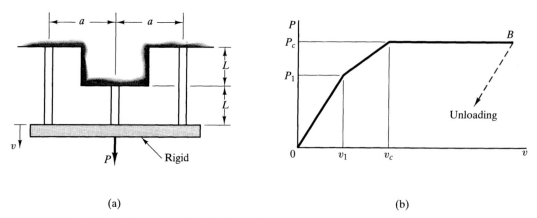

FIGURE 11.2-2 (a) Load P is supported by three bars of the same material and cross-sectional area. (b) Relation between load P and its displacement v.

Initially, when all vertical bars are linearly elastic, P and v are linearly related. Dimensions in Fig. 11.2-2a show that when outer bars have axial strain ϵ, the middle bar has axial strain 2ϵ. Accordingly, when axial stress σ_Y is reached in the middle bar, stress $\sigma_Y/2$ exists in each outer bar. Summation of vertical forces on the horizontal rigid bar provides P_1, the value of P at first yield.

At first yield:
$$\begin{cases} P_1 = \sigma_Y A + 2\left(\dfrac{\sigma_Y}{2}A\right) = 2\sigma_Y A \\ v_1 = \epsilon_Y L = \dfrac{\sigma_Y}{E}L \end{cases} \quad (11.2\text{-}1)$$

Displacement v_1 is calculated from the elastic stress-strain relation, with the middle bar on the verge of yielding. Coordinates (P_1, v_1) at the end of the initial straight line segment in Fig. 11.2-2b have now been determined.

When P reaches a value that initiates yielding in the outer bars, axial stress in the middle bar is still σ_Y. All three bars are now plastic, and collapse load P_c has been reached. Summation of vertical forces provides P_c.

Fully plastic:
$$\begin{cases} P_c = 3\sigma_Y A \\ v_c = \epsilon_Y(2L) = 2\dfrac{\sigma_Y}{E}L \end{cases} \quad (11.2\text{-}2)$$

Displacement v_c is calculated from an outer bar, when it is on the verge of yielding. Since all bars are now plastic, no further increase of P is possible. An attempt to increase P to even slightly more than P_c produces deflections that are infinite for our idealized material, but are in reality limited by strain hardening.

If only the plastic collapse load P_c is of interest, P_1 need not be calculated; we can proceed directly to P_c. Rather than summing forces as in Eq. 11.2-2, we can determine P_c by a virtual work argument, which is not advantageous for the present problem but is valuable in the analysis of beams and frames. For a virtual work solution we consider a small vertical displacement v, imagined to occur *after* the plastic collapse load has been applied, and calculate work done by external loads and by internal forces. Total virtual work must vanish if static equilibrium is to prevail. Thus

$$P_c v - (3\sigma_Y A)v = 0 \quad \text{hence} \quad P_c = 3\sigma_Y A \quad (11.2\text{-}3)$$

Displacement v cancels; its value does not matter and cannot be determined. Equations 11.2-2 and 11.2-3 provide the same value of P_c.

Remarks. Consider the application of a safety factor, say $SF = 2$. According to elastic analysis, a working load $P_w = P_1/2 = \sigma_Y A$ is allowed. According to plastic collapse analysis, a working load $P_w = P_c/2 = 1.5\sigma_Y A$ is allowed, a 50% increase in P_w. In this example, since P_w is less than the load that initiates yielding, plastic collapse analysis with $SF = 2$ provides a working load that produces no plastic action.

If load P_c is removed, unloading proceeds elastically, leaving residual stresses in the bars and a residual displacement. Residual stresses are calculated by superposing on plastic stresses the *elastic* stresses associated with reversed load P_c (in the manner described for a torsion problem in Section 1.9). For the present problem, linear elastic analysis under downward load P provides axial stresses $\sigma_o = P/4A$ in the outer bars and $\sigma_m = P/2A$ in the middle bar. (This type of analysis is treated in elementary mechanics of materials; the reader may fill in details.) Residual stresses are plastic stresses plus the foregoing elastic stresses with $P = -P_c$. Thus, in each outer bar, residual stress is $\sigma_{or} = \sigma_Y + (-3\sigma_Y/4) = \sigma_Y/4$. In the middle bar, residual stress is $\sigma_{mr} = \sigma_Y + (-3\sigma_Y/2) = -\sigma_Y/2$. As for displacement, removal of downward load P_c produces elastic recovery in the amount $v = L(3\sigma_Y/2)/E$. Residual displacement cannot be determined without knowledge of where point B lies on the horizontal plateau in Fig. 11.2-2b.

If load P is reapplied after unloading, it creates elastic stresses that are superposed on residual stresses. If the reapplied P acts downward, it can reach P_c before yielding begins again (in all bars simultaneously). If the reapplied P acts upward it can reach only $\sigma_Y A$ before yielding begins again (in the middle bar). Therefore P has an elastic range of $4\sigma_Y A$. This range of P is the same as its original elastic range, where $-P_1 < P < P_1$ was possible without yielding. Accordingly, plastic straining of the structure has shifted the location of the elastic range without changing its size. This behavior can be regarded as the structural analogue of the Bauschinger effect. Here macroscopic members have taken the place of material microstructure.

11.3 PLASTIC ACTION IN BENDING

We consider pure bending of a prismatic beam—whose material behaves as depicted in Fig. 11.3-1a—and develop relations between bending moment, depth of yielding, and curvature of the beam. We assume that there are no initial stresses from manufacture or previous loading and exclude unsymmetric bending from the analysis.

Figure 11.3-1b shows stress and strain distributions when outer layers of a beam of rectangular cross section have yielded. Distance from the neutral axis to the elastic-plastic boundary is ηc. Thus the beam is entirely elastic when $\eta > 1$. Yielding impends, at bending moment M_Y, when $\eta = 1$. Fully plastic bending moment M_{fp} is applied when $\eta = 0$. For $M_Y < M < M_{fp}$, the depth of yielding is $(1 - \eta)c$. For a solid rectangular cross section of dimensions b by $2c$, bending moment for an arbitrary value of η in the range $0 \leq \eta \leq 1$ is

$$M = 2\left[\frac{\sigma_Y}{2}b(\eta c)\frac{2\eta c}{3} + \sigma_Y b(1-\eta)c\frac{(1+\eta)c}{2}\right] = \sigma_Y bc^2\left(1 - \frac{\eta^2}{3}\right) \quad (11.3\text{-}1)$$

This equation is written by expressing the moment about the neutral axis of forces produced by linear and constant stress distributions on a cross section, as explained in texts on elementary mechanics of materials. The "shape factor" f is defined as $f = M_{fp}/M_Y$. For the solid rectangular cross section,

$$\left.\begin{array}{l}\eta = 1: \quad M = M_Y = 2\sigma_Y bc^2/3 \\ \eta = 0: \quad M = M_{fp} = \sigma_Y bc^2\end{array}\right\} \quad f = \frac{M_{fp}}{M_Y} = 1.50 \quad (11.3\text{-}2)$$

By definition, the elastic section modulus is $S = I/c$ (hence $S = M_Y/\sigma_Y$), and the plastic section modulus is $Z = M_{fp}/\sigma_Y$. In terms of dimensions b and c of the rectangular cross section,

$$S = \frac{I}{c} = \frac{M_Y}{\sigma_Y} = \frac{2bc^2}{3} \qquad Z = \frac{M_{fp}}{\sigma_Y} = bc^2 \quad (11.3\text{-}3)$$

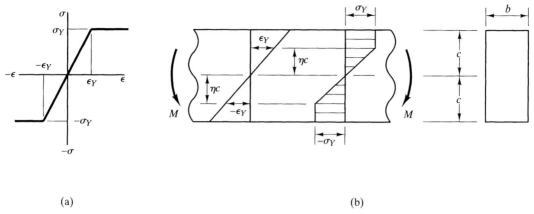

FIGURE 11.3-1 (a) Assumed stress-strain relation. (b) Distributions of strain and stress in a beam whose cross section has two axes of symmetry.

We see that an alternative definition of shape factor is Z/S. For any shape of cross section, $f = M_{fp}/M_Y$ and also $f = Z/S$. The numerical value of f depends on the geometry of the cross section.

Beam curvature κ is the reciprocal of radius of curvature ρ. Curvature κ can be expressed in terms of η by the procedure explained for linearly elastic conditions in Section 1.6. Here we allow plastic action, and do not require that the cross section be rectangular. Briefly, the procedure is as follows. Consider two parallel cross sections a distance dx apart. After M is applied, the two cross sections include angle $d\theta$, and the beam axis has radius of curvature ρ. For any value of η,

$$d\theta = \frac{dx}{\rho} \quad \text{and} \quad d\theta = \frac{\epsilon_Y \, dx}{\eta c} \quad \text{hence} \quad \kappa = \frac{1}{\rho} = \frac{\epsilon_Y}{\eta c} \quad (11.3\text{-}4)$$

where $\epsilon_Y = \sigma_Y/E$.

Moment-curvature relations for selected cross sections are plotted in Fig. 11.3-2a. When yielding begins, the curvature is $\kappa = \kappa_Y$ (for which $\eta = 1$ in Eq. 11.3-4). For $\eta < 1$, the curve that represents a rectangular cross section is a plot of $M/\sigma_Y bc^2$ versus $1/\eta$ according to Eq. 11.3-1. If $\eta = 1/4$, so that κ is four times its value when yielding begins, then $M = 0.98 M_{fp}$ in a rectangular cross section. We see that curvature need not be large for M to closely approach M_{fp}. The physical reason is that stresses in a small elastic core are not only smaller than σ_Y but also have small moment arms about the neutral axis.

Any shape of cross section can be analyzed to provide a relation analogous to Eq. 11.3-1. Note that when M_{fp} is applied, the neutral axis has equal areas on either side of it, so that forces parallel to the beam axis sum to zero on a cross section. This consideration is important when calculating M_{fp} for a cross section that has but one axis of symmetry (such as the triangular cross section in Fig. 11.3-2b), or for a cross section that has no axis of symmetry. In these cases the neutral axis may not be centroidal, or even parallel to the bending moment vector.

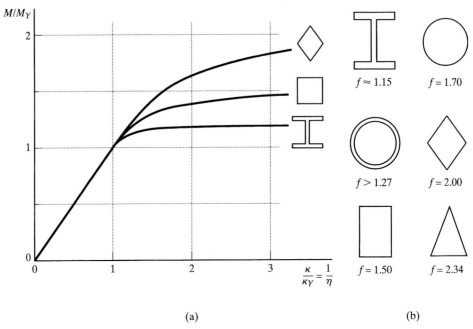

FIGURE 11.3-2 (a) Moment-curvature relations in pure bending. The material is elastic, then perfectly plastic. (b) Shape factors f in pure bending. The neutral axis of each cross section is horizontal on the paper.

For I and wide-flange cross sections, shape factor f is little greater than unity. The obvious reason is that only a small depth of yield is needed to produce stress σ_Y throughout the flanges, which contain almost all moment-carrying material. Further bending spreads yielding into the web, but the web develops comparatively little bending moment.

Springback. When loads that produce yielding are removed, some of the total deformation disappears as the structure "springs back" elastically. An illustration of this action appears in Fig. 11.3-3, where a bar is bent over a cylindrical mandrel to mean radius R by a moment M that causes plastic action. Release of M allows a springback change in curvature. Superposition of loading and unloading, as shown, results in an unloaded bar with residual stresses and residual curvature ρ_{res}. The change in curvature due to unloading, $\Delta\kappa$, is the final curvature minus the original curvature.

$$\Delta\kappa = \frac{1}{\rho_{\text{res}}} - \frac{1}{R} \quad \text{where} \quad \Delta\kappa = -\frac{M}{EI} \quad (11.3\text{-}5)$$

Here $\Delta\kappa$ is calculated from the elastic formula, and is negative because κ decreases. (In using EI for flexural stiffness, we presume that the bar is not so wide that it behaves like a plate, whose flexural stiffness is somewhat greater than EI; see Chapter 12.) Even if loading causes all material to yield, springback is entirely elastic if shape factor f does not exceed 2. Equation 11.3-5 can be used to determine the mean radius R to which a bar must be bent in order to achieve a final mean radius ρ_{res}.

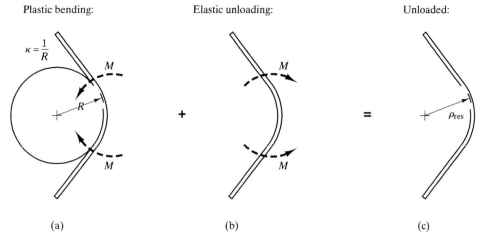

FIGURE 11.3-3 Plastic bending, followed by springback upon unloading.

Consider the special case of a prismatic member of solid rectangular cross section of width b and depth $2c$, initially stress free, whose material has yield stress σ_Y and no strain hardening. Let it be bent to the fully plastic condition by moment M_{fp} and then unloaded. From Eqs. 11.3-2 and 11.3-5, with elastic unloading,

$$\text{Solid rectangular cross section:} \quad \Delta\kappa = -\frac{3\sigma_Y}{2Ec} \qquad (11.3\text{-}6)$$

Thus we see that springback is reduced by a reduction in yield stress, and increased by a reduction in modulus or a reduction in thickness.

11.4 PLASTIC HINGES IN BEAMS

Zone of Yielding. The prismatic beam of Fig. 11.4-1a carries midspan load P_c, where subscript c means "collapse." Fully plastic moment M_{fp} prevails at midspan. Let us assume that the material does not strain harden and that deflections are small enough that the original geometry can be used in analysis. Statics dictates that the spanwise distribution of bending moment M in Fig. 11.4-1a is symmetric about midspan and varies linearly with x; in the right half it is $M = M_{fp}(L - 2x)/L$. At what distance from midspan is M equal to M_Y, the bending moment at which yielding begins? It depends on shape factor f. Letting x_Y be the desired distance,

$$M_Y = M_{fp}\left(\frac{L - 2x_Y}{L}\right) \quad \text{hence} \quad x_Y = \frac{L}{2}\left(1 - \frac{1}{f}\right) \qquad (11.4\text{-}1)$$

where $f = M_{fp}/M_Y$. Accordingly, cross sections are partially yielded over a central span $2x_Y = L(1 - 1/f)$, as shown in Fig. 11.4-1a. Only the cross section at $x = 0$ is completely yielded. A similar argument applies to a simply supported beam under uniform loading with $M = M_{fp}$ at midspan, for which the span of yielded cross sections is shown in Fig. 11.4-1b.

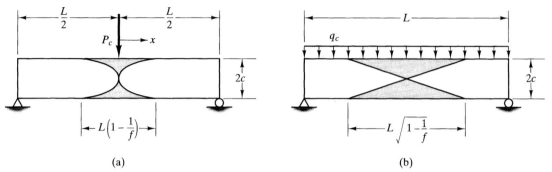

FIGURE 11.4-1 Yielded zones in uniform simply supported beams carrying collapse loads P_c or q_c. The shape factor of the cross section is f. Parabolic and straight-line envelopes shown pertain to a rectangular cross section ($f = 1.50$).

The shape of the elastic-plastic boundary can also be determined. For example, let the beam in Fig. 11.4-1a have a rectangular cross section, for which $f = 1.50$. From Eqs. 11.3-1 and 11.3-2, $M = M_{fp}(1 - \eta^2/3)$. With $M = M_{fp}$ at midspan, bending moment in the right half of the beam is $M = M_{fp}(L - 2x)/L$. Accordingly, for concentrated center load P_c,

$$M_{fp}\left(1 - \frac{\eta^2}{3}\right) = M_{fp}\left(\frac{L - 2x}{L}\right) \quad \text{hence} \quad x = \frac{L\eta^2}{6} \qquad (11.4\text{-}2)$$

This equation describes the parabola shown in Fig. 11.4-1a. For the case of uniform loading q_c on a beam of rectangular cross section, a similar argument establishes the straight-line elastic-plastic boundary shown in Fig. 11.4-1b.

Deflections. If a deflected beam has small slope, its curvature $\kappa = \kappa(x)$ can be approximated by the usual expression $\kappa = d^2v/dx^2$, where $v = v(x)$ is lateral deflection. However, the elementary formula $d^2v/dx^2 = M/EI$ is not applicable to portions of a beam where cross sections are partially plastic. Consider, for example, a uniform cantilever beam of rectangular cross section, Fig. 11.4-2a. For this beam, $M = P(L - x)$ in

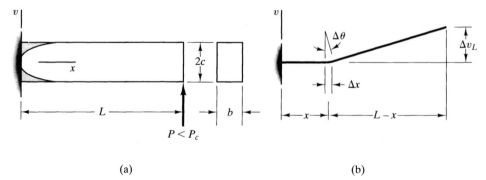

FIGURE 11.4-2 (a) Cantilever beam, with incomplete yielding at the fixed-end cross section. (b) The contribution to tip deflection v_L from rotation $\Delta\theta$ in a length increment Δx.

Eq. 11.3-1. Equations 11.3-1 and 11.3-4 provide an expression for curvature in the partially plastic span, where $0 \leq \eta \leq 1$.

$$\frac{d^2v}{dx^2} = \frac{\epsilon_Y}{\eta c} \quad \text{in which} \quad \eta = \left[3 - \frac{3P(L-x)}{M_{fp}}\right]^{1/2} \qquad (11.4\text{-}3)$$

where $M_{fp} = \sigma_Y bc^2$. In the entirely elastic span, $d^2v/dx^2 = P(L-x)/EI$. After integration of both equations for d^2v/dx^2, the two solutions for $v = v(x)$ must be matched in v and dv/dx where the elastic span meets the partially plastic span.

The same problem can be attacked numerically using an analysis suggested by Fig. 11.4-2b. A contribution to total tip deflection v_L is Δv_L, produced by rotation $\Delta\theta$. Equation 11.3-4 provides an expression for $\Delta\theta$. Thus

$$v_L = \sum \Delta v_L \quad \text{where} \quad \Delta v_L = (L-x)\,d\theta = (L-x)\frac{\epsilon_Y \Delta x}{\eta c} \qquad (11.4\text{-}4)$$

The summation corresponds to spanning the beam with length increments Δx, each of which displays a different η and a different x.

Plastic Hinge. Equation 11.4-3 shows that curvature approaches infinity as η approaches zero. Beam deflection is then dominated by a small zone of high curvature, as shown for a simply supported prismatic beam in Fig. 11.4-3a. When M_{fp} exists in the central cross section of this beam, collapse load P_c has been reached. Then, until strain hardening develops, deflection takes place as the central cross section deforms plastically, without any increase in load. The two halves of the beam maintain their deformed shapes as they rotate. In other words the central cross section has become a *plastic hinge*.

If the beam in Fig. 11.4-3a has a rectangular cross section, a plot of central load versus central deflection resembles the middle curve in Fig. 11.3-2a. Experiments [11.8] show that the load does not stay at P_c as deflection increases, but gradually rises due to strain hardening. Even so, although load P_c is calculated with strain hardening ignored, it is the load at which deflections increase greatly for little increase in load. P_c is therefore a practical measure of load-carrying capacity.

By isolating half the beam in Fig. 11.4-3a as a free-body diagram and taking moments about the center, we obtain $(P_c/2)(L/2) - M_{fp} = 0$, from which the collapse load is $P_c = 4M_{fp}/L$. The elastic limit load is $P_Y = 4M_Y/L$. The ratio P_c/P_Y is $M_{fp}/M_Y = f$, the shape factor. For a statically determinate structure, it is always the case that $P_c/P_Y = f$. We will see that for a statically *in*determinate structure, $P_c/P_Y > f$.

We have tacitly assumed that when transverse shear force is present, it does not reduce the bending moment capacity M_{fp}. This simplification is acceptable for ordinary beams, which are slender. Similarly, in a slender column, axial force may not appreciably reduce the bending moment capacity. (Combined loadings are discussed in Sections 11.8 and 11.9.)

Virtual Work. A collapse load can be determined by equilibrium considerations, as in the preceding paragraph. However, it is often easier to use a virtual work argument to analyze beams and frames, especially if the problem is statically indeterminate.

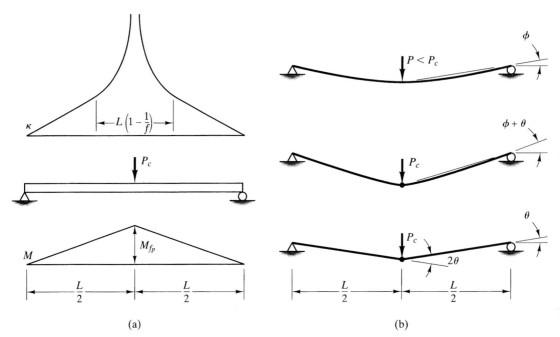

FIGURE 11.4-3 (a) Curvature and bending moment in a uniform beam with a plastic hinge. (b) Deformations associated with virtual work arguments. The dot beneath load P_c represents a plastic hinge.

As applied to the beam in Fig. 11.4-3, the virtual work argument proceeds as follows. With collapse load P_c applied, allow a virtual displacement, here described by angle θ. Angle θ represents a small and imaginary *additional* displacement that occurs *after* the collapse load has been reached. All virtual displacement is produced by rotation of the plastic hinge. Deformation previously accumulated does not change during this virtual displacement and can be ignored. As shown in the lowermost part of Fig. 11.4-3b, for collapse analysis we can regard halves of the beam as rigid bars that are hinged at the center. Work is done by load P_c, and work is absorbed by the plastic hinge. Physically, the hinge acts as a "rusty" hinge, as if absorbing energy by friction in the manner of the block shown in Fig. 11.2-1b. Displacement of load P_c is $(L/2)\theta$ and rotation of the hinge is 2θ. Total virtual work must vanish; therefore

$$P_c\left(\frac{L}{2}\theta\right) - M_{fp}(2\theta) = 0 \quad \text{hence} \quad P_c = \frac{4M_{fp}}{L} \qquad (11.4\text{-}5)$$

Angle θ cancels, and we obtain the same P_c as obtained previously by equilibrium considerations.

11.5 COLLAPSE ANALYSIS OF BEAMS

Statically determinate beam problems are trivial in that the collapse load is simply the shape factor times the load that initiates yielding. In a statically determinate beam or frame, a single plastic hinge forms where bending moment is largest. A single hinge is sufficient to cause collapse because the structure has become a mechanism. In the present context, "mechanism" can be taken to mean a linkage of rigid bars connected by hinges. Plastic hinges are "rusty" hinges that rotate with constant moment. They are analogous to the block in Fig. 11.2-1b that slides with constant friction force.

In a statically *in*determinate beam or frame, the collapse load is greater than the shape factor times the load that initiates yielding. Hinge locations are often not immediately obvious. Usually, the number of hinges needed for plastic collapse is one greater than the degree of static indeterminacy (exceptions are noted at the end of the present section).

When a collapse mechanism has been identified, the corresponding load is easily determined by the virtual work method. For collapse analysis it is not necessary to do an elastic analysis or to know the order in which hinges form. The virtual work procedure is explained by beam examples in the present section. In these examples, beams have bending moment capacity M_{fp} throughout their lengths.

EXAMPLE 1

The uniform beam in Fig. 11.5-1a is fixed at A and simply supported at C. What is the load P_c that causes collapse by development of plastic hinges?

Although not needed to answer the question posed, the elastic bending moment diagram, for the load P_Y at which yielding begins, is shown in Fig. 11.5-1b (linear elastic analysis shows that bending moments are $-3P_YL/16$ at A and $5P_YL/32$ at B). We see that yielding begins at A, at load $P_Y = 16M_Y/3L$. Simply increasing the load until the bending moment at A reaches M_{fp} is not sufficient to cause collapse: The beam is statically indeterminate to the first degree; *two* plastic hinges are needed in order to create a mechanism. These hinges appear at A and at B. Collapse load P_c can be calculated by considering small displacements that take place after load P_c is reached, ignoring deformations previously accumulated. The geometry of the collapse mechanism is shown in Fig. 11.5-1c, where dots represent plastic hinges. In moving downward, P_c does work that is absorbed by plastic hinges at A and B as they undergo the respective rotations θ and 2θ. Net virtual work is zero. Thus

$$P_c\left(\theta\frac{L}{2}\right) - M_{fp}\theta - M_{fp}(2\theta) = 0 \qquad P_c = \frac{6M_{fp}}{L} \qquad (11.5\text{-}1)$$

This result is independent of the linearly elastic solution. The ratio P_c/P_Y is $(9/8)M_{fp}/M_Y = 1.125f$, where f is the shape factor of the cross section. Note that if P_c and L are prescribed, Eq. 11.5-1 can be used to determine the required bending moment capacity M_{fp}.

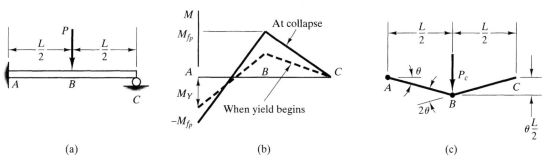

FIGURE 11.5-1 (a) Uniform propped cantilever beam with concentrated load P. (b) Bending moment diagrams. (c) Collapse mechanism, with plastic hinges at A and B.

A calculated elastic limit load P_Y may not be realistic, due to residual stresses in the beam, settlement of supports, or construction errors. These factors may influence the location of initial yielding and the order in which hinges form, but have practically no effect on collapse load P_c.

EXAMPLE 2

Determine collapse load q_c for the uniform propped cantilever beam shown in Fig. 11.5-2a.

Again two plastic hinges are needed in order to create a mechanism. One hinge is at A, but the location of hinge B is not obvious. Let hinge B be a distance a from A as shown in Fig. 11.5-2b, where a is as yet unknown. For small deflections, geometric relations are

$$a + b = L \qquad \lambda = \theta + \phi \qquad \phi = \theta a/b \tag{11.5-2}$$

Concentrated equivalents for distributed load in spans AB and BC are $q_c a$ and $q_c b$ respectively. The virtual work equation is

$$q_c a \left(\theta \frac{a}{2}\right) + q_c b \left(\phi \frac{b}{2}\right) - M_{fp}\theta - M_{fp}\lambda = 0 \tag{11.5-3}$$

From Eqs. 11.5-2 and 11.5-3,

$$q_c = \frac{2M_{fp}(2L - a)}{La(L - a)} \tag{11.5-4}$$

Different values of a result in different collapse mechanisms. We might assume that $a = L/2$. This assumption, however, is slightly incorrect, so Eq. 11.5-4 then yields an approximate collapse load:

$$\text{Approximate:} \quad q_c = \frac{12M_{fp}}{L^2} \qquad \text{for} \qquad a = \frac{L}{2} \tag{11.5-5}$$

Placing a hinge where it does not want to be acts as a constraint on the mathematical model, thereby stiffening it and raising the calculated collapse load. A plastic collapse mechanism is analogous to a buckling mode of a column, in that of all possible mechanisms or modes, the one that actually occurs is the one for which the collapse or buckling load is smallest. In other words, as soon as a structure can collapse, it will; it does not await the higher load associated with a misplaced plastic hinge.

The correct mechanism minimizes the collapse load. Accordingly, we write $dq_c/da = 0$, solve for a, and insert this value of a into Eq. 11.5-4. Thus

$$\text{Exact:} \quad q_c = \frac{11.657 M_{fp}}{L^2} \qquad \text{for} \qquad a = 0.586L \tag{11.5-6}$$

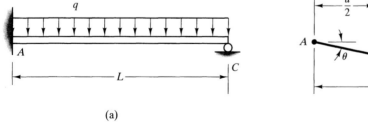
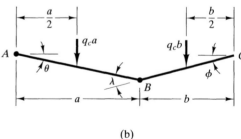

(a)

(b)

FIGURE 11.5-2 (a) Uniform propped cantilever beam with uniform load q. (b) Collapse mechanism, with plastic hinges at A and B.

Note that in the approximate solution, Eq. 11.5-5, distance a has a greater percentage error than load q_c. This difference should be expected, since the curve that represents the q_c versus a relation is horizontal at the minimum, where $a = 0.586L$.

In this example the ratio of q_c to the load that initiates yielding, $q_Y = 8M_Y/L^2$, is $q_c/q_Y = 1.46f$, where f is the shape factor of the cross section.

Remarks. Equilibrium considerations can be used to determine whether a proposed collapse mechanism is possible. Let us do so for the mechanism of Eq. 11.5-5. First we isolate the right half of the beam as a free body diagram, with q_c from Eq. 11.5-5 applied and with $M = M_{fp}$ at B; thus we obtain $R_C = 5M_{fp}/L$ as the support reaction at C in Fig. 11.5-2. Then we calculate the largest bending moment between B and C; it turns out to be $1.04M_{fp}$, a distance $L/12$ to the right of B. But M_{fp} is the largest bending moment possible. Thus we know that the proposed collapse mechanism is not possible; that is, $a = L/2$ is not correct.

An assumed mechanism provides a collapse load greater than the exact collapse load or at best equal to it: this is the "upper bound theorem" (see Section 11.7). Of all possible collapse mechanisms, the mechanism that provides the smallest collapse load is the mechanism that actually occurs. In practice one examines only likely or plausible mechanisms, which may be few in number. With these thoughts in mind, we consider the following example.

EXAMPLE 3

Two equal loads P are applied to a uniform propped cantilever beam, as shown in Fig. 11.5-3a. Determine the value P_c that causes plastic collapse.

Two plastic hinges are needed to create a mechanism. Three plausible mechanisms are shown in the latter parts of Fig. 11.5-3. In each case the bending moment diagram consists of three straight line segments. A hinge can appear only where segments meet because only there will bending moment be maximum or minimum. For the mechanism in Fig. 11.5-3b, with plastic hinges at A and B,

$$P_c\left(2\theta\frac{L}{3} + \theta\frac{L}{3}\right) - M_{fp}(2\theta + 3\theta) = 0 \qquad P_c = \frac{5M_{fp}}{L} \qquad (11.5\text{-}7)$$

For the mechanism in Fig. 11.5-3c, with plastic hinges at A and C,

$$P_c\left(\theta\frac{L}{3} + 2\theta\frac{L}{3}\right) - M_{fp}(\theta + 3\theta) = 0 \qquad P_c = \frac{4M_{fp}}{L} \qquad (11.5\text{-}8)$$

Finally, for the mechanism in Fig. 11.5-3d, with plastic hinges at B and C,

$$P_c\left(\theta\frac{L}{3}\right) - M_{fp}(\theta + 2\theta) = 0 \qquad P_c = \frac{9M_{fp}}{L} \qquad (11.5\text{-}9)$$

The smallest of the three loads is $P_c = 4M_{fp}/L$, from Eq. 11.5-8, which is the correct plastic collapse load.

One can use the value of P_c for a mechanism and equations of statics to determine bending moment at any location. When using P_c for the mechanism of Fig. 11.5-3c, we find that the

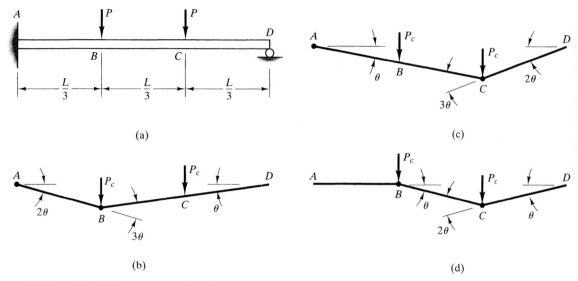

FIGURE 11.5-3 (a) Uniform propped cantilever beam with equal concentrated loads P. (b, c, d) Plausible collapse mechanisms.

bending moment is at or below M_{fp} in magnitude throughout the beam, as is required. We conclude that this mechanism is correct. (An understanding of the bound theorems in Section 11.7 helps in understanding this conclusion.)

Number of Hinges. When *complete collapse* occurs, the number of hinges is one greater than the degree of static indeterminacy. Other possible modes of collapse are shown in Fig. 11.5-4. Figure 11.5-4a shows an example of *partial collapse*. The continuous beam is statically indeterminate to the second degree, but the collapse mechanism consists of only two hinges, at C and D. Portion ABC remains elastic and statically indeterminate. Figure 11.5-4b shows an example of *over collapse* (also called "overcomplete collapse"). The beam is statically indeterminate to the second degree. It is

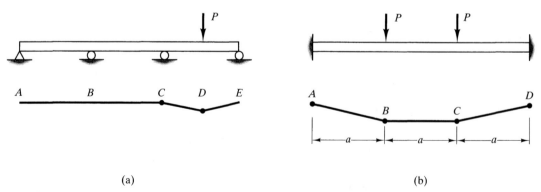

FIGURE 11.5-4 The respective beams display partial collapse and over collapse.

reasonable to expect that equal loads P will cause hinges to appear simultaneously at A and D, and simultaneously at B and C, for a total of four hinges in the collapse mechanism. However, only three hinges are needed. One can easily show that three plausible mechanisms (plastic hinges at $ABCD$, at ABD, or at ACD) all result in the same collapse load P_c.

11.6 COLLAPSE ANALYSIS OF PLANE FRAMES AND ARCHES

As is the case for a beam, the collapse load of a frame or an arch can be obtained by determining the plastic collapse load associated with each possible mechanism, and selecting the smallest. Calculation procedures are the same as for a beam, but mechanism geometry is often more complicated. In what follows, M_{fp} designates bending moment capacity of a member. In practice, members are often stronger than connections between them, so M_{fp} may then designate the moment capacity of a connection.

EXAMPLE 1

The plane frame in Fig. 11.6-1a is fixed at A and pinned at E. Bending moment capacity is M_{fp} throughout the frame. Determine the value of P for plastic collapse.

The frame is statically indeterminate to the second degree. Three plausible mechanisms, each containing three hinges, are shown in the latter parts of Fig. 11.6-1. We consider the mechanisms in order, and apply the virtual work argument. In the mechanism of Fig. 11.6-1b, vertical load $2P_c$ does no work, so

$$P_c(2L\theta) - M_{fp}(\theta + \theta + \theta) = 0 \qquad P_c = \frac{3M_{fp}}{2L} \qquad (11.6\text{-}1)$$

In the mechanism of Fig. 11.6-1c, columns remain essentially vertical because angle θ is small. Horizontal load P_c does no work, so

$$2P_c(L\theta) - M_{fp}(\theta + 2\theta + \theta) = 0 \qquad P_c = \frac{2M_{fp}}{L} \qquad (11.6\text{-}2)$$

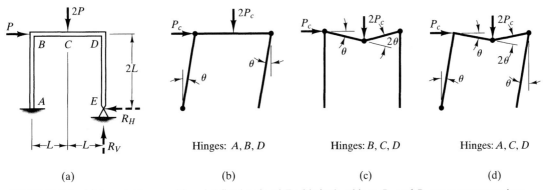

Hinges: A, B, D Hinges: B, C, D Hinges: A, C, D

(a)   (b)   (c)   (d)

FIGURE 11.6-1 (a) A portal frame, with end A fixed and end E a frictionless hinge. R_H and R_V are support reactions. (b, c, d) Plausible collapse mechanisms. Dots represent plastic hinges.

Both loads do work in the mechanism of Fig. 11.6-1d, so

$$P_c(2L\theta) + 2P_c(L\theta) - M_{fp}(\theta + 2\theta + 2\theta) = 0 \qquad P_c = \frac{5M_{fp}}{4L} \qquad (11.6\text{-}3)$$

The latter result for P_c, the smallest of the three, is the actual collapse load.

One can check that the result of Eq. 11.6-3 is indeed correct by evaluating bending moments at nonhinge locations. In terms of forces R_H and R_V in Fig. 11.6-1a, bending moments at lettered locations are

$$M_A = -2R_V L + 4P_c L \qquad M_B = -2R_V L + 2P_c L + 2R_H L$$
$$M_C = -R_V L + 2R_H L \qquad M_D = 2R_H L \qquad (11.6\text{-}4)$$

in which positive M is associated with tensile stress on the outside of the frame. The bending moment diagram consists of straight line segments between lettered locations. Elimination of R_H and R_V reduces Eqs. 11.6-4 to

$$M_A - 2M_C + 2M_D = 4P_c L \qquad (11.6\text{-}5a)$$

$$M_B - 2M_C + M_D = 2P_c L \qquad (11.6\text{-}5b)$$

For the mechanism of Fig. 11.6-1d, $M_A = M_{fp}$, $M_C = -M_{fp}$, $M_D = M_{fp}$, and $P_c = 5M_{fp}/4L$. Equation 11.6-5a reduces to $1 = 1$, and Eq. 11.6-5b yields $M_B = -M_{fp}/2$. Thus we know that this mechanism is physically possible because $|M| \le M_{fp}$ throughout the frame. For the mechanism in Fig. 11.6-1b, $M_A = M_{fp}$, $M_B = -M_{fp}$, $M_D = M_{fp}$, and $P_c = 3M_{fp}/2L$; Eqs. 11.6-5 therefore yield $M_c = -1.5M_{fp}$, so we know this mechanism is not possible. Similarly, in Fig. 11.6-1c, we obtain $M_A = 4M_{fp}$, which is not possible. Thus we confirm that the mechanism of Fig. 11.6-1d, which provides the smallest P_c, is the only mechanism of the three that is possible.

A sketch of a collapse mechanism may show displacements so greatly exaggerated that one may be tempted to install unnecessary hinges. For example, in Fig. 11.6-1c the base of each column seems to invite installation of a hinge because it appears that columns must rotate inward. Actual column rotations are negligible in this mechanism; therefore, it would be incorrect to install more hinges than shown in Fig. 11.6-1c.

EXAMPLE 2

Loads P, Q, and R act on the uniform frame in Fig. 11.6-2a. For the mechanism shown in Fig. 11.6-2b, with plastic hinges at B, C, D, and E, express the collapse load in terms of dimensions shown.

The main task is to relate rotations of links BC, CD, and DE, so that hinge rotations can be expressed in terms of a single parameter, such as rotation θ of link BC. To do so we use the instantaneous center, which is discussed in elementary texts about dynamics. Points C and D have displacements Δ_C and Δ_D in the mechanism shown. Therefore point I is the instantaneous center of rotation for link CD. Accordingly

$$\Delta_C = \ell_1 \theta = \ell_2 \phi \qquad \text{hence} \qquad \phi = \frac{\ell_1}{\ell_2}\theta = \frac{a}{b}\theta \qquad (11.6\text{-}6)$$

where ϕ is the rotation of link CD. As for rotation α of link DE,

$$\Delta_D = h\alpha = s\phi \qquad \text{hence} \qquad \alpha = \frac{s}{h}\phi = \frac{r(a+b)}{ah}\phi = \frac{r(a+b)}{bh}\theta \qquad (11.6\text{-}7)$$

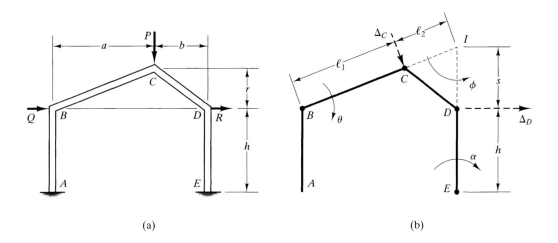

FIGURE 11.6-2 (a) A gable frame. (b) Rotations θ, ϕ, and α of links in a particular mechanism. Dots represent plastic hinges.

Fully plastic moment M_{fp} appears at B, C, D, and E. The virtual work equation for this mechanism, which does not involve the load at B, is

$$P_c(a\theta) + R_c(h\alpha) - M_{fp}[\theta + (\theta + \phi) + (\phi + \alpha) + \alpha] = 0 \tag{11.6-8}$$

Upon substitution of Eqs. 11.6-6 and 11.6-7 into Eq. 11.6-8, angle θ cancels. Thus we obtain an equation that relates loads P_c and R_c to M_{fp} and dimensions. When the ratio P_c/R_c is prescribed, loads P_c and R_c at collapse are determined. Other collapse mechanisms can be analyzed in similar fashion.

Arches. The uniform arch in Fig. 11.6-3a is supported by frictionless pins at A and D. There is one redundant reaction. Therefore, at collapse, two plastic hinges appear. One is at the load, point B. The other is at point C. To locate point C, we recall from statics that when a body in equilibrium is acted upon by three forces, the forces must be either parallel or concurrent. In Fig. 11.6-3 the point of concurrency is at such a location on the line of action of P_c that bending moments at B and C both have magnitude M_{fp}. Since a small error in the location of plastic hinge C causes a much smaller error in the calculated value of P_c, point C might be located by scaling an accurate drawing. Similar work with a drawing might serve to locate the instantaneous center of rotation for link BC, so that link rotations can be expressed by equations similar to Eqs. 11.6-6 and 11.6-7.

The arch in Fig. 11.6-3b has fixed supports at A and D. At collapse, plastic hinges appear at A, B, C, and D. If the arch is uniform, C is located at the midpoint of arc BD. This conclusion is reached by drawing a free-body diagram of the arch between B and D. This portion of the arch is free of external load. One discovers that the resultant

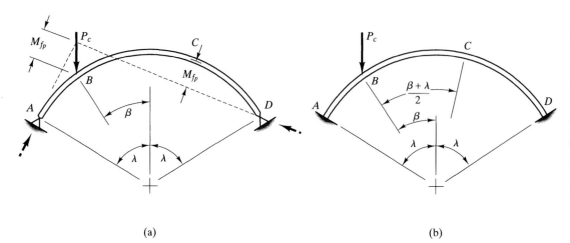

FIGURE 11.6-3 (a) Uniform two-hinged circular arch. Dashed lines indicate support reactions and their lines of action. (b) Hingeless circular arch.

force and moment at B also appear at D, but oppositely directed; therefore the bending moment diagram of arc BD is symmetric about its midpoint.

Deflections. Deflections may be of interest for two reasons. First, design criteria may impose deflection limits. Second, one must know that deflections are indeed small if collapse analysis is to be based on the assumption that original geometry is essentially unchanged by the development of plastic hinges (as is assumed in the present chapter).

If the working load (rather than the collapse load) produces no yielding, deflections at the working load can be calculated by standard elastic analysis methods. If yielding is present, there appears to be no simple calculation procedure for deflections. When the next-to-last plastic hinge is fully formed, a portion of the structure is still elastic, and elastic analysis methods are applicable to this portion. However, for such an analysis one must know which hinge is last to form, and ignore the effect of incomplete yielding at this hinge. If deflections under a given elastic load P_e are known, one can *estimate* deflections at the collapse load P_c by multiplying elastic deflections by the ratio $(P_c/P_e)f$, where f is the shape factor of the cross section. The estimate is sufficiently unreliable that it is scarcely worth asking what to do if f is not the same at all plastic hinges.

Fortunately, computational methods can analyze a structure from the initiation of loading to collapse, allowing hinges to form when and where required, and keeping track of displacements along the way. Figure 11.6-4 shows results from such an analysis. This particular analysis uses the simplifying assumption that $f = 1$: that is, as load increases, there is no yielding at a potential hinge location until M_{fp} is reached there, whereupon a plastic hinge is instantly present. Displacement u_A is the horizontal displacement of point A at the instant prior to formation of a hinge. One can verify that the largest load, $P = 4.185$ kN, agrees with the collapse load calculated by hand, based on the mechanism of seven plastic hinges and the assumption that calculations can be based on the original geometry of the frame. However, we see that horizontal deflection u_A becomes large

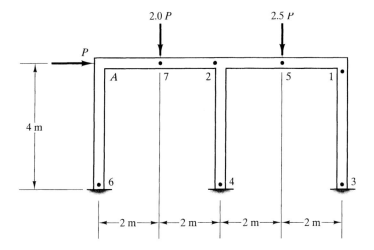

FIGURE 11.6-4 A plane frame, statically indeterminate to the sixth degree. The tabulation, from computer analysis, shows the order in which hinges form and the horizontal displacement of point A. (*Courtesy of Professor C. K. Wang, University of Wisconsin, Madison.*)

enough that vertical loads on column tops appreciably change their moment arms about points 3, 4, and 6. Therefore results must be regarded as approximate.

11.7 GENERAL THEOREMS

The following statements apply to a material that does not strain harden. The statements are made without proofs, which can be found in books devoted to plastic analysis.

Upper Bound Theorem. If a collapse load is calculated by the virtual work method, in which external loads and plastic action do zero net work in a compatible displacement pattern, then the calculated collapse load either is too large or is exact.
  Here the word "exact" pertains not to the physical structure but to its mathematical model, in which geometry may be somewhat idealized and small deflections can be assumed. The upper bound theorem implies that when all possible collapse mechanisms are considered, the mechanism that provides the smallest collapse load is correct. This method is used in the example problems of Sections 11.5 and 11.6.

Lower Bound Theorem. If a stress field satisfies equilibrium conditions and does not exceed the yield criterion at any point, then the load supported by the stress field is less than, or at most equal to, the exact collapse load.
  If only bending moments need be considered in the plastic collapse analysis of beams and frames, the lower bound theorem says that if a bending moment diagram satisfies equilibrium conditions and nowhere exceeds M_{fp} in magnitude, then the

associated load either is too small or is exact. As an example, consider a propped cantilever beam with load P at midspan, Figs. 11.5-1a and 11.7-1a. We can start with the *elastic* bending moment diagram, which satisfies equilibrium conditions. The elastic bending moment at B is 5/6 its magnitude at A. Now we scale the bending moment at all locations by a factor such that $M_A = -M_{fp}$. Then $M_B = 5M_{fp}/6$ as shown in Fig. 11.7-1b. With R the support reaction at C, we write moment equilibrium equations about points A and B in Figs. 11.7-1c and 11.7-1d, then eliminate R.

$$\left. \begin{array}{c} RL - P_c\dfrac{L}{2} + M_{fp} = 0 \\ \\ R\dfrac{L}{2} - \dfrac{5M_{fp}}{6} = 0 \end{array} \right\} \quad \text{hence} \quad P_c = \dfrac{16M_{fp}}{3L} \qquad (11.7\text{-}1)$$

This value of P_c is indeed smaller than the exact value $P_c = 6M_{fp}/L$.

The lower bound theorem is attractive because it provides a conservative result. However, it is usually more difficult to apply than the upper bound theorem, simply because it is harder to construct an equilibrium stress field than a compatible mechanism.

Corollary. Collapse load cannot be increased by decreasing the strength of any portion of the structure, nor decreased by increasing the strength of any portion.

This corollary is stated in the negative because a converse is not necessarily true. For example, in Fig. 11.7-1a, let portion BC be strengthened, while the bending moment capacity of portion AB remains M_{fp}. Plastic hinges at A and B will still form at moment M_{fp}, so the collapse load will not change.

Uniqueness. If the stress distribution referred to in the lower bound theorem is associated with a mechanism, then the applied load is the plastic collapse load.

The foregoing statement means that the smallest upper bound coincides with the largest lower bound. Without referring to the uniqueness theorem by name, we have used it before (following Eq. 11.5-9 and following Eq. 11.6-5) to argue that a collapse load calculated from an assumed mechanism is in fact correct. A corollary is that

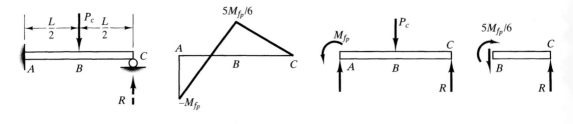

FIGURE 11.7-1 (a) Uniform propped cantilever beam with a concentrated load. (b) Elastic bending moment diagram. (c, d) Free body diagrams.

if the stress (or bending moment) field associated with a mechanism exceeds the yield criterion at any point, then that mechanism is not the correct collapse mechanism. This argument is used in remarks that follow Example 2 of Section 11.5 and in Example 1 of Section 11.6.

Statements in the present section make no reference to the state of stress before the collapse mechanism forms. Therefore, for small deflections and zero strain hardening, a collapse load is not affected by misalignment of components, support settlement, slip at connections, residual stress, or thermal stress.

11.8 INTERACTION: BENDING AND AXIAL FORCE

An *interaction formula* describes the relation among two or more loads when their combination causes failure, where "failure" might be defined as the onset of yielding, reaching a fully plastic condition, or perhaps something else. The loads cited might be separate forces on a frame, as in Fig. 11.6-2a, where failure is defined as the development of enough plastic hinges to create a mechanism. Or, the loads might be force and moment applied to a single cross section. In the present discussion we consider the latter. Axial force N and bending moment M are simultaneously applied to a cross section of a prismatic member. We ask how N and M are related for the onset of yielding and for full plasticity.

Onset of Yielding. For simplicity of explanation, we consider a prismatic member whose cross section has two axes of symmetry. Let the member be loaded by centroidal axial force N and bending moment M. If conditions are linearly elastic and the largest stress is yield stress σ_Y (Fig. 11.8-1a), then

$$\frac{N}{A} + \frac{Mc}{I} = \sigma_Y \quad \text{or} \quad \frac{N}{\sigma_Y A} + \frac{M}{\sigma_Y I/c} = 1 \qquad (11.8\text{-}1)$$

We recognize $\sigma_Y A$ as the axial force that would initiate yielding if it were to act alone, and $\sigma_Y I/c$ as the bending moment that would initiate yielding if it were to act alone. For these loads we adopt the symbols N_o and M_o; that is, $N_o = \sigma_Y A$ and $M_o = \sigma_Y I/c$ are separately applied failure loads in the present example, where failure is regarded as the onset of yielding. For load ratios we adopt the notation

$$R_N = \frac{N}{N_o} \quad \text{and} \quad R_M = \frac{M}{M_o} \qquad (11.8\text{-}2)$$

Thus Eq. 11.8-1 becomes the interaction formula

$$\text{Onset of yielding:} \quad R_N + R_M = 1 \qquad (11.8\text{-}3)$$

An interaction formula expresses a limiting condition. Here it is the onset of yielding. If loads are such that $R_N + R_M < 1$, then yielding does not impend.

Loads N and M need not have the directions shown in Fig. 11.8-1. To allow for reversed (negative) values of N and M, Eq. 11.8-3 can be written in the form

368 Chapter 11 Plasticity in Structural Members and Collapse Analysis

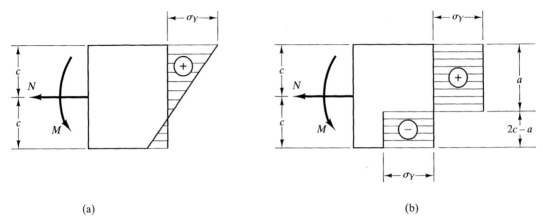

FIGURE 11.8-1 Member loaded by centroidal axial force N and bending moment M. (a) Onset of yielding. (b) Fully plastic condition.

$|R_N|+|R_M|=1$. When plotted, this relation provides straight-line interaction relations in all four quadrants (Fig. 11.8-2a).

Full Plasticity. Let loads N and M be sufficient to cause yielding throughout the cross section (Fig. 11.8-1b). The resulting interaction formula describes a curve rather than a straight line. The shape of the curve depends on the shape of the cross section. Equations that follow pertain to a *rectangular cross section* and a material without strain hardening.

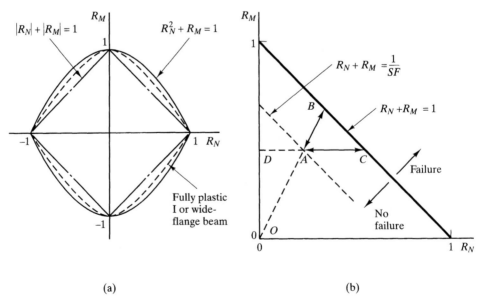

FIGURE 11.8-2 (a) Interaction curves for combined axial and bending loads. (b) Information provided by a straight-line interaction curve.

In Fig. 11.8-1b, N and M can be expressed in terms of σ_Y and dimensions by using equations of static equilibrium. Summing axial forces and taking moments about a convenient point, such as a point on the line of action of N, we obtain

$$N = 2\sigma_Y b(a - c) \quad \text{and} \quad M = \sigma_Y ab(2c - a) \tag{11.8-4}$$

where b is the width of the rectangular cross section. Elimination of a between these two equations provides

$$\left(\frac{N}{2\sigma_Y bc}\right)^2 + \frac{M}{\sigma_Y bc^2} = 1 \tag{11.8-5}$$

Now $N_o = 2\sigma_Y bc$ and $M_o = \sigma_Y bc^2$ are loads that would produce a fully plastic condition if each acted alone. These are separately applied failure loads in the present example, where failure is regarded as reaching a fully plastic condition. With $R_N = N/N_o$ and $R_M = M/M_o$, Eq. 11.8-5 becomes

$$\text{Fully plastic:} \quad R_N^2 + R_M = 1 \tag{11.8-6}$$

This interaction formula, generalized to allow for positive and negative N and M, is plotted in Fig. 11.8-2a. A combination of R_N and R_M that plots as a point within the envelope is "safe" in that the fully plastic capacity is not exceeded. The straight-line envelope might be adopted as a simple and conservative approximation of the curved envelope.

Safety Factor. Let the same safety factor SF be applied to each load in the straight-line interaction formula. Thus, Eq. 11.8-3 becomes

$$\frac{(SF)N}{N_o} + \frac{(SF)M}{M_o} = 1 \quad \text{or} \quad R_N + R_M = \frac{1}{SF} \tag{11.8-7}$$

This equation is plotted in Fig. 11.8-2b. Distance AB represents reserve strength for proportional loading (in which the ratio N/M is maintained as both change). We see that reserve strength is greater for nonproportional loading (distance AC).

A safety factor can be interpreted graphically as a ratio of lengths. For proportional loading in Fig. 11.8-2b it is $SF = OB/OA$, as is most easily seen for the special case of N or M equal to zero in Eq. 11.8-7.

Remarks. The general definition of a load ratio is [11.9]

$$R_i = \frac{\text{applied loading of type } i}{\text{critical loading of type } i \text{ when it acts alone}} \tag{11.8-8}$$

Critical loading refers to loading applied at failure, whether failure is defined as the onset of yielding, the fully plastic condition, buckling, or something else. An interaction curve always intersects the R_i axis at $R_i = 1$, regardless of how failure is defined.

An interaction formula can be derived from a failure criterion in some situations. For example, Eqs. 11.8-3 and 11.8-6 are consistent with both the maximum shear stress failure criterion and the von Mises failure criterion. Indeed, equations that describe failure criteria themselves can be regarded as interaction relations (Sections 3.2 and

3.3). Other situations are better suited to experiment than to analytical determination. Examples include fatigue (Fig. 3.6-4) and local buckling [11.9].

EXAMPLE

Determine collapse load P_c for the angle frame in Fig. 11.8-3. Assume that fully plastic conditions develop at the fixed base, and use a straight-line formula to approximate the interaction relation.

Let M_c be the bending moment at the base when fully plastic conditions develop. For a small rotation θ about the base, the virtual work equation is

$$P_c(L\theta) - M_c\theta = 0 \tag{11.8-9}$$

With $N_o = 2\sigma_Y bc$, $M_o = \sigma_Y bc^2$, and $SF = 1$ in Eq. 11.8-7,

$$\frac{P_c}{2\sigma_Y bc} + \frac{M_c}{\sigma_Y bc^2} = 1 \tag{11.8-10}$$

Elimination of M_c between Eqs. 11.8-9 and 11.8-10 provides

$$P_c = \frac{\sigma_Y bc^2}{L + 0.5c} \tag{11.8-11}$$

As a partial check on this result, note that if $L \ll c$, then $P_c = 2\sigma_Y bc$, as is reasonable. The other extreme, $L \gg c$, appears in ordinary structural frames. Then Eq. 11.8-11 reduces to $P_c = \sigma_Y bc^2/L$, which is $P_c = M_{fp}/L$ in the notation of preceding sections. Thus we justify our neglect of axial forces in the frame problems discussed in Section 11.6. Note that if $H \gg L$, lateral deflection will cause load P to exert a moment about the base appreciably larger than PL. Load P_c from the foregoing linear analysis is then an overestimate of the actual collapse load. This complication is discussed in [11.10].

11.9 INTERACTION: SOME ADDITIONAL RELATIONS

Interaction formulas have been proposed for a great many situations. Some formulas are curves fitted to experimental data, and some relate three or more kinds of loads. What follows is a very brief sampling.

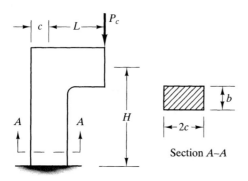

FIGURE 11.8-3 Angle frame of rectangular cross section.

Section 11.9 Interaction: Some Additional Relations

Shear and Bending. Approximate analyses and experiments have provided several formulas and plots that relate transverse shear force V and bending moment M in the fully plastic condition [11.1, 11.4, 11.5]. Representative plots are shown in Fig. 11.9-1a, where $R_M = M/M_o$ and $R_V = V/V_o$, and M_o and V_o each produce a fully plastic condition when acting alone. From Fig. 11.9-1a we draw the useful conclusion that for I and wide-flange sections commonly used in engineering structures, transverse shear force is unlikely to be large enough to significantly reduce bending moment capacity.

Bending and Torsion. Let M and T represent applied bending moment and torque, and let M_o and T_o represent failure values of these quantities when each acts alone. With $R_M = M/M_o$ and $R_T = T/T_o$, the interaction formula for fully plastic conditions in a circular cross section has the form

$$R_M^2 + R_T^2 = 1 \quad \text{or} \quad R_M^2 + R_T^2 = \frac{1}{(SF)^2} \tag{11.9-1}$$

The latter formula contains a safety factor SF applied to both M and T. This relation is plotted in Fig. 11.9-1b. The first of Eqs. 11.9-1 can be derived using the maximum shear stress failure criterion. Experiments show that this equation is also applicable to thin-walled tubes of circular cross section that fail by local buckling under combined bending and torsion [11.9].

Collapse problems that involve M and T in combination are often difficult and require use of the "normality condition" of plasticity theory [11.6]. Fortunately, for I and wide-flange sections, bending stiffness is so much greater than torsional stiffness that even for structures in which members twist, members usually do not develop enough torque for it to affect bending moment capacity.

Torsion and Axial Force. For thin-walled tubes of circular cross section that fail by local buckling, the following formula provides a reasonable fit to experimental data [11.9]:

$$R_N + R_T^3 = 1 \quad \text{for} \quad -0.7 \leq R_N < 1.0 \tag{11.9-2}$$

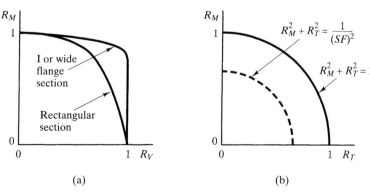

FIGURE 11.9-1 (a) Interaction of bending moment and transverse shear force. (b) Interaction of bending moment and torsion.

where, in the axial force ratio $R_N = N/N_o$, N_o is a *negative* quantity that represents failure by local buckling in pure compression. We see that tensile force ($N < 0$ in the present sign convention) raises the torque required for failure.

PROBLEMS

In the following problems, assume that deflections are small and that the material does not strain harden. Ignore interaction effects except in problems related to Sections 11.8 and 11.9.

11.2-1. Each bar of the truss shown yields at an axial load of 24 kN.
  (a) Determine the value of load P for collapse (that is, determine the value of P at which displacements begin to increase greatly for small increases in P).
  (b) Determine the collapse load if bar AD is removed before loading begins.

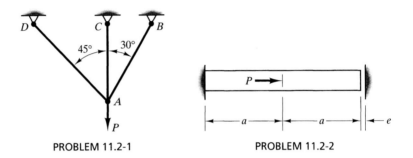

PROBLEM 11.2-1   PROBLEM 11.2-2

11.2-2. The uniform bar shown has cross-sectional area A and is made of the material depicted in Fig. 11.2-1a. When load P is zero, gap e measures $0.7\sigma_Y a/E$. Determine and plot the relation between load P and its displacement in the manner of Eqs. 11.2-1 and 11.2-2 and Fig. 11.2-2b.

11.2-3. Bar BC in the sketch is rigid. The other three bars are slender links, pin-connected at their ends, with cross-sectional areas A and $2A$ as shown. All links are the same length and are made of the same material, with yield stress σ_Y. Determine load P_c for plastic collapse in terms of A and σ_Y. Suggestions: Consider three plausible collapse mechanisms; for each, calculate forces and stresses in the links.

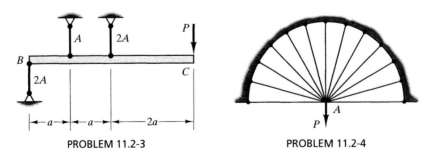

PROBLEM 11.2-3   PROBLEM 11.2-4

11.2-4. The structure shown consists of uniform and uniformly spaced bars, each pin-connected to point A and to the semicircular support. There are $2n + 1$ identical bars, where n is "large."

(a) Determine the value of P for which yielding begins, in terms of n, σ_Y, and cross-sectional area A of each bar.

(b) Determine collapse load P_c in terms of the same quantities. Explain why assumptions you may make about the geometry of deformation are appropriate.

11.3-1. Verify numerical factors in Fig. 11.3-2b for the following cross sections.
  (a) Solid circle.
  (b) Hollow circle.
  (c) Diamond.
  (d) Triangle.

11.3-2. Determine shape factor f for the cross section shown, when it is bent in the usual way (with neutral axis parallel to the flanges). Dimensions are in centimeters.

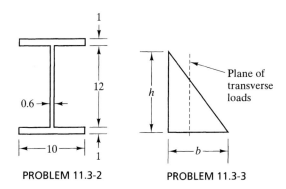

PROBLEM 11.3-2   PROBLEM 11.3-3

11.3-3. A beam has the unsymmetric triangular cross section shown. Determine M_{fp} in terms of b, h, and σ_Y.

11.3-4. For the beam cross section in Fig. 11.3-1, let $b = 20$ mm, $2c = 60$ mm, and $E = 200$ GPa. Imagine that the beam is not homogeneous: Let $\sigma_Y = 300$ MPa for top and bottom layers each 9 mm deep, and $\sigma_Y = 400$ MPa for the 42 mm core that straddles the neutral axis. As M increases monotonically from zero, for what ranges of M does the moment-curvature relation plot as a straight line?

11.3-5. Show that if $Z/S \le 2$, then unloading from fully plastic bending produces no renewed yielding as the load approaches zero. Do not consider unsymmetric cross sections.

11.3-6. (a) Let a beam be unloaded after M_{fp} has been applied. For a beam of rectangular cross section, Fig. 11.3-1, determine the pattern of residual stress and show it on a sketch.

(b) Show that the residual stress pattern is self-equilibrating. That is, show that it produces zero resultant force and zero resultant moment.

(c) The beam is now reloaded by bending moment M opposite in direction to the original M_{fp}. How large can M become before yielding begins again? How large can M become if yielding is allowed?

11.3-7. Repeat Problem 11.3-6a for the following cross sections in Fig. 11.3-2b.
  (a) Solid circle (for which $f = 1.70$).
  (b) Diamond (for which $f = 2.00$).

11.3-8. For a beam of rectangular cross section, Fig. 11.3-1, let $b = 20$ mm, $2c = 60$ mm, $E = 200$ GPa, and $\sigma_Y = 280$ MPa. The beam is bent as shown in Fig. 11.3-1 until two-thirds of the material has yielded.

(a) What bending moment is applied, and what is the radius of curvature?

(b) If the bending moment is removed, what is the radius of curvature?

11.3-9. A straight wire is of square cross section, 2.5 mm on a side, with $E = 200$ GPa and $\sigma_Y = 500$ MPa. The wire is to be wound on a mandrel of such radius R that when the mandrel is removed the wire has become a closely coiled helical spring of mean radius 25 mm.

(a) Determine R. Assume that winding makes the wire fully plastic.

(b) Again determine R, this time without the simplifying assumption of part (a).

11.3-10. Imagine that a tape is wrapped on a spool of radius R, such that each layer of tape lies over the layer below, thus forming a tight spiral when viewed parallel to the axis of the spool. Now let the "tape" be a wire that was initially straight, of square cross section $2c$ units on a side, with elastic modulus E and yield stress σ_Y. When the wrapping force is released, the spiral opens up a bit.

(a) Determine the equation of the final spiral, in the form $1/\rho_{res} = f(\alpha)$, in terms of R, c, E, σ_Y, and an angle α. Assume that $R \gg c$.

(b) How many layers could thus be wrapped before the remainder of the wire would return to its straight shape when the wrapping force is released?

11.4-1. (a) The span of yielded material in Fig. 11.4-1b is $L\sqrt{1 - 1/f}$. Derive this result.

(b) Show that if the cross section is rectangular, the plastic material is bounded by straight lines, as shown in Fig. 11.4-1b.

11.4-2. Distributed loading q varies linearly on a prismatic beam in the manner shown, reaching intensity q_o at each end. Determine the spanwise extent of the yielded zone, and the shape of the elastic-plastic boundary if the cross section is rectangular (as done for other loadings in Fig. 11.4-1).

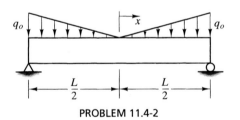

PROBLEM 11.4-2

11.5-1. Use equilibrium considerations, rather than the virtual work method, to obtain the value of P_c given by Eq. 11.5-1.

11.5-2. Solve for collapse load q_c in Fig. 11.5-2a by setting $M_A = -M_{fp}$, locating the other plastic hinge at the point of maximum bending moment, and applying equilibrium considerations.

11.5-3. For each mechanism in Fig. 11.5-3, what are the bending moments at A, B, C, and D in terms of M_{fp}? Sketch the bending moment diagram for each mechanism.

11.5-4. In each of the beams shown, bending moment capacity is M_{fp} throughout except as noted. Apply the method of virtual work to plausible collapse mechanisms, and determine collapse load P_c or M_{oc} (concentrated force or moment) or q_c (distributed load). Express answers in terms of M_{fp} and dimensions shown.

11.5-5. For the individual parts of Problem 11.5-4, show that when the correct collapse mechanism prevails, bending moments at possible critical locations are at or below the bending moment capacity.

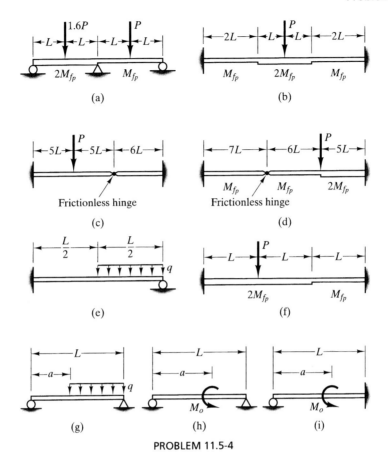

PROBLEM 11.5-4

11.5-6. Distributed loading on the uniform beam shown varies linearly, reaching intensity q_L at $x = L$. Determine q_{Lc}, the collapse value of q_L, in terms of L and M_{fp} by the following methods.
  (a) Use equilibrium considerations. Locate the intermediate plastic hinge at the point of maximum bending moment.
  (b) Use virtual work. Make two approximate calculations: Place the intermediate plastic hinge at $x = L/2$, then at $x = 2L/3$.

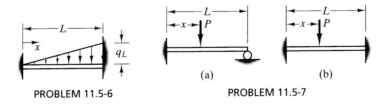

PROBLEM 11.5-6

PROBLEM 11.5-7

11.5-7. For each of the uniform beams shown, obtain an expression for collapse load P_c in terms of x, L, and M_{fp}. Also determine location x in the range $0 \le x \le L$ for which P_c is least, and the value of that P_c.

11.5-8. Beam AD and wire BC in the sketch are made of the same material, whose yield stress is σ_Y. The beam has a square cross section, a units on a side. What should be the diameter of the wire, in terms of a and L, if two collapse mechanisms are to be equally likely?

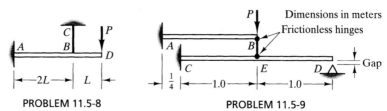

PROBLEM 11.5-8

PROBLEM 11.5-9

11.5-9. For beams AB and CD shown, let $EI = 40$ kN·m², shape factor $f = 1$, and $M_{fp} = 1000$ N·m. Assume that link BE is rigid, and that M_{fp} is reached in both beams before the gap closes. Qualitatively plot the relation between load P and its displacement v. Provide numerical values of P at points where the P versus v plot changes slope.

11.5-10. For the beam shown, assume that a first collapse mechanism appears before the gap closes. Qualitatively plot the relation between load quantity qL^2/M_{fp} and vertical displacement v of point C. Provide numerical values of qL^2/M_{fp} at points where the qL^2/M_{fp} versus v plot changes slope.

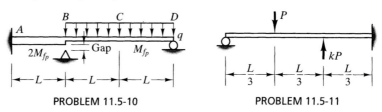

PROBLEM 11.5-10

PROBLEM 11.5-11

11.5-11. (a) For the uniform beam shown, determine P for collapse in terms of M_{fp}, L, and k, where k is a positive number. The simple support at the left can supply either upward or downward force. Using numerical values, plot $P_c L/M_{fp}$ versus k.

(b) Consider P_c when both loads are the same ($k = 1$). Can these two loads be applied in arbitrary order, rather than simultaneously? Explain.

11.5-12. The sketch represents the top view of a grillage in the horizontal plane, built by welding uniform members together at connection points A and B. Dots at member ends represent simple supports. Assume that all members have bending moment capacity M_{fp} and negligible torsional resistance. Apply a downward load P at A.

(a) Let $L = s$. Determine the collapse value of P.

(b) If now $L \neq s$, determine the ratio s/L such that two collapse mechanisms are equally likely.

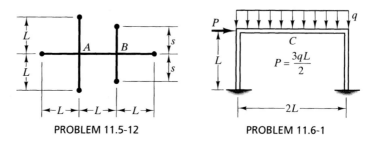

PROBLEM 11.5-12

PROBLEM 11.6-1

11.6-1. Determine the collapse value of load q for the plane frame shown. Let $P = 3qL/2$. Consider a mechanism in which a plastic hinge appears at C, and loads P and q both do work. Bending moment capacity is M_{fp} throughout the frame.
  (a) Assume that C is at the middle of the horizontal member.
  (b) Place C at its proper location.
  (c) Does some other collapse mechanism actually prevail? What is the correct q_c?

11.6-2. Let the frame in Fig. 11.6-2a have bending moment capacity M_{fp} throughout. Determine the collapse value of load P in terms of M_{fp} and L for the following arrangements of loading and dimensions. Load P at C is present in both cases.
  (a) $Q = P/3, R = 0, a = L, b = L, h = L, r = L/3$.
  (b) $Q = 0, R = P/2, a = 4L, b = 2L, h = 3L, r = 2L$.

11.6-3. Determine the collapse value of P_c for each of the two arches in Fig. 11.6-3. Let $\lambda = 60°$ and $\beta = 30°$. Throughout each arch, the bending moment capacity is M_{fp}.

11.6-4. Using the seven-hinge mechanism shown in Fig. 11.6-4, calculate collapse load P_c.

11.6-5. In each of the plane structures shown, bending moment capacity is M_{fp} throughout, except as noted. Apply the method of virtual work to plausible collapse mechanisms, and determine collapse loads P_c in terms of M_{fp} and dimensions shown.

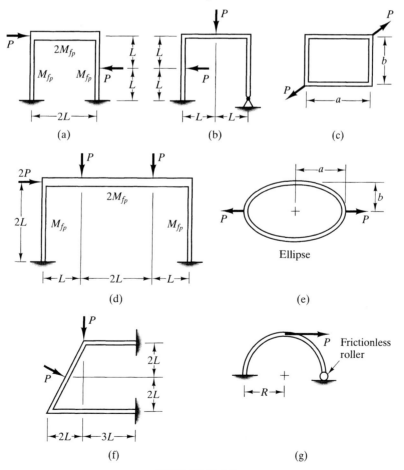

PROBLEM 11.6-5

11.6-6. For the individual parts of Problem 11.6-5, show that when the correct collapse mechanism prevails, bending moments at possible critical locations are at or below the bending moment capacity.

11.6-7. Let the frame in Fig. 11.6-2a have the same bending moment capacity M_{fp} throughout. Adopt the following dimensions: $a = b = h = 2L$, $r = L$. Omit load R. Consider three mechanisms: one in which load P does not move, one in which load Q does not move, and one in which both loads move. In each case, express collapse load P_c and/or Q_c in terms of M_{fp} and L. Plot lines that represent these expressions on axes labeled $P_c L/M_{fp}$ and $Q_c L/M_{fp}$. Identify the "safe" region of the plot.

11.7-1. For the following problems, obtain lower-bound solutions for collapse load by setting the largest bending moment in a linearly elastic solution equal to the bending moment capacity. For nonuniform members, assume that EI is directly proportional to the bending moment capacity shown.
  (a) Problem 11.5-4c.
  (b) Problem 11.5-4d.
  (c) Problem 11.5-4e.
  (d) Problem 11.6-5e (let $a = b$).

11.8-1. A beam of square cross section is to be made of a material whose yield stress is 360 MPa. If full plasticity is allowed, what must be the size of the cross section if axial force 130 kN and bending moment 600 N·m are to be simultaneously applied?

11.9-1. A member has a circular cross section and is made of a brittle material. It is loaded in combined bending and torsion. Use the maximum principal stress failure criterion to derive an interaction formula.

11.9-2. At a point where a plane state of stress is described by one normal stress σ and one shear stress τ, criteria of ductile failure (Chapter 3) provide the yield criterion

$$\left(\frac{\sigma}{\sigma_Y}\right)^2 + C^2\left(\frac{\tau}{\sigma_Y}\right)^2 = 1$$

where $C^2 = 4$ for the maximum shear stress yield criterion and $C^2 = 3$ for the von Mises yield criterion. Let a member of circular cross section carry both bending moment M and torque T.
  (a) If failure is defined as the onset of yielding, what is the interaction formula that relates R_M and R_T?
  (b) Can the formula obtained in part (a) be applied to a noncircular cross section? Explain.
  (c) Let $\sigma_Y = 300$ MPa, $M = 500$ N·m, and $T = 400$ N·m. For a safety factor of 2 against the onset of yielding, what must be the radius of a solid cross section? Use the maximum shear stress yield criterion.
  (d) Repeat part (c), but use the von Mises yield criterion.
  (e) Show that the interaction formula is $R_M^2 + R_T^2 = 1$ if the cross section is fully plastic. Assume that σ and τ are constant over a fully yielded cross section.
  (f) Let $M = 0.9 M_o$, where M_o is the fully plastic failure load under pure bending. By what percentage has the presence of M reduced the torque that can be carried?

11.9-3. In the opening remarks of Problem 11.9-2, replace bending moment M on the circular cross section by axial force N. Then determine the interaction formula if failure is defined by (a) the onset of yielding, and (b) the fully plastic condition.

11.9-4. Determine the safety factor for proportional loading of a thin-walled tube of circular cross section in the following situations.
  (a) $R_M = 0.5$ and $R_T = 0.5$ in combined bending and twisting.
  (b) $R_N = -0.5$ and $R_T = 0.5$ in combined stretching and twisting.

CHAPTER 12

Plate Bending

12.1 INTRODUCTION. ASSUMPTIONS AND LIMITATIONS

Plates are frequently used as structural elements. Concrete floors and paving are common examples, although concrete plates are usually called *slabs*. Like a beam, a plate carries lateral loads by bending. At a given cross section, a typical beam acquires a single bending moment and a single transverse shear force. A typical plate acquires two bending moments, a twisting moment, and two transverse shear forces. An unloaded beam may be straight or curved, but by definition an unloaded plate is flat (a curved geometry would make it a shell). Although a plate may be loaded by in-plane loads only, as discussed in Chapter 7, the present chapter deals with lateral loads, which produce bending.

In Fig. 12.1-1a, the xy plane is the plate *midsurface*, halfway between top and bottom plate surfaces. The z axis is shown positive downward, as is commonly done. The coordinate system is right-handed. Normal stresses σ_x and σ_y result from bending, and shear stress τ_{xy} results from twisting. These three stresses act parallel to the midsurface. Stresses τ_{yz} and τ_{zx} are transverse shear stresses. All these stresses vary with z, in accordance with assumptions noted in what follows. Distributed lateral load $q = q(x,y)$ is arbitrary. If q is constant, it represents uniform lateral pressure. Concentrated forces

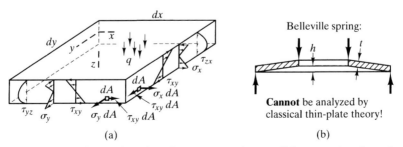

FIGURE 12.1-1 Stresses in a plate, shown on xz- and yz-parallel cross sections. Lateral load q has units of pressure. (b) Cross section of a shallow conical shell, loaded around its inner and outer edges.

and moments may also be applied. Supports, not shown in Fig. 12.1-1a, include fixed and simple supports, in direct analogy to supports of beams.

Assumptions and Restrictions. The following assumptions are used in classical thin-plate theory. (With few changes in wording, they apply also to elementary beam theory.)

1. Before loading, the midsurface is flat.
2. Plate thickness t is small in comparison with spanwise dimensions of the plate.
3. The material is homogeneous, isotropic, and linearly elastic.
4. A line normal to the midsurface before loading remains normal to the midsurface after loading.
5. Stress σ_z (normal to the midsurface) is negligible in comparison with σ_x and σ_y.
6. Lateral deflections are small in comparison with the plate thickness, and slopes of the midsurface are small in comparison with unity.
7. The midsurface is a neutral surface; that is, strains ϵ_x, ϵ_y, and γ_{xy} are zero at $z = 0$.

Assumption 2 is usually taken to mean that t should be no more than about 1/10 the smallest plate dimension in the xy plane. A plate with larger t may have significant deformation due to transverse shear and is known as a *thick plate*. Assumptions 4, 6, and 7 dictate that the deformation state of a plate is completely defined by the z-direction deflection w of its midsurface, where $w = w(x,y)$. Assumption 4 also says that transverse shear strains γ_{yz} and γ_{zx} are zero. This assumption is known as the *Kirchhoff approximation*. A plate for which it is satisfactory is usually called a *thin plate*. Assumptions 3, 4, and 7 dictate the linear distributions of σ_x, σ_y, and τ_{xy} seen in Fig. 12.1-1a.

Assumptions 2, 6, and 7 are related. If a plate is extremely thin it has very little bending stiffness, and lateral load makes it sag like a wet blanket. Load is then supported mainly by membrane stresses, which act tangent to the deflected midsurface, and assumptions 6 and 7 are violated. Such a plate is properly called a *membrane* and is analyzed as such. However, even when most load is supported by bending action, a plate whose maximum lateral deflection is approximately equal to its thickness may violate assumption 7 to such an extent that classical plate theory must be extended to account for membrane stresses. The reason is that the deflected shape $w = w(x,y)$ is usually neither cylindrical nor conical. Therefore it cannot be obtained from the initially flat configuration without straining the midsurface and thus producing membrane stresses. Significant membrane stresses make a plate problem *nonlinear*; that is, deflections and stresses are not directly proportional to lateral load. Examples are discussed in Section 12.6.

Figure 12.1-1b shows the cross section of a washer-shaped solid called a "Belleville spring," loaded by line loads around inner and outer edges. Superficially, it looks like a circular plate, but it violates assumption 1 because h is not zero. Membrane stresses develop, and stresses and deflections are not directly proportional to load if h is comparable to t. This structure must be analyzed by shell theory, or by a modified plate theory that accounts for initial deflections and includes nonlinear terms.

In summary, we have categorized plates as follows: (1) thick, where transverse shear deformation is significant; (2) thin, and with negligible membrane stresses;

Section 12.2 Governing Equations in Cartesian Coordinates

(3) thin, but nonlinear because of appreciable membrane stresses. Category (2) is the most restrictive, but permits the simplest analysis. Except for examples in Section 12.6 that show the effect of large deflection, discussion in the present chapter is limited to category (2). This category may be called "Kirchhoff theory" or "classical thin-plate theory." It is based on the foregoing seven assumptions. We consider only comparatively simple cases of loading, geometry, and support conditions. Results are available for many additional cases [1.6, 1.7, 12.1]. Complicated situations, including varying thickness, anisotropy, and elastic support may be most easily addressed by computational methods.

12.2 GOVERNING EQUATIONS IN CARTESIAN COORDINATES

The following formulation applies to a plate for which assumptions 1 through 7 of Section 12.1 are acceptable. Ingredients of the formulation are equilibrium equations, strain-displacement relations, and stress-strain relations.

First, consider bending moments M_x and M_y, twisting moment M_{xy}, and transverse shear forces Q_x and Q_y. They are associated with stresses shown in Fig. 12.1-1a. For example, moment increment dM_x is $dM_x = z(\sigma_x \, dA)$, where M_x is shown in Fig. 12.2-1. By definition, the M's and Q's are moments and forces *per unit length* in the xy plane. Therefore $dA = (1)dz$, and

$$M_x = \int_{-t/2}^{t/2} \sigma_x z \, dz \qquad M_y = \int_{-t/2}^{t/2} \sigma_y z \, dz \qquad M_{xy} = \int_{-t/2}^{t/2} \tau_{xy} z \, dz$$

$$Q_x = \int_{-t/2}^{t/2} \tau_{zx} \, dz \qquad Q_y = \int_{-t/2}^{t/2} \tau_{yz} \, dz \tag{12.2-1}$$

The M's have dimensions [force·length/length] or simply [force]; the Q's have dimensions [force/length]. In general, all are functions of coordinates x and y. Their positive directions are as shown in Fig. 12.2-1. These directions are consistent with stress directions shown in Fig. 12.1-1a.

Stresses associated with the M's vary linearly with z. For example, if $\bar{\sigma}_x$ is the value of σ_x at $z = t/2$, then $\sigma_x = 2\bar{\sigma}_x z/t$. Hence Eq. 12.2-1 yields $\bar{\sigma}_x = 6M_x/t^2$. This formula can be regarded as the result of applying the flexure formula $\sigma = Mc/I$ to a unit

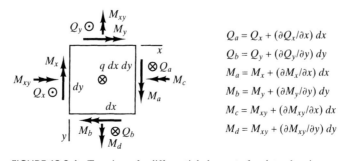

FIGURE 12.2-1 Top view of a differential element of a plate, showing forces and moments per unit length. Symbols ⊗ and ⊙ indicate arrows in the $+z$ (downward) and $-z$ (upward) directions, respectively.

width of plate, so that $I = t^3/12$. In summary, lateral load on a plate produces the following stresses at plate surfaces $z = \pm t/2$.

$$\overline{\sigma}_x = \pm \frac{6M_x}{t^2} \qquad \overline{\sigma}_y = \pm \frac{6M_y}{t^2} \qquad \overline{\tau}_{xy} = \pm \frac{6M_{xy}}{t^2} \tag{12.2-2}$$

If moments have directions shown in Fig. 12.2-1, stresses are positive at $z = t/2$ (directed as shown in Fig. 12.1-1a). Transverse shear stresses τ_{yz} and τ_{zx} can usually be ignored because they are relatively small. They vary quadratically with z and have maximum magnitudes $1.5 Q_y/t$ and $1.5 Q_x/t$ at $z = 0$, as shown in Fig. 12.1-1a.

Figure 12.2-1 shows a typical differential element of a plate. If the element is to be in static equilibrium, three equations must be satisfied: forces must sum to zero in the z direction, and moments must sum to zero about the x axis and about the y axis. In general, forces and moments per unit length are not quite the same on opposite edges of the element, so increments such as $(\partial Q_x/\partial x)dx$ appear. Here $\partial Q_x/\partial x$ is *rate* of change and $(\partial Q_x/\partial x)dx$ is *amount* of change. Forces $Q_x \, dy$ and $Q_y \, dx$ act upward in Fig. 12.2-1; forces $Q_a \, dy$, $Q_b \, dx$, and $q \, dx \, dy$ act downward. Contributions to moment about the x axis in Fig. 12.2-1 are $(M_{xy} - M_c)dy$, $(M_y - M_b)dx$, $(Q_b \, dx)dy$, and higher-order terms associated with Q_x, Q_a, and q. Pursuing these arguments, substituting for Q_a and so on, and omitting higher-order terms, we obtain the equilibrium equations

$$\frac{\partial Q_x}{\partial x} + \frac{\partial Q_y}{\partial y} + q = 0 \qquad \frac{\partial M_{xy}}{\partial x} + \frac{\partial M_y}{\partial y} = Q_y \qquad \frac{\partial M_x}{\partial x} + \frac{\partial M_{xy}}{\partial y} = Q_x \tag{12.2-3}$$

Substitution of the latter two equations into the first yields an equilibrium equation that relates moments.

$$\frac{\partial^2 M_x}{\partial x^2} + 2\frac{\partial^2 M_{xy}}{\partial x \, \partial y} + \frac{\partial^2 M_y}{\partial y^2} + q = 0 \tag{12.2-4}$$

Figure 12.2-2 shows the displacement of a differential element as viewed parallel to the y axis. Right angles are preserved because transverse shear deformation is ignored. A

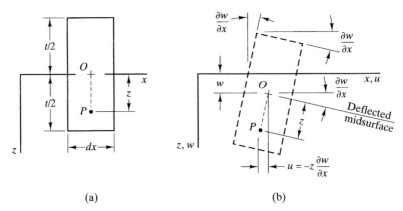

FIGURE 12.2-2 Differential slice of a plate of thickness t. (a) Position before loading. (b) Position after loading, shown dashed, with transverse shear deformation neglected.

typical point P has x-direction displacement $u = -z(\partial w/\partial x)$. Similarly, a view parallel to the x axis shows that P has y-direction displacement $v = -z(\partial w/\partial y)$. Strains parallel to the midsurface are $\epsilon_x = \partial u/\partial x$, $\epsilon_y = \partial v/\partial y$, and $\gamma_{xy} = \partial u/\partial y + \partial v/\partial x$ (see Eqs. 7.2-3). Hence, strains are related to curvatures by the expressions

$$\epsilon_x = -z\frac{\partial^2 w}{\partial x^2} \qquad \epsilon_y = -z\frac{\partial^2 w}{\partial y^2} \qquad \gamma_{xy} = -2z\frac{\partial^2 w}{\partial x\, \partial y} \qquad (12.2\text{-}5)$$

These strains may be substituted into the stress-strain relation for isotropy and linear elasticity, Eqs. 7.1-3, and the resulting stresses substituted into Eqs. 12.2-1. These manipulations yield the moment expressions

$$M_x = -D\left(\frac{\partial^2 w}{\partial x^2} + \nu\frac{\partial^2 w}{\partial y^2}\right) - \frac{E\alpha}{1-\nu}\int_{-t/2}^{t/2} Tz\, dz$$

$$M_y = -D\left(\frac{\partial^2 w}{\partial y^2} + \nu\frac{\partial^2 w}{\partial x^2}\right) - \frac{E\alpha}{1-\nu}\int_{-t/2}^{t/2} Tz\, dz \qquad (12.2\text{-}6)$$

$$M_{xy} = -(1-\nu)D\frac{\partial^2 w}{\partial x\, \partial y} \qquad \text{where} \qquad D = \frac{Et^3}{12(1-\nu^2)}$$

where ν is Poisson's ratio, α is the coefficient of thermal expansion, and T is temperature change, usually measured relative to a uniform reference temperature at which the plate is free of stress. Assumption 7 of Section 12.1 is satisfied if $T = T(x,y,z)$ has a form such that the integral $T\, dz$ through the plate thickness is zero. In writing Eqs. 12.2-6, we have assumed that E, ν, and α are independent of temperature and have used the relation $G = 0.5E/(1+\nu)$. D is called *flexural rigidity*. It has dimensions [force·length], and is analogous to flexural stiffness EI of a beam. Indeed, if a beam has depth t and unit width, and Poisson's ratio is zero, then D is equal in magnitude to EI.

For an isotropic plate of uniform thickness, D is independent of x and y, and substitution of Eqs. 12.2-6 into Eq. 12.2-4 yields

$$\frac{\partial^4 w}{\partial x^4} + 2\frac{\partial^4 w}{\partial x^2 \partial y^2} + \frac{\partial^4 w}{\partial y^4} = \frac{1}{D}\left[q - \frac{E\alpha}{1-\nu}\int_{-t/2}^{t/2}\left(\frac{\partial^2 T}{\partial x^2} + \frac{\partial^2 T}{\partial y^2}\right)z\, dz\right] \qquad (12.2\text{-}7)$$

which is the governing equation for the thin plate problem. The solution to a given plate problem is obtained by devising a function $w = w(x,y)$ that simultaneously satisfies Eq. 12.2-7 and boundary conditions of the given problem (boundary conditions are described in the following section). Equation 12.2-7, without temperature terms, was developed between 1810 and 1820, with most credit going to Sophie Germain, Lagrange, and Navier.

12.3 UNIFORM BENDING. BOUNDARY CONDITIONS

By "uniform bending" we mean states in which M_x, M_y, and M_{xy} are independent of x and y. These states are considered in order to illustrate thin-plate behavior, especially how it differs from beam behavior. Deflections shown in sketches are greatly exaggerated. For simplicity, plates are shown as rectangular. The basic plate equations contain no such shape restriction, but solutions are most easily determined for rectangular plates and circular plates.

Bending to a Cylindrical Surface. Lateral displacement $w = -c_1 x^2$, where c_1 is a positive constant, provides the cylindrical (or shallow parabolic) surface shown in Fig. 12.3-1a. With this w, Eqs. 12.2-6 yield the uniform moment field

$$M_x = 2c_1 D \qquad M_y = 2\nu c_1 D \qquad M_{xy} = 0 \qquad (12.3\text{-}1)$$

We see that the expected moments M_x must be accompanied by moments $M_y = \nu M_x$ in order to produce a cylindrical surface. Except at the midsurface, where stresses are zero, the state of stress at any point is biaxial, with $\sigma_y = \nu \sigma_x$.

If x-parallel edges are not loaded, so that M_y and σ_y are zero there, these edges respond by curling (Fig. 12.3-1b). The amount of curling is directly proportional to ν and is slight; displacement w along x-parallel edges is only about $t/10$ when $\nu = 1/3$ [12.2]. So long as $b \gg t$, the central portion of the plate remains a cylindrical surface, with $M_y = \nu M_x$. If $b < t$, we have a beam, not a plate: Then for $w = -c_1 x^2$ the applied bending moment is $2c_1 EI$, and $\sigma_y \approx 0$ throughout. In a beam, the tendency for sideways strain associated with the Poisson effect is almost unrestrained (Fig. 12.3-1c). The deformed shape of top and bottom surfaces of the beam is called "anticlastic" and may easily be demonstrated by bending a rectangular rubber eraser.

Pure Twist. Lateral displacement $w = -c_2 xy$, where c_2 is a positive constant, provides the saddle-shaped surface shown in Fig. 12.3-2a. With this w, Eqs. 12.2-6 yield the uniform moment field

$$M_x = 0 \qquad M_y = 0 \qquad M_{xy} = (1 - \nu)c_2 D \qquad (12.3\text{-}2)$$

At any point not on the midsurface, the state of stress is pure shear. Therefore, as shown in Fig. 12.3-2b, a plate element whose sides are at 45° to x and y axes carries principal stresses, of equal magnitude and opposite sign. These principal stresses are associated with the bending moment loading shown in Fig. 12.3-2c, which also produces lateral displacement $w = -c_2 xy$.

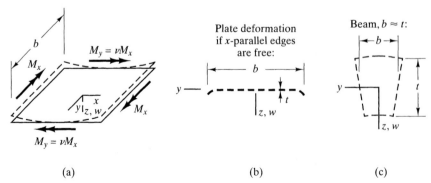

FIGURE 12.3-1 (a) Bending to a cylindrical surface, shown by dashed lines. (b) Deformed neutral surface if moments M_y are removed from x-parallel edges. (c) Analogous deformation of an initially rectangular cross section of a beam.

Moments differ from stresses only by a constant, as is shown, for example, by Eqs. 12.2-2. Accordingly, transformation equations for stress are immediately adaptable to moments. Thus, in Fig. 1.10-1, the respective stresses σ_x, σ_y, and τ_{xy} can be replaced by M_x, M_y, and M_{xy} in order to determine M_n, M_s, and M_{ns}. In the Mohr circle representation for moments in a plate, the center of the circle is on the axis along which M_x and M_y are plotted, and M_{xy} is plotted along the other axis.

Boundary Conditions. Boundary conditions may also be called edge or support conditions. They include *simply supported*, *clamped*, and *free*, as one might expect from similar support conditions in beam theory. In practice, simply supported conditions, and especially clamped conditions, are difficult to enforce. Slight relaxation or yielding at nominally clamped edges may significantly increase deflections and stresses elsewhere.

Boundary conditions can be applied along any edge, straight or curved. For brevity, we limit our examples to a straight edge parallel to the y axis, such as the edge of length b in Fig. 12.3-1. If this edge is *simply supported,* it does not deflect, and $\sigma_x = 0$ along the edge. Therefore

$$w = 0 \quad \text{and} \quad M_x = 0 \tag{12.3-3}$$

If a y-parallel edge is *clamped*, it neither deflects nor rotates. Therefore

$$w = 0 \quad \text{and} \quad \frac{\partial w}{\partial x} = 0 \tag{12.3-4}$$

If a y-parallel edge is *free*, the expected conditions are zero values of M_x, M_{xy}, and Q_x. In 1850, Kirchhoff showed mathematically that three conditions are too many for thin-plate theory, and that the appropriate conditions for a y-parallel free edge are

$$M_x = 0 \quad \text{and} \quad V_x = 0 \quad \text{where} \quad V_x = Q_x + \frac{\partial M_{xy}}{\partial y} \tag{12.3-5}$$

The latter equation was explained in a physical way by Kelvin in 1870, as follows. Consider an edge, such as the y-parallel edge that extends from the corner circled in

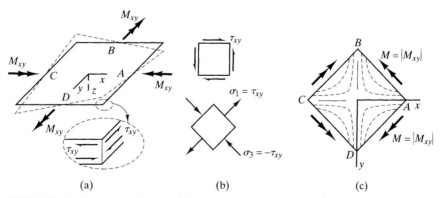

FIGURE 12.3-2 Pure twist of a plate. (a) Twisting moments and deflected shape. Inset shows stresses at a corner. (b) State of stress at a typical point above the midsurface. (c) Equivalent loading on portion $ABCD$. Dashed lines are contour lines of w.

Fig. 12.3-2a. Along a typical length dy, M_{xy} provides a couple $M_{xy}\,dy$. Statically equivalent couple-forces are shown in Fig. 12.3-3a: M_{xy} in a span dy, $M_{xy}+dM_{xy}$ in the next span dy, and $M_{xy}-dM_{xy}$ in the preceding span dy. Where adjacent spans dy meet, along a thickness-direction line, upward and downward forces have the downward force resultant $dM_{xy}=(\partial M_{xy}/\partial y)dy$, as shown in Fig. 12.3-3b. These force resultants are separated by distance dy, and so provide a force per unit length $\partial M_{xy}/\partial y$ (Fig. 12.3-3c). Transverse shear force Q_x may also be present, so net force per unit length is $V_x=Q_x+\partial M_{xy}/\partial y$, which must vanish if the edge is free. Along an x-parallel free edge, boundary conditions are $M_y=0$ and $V_y=0$, where $V_y=Q_y+\partial M_{xy}/\partial x$.

At the corner in Fig. 12.3-3, force M_{xy} on the y-parallel edge is left over by the preceding argument. Another force M_{xy} is contributed by the adjacent x-parallel edge. Thus there appears the concentrated corner force $R=2M_{xy}$. In a state of pure twist, Fig. 12.3-2, M_{xy} is constant, so $\partial M_{xy}/\partial x$ and $\partial M_{xy}/\partial y$ are zero. Then the argument of Fig. 12.3-3 provides only forces R at corners, acting upward at one pair of opposite corners and downward at the other pair. This loading is an alternative to the loading shown in Fig. 12.3-2. According to Saint-Venant's principle, the two loadings produce stress states that are indistinguishable at distances greater than about t units from an edge.

Uniform Thermal Loading. Let $T=2T_o z/t$, which is a linear variation of temperature through the plate, from $-T_o$ on one surface to $+T_o$ on the other, but independent of x and y. Also, let $q=0$. Then, if all edges are clamped, Eqs. 12.2-7 and 12.3-4 are satisfied if $w=0$ throughout the plate. For $w=0$, Eqs. 12.2-6 yield

$$M_x = M_y = -\frac{E\alpha t^2 T_o}{6(1-\nu)} \quad \text{and} \quad M_{xy}=0 \qquad (12.3\text{-}6)$$

These moments prevail at the clamped edges and everywhere else in the plate. Equations 12.2-2 yield surface stresses

$$\bar{\sigma}_x = \bar{\sigma}_y = \pm\frac{E\alpha T_o}{1-\nu} \quad \text{and} \quad \bar{\tau}_{xy}=0 \qquad (12.3\text{-}7)$$

compressive on the heated side, tensile on the cooled side.

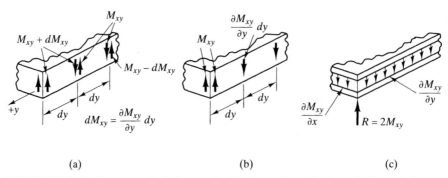

(a)   (b)   (c)

FIGURE 12.3-3 Replacement of twisting couples $M_{xy}\,dy$ on a free edge by statically equivalent transverse forces.

Next, again with $T = 2T_o z/t$, consider the case of all edges free. Equations 12.2-6, 12.2-7, and 12.3-5 are satisfied by a moment-free and stress-free state with bending to a spherical (or shallow paraboloidal) surface, namely $w = c_3(x^2 + y^2)$, where c_3 is a constant. From the first (or the second) of Eqs. 12.2-6, with zero moment because there is zero stress, we obtain $c_3 = -\alpha T_o/t$. In a *circular* plate with $T = 2T_o z/t$, a state of bending to a spherical surface without stress appears whether the edge is free or simply supported. (Remarks near the end of Section 7.10 lead us to anticipate deformation without stress when T is linear in z and there are no supports.)

12.4 SERIES SOLUTION FOR RECTANGULAR PLATES. SELECTED RESULTS

Consider first a plate loaded by uniform lateral pressure q, with all edges simply supported. Let the plate be square ($a = b$ in Fig. 12.4-1a). At center C, bending moment M per unit length can be approximated easily by use of elementary statics, as follows. Cut the plate in half along a diagonal, Fig. 12.4-1b. Assume that M has the uniform value M_{ave} along a diagonal. Summing moments about the diagonal in Fig. 12.4-1b, with $h = a/\sqrt{2}$, we obtain

$$M_{\text{ave}}(a\sqrt{2}) + \frac{qa^2}{2}\left(\frac{h}{3}\right) - 2\frac{qa^2}{4}\left(\frac{h}{2}\right) = 0 \quad \text{hence} \quad M_{\text{ave}} = \frac{qa^2}{24} \quad (12.4\text{-}1)$$

The same M_{ave} appears about the other diagonal. Neither diagonal carries twisting moment because diagonals are axes of symmetry for a square plate. Transformation considerations noted in text following Eq. 12.3-2 show that M is the same in all directions at the center. If $\nu = 0.3$, M_{ave} is 13% lower than the correct M at the center.

Simply Supported Plates: Classical Solution. Let the plate have uniform thickness, and let distributed lateral load q have the form

$$q = q_o \sin\frac{m\pi x}{a} \sin\frac{n\pi y}{b} \quad (12.4\text{-}2)$$

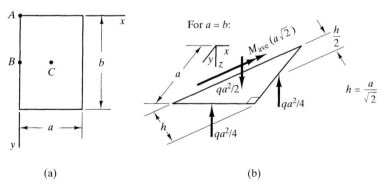

(a)   (b)

FIGURE 12.4-1 (a) Rectangular plate of aspect ratio a/b. (b) Diagram for calculation of average bending moment per unit length along a diagonal of a simply supported square plate.

where m and n are positive integers. This loading has m and n half-waves in x and y directions respectively, and maximum intensity q_o. We tentatively assume that lateral deflection has the form $w = w_o \sin(m\pi x/a) \sin(n\pi y/b)$, where w_o is a constant. This form satisfies the boundary conditions of zero deflection and zero bending moment along all four edges, as required for a simply supported plate. Substituting expressions for q and w into Eq. 12.2-7 with $T = 0$, we obtain w_o in terms of q_o, D, m, n, a, and b. Thus, it turns out that w_o is indeed independent of x and y, as assumed. The expression for w is

$$w = \frac{q_o}{\pi^4 D \left(\frac{m^2}{a^2} + \frac{n^2}{b^2}\right)^2} \sin \frac{m\pi x}{a} \sin \frac{n\pi y}{b} \qquad (12.4\text{-}3)$$

In 1820, Navier generalized the foregoing analysis by writing lateral load q in the form

$$q = \sum_{m=1}^{\infty} \sum_{n=1}^{\infty} q_{mn} \sin \frac{m\pi x}{a} \sin \frac{n\pi y}{b} \qquad (12.4\text{-}4)$$

where m and n are positive integers, and q_{mn} depends on m and n but not on x or y. Various loadings—uniform, linearly varying, and others—can be represented by Eq. 12.4-4 by appropriate choice of q_{mn}. For example, with $q_{mn} = 16q/(\pi^2 mn)$ and m and n restricted to odd integers, Eq. 12.4-4 provides a uniform lateral load of intensity q [12.1]. Lateral deflection w is written as the double sum of $w_{mn} \sin(m\pi x/a) \sin(n\pi y/b)$, and w_{mn} is obtained by substituting expressions for w and q into Eq. 12.2-7 and equating coefficients of like sine products. The resulting expression for w is

$$w = \sum_{m=1}^{\infty} \sum_{n=1}^{\infty} \frac{q_{mn}}{\pi^4 D \left(\frac{m^2}{a^2} + \frac{n^2}{b^2}\right)^2} \sin \frac{m\pi x}{a} \sin \frac{n\pi y}{b} \qquad (12.4\text{-}5)$$

Moments and stresses can be obtained by applying Eqs. 12.2-6 and 12.2-2 to Eq. 12.4-5. The series for q need not be convergent in order to produce convergent series for w, M_x, and M_y. (Use of trigonometric series for *beam* problems is discussed in Section 4.12.)

Some results specific to a simply supported and uniformly loaded square plate with Poisson's ratio 0.3 are as follows [12.1]. Points A, B, and C are shown in Fig. 12.4-1a.

$$\text{For } a = b \text{ and } \nu = 0.3 \begin{cases} w_C = 0.00406 \dfrac{qa^4}{D} & (M_x)_C = 0.0479 qa^2 \\ (M_{xy})_A = -0.0325 qa^2 & (V_x)_B = 0.420 qa \end{cases} \qquad (12.4\text{-}6)$$

At C, $M_x = M_y$ and $M_{xy} = 0$. At each corner, a force R appears, as shown in Fig. 12.3-3c. In the present case, R acts in the same direction as q and has magnitude $0.065 qa^2$. At B, the respective contributions of Q_x and $\partial M_{xy}/\partial y$ to the net line load V_x applied by the support are $0.338 qa$ and $0.082 qa$.

For a simply supported and uniformly loaded rectangular plate with $a \ll b$, the deformed shape is cylindrical except near edges $y = 0$ and $y = b$. Accordingly, most

x-parallel strips of the plate deform like simply supported beams, except that moment $M_y = \nu M_x$ is also present. From beam theory, using D rather than EI, we obtain

$$\text{For} \quad a \ll b: \quad w_C \approx \frac{5qa^4}{384D} \quad (M_x)_C \approx \frac{qa^2}{8} \quad (V_x)_B \approx \frac{qa}{2} \quad (12.4\text{-}7)$$

If $b = 3a$, Eqs. 12.4-7 err by 6.5% for w_C and 5.1% for $(M_x)_C$ (both high) and by 1.0% for $(V_x)_B$ (low).

A simply supported square plate of uniform thickness with concentrated center load P has center deflection $w_C = 0.01160 Pa^2/D$. According to classical thin-plate theory, bending moments beneath a concentrated load are infinite, regardless of how the plate is shaped or supported. However, assumptions 4, 5, and 7 of Section 12.1 are not valid close to a load that is concentrated or nearly so. Approximate formulas are available for stresses near such a load. Details, as well as results for many other plate problems, may be found in [1.6, 1.7, 12.1].

Other Support Conditions. Rectangular plates with clamped edges, or with a mixture of support conditions, can also be analyzed by series methods. None of these solutions are as easily obtained as Eq. 12.4-5. If Fig. 12.4-1 represents a square plate of uniform thickness with all edges clamped, carrying uniform load q, and with $\nu = 0.3$, then [12.1]

$$w_C = 0.00126 \frac{qa^4}{D} \quad (M_x)_B = -0.0513 qa^2 \quad (M_x)_C = 0.0231 qa^2 \quad (12.4\text{-}8)$$

We see that the magnitude of bending moment at the middle of a clamped support is about twice the central bending moment. In this way the plate resembles a uniformly loaded beam with clamped ends. If load q on the square clamped plate is replaced by concentrated center force P, then, for $\nu = 0.3$ [12.1],

$$w_C = 0.00560 \frac{Pa^2}{D} \quad (M_x)_B = -0.1257 P \quad (12.4\text{-}9)$$

Bending moments are infinite at load P according to classical thin plate theory.

12.5 GOVERNING EQUATIONS FOR AXISYMMETRIC PLATES

Polar coordinates $r\theta$ are convenient for analysis of a circular plate. The problem is *axisymmetric* if nothing varies with θ: that is, if geometry, supports, elastic properties, and loads depend only on radial coordinate r. Again we take z as positive downward. The following development for axisymmetric plates parallels that in Section 12.2, and is also based on assumptions stated in Section 12.1.

Shear stresses $\tau_{r\theta}$ and $\tau_{\theta z}$ are zero because every diameter is an axis of symmetry. The only nonzero stresses are those shown in Fig 12.5-1a. Associated bending moments and transverse shear force per unit length are

$$M_r = \int_{-t/2}^{t/2} \sigma_r z \, dz \quad M_\theta = \int_{-t/2}^{t/2} \sigma_\theta z \, dz \quad Q_r = \int_{-t/2}^{t/2} \tau_{rz} \, dz \quad (12.5\text{-}1)$$

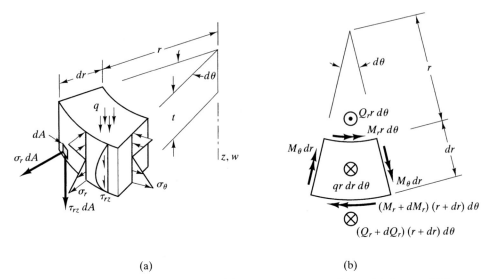

FIGURE 12.5-1 Axisymmetric plate problem. (a) Stresses that act on a differential element. (b) Forces and moments associated with stresses, viewed from above. Symbols ⊗ and ⊙ indicate arrows in the $+z$ (downward) and $-z$ (upward) directions, respectively.

The largest normal stresses appear on plate surfaces. These stresses, which we call $\bar{\sigma}_r$ and $\bar{\sigma}_\theta$, are related to bending moments by the equations

$$\bar{\sigma}_r = \pm \frac{6M_r}{t^2} \qquad \bar{\sigma}_\theta = \pm \frac{6M_\theta}{t^2} \qquad (12.5\text{-}2)$$

The argument that leads to these equations is as stated in text preceding Eqs. 12.2-2, with x and y replaced by r and θ. Transverse shear stress τ_{rz} is largest at the midsurface, where it has magnitude $1.5Q_r/t$. It is usually small in comparison with flexural stresses.

In general, M_r, M_θ, and Q_r are functions of r, and so have increments dM_r, dM_θ, and dQ_r over a distance dr. Forces and moments per unit length, when multiplied by lengths over which they act, provide the net forces and moments shown in Fig. 12.5-1b. Relative to a radial line through the center of the element, each moment $M_\theta\,dr$ has a θ-direction component proportional to $\sin(d\theta/2)$. The sum of moments about any line must vanish. Choosing the line tangent to the arc at radius $r + dr$, we obtain the equilibrium equation

$$(M_r + dM_r)(r + dr)d\theta - M_r r\,d\theta - (Q_r r\,d\theta)dr \\ - 2\left(M_\theta\,dr\sin\frac{d\theta}{2}\right) + qr\,dr\,d\theta\frac{dr}{2} = 0 \qquad (12.5\text{-}3)$$

Some terms cancel, and $\sin(d\theta/2) \approx d\theta/2$. Compared with terms that remain, the term that contains distributed lateral load q is of higher order and can be discarded. Thus Eq. 12.5-3 reduces to

$$\frac{M_r - M_\theta}{r} + \frac{dM_r}{dr} = Q_r \qquad \text{or} \qquad \frac{d}{dr}(rM_r) - M_\theta = rQ_r \qquad (12.5\text{-}4)$$

Section 12.5 Governing Equations for Axisymmetric Plates

An arbitrary point not on the midsurface has displacements u radially and w laterally. As described in connection with Fig. 12.2-2, but with x replaced by r, radial displacement of an arbitrary point is $u = -z(dw/dr)$. Relations between strains and lateral displacement w are

$$\epsilon_r = \frac{du}{dr} = -z\frac{d^2w}{dr^2} \quad \text{and} \quad \epsilon_\theta = \frac{u}{r} = -\frac{z}{r}\frac{dw}{dr} \tag{12.5-5}$$

The argument that provides the formula $\epsilon_\theta = u/r$ is stated in connection with Fig. 1.4-2b. Strains of Eqs. 12.5-5 may be substituted into the stress-strain relation for isotropy and linear elasticity, Eqs. 7.1-3, and the resulting stresses substituted into Eqs. 12.5-1. These manipulations yield the moment expressions

$$M_r = -D\left(\frac{d^2w}{dr^2} + \frac{\nu}{r}\frac{dw}{dr}\right) - \frac{E\alpha}{1-\nu}\int_{-t/2}^{t/2} Tz\,dz \tag{12.5-6a}$$

$$M_\theta = -D\left(\frac{1}{r}\frac{dw}{dr} + \nu\frac{d^2w}{dr^2}\right) - \frac{E\alpha}{1-\nu}\int_{-t/2}^{t/2} Tz\,dz \tag{12.5-6b}$$

where $D = Et^3/[12(1-\nu^2)]$. Thermal terms are as described following Eq. 12.2-6.

For an isotropic plate of uniform thickness, D is independent of r, and substitution of Eqs. 12.5-6 into Eq. 12.5-4 yields

$$\frac{d^3w}{dr^3} + \frac{1}{r}\frac{d^2w}{dr^2} - \frac{1}{r^2}\frac{dw}{dr} = -\frac{1}{D}\left[Q_r + \frac{E\alpha}{1-\nu}\int_{-t/2}^{t/2}\frac{dT}{dr}z\,dz\right] \tag{12.5-7}$$

This equation can also be written in the following form, which is more convenient for integration.

$$\frac{d}{dr}\left[\frac{1}{r}\frac{d}{dr}\left(r\frac{dw}{dr}\right)\right] = -\frac{1}{D}\left[Q_r + \frac{E\alpha}{1-\nu}\int_{-t/2}^{t/2}\frac{dT}{dr}z\,dz\right] \tag{12.5-8}$$

To integrate Eq. 12.5-8, load terms Q_r and T must be known. An expression for Q_r in terms of a distributed lateral load of intensity q is obtained by summing z-direction forces that act on a central disk of radius r:

$$2\pi r Q_r + \int_0^{2\pi}\int_b^r qr_r\,dr_r\,d\theta = 0 \quad \text{yields} \quad Q_r = -\frac{1}{r}\int_b^r qr_r\,dr_r \tag{12.5-9}$$

where b is the radius of a central hole (if present), and r_r is a radius in the range $b < r_r < r$. If instead of q the load is central force P in the $+z$ direction, the equilibrium argument yields $Q_r = -P/2\pi r$.

Boundary Conditions. Boundary conditions for an axisymmetric circular plate are as follows. These conditions pertain to the (outer) edge of a solid plate, or to inner and outer edges of a plate with a central hole.

1. Around a simply supported edge, $w = 0$ and $M_r = 0$.
2. Around a clamped edge, $w = 0$ and $dw/dr = 0$.
3. Around a free edge, $Q_r = 0$ and $M_r = 0$.

The latter condition is less complicated than Eq. 12.3-5 because there is no twisting moment $M_{r\theta}$ in an axisymmetric state.

12.6 SOLUTION OF SIMPLE AXISYMMETRIC PROBLEMS

The procedure here described is that of integrating Eq. 12.5-8, followed by evaluation of constants of integration. The resulting expression for $w = w(r)$ then provides moments according to Eqs. 12.5-6 and surface stresses according to Eq. 12.5-2. Here we consider plates of uniform thickness, and only mechanical loads ($T = 0$ in Eqs. 12.5-6, 12.5-7, and 12.5-8).

Edge Moment Loading. The plate of Fig. 12.6-1a is simply supported around its edge $r = a$, where moment M_a per unit of circumferential length is applied. Since $q = 0$, Eq. 12.5-9 yields $Q_r = 0$. One integration of Eq. 12.5-8 yields

$$\frac{1}{r}\frac{d}{dr}\left(r\frac{dw}{dr}\right) = C_1 \quad \text{or} \quad \frac{d}{dr}\left(r\frac{dw}{dr}\right) = C_1 r \quad (12.6\text{-}1)$$

where C_1 is a constant of integration. The second and third integrations yield, respectively,

$$\frac{dw}{dr} = \frac{C_1 r}{2} + \frac{C_2}{r} \quad \text{and} \quad w = \frac{C_1 r^2}{4} + C_2 \ln r + C_3 \quad (12.6\text{-}2)$$

To evalute integration constants C_1, C_2, and C_3, we first note that since there is no central hole, w and dw/dr must remain finite at $r = 0$. Therefore $C_2 = 0$. Two additional conditions apply at $r = a$: that $w = 0$ and $M_r = M_a$. Thus, from Eqs. 12.6-2 and 12.5-6,

$$0 = \frac{C_1 a^2}{4} + C_3 \quad \text{and} \quad M_a = -\frac{(1 + \nu)D}{2} C_1 \quad (12.6\text{-}3)$$

Solving for C_1 and C_3, and substituting them into Eq. 12.6-2, we obtain

$$w = \frac{M_a(a^2 - r^2)}{2(1 + \nu)D} \quad (12.6\text{-}4)$$

Equation 12.6-4 describes bending to a spherical (or shallow paraboloidal) surface, with a uniform moment field $M_r = M_\theta = M_a$ throughout the plate.

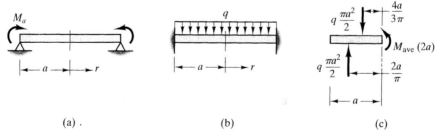

FIGURE 12.6-1 (a) Simply supported circular plate with edge moment M_a (uniform around the entire circumference). (b) Clamped circular plate wih uniformly distributed lateral load q. (c) Free body diagram of half a simply supported circular plate.

Uniform q, Clamped Edge. Uniform load q acts downward, and the plate has no central hole (Fig. 12.6-1b). Equation 12.5-9 yields $Q_r = -qr/2$, and integration of Eq. 12.5-8 yields

$$w = \frac{qr^4}{64D} + \frac{C_1 r^2}{4} + C_2 \ln r + C_3 \tag{12.6-5}$$

Again $C_2 = 0$ so that w remains finite at $r = 0$. Constants C_1 and C_3 are evaluated from the conditions that at $r = a$, $w = 0$ and $dw/dr = 0$. Equation 12.6-5 becomes

$$w = \frac{q(a^2 - r^2)^2}{64D} \tag{12.6-6}$$

Then, from Eqs. 12.5-6, we obtain bending moments per unit length

$$M_r = \frac{q}{16}[(1+\nu)a^2 - (3+\nu)r^2] \quad M_\theta = \frac{q}{16}[(1+\nu)a^2 - (1+3\nu)r^2] \tag{12.6-7}$$

The bending moment of largest magnitude is M_r at the edge, $r = a$, where $M_r = -qa^2/8$. At the center, where $r = 0$, $M_r = M_\theta = (1+\nu)qa^2/16$.

Uniform q, Simply Supported Edge. The solution for this case is easily obtained by superposition, using the foregoing two examples. All we need do is remove edge moment $M_r = -qa^2/8$ from the clamped plate (Fig. 12.6-1b) by applying an equal edge moment in the opposite sense (Fig. 12.6-1a). Accordingly we set $M_a = qa^2/8$ in Eq. 12.6-4, then add Eqs. 12.6-4 and 12.6-6, with the result

$$w = \frac{q(a^2 - r^2)}{64D}\left[\frac{5+\nu}{1+\nu}a^2 - r^2\right] \tag{12.6-8}$$

Similarly, adding $qa^2/8$ to the moments of Eqs. 12.6-7, we obtain for the simply supported and uniformly loaded circular plate

$$M_r = \frac{q}{16}[(3+\nu)(a^2 - r^2)] \quad M_\theta = \frac{q}{16}[(3+\nu)a^2 - (1+3\nu)r^2] \tag{12.6-9}$$

As a check, an estimate of the maximum bending moment (at $r = 0$) can be obtained from an equilibrium argument, like that used in connection with Fig. 12.4-1b. Half of the plate is isolated by making a cut along a diameter, as shown in a view parallel to the cut in Fig. 12.6-1c. Resultant forces $q(\pi a^2/2)$ come from pressure on top and line load provided by the support. Equilibrium of moments about the diametral cut yields

$$M_{\text{ave}}(2a) - q\frac{\pi a^2}{2}\left(\frac{2a}{\pi} - \frac{4a}{3\pi}\right) = 0 \quad \text{hence} \quad M_{\text{ave}} = \frac{qa^2}{6} \tag{12.6-10}$$

(Also, integration of $M_\theta \, dr$ for $0 < r < a$ yields $qa^3/6$, or an average M_θ of $qa^2/6$.) If $\nu = 0.3$, M_{ave} is 19% lower than the correct bending moment at the center.

Large Deflections. If lateral deflection is not small, it becomes necessary to account for the effect of midsurface stretching, which we have thus far ignored. The effect introduces nonlinearities, so that lateral deflection is not directly proportional to lateral load. In what follows we discuss the nature of the problem and provide selected results. Details of analysis and further results appear in other references [1.6, 4.1, 12.1].

Physically, what happens is that lateral load is not resisted entirely by bending. Some is resisted by membrane action. Therefore lateral deflection is less than predicted by classical plate theory, which accounts only for bending resistance. Stresses come not only from bending, but from membrane stresses as well. For membrane action to be negligibly small, lateral deflection w must nowhere exceed roughly half the plate thickness, the exact amount being dependent on the accuracy required and the particular circumstances of plate shape, loading, and support.

Figure 12.6-2a shows that small to moderately large lateral displacement produces radial strain $\epsilon_1 = (dw/dr)^2/2$ at the midsurface. Radial displacement u at the midsurface contributes additional radial strain $\epsilon_2 = du/dr$. Total radial strain at the midsurface is $\epsilon_{rm} = \epsilon_1 + \epsilon_2 = (dw/dr)^2/2 + du/dr$. Total circumferential strain at the midsurface is simply $\epsilon_{\theta m} = u/r$. Points *not* on the midsurface have strains ϵ_{rm} and $\epsilon_{\theta m}$ *plus* flexural strains stated in Eq. 12.5-5. If lateral deflections are very large, the expression for ϵ_{rm} must contain the additional term $(du/dr)^2/2$ [12.1].

Approximate solutions for lateral deflection and largest normal stress can be obtained by truncating large-deflection series solutions at a finite number of terms. For example, consider a uniformly loaded circular plate whose edge is simply supported ($w = 0$, but rotation dw/dr and radial displacement u both unrestrained). With w_0 the maximum lateral deflection (at $r = 0$) and σ_0 the maximum stress on the plate surface at $r = 0$, including both bending and membrane contributions, from [12.1] with $\nu = 0.3$,

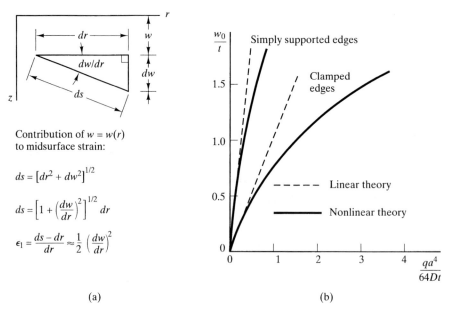

FIGURE 12.6-2 Large deflection of a circular plate. (a) Contribution of lateral deflection w to radial strain at the midsurface. (b) Center deflection w_0 in a uniformly loaded plate.

$$w_0 \approx \frac{qa^4}{64D}\left[\frac{4.08}{1 + 0.262\left(\dfrac{w_0}{t}\right)^2}\right] \qquad \sigma_0 \approx \frac{Et^2}{a^2}\left[1.78 + 0.295\frac{w_0}{t}\right]\frac{w_0}{t} \qquad (12.6\text{-}11)$$

When load, dimensions, and elastic properties are prescribed, one obtains w_0/t from the first equation, then σ_0 from the second equation.

Deflection versus load is plotted in dimensionless form in Fig. 12.6-2b, for both simply supported and clamped edges. Here the clamped edge condition means that $u = w = dw/dr = 0$ at the edge. Nonlinearity is less pronounced with simply supported edges because radial stress vanishes at the edge, $r = a$. However, nonlinearity is still present because the deformed shape is a shallow dish, which cannot be obtained from an initially flat plate without straining its midsurface. Since a simple support provides no radial restraint, circumferential membrane stresses must be self-equilibrating on any diametral cross section. Accordingly, membrane tensile stress near the center of a simply supported plate must be accompanied by membrane compressive stress near the support. Compressive stress raises the possibility of buckling, with portions of the plate edge lifting off the support.

12.7 SELECTED SOLUTIONS FOR AXISYMMETRIC PLATES

The following short catalog pertains to axisymmetric conditions in homogeneous, isotropic, and linearly elastic circular plates of uniform thickness. Poisson's ratio is taken as $\nu = 0.3$. Results are based on classical small-deflection theory, which predicts infinite bending moment and infinite stress beneath a concentrated load. More realistic stresses can be determined from special formulas [1.6, 12.1].

NOTATION: cl = clamped edge, ss = simply supported edge

Case 1: center force P, cl.

At $r = 0$: $\quad w = 0.0199\dfrac{Pa^2}{D}$

$\quad M_r = M_\theta = \infty$

At $r = a$: $\quad M_r = -0.0796P$

$\quad M_\theta = -0.0239P$

Case 2: center force P, ss.

At $r = 0$: $\quad w = 0.0505\dfrac{Pa^2}{D}$

$\quad M_r = M_\theta = \infty$

At $r = a$: $\quad \dfrac{dw}{dr} = -0.0612\dfrac{Pa}{D}$

$\quad M_\theta = 0.0557P$

Case 3: uniform pressure q, cl.

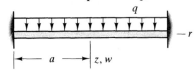

At $r = 0$: $w = 0.0156 \dfrac{qa^4}{D}$

$M_r = M_\theta = 0.0813 qa^2$

At $r = a$: $M_r = -0.125 qa^2$

$M_\theta = -0.0375 qa^2$

Case 4: uniform pressure q, ss.

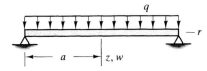

At $r = 0$: $w = 0.0637 \dfrac{qa^4}{D}$

$M_r = M_\theta = 0.206 qa^2$

At $r = a$: $\dfrac{dw}{dr} = -0.0962 \dfrac{qa^3}{D}$

$M_\theta = 0.0875 qa^2$

Case 5: moments M_1 and M_2 distributed around edges, ss.

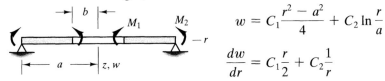

$w = C_1 \dfrac{r^2 - a^2}{4} + C_2 \ln \dfrac{r}{a}$

$\dfrac{dw}{dr} = C_1 \dfrac{r}{2} + C_2 \dfrac{1}{r}$

$C_1 = -\dfrac{2(a^2 M_2 - b^2 M_1)}{(1+\nu)(a^2 - b^2) D}$ $\quad C_2 = -\dfrac{a^2 b^2 (M_2 - M_1)}{(1-\nu)(a^2 - b^2) D}$

If $b = 0$, then $M_r = M_\theta = M_2$ throughout the plate, and, with $\nu = 0.3$,

At $r = 0$: $w = 0.385 \dfrac{M_2 a^2}{D}$ $\quad$ At $r = a$: $\dfrac{dw}{dr} = -0.769 \dfrac{M_2 a}{D}$

Case 6: line load V (force per unit length), ss.

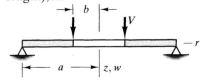

$w = g_1 \dfrac{Va^3}{D}$

$\dfrac{dw}{dr} = g_2 \dfrac{Va^2}{D}$

$M_{\max} = M_{\theta b} = g_3 V a$

b/a	0.1	0.3	0.5	0.7	0.9
g_1 at $r = b$	0.0364	0.1266	0.1934	0.1927	0.0938
g_2 at $r = b$	−0.0371	−0.2047	−0.4262	−0.6780	−0.9532
g_2 at $r = a$	−0.0418	−0.1664	−0.3573	−0.6119	−0.9237
g_3 at $r = b$	0.3374	0.6210	0.7757	0.8814	0.9638

Case 7: uniform pressure q, ss.

$$w = g_4 \frac{qa^4}{D}$$

$$\frac{dw}{dr} = g_5 \frac{qa^3}{D}$$

$$M_{\max} = M_{\theta b} = g_6 qa^2$$

b/a	0.1	0.3	0.5	0.7	0.9
g_4 at $r=b$	0.0687	0.0761	0.0624	0.0325	0.0048
g_5 at $r=b$	−0.0436	−0.1079	−0.1321	−0.1130	−0.0491
g_5 at $r=a$	−0.0986	−0.1120	−0.1201	−0.1041	−0.0477
g_6 at $r=b$	0.3965	0.3272	0.2404	0.1469	0.0497

12.8 SUPERPOSITION SOLUTIONS FOR AXISYMMETRIC PLATES

In the present context, "superposition" means solving a problem by regarding it as the sum of two or more component problems. The motivation is to obtain a solution by making use of existing solutions that are conveniently available. The same procedure is commonly used in beam problems, where a complicated case is treated as the sum of simpler cases, for each of which the solution is known. We have already used superposition in Section 12.6, where solutions for the problems of Figs. 12.6-1a and 12.6-1b are combined to solve the problem of a simply supported and uniformly loaded circular plate (Eqs. 12.6-8 and 12.6-9). In the following additional examples, a verbal summary of the solution procedure is presented, followed by detailed calculations. Poisson's ratio is taken as 0.3 in all cases.

EXAMPLE 1

The annular plate in the upper part of Fig. 12.8-1a is simply supported around its inner edge and is loaded by uniform lateral pressure q. Determine the maximum deflection.

The given problem can be regarded as the sum of the latter two problems in Fig. 12.8-1a, which are Cases 6 and 7 of Section 12.7. Thus, we imagine that line load V in Case 6 is reversed and is made large enough to carry the entire load produced by q. Combining this case with Case 7 unloads the outer edge. At this time the inner edge is as far above the outer edge as the outer edge in the upper part of Fig. 12.8-1a is below the inner edge.

Line load V at $r = 0.7a$ that will support all the force produced by q has magnitude

$$2\pi(0.7a)V = \pi[a^2 - (0.7a)^2]q \quad \text{hence} \quad V = 0.3643qa \quad (12.8\text{-}1)$$

Part of the deflection of the inner edge relative to the outer edge comes from Case 6, with $V = -0.3643qa$. An additional contribution comes from Case 7. Thus, by use of tabulated data and superposition, we obtain the net inner-edge deflection w_b, measured relative to the outer edge:

$$w_b = 0.0325 \frac{qa^4}{D} + 0.1927 \frac{(-0.3643qa)a^3}{D} = -0.0377 \frac{qa^4}{D} \quad (12.8\text{-}2)$$

The required result—the (downward) displacement of the outer edge relative to the inner edge—is the magnitude of w_b.

EXAMPLE 2

The annular plate in the upper part of Fig. 12.8-1b is clamped around its outer edge and is loaded by uniform line load V around its inner edge. Determine the maximum flexural stress.

The given problem can be regarded as the sum of the latter two problems in Fig. 12.8-1b, which are Cases 5 and 6 of Section 12.7. Moment M_2 must have such magnitude that the outer edge does not rotate under the combined action of V in Case 6 and M_2 in Case 5. Thus we determine M_2 in terms of V and can then investigate stresses, again using tabulated data for Cases 5 and 6.

From Case 6, $dw/dr = -0.3573Va^2/D$ at $r = a$ if the edge at $r = a$ is simply supported. Adding this rotation to dw/dr from Case 5 and equating the total to zero, we have

$$-0.3573\frac{Va^2}{D} - 2.0513\frac{M_2}{D}\frac{a}{2} - 0.4762a^2\frac{M_2}{D}\frac{1}{a} = 0 \qquad (12.8\text{-}3)$$

from which $M_2 = -0.2379Va$. This moment is shown acting in its actual direction in the lowest part of Fig. 12.8-1b. M_2 may be the bending moment of largest magnitude, but we must also consider M_θ at $r = b$. At $r = b$, with $T = 0$ and $M_r = 0$, Eq. 12.5-6a yields $(d^2w/dr^2)_b = -(\nu/b)(dw/dr)_b$. Equation 12.5-6b then yields, at $r = b$,

$$M_{\theta b} = -\frac{1-\nu^2}{b}D\left(\frac{dw}{dr}\right)_b \qquad (12.8\text{-}4)$$

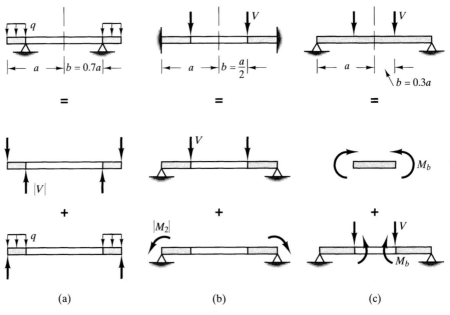

(a)   (b)   (c)

FIGURE 12.8-1 Three problems of axisymmetric plates, each solved by superposing solutions of other problems.

Section 12.9   Plastic Collapse Analysis of Plates   399

At $r = b$, combining Cases 5 and 6, we have

$$\left(\frac{dw}{dr}\right)_b = -2.0513\frac{M_2}{D}\frac{a/2}{2} - 0.4762a^2\frac{M_2}{D}\frac{1}{a/2} - 0.4262\frac{Va^2}{D} \quad (12.8\text{-}5)$$

With $M_2 = -0.2379Va$, $b = a/2$, and $\nu = 0.3$, Eqs. 12.8-4 and 12.8-5 yield

$$\left(\frac{dw}{dr}\right)_b = -0.07762\frac{Va^2}{D} \quad \text{and} \quad M_{\theta b} = 0.1413Va \quad (12.8\text{-}6)$$

So M_2 at $r = a$ is after all the bending moment of largest magnitude. From Eq. 12.5-2, the largest tensile stress is $\bar{\sigma}_r = 6|M_2|/t^2$ on top of the plate at $r = a$.

EXAMPLE 3

The solid plate in the upper part of Fig. 12.8-1c is simply supported and is loaded by uniform line load V around a circle of radius $b = 0.3a$. Determine the maximum deflection.

The given problem can be regarded as the sum of the latter two problems in Fig. 12.8-1c, which are described by Eq. 12.6-4 and Cases 5 and 6 of Section 12.7. Moment M_b must have such magnitude that the inner solid plate and the outer annular plate have the same rotation at $r = b$ when M_b and V are applied to the component cases as shown. When M_b is known, center deflection can be calculated by adding deflections of the component cases.

From Eq. 12.6-4, with M_a replaced by M_b, $\nu = 0.3$, and $r = 0.3a$, we obtain $dw/dr = -0.2308 M_b a/D$ at the edge of the inner solid plate. In Case 5 of Section 12.7, with $M_1 = M_b$, $M_2 = 0$, $\nu = 0.3$, and $b = 0.3a$, constants are $C_1 = 0.1522 M_b/D$ and $C_2 = 0.1413 M_b a^2/D$. Load V also contributes to dw/dr in the outer annular plate (Case 6, with $b = 0.3a$). To match slopes where plates meet, we write

$$-0.2308\frac{M_b a}{D} = 0.1522\frac{M_b}{D}\frac{0.3a}{2} + 0.1413\frac{M_b a^2}{D}\frac{1}{0.3a} - 0.2047\frac{Va^2}{D} \quad (12.8\text{-}7)$$

from which $M_b = 0.2825 Va$. This bending moment prevails throughout the region $0 < r < 0.3a$ and is the largest bending moment in the entire plate. The deflection at $r = 0$ is $w_{\max}$. It is the algebraic sum of w contributions from Eq. 12.6-4, Case 5, and Case 6. With $\nu = 0.3$,

$$w_{\max} = \frac{M_b (0.3a)^2}{2.6D} + 0.1522\frac{M_b}{D}\frac{(0.3a)^2 - a^2}{4}$$

$$+ 0.1413\frac{M_b a^2}{D}\ln 0.3 + 0.1266\frac{Va^3}{D} \quad (12.8\text{-}8)$$

With $M_b = 0.2825 Va$, Eq. 12.8-8 yields $w_{\max} = 0.0785 Va^3/D$.

12.9   PLASTIC COLLAPSE ANALYSIS OF PLATES

Plastic collapse analysis provides a way to estimate the load-carrying capacity of reinforced concrete slabs and ductile metal plates. The load that causes plastic collapse creates zones of plastic material such that subsequent deflection of the plate is associated with deformation of material in these zones, with little or no enlargement of the zones or increase in externally applied load. This kind of behavior in beams is described in Chapter 11, whose early sections the reader may wish to review.

Chapter 12 Plate Bending

In plastic collapse analysis it is not necessary to use the initial elastic part of the stress-strain relation. We will assume that the uniaxial stress-strain relation is flat-topped at a definite yield stress σ_Y (Fig. 11.2-1). We will also assume that deflections are small, even in the plastic collapse mode. With these assumptions we ignore possible strain hardening of the material and possible membrane stresses due to large deflections (Section 12.6). Both of these effects increase the load-carrying capacity. Therefore our analyses are likely to underestimate the actual collapse load.

Imagine that, on some cross section of an isotropic and ductile plate, all material through the plate thickness has yielded, with σ_Y the normal stress of greatest magnitude, as shown in Fig. 12.9-1a. The associated "fully plastic" bending moment per unit length, M_{fp}, can be calculated from Eq. 12.2-1, with $\sigma = \sigma_Y$ for $z > 0$ and $\sigma = -\sigma_Y$ for $z < 0$. This calculation appears in Fig. 12.9-1a. By using a yield criterion, M_{fp} can be related to principal moments M_1 and M_2, where $M_1 \geq M_2$. Principal moments are maximum and minimum moments, and act on a plate element of such orientation that twisting moment is zero (Fig. 12.9-1c). The maximum shear stress yield criterion, Eq. 3.3-1, assumes the following form when applied to plate bending. Failure by plastic flow is predicted when

$$\max(|M_1|, |M_2|, M_1 - M_2) - M_{fp} \geq 0 \quad \text{where} \quad M_{fp} = \frac{t^2}{4}\sigma_Y \quad (12.9\text{-}1)$$

That is, the principal moment of greatest magnitude, or the difference of principal moments if they are of opposite sign, is equal to M_{fp} when yielding spans the entire plate thickness. (As examples of principal moments, $M_1 = M_x$ and $M_2 = M_y$ in Fig. 12.3-1, and $M_1 = M_{xy}$ and $M_2 = -M_{xy}$ in Fig. 12.3-2.) The von Mises yield criterion, Eq. 3.3-2, can also be stated in terms of moments in a plate, but collapse analysis is much simpler when Eq. 12.9-1 is used.

Rectangular Plates. Let a square plate be simply supported on all four edges and loaded by uniform lateral pressure q (Fig. 12.9-2a). When q reaches its plastic collapse value q_c, bands of yielded material called "yield lines" appear along diagonals of the plate. Due to symmetry, no shear stress is present along a diagonal. Hence $M_{xy} = 0$

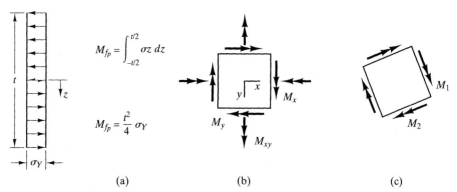

FIGURE 12.9-1 (a) Stress distribution through the plate thickness that provides fully plastic bending moment M_{fp}. (b) Plate element in xy coordinates. (c) Principal element and principal bending moments.

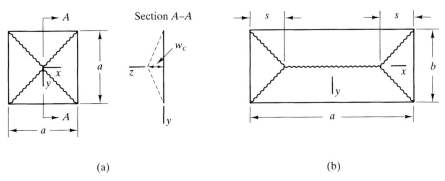

FIGURE 12.9-2 Square and rectangular plates. Irregular lines denote yield lines.

along diagonals, and moment vectors parallel and normal to diagonals are principal moments M_1 and M_2. They have the same sign, since the deformed shape of the plate is not saddle-shaped as it is in Fig. 12.3-2. Therefore the moment vector parallel to a diagonal is $M_1 = M_{fp}$.

An upper bound on collapse load q_c can be obtained from a virtual work argument, using an assumed collapse mechanism, as described for beams and frames in Chapter 11. Consider a simply supported square plate of uniform thickness, and the collapse mechanism shown in Fig. 12.9-2a. Elastic deformation accumulated prior to collapse is not shown: Center deflection w_c is regarded as a virtual displacement that occurs *after* M_{fp} has developed along yield lines. Thus each of four triangular portions of the plate rotates as a rigid body about its edge support, through the (small) angle $\theta = w_c/(a/2)$. Force resultant $F = q_c a^2/4$ on a triangular portion acts at its centroid, which has displacement $w_c/3$, and so does work $F(w_c/3)$. Work absorbed by plastic moments M_{fp} is most easily calculated by projecting moment resultants on axes of rotation. For a single triangular portion, the projected moment is $M_{fp}a$, which is associated with rotation θ. For equilibrium, net virtual work must vanish. Thus, summimg over all four triangular portions, we obtain the upper bound solution for collapse pressure q_c.

$$4\left[\frac{q_c a^2}{4}\left(\frac{w_c}{3}\right)\right] - 4\left[M_{fp}a\left(\frac{w_c}{a/2}\right)\right] = 0 \quad \text{hence} \quad q_c = \frac{24 M_{fp}}{a^2} \quad (12.9\text{-}2)$$

In Chapter 11, calculation of work absorbed by M_{fp} is accomplished by multiplying M_{fp} by the relative rotation between members joined by a plastic hinge. In the present plate example, rotation θ has component $\theta/\sqrt{2}$ parallel to a diagonal. Relative rotation across a diagonal is twice this amount. Net moment along a half-diagonal is $M_{fp}(\sqrt{2}a/2)$. The reader can easily show that this viewpoint again leads to $q_c = 24 M_{fp}/a^2$.

A lower bound for q_c is obtained from a moment field that satisfies the equilibrium equation, Eq. 12.2-4, and nowhere exceeds the yield citerion (see Section 11.7). That is, for the maximum shear stress criterion, the left side of Eq. 12.9-1 cannot be positive at any point on the plate. These requirements are met by the *elastic* solution with the maximum bending moment increased to M_{fp}. Thus, from Eq. 12.4-6, a lower bound is $q_c = M_{fp}/(0.0479a^2) = 20.9 M_{fp}/a^2$.

Imagine next that the simply supported plate is not square (Fig. 12.9-2b). Various collapse mechanisms might be investigated. Of these, the mechanism that provides the smallest q_c is most nearly correct. A mechanism like that for the square plate would have four yield lines, each extending from the center of the plate to a corner. An alternative mechanism, shown in Fig. 12.9-2b, is more realistic. Distance s is unknown. A convenient value such as $s = b/2$ might be assumed, but the correct s makes q_c a minimum.

If all edges are clamped, yield lines also appear along edges, with the sign of M_{fp} opposite to its sign in the yield lines shown in Fig. 12.9-2. For a square plate, the effect of these added yield lines is to change the multiplier of the second bracketed term in Eq. 12.9-2 from 4 to 8. Therefore $q_c = 48M_{fp}/a^2$ is the (upper bound) solution associated with this mechanism in a clamped square plate.

Let q_Y be the uniformly distributed lateral load that initiates yielding, and let us calculate the ratio q_c/q_Y. The bending moment that initiates yielding on plate surfaces is $M_Y = 2M_{fp}/3$ (see Eq. 12.9-1, and Eq. 12.2-2 with, say, $\bar{\sigma}_x = \sigma_Y$ and $M_x = M_Y$). For a simply supported square plate, from Eq. 12.4-6, $q_Y = M_Y/(0.0479a^2) = 13.9M_{fp}/a^2$. Therefore, using the upper-bound collapse load, $q_c/q_Y = 24/13.9 = 1.73$. Similarly, for a clamped square plate, from Eq. 12.4-8, $q_Y = M_Y/(0.0513a^2) = 13.0M_{fp}/a^2$. Therefore $q_c/q_Y = 48/13.0 = 3.69$.

Circular Plates. Consider a simply supported circular plate of uniform thickness, loaded uniformly by lateral pressure q. A possible collapse mechanism is shown in Fig. 12.9-3a. This "conical mode" is a virtual displacement that appears after collapse load q_c is reached; it does not include previously accumulated elastic deformation. In effect, every radial line is a yield line. We will assume, subject to verification later, that $M_\theta = M_{fp}$. In Fig. 12.9-3b, a typical sector rotates about the support at $r = a$. Work absorbed by plastic action is equal to the rotation times the moment projected on arc length $a\,d\theta$. A virtual work equation analogous to Eq. 12.9-2 provides the upper bound solution.

$$q_c \frac{a^2\,d\theta}{2}\left(\frac{w_c}{3}\right) - M_{fp} a\,d\theta\left(\frac{w_c}{a}\right) = 0 \quad \text{hence} \quad q_c = \frac{6M_{fp}}{a^2} \quad (12.9\text{-}3)$$

Next we obtain a lower bound solution for q_c. From this discussion we will also show that $M_\theta = M_{fp}$, as assumed in the upper bound solution. Transverse shear force at

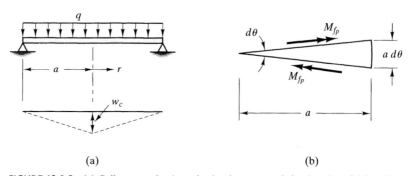

FIGURE 12.9-3 (a) Collapse mechanism of a simply supported circular plate. (b) Bending moments that act on a typical sector of the plate.

arbitrary r is $Q_r = -(q_c \pi r^2 / 2\pi r) = -q_c r/2$. With $M_\theta = M_{fp}$, the equilibrium equation (Eq. 12.5-4) becomes

$$\frac{d}{dr}(rM_r) = M_{fp} - \frac{q_c r^2}{2} \quad \text{hence} \quad M_r = M_{fp} - \frac{q_c r^2}{6} + \frac{C_1}{r} \quad (12.9\text{-}4)$$

where C_1 is a constant of integration. M_r remains finite at $r = 0$; therefore $C_1 = 0$. Also, because the plate is simply supported, $M_r = 0$ at $r = a$; therefore

$$q_c = \frac{6M_{fp}}{a^2} \quad \text{and} \quad M_r = M_{fp}\left(1 - \frac{r^2}{a^2}\right) \quad (12.9\text{-}5)$$

Since upper and lower bound solutions for q_c agree, the collapse load $q_c = 6M_{fp}/a^2$ is exact (within the limits of assumptions made). With $M_\theta = M_{fp}$, the second of Eqs. 12.9-5 shows that M_r is everywhere less than M_θ and both are positive. Accordingly Eq. 12.9-1 provides $M_\theta = M_{fp}$, as was assumed in the upper bound solution.

If a circular plate of uniform thickness has a clamped edge, an upper bound solution is easily obtained. All that is needed is to add moment M_{fp} along the edge of length $a\,d\theta$ in Fig. 12.9-3b. This addition doubles the virtual work absorbed, and we obtain $q_c = 12M_{fp}/a^2$ for a uniformly loaded clamped plate. A lower bound solution is not so simple. Over the outer portion of the plate, M_θ and M_r have opposite sign (Fig. 12.9-4). The yield criterion is $M_\theta = M_{fp}$ for $0 < r < b$ and $M_\theta - M_r = M_{fp}$ for $b < r < a$. The inner portion $0 < r < b$ is in effect a simply supported plate of radius b. Therefore $q_c = 6M_{fp}/b^2$ (from Eq. 12.9-5). An expression for q_c in the outer portion $b < r < a$ is obtained by integrating Eq. 12.5-4. The two q_c expressions must be equal, thus providing $b = 0.730a$, and finally $q_c = 11.3M_{fp}/a^2$ as the lower bound solution for a clamped plate. This equation is in fact the exact solution (within the limits of assumptions made).

Remarks. In the foregoing examples we have assumed that the material is isotropic. This assumption is approximately true for a concrete slab if it is "isotropically reinforced," since the plastic condition is associated with yielding of reinforcement bars. Another consideration for concrete is that a localized or concentrated load may cause "punching failure" associated with large transverse shear stress. In this context, localized loads include forces applied by column supports.

Useful rules for upper bound calculations include the following. Yield lines are usually *straight* lines, along which M_{fp} is constant. In the mechanism that provides the smallest upper bound, transverse shear force and twisting moment are zero along yield

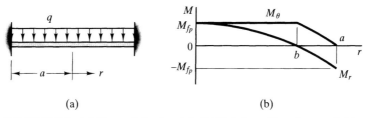

FIGURE 12.9-4 (a) Clamped circular plate. (b) Variation with r of moments M_r and M_θ when collapse load q_c is applied.

lines. Yield lines lie along fixed edges and pass over column supports. Yield lines form a symmetric pattern if there is symmetry of geometry, supports, elastic properties, and loading.

For more complete discussion of plastic action in plates, the reader is urged to consult additional references, such as [11.4, 12.3–12.5].

PROBLEMS

12.2-1. At a certain location in a 6 mm-thick plate, $M_x = 90$ N·mm/mm, $M_y = 50$ N·mm/mm, and $M_{xy} = 35$ N·mm/mm. At this location, what are the principal (maximum and minimum) bending moments and the maximum twisting moment? On the plate surface, what is the maximum shear stress on planes normal to the surface and what is the maximum shear stress on any plane?

12.2-2. Consider a square plate with edges parallel to x and y axes, and loaded only by moments applied along edges. For each of the following relations, describe the edge moments applied. M, M_A, and M_B are constants.
(a) $2D(1+\nu)w = M(x^2 + y^2)$
(b) $2D(1-\nu^2)w = (M_A - \nu M_B)x^2 + (M_B - \nu M_A)y^2$
(c) $2D(1-\nu)w = M(y^2 - x^2)$

12.2-3. If lateral deflection w is a known function of x and y, what are the maximum slope $\partial w/\partial n$ and the direction of axis n with respect to the x axis, in terms of $\partial w/\partial x$ and $\partial w/\partial y$?

12.3-1. A rectangular plate of dimensions a by b carries uniformly distributed lateral load q. Edges of length b are clamped; edges of length a are free. If $a \gg b$, what are the lateral deflection, principal stresses, and principal strains at the middle of the lower surface, in terms of q, a, b, E, t, and ν?

12.3-2. Rectangular plate $ABCD$ serves as a lid over an opening. It is hinged along AB and is raised by force F applied at C. The builder has attached edge beams as shown, making the lid a shallow box, in hopes of greatly reducing bending and twisting deformations when force F is applied. Will the builder's hopes be realized? Explain.

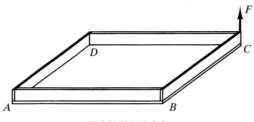

PROBLEM 12.3-2

12.4-1. Obtain M_{ave} of Eq. 12.4-1 by considering the equilibrium of one-eighth of a square plate, such as triangle ABC in Fig. 12.4-1a.

12.4-2. A concentrated lateral force P is applied at an arbitrary location on a simply supported rectangular plate of dimensions a by b. Assume that q_{mn} in Eq. 12.4-5 is known for this loading. Imagine now a triangular plate with all three edges simply supported and a right angle between adjacent edges of lengths a and b. Without writing equations, explain how $w = w(x,y)$ for the triangular plate can be obtained by superposing two solutions for the rectangular plate.

12.4-3. Specialize Eq. 12.4-5 to the case of a uniformly loaded square plate. Use the first series term ($m = n = 1$) to obtain approximations for (a) deflection, and (b) bending moment at the center of a plate with simple supports. In (b), let $\nu = 0.3$. Express answers in terms of q, a, and D. What are percentage errors of these results?

12.4-4. Repeat Problem 12.4-3 using the first three nonzero series terms for which $m + n$ is smallest.

12.4-5. For the linearly varying distributed load $q = q_a x/a$, q_{mn} in Eq. 12.4-4 is $q_{mn} = 8q_a(-1)^{m+1}/\pi^2 mn$, where $m = 1, 2, 3, \ldots$ and $n = 1, 3, 5, \ldots$. Use the first three nonzero series terms for which $m + n$ is smallest to obtain approximations for (a) deflection, and (b) bending moment at the center of a square plate with simple supports. In (b), let $\nu = 0.3$.

12.4-6. For a concentrated load P at the center of a square plate, q_{mn} in Eq. 12.4-4 is $q_{mn} = (4P/a^2) \sin(m\pi/2) \sin(n\pi/2)$. Use the first three nonzero series terms for which $m + n$ is smallest to approximate the center deflection of a square plate with simple supports.

12.4-7. A simply supported square plate carries uniformly distributed lateral load q. Assume that along an edge, net shear force V plots as a half-ellipse, being maximum at the center of an edge (see Eq. 12.4-6) and zero at corners. Then calculate the net upward force provided by all support forces, and compare it with the total lateral load applied.

12.4-8. A simply supported square plate of uniform thickness, 300 mm on a side, must carry uniform lateral pressure $q = 0.8$ MPa without deflecting more than one-third its thickness and without exceeding a maximum normal stress of 280 MPa. What is the minimum permissible thickness? Let $E = 200$ GPa and $\nu = 0.3$.

12.5-1. (a) Derive Eq. 12.5-4 by taking moments about a line tangent to the arc of radius r in Fig. 12.5-1b.

(b) By summing z-direction forces in Fig. 12.5-1b, obtain an expression for $d(rQ_r)/dr$.

(c) Combine Eq. 12.5-8 with the expression obtained in part (b), to obtain a fourth-order equation analogous to Eq. 12.2-7.

12.5-2. A circular plate has radius a and no hole. Consider the lateral displacement field $w = (9a^4 - 10a^2r^2 + r^4)c/D$, where c is a constant. How is the plate supported, and what is the nature of the loading? Let $\nu = 0$.

12.6-1. (a) A circular plate with its edge clamped carries distributed lateral load, whose intensity varies linearly with radius r, from zero at $r = 0$ to q_a at edge $r = a$. Derive the expression for $w = w(r)$. Also determine the center deflection.

(b) Combine the answer to part (a) with one other case to obtain the center deflection due to a distributed load that varies linearly with r, from q_a at $r = 0$ to zero at $r = a$.

12.6-2. In the following problems (a) and (b) of circular plates, obtain the expression for $w = w(r)$ by integrating the governing differential equation and imposing appropriate boundary conditions.

(a) Uniformly loaded, simply supported edge.

(b) Case 5 of Section 12.7.

12.6-3. The following loadings on circular plates lack axial symmetry. Use Maxwell's reciprocal theorem to determine the center deflection. Each plate shown is simply supported at its outer radius a. For simplicity, let $\nu = 0$. A useful formula: for a concentrated center force P and $\nu = 0$, $16\pi Dw = P[3(a^2 - r^2) + 2r^2 \ln(r/a)]$.

(a) Distributed lateral pressure whose intensity varies linearly with Cartesian coordinate x, from zero at $x = 0$ to q_m at $x = 2a$.

(b) Concentrated lateral force F at radius b.

(c) Lateral line load of uniform intensity f (force per unit length) along a diameter.

(d) Concentrated moment M_b at radius b.

(e) Uniform edge moment M_a per unit length applied over arc α at $r = a$.

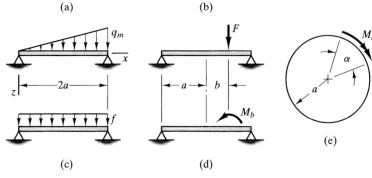

PROBLEM 12.6-3

12.6-4. Imagine that the sketch shows a uniform beam, not a plate. Load q is force per unit length. The cross section is rectangular, of dimensions b by t. Simple supports allow rotation but no lateral or axial displacement at the ends. Obtain an expression for w_0/t analogous to Eq. 12.6-11. Suggestions: Assume that $q = q_s + q_b$, where q_s is load supported by "string tension" T and q_b is load supported by bending. Also assume that $w = 4w_0 x(L-x)/L^2$ in order to calculate T. (Why is the assumption $q = q_s + q_b$ an approximation?)

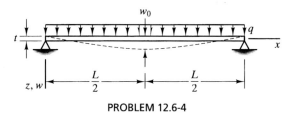

PROBLEM 12.6-4

12.6-5. A circular plate has the following dimensions and properties: $a = 250$ mm, $t = 5$ mm, $E = 200$ GPa, and $\nu = 0.3$. The boundary is simply supported. A uniformly distributed lateral load $q = 0.08$ MPa is applied. What are the center deflection and the largest normal stress? Consider both linear and nonlinear analyses.

12.6-6. A circular plate, 10 mm in diameter and having a simply supported edge, is proposed as a control device: Uniform lateral pressure 0.06 MPa is to produce center deflection w_0 such that electrical contact is made. In trial designs, the effect of w_0 is to be studied. What thickness t should the plate have, and what is the maximum normal stress on the plate surface at its center, if (a) $w_0 = 0.08$ mm, and (b) $w_0 = 0.16$ mm? For comparison, also calculate thickness and stress according to linear plate theory. Let $E = 200$ GPa and $\nu = 0.3$.

12.7-1. Obtain the bending moment at the center of the circular plate in Case 3 of Section 12.7 by superposition, using data from Cases 4 and 5.

12.7-2. (a) Obtain the solution to Problem 12.6-3(a) by a superposition argument rather than by the reciprocal theorem.

(b) What line load alternative to force F in Problem 12.6-3(b) produces the same center deflection? Suggestion: Consider first two forces $F/2$ at opposite ends of a diameter.

12.8-1. Determine the maximum stress σ_r in the circular plate shown if uniformly distributed edge moment M_a is large enough to make the center deflection 40% of what it would be if M_a were absent. Let $\nu = 0.3$. Express σ_r in terms of q, a, and t.

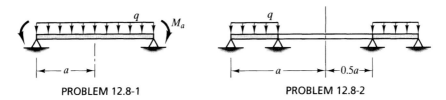

PROBLEM 12.8-1

PROBLEM 12.8-2

12.8-2. (a) In the simply supported annular plate shown, what fraction of the total load is carried by the inner support? Let $\nu = 0.3$.

(b) If each support is to carry half the load, which support should be lower, and how much lower? Express the result in terms of q, a, and D.

12.8-3. (a) A circular plate is uniformly loaded. There is a simple support along the edge (at $r = a$) and a point support at the center (at $r = 0$). What percentage of the total load is carried by the point support? Let $\nu = 0.3$.

(b) Repeat part (a), but change the simple support at $r = a$ to a clamped edge.

12.8-4. A thin flat plate rests on a rigid horizontal surface. When an upward force P is applied at a location distant from an edge, a circular portion of the plate lifts off the horizontal surface. Determine the radius of this portion in terms of P and the weight density per unit area of the plate. Let $\nu = 0.3$.

12.8-5. The annular and solid plates shown have the same thickness t and share a simple support (that is, they have a moment-free connection). The lower plate is point-supported at its center. Obtain an expression for the deflection at $r = 0.5a$ due to uniform load q in terms of q, a, and D. Let $\nu = 0.3$.

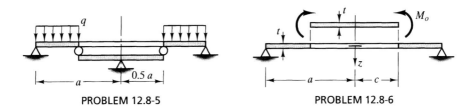

PROBLEM 12.8-5

PROBLEM 12.8-6

12.8-6 A solid plate of radius c is loaded by distributed edge moment M_o, then inserted into the annular plate shown and welded to it with M_o still applied. Then M_o is released. Relative to the $z = 0$ plane, what is the center deflection in terms of M_o, a, and D? What is the largest bending moment in the annular plate? Let $\nu = 0.3$.

12.8-7. In each of the two cases shown, imagine that loads, dimensions, and material properties are known. Write, but do not solve, sufficient equations to determine M_r at $r = a$. Let $\nu = 0.3$.

(a) Centers of the two circular plates are connected by a linear spring. The plates have the same E and t.

(b) The annular plate is uniformly loaded.

(c) Without writing equations, explain how to determine the deflection of largest magnitude in part (a).

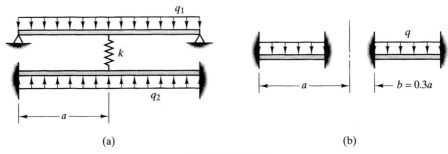

PROBLEM 12.8-7

12.8-8. The annular plate shown is loaded by opposing line loads, uniformly distributed around circles of radii a and $0.7a$. Each line load totals P units of force. Express the relative approach of these loads in terms of P, D, and a. Let $\nu = 0.3$.

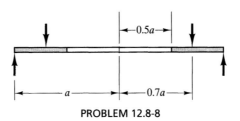

PROBLEM 12.8-8

12.8-9. In each of the axisymmetric plates shown, determine the deflection of largest magnitude and bending moment M_r at fixed supports. Let $\nu = 0.3$.

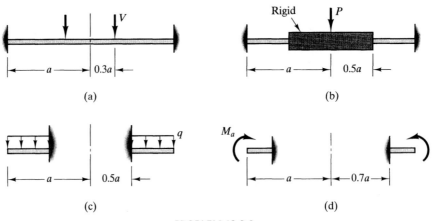

PROBLEM 12.8-9

12.9-1. **(a)** Show how Eq. 12.9-1 is obtained from Eq. 3.3-1.
(b) Use the von Mises yield criterion, Eq. 3.3-2, to develop a yield expression analogous to Eq. 12.9-1. (The expression involves M_x, M_y, M_{xy}, and M_{fp}.)

12.9-2. For the simply supported rectangular plate in Fig. 12.9-2b, obtain an upper bound solution for uniformly distributed collapse load q_c by using the mechanism in which yield lines lie along diagonals (that is, set $s = a/2$ in Fig. 12.9-2b). Then evaluate this expression in terms of M_{fp} and b for the case $a/b = 2$.

12.9-3. (a) For the simply supported rectangular plate in Fig. 12.9-2b, obtain an upper bound expression for collapse load q_c in terms of M_{fp}, a, b, and s.

(b) Let $a/b = 2$, and assume that $s = b/2$. Determine q_c in terms of M_{fp} and b.

(c) Obtain an expression for the s that minimizes q_c. For $a/b = 2$, determine this s in terms of b, and express the associated q_c in terms of M_{fp} and b.

12.9-4. Along all four edges of a uniformly loaded rectangular plate, support conditions are changed from simply supported to clamped. By what factor is the collapse load increased by this change?

12.9-5. The square plate shown is uniformly loaded, simply supported along one edge (ss), clamped along one edge (cl), and free along two adjacent edges. Obtain an upper bound expression for collapse load q_c in terms of M_{fp} and a.

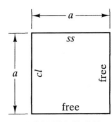

PROBLEM 12.9-5

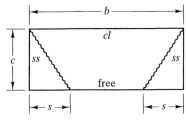

PROBLEM 12.9-6

12.9-6. (a) The rectangular plate shown has one clamped edge (cl) and two simply supported edges (ss). The remaining edge is free. Use the yield line mechanism shown to determine the uniformly distributed collapse load q_c in terms of M_{fp}, b, c, and s.

(b) For the case $b = c$, determine s and evaluate q_c in terms of M_{fp} and b.

(c) If c/b is large enough, the two yield lines may meet to form a Y pattern. With this potential pattern in mind, improve the upper bound estimate of q_c if possible. Consider the case $b = c$.

12.9-7. Each of the following plates is loaded by concentrated lateral force P at its center. In each case obtain an upper bound for collapse load P_c. Assume that collapse does not occur by the load "punching through" the plate.

(a) Circular plate, simply supported edge.

(b) Square plate, simply supported edges.

(c) Circular plate, clamped edge.

(d) Square plate, clamped edges.

(e) Let lateral force P act anywhere on a uniform solid plate with a clamped edge. In view of the results of parts (c) and (d), what is an upper bound formula for P_c?

12.9-8. Derive the lower bound formula $q_c = 11.26 M_{fp}/a^2$ for a uniformly loaded clamped circular plate (see Fig. 12.9-4).

12.9-9. Let q_Y be the uniformly distributed lateral load that initiates yielding of a plate. If the plate is circular, what is q_c/q_Y if the edge is (a) simply supported, or (b) clamped? State any assumptions you may make.

12.9-10. A simply supported circular plate of radius a is loaded by lateral force P, which is uniformly distributed around a circle of radius c, where $0 < c < a$. Obtain an upper bound formula for the collapse load P_c in terms of M_{fp}, a, and c.

12.9-11. An annular plate has outer radius a, inner (hole) radius b, and is uniformly loaded. Edge $r = a$ is simply supported; edge $r = b$ is free. Express collapse load q_c in terms of M_{fp}, a, and b. Obtain (a) an upper bound, and (b) a lower bound.

12.9-12. Repeat Problem 12.9-11, but let the annular plate be simply supported at $r = b$ and free at $r = a$. Suggestion: Take care with the yield criterion in part (b).

12.9-13. If the support condition is changed from simply supported to clamped, what is an upper bound for q_c in (a) Problem 12.9-11, and (b) Problem 12.9-12?

CHAPTER 13

Shells of Revolution with Axisymmetric Loads

13.1 INTRODUCTION

A shell forms a curved suface in space. Eggshells and water tanks are common examples. A shell is defined by its thickness and the shape of its midsurface, which lies midway between inner and outer surfaces. This chapter is restricted to *shells of revolution*, for which the midsurface is generated by revolving a plane curve about an axis. As examples, a cylinder is generated by revolving a straight line about a parallel line; a sphere is generated by revolving a circle about a diameter. In this chapter we consider only *thin* shells. A shell is usually regarded as thin if $R/t > 20$ throughout the shell, where R is the smallest midsurface radius of curvature and t is the shell thickness. For an eggshell, R/t is roughly 50. For constructed shells, R/t may approach 1000.

Membrane stresses are defined as stresses directed tangent to the midsurface and constant through the shell thickness. A state of membrane stress, with little bending, can be approached if the shell is thin. A membrane state is an efficient way to carry load. For example, consider Fig. 13.1-1. Plate and hemispherical shell carry the same vertical force, namely $F = \pi R^2 p$. The plate carries F by bending; the shell carries F by membrane action. Stress σ_{max} in the plate comes from Case 4 of Section 12.7, with $q = p$; stress σ in the shell comes from the elementary pressure vessel formula $\sigma = pR/2t$. The ratio of these stresses is $|\sigma_{max}/\sigma| = 2.48R/t$. Accordingly, for a given allowable stress and (say) $R/t = 100$, the hemispherical shell can carry 248 times as much load as a plate of the same radius. *A shell is strong mainly because of its shape*, not because of the strength of its material, which explains why most people cannot break an egg by squeezing it along its axis between thumb and forefinger, though it is easy to break an eggshell fragment by bending it.

Bending stresses, or *flexural stresses*, are stresses that vary linearly through the thickness, as they do in a plate. They are usually localized, and arise near changes in geometry and near some types of loads and supports. Bending stresses are almost impossible to avoid in practical shells, so they must be calculated for design purposes. Fortunately, for thin shells of revolution with axisymmetric loads and supports, bending stresses can usually be calculated without much difficulty.

412 Chapter 13   Shells of Revolution with Axisymmetric Loads

Circular plate:

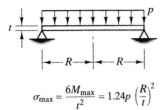

$$\sigma_{max} = \frac{6M_{max}}{t^2} = 1.24p\left(\frac{R}{t}\right)^2$$

Hemispherical shell:

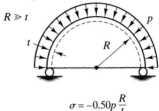

$$\sigma = -0.50p\frac{R}{t}$$

FIGURE 13.1-1 Uniform pressure p applied to a circular plate of radius R and to a thin hemispherical shell, also of radius R.

How does a shell differ from a *plane* curved structure such as an arch? Bending is absent in an arch only if the load distribution is suited to the shape of the arch. The same arch will bend under any other distribution of loading. In contrast, a given shell of revolution can carry a variety of distributed loads with practically no bending (except near supports).

Restrictions of the present chapter are as follows. Shells are thin, and axisymmetric (that is, they are shells of revolution). Loads, supports, and material properties are also axisymmetric. Loads are static. For bending theory, we also assume that the material is homogeneous, isotropic, and linearly elastic. Deflections are assumed to be small enough that linear theory is adequate. Buckling is excluded. (As rough approximations, if buckling occurs it will be elastic rather than plastic if $R/t > 400$, and local buckling is expected when membrane stress is $\sigma = -0.2Et/R$ in either an axially compressed cylindrical shell or a spherical shell under external pressure [1.6, 1.9].)

Membrane stresses are discussed first, then bending stresses that are superposed on membrane stresses. Finally we consider the reinforcing ring, which may be present where shells of different geometry are connected.

13.2   GEOMETRY AND TERMINOLOGY

Figure 13.2-1a shows the midsurface of a shell of revolution. The line of intersection of the midsurface and a plane that contains the axis of revolution is called a *meridian*. The line of intersection of the midsurface and a plane normal to the axis is called a *parallel* (or a *parallel circle* or a *circumference*). Figure 13.2-1a also shows internal forces N_θ and N_ϕ. These variables are membrane forces per unit of meridional or circumferential length, respectively, whose calculation we consider in subsequent sections.

At a point on the midsurface of any shell there are two principal radii of curvature. Both are directed normal to the midsurface. They are radii of mutually perpendicular infinitesimal arcs in the midsurface. One radius is the largest, the other the smallest, of the radii of all possible arcs in the midsurface at the point in question. For a shell of revolution, we call these radii R_θ and R_ϕ (Fig. 13.2-1b). Both lie in a plane that contains the axis, and in general both vary from point to point along a meridian. The center for R_θ always lies on the axis. The center for R_ϕ does not, unless the shell is spherical. Indeed, the two centers may lie on opposite sides of the midsurface; then R_θ and R_ϕ have opposite sign. A sign convention may be chosen arbitrarily. We elect to say

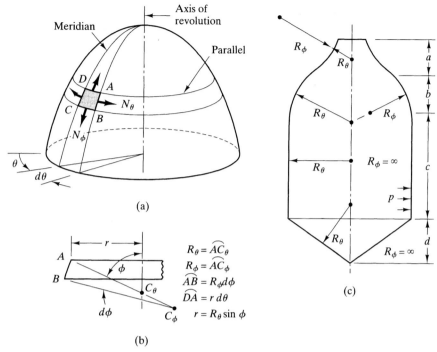

FIGURE 13.2-1   Geometry of a shell of revolution. (a) Terminology. (b) Cross section containing meridian AB, showing arc centers C_θ and C_ϕ. (c) Examples of principal radii.

that the positive side of a shell is the side on which pressure acts, and that a radius of curvature is positive if the arrowhead on the midsurface is on the positive side.

Examples appear in Fig. 13.2-1c. In span a, $R_\theta > 0$ and $R_\phi < 0$. In span b, R_θ and R_ϕ are both positive. In span c, where the shell is cylindrical, R_θ is constant and R_ϕ is infinite. In span d, where the shell is conical, R_θ is a function of distance from the apex of the cone and R_ϕ is infinite. In spans a and b the shell is called "doubly curved." In spans c and d it is called "singly curved."

13.3   MEMBRANE FORCES AND STRESSES

Membrane forces are N_ϕ in the meridional direction and N_θ in the circumferential direction (Fig. 13.2-1a). By definition, these are forces *per unit length*. In general, N_ϕ and N_θ vary along a meridian. They do not vary around a parallel, nor is there any shear force $N_{\phi\theta}$, because the situation is axisymmetric. N_ϕ and N_θ yield membrane stresses when divided by shell thickness.

$$\sigma_\phi = \frac{N_\phi}{t} \quad \text{and} \quad \sigma_\theta = \frac{N_\theta}{t} \tag{13.3-1}$$

Two equilibrium equations suffice to determine N_ϕ and N_θ. Thus, analysis for membrane forces in a shell of revolution is a *statically determinate problem*. Material properties do not matter. Aside from symmetry about the axis, we require only that the shape of the shell not change much when load is applied. Thus we exclude a balloon, unless we know what its inflated shape will be. As for signs, we will say that N_ϕ and N_θ are positive if tensile.

One of the two equilibrium equations can be written by considering net force normal to a differential element. Element $ABCD$ of Fig. 13.2-1a is shown again in Fig. 13.3-1a. Here p is an internal pressure, which may vary along a meridian. Forces per unit length N_ϕ and N_θ become ordinary forces when multiplied by lengths ds_θ and ds_ϕ on which they act. These forces subtend small angles $d\phi$ and $d\theta$, and therefore have components along the shell normal proportional to $2\sin(d\phi/2) \approx d\phi$ and $2\sin(d\theta/2) \approx d\theta$. The sum of normal forces is

$$p\,ds_\phi\,ds_\theta - (N_\phi ds_\theta)d\phi - [(N_\theta ds_\phi)d\theta]\sin\phi = 0 \qquad (13.3\text{-}2)$$

The force in brackets acts normal to the axis; it is multiplied by $\sin\phi$ to obtain its component normal to the element. Terms associated with changes in N_ϕ and ds_θ over distance ds_ϕ have been omitted because they are of higher order. Now $ds_\theta = r\,d\theta = (R_\theta \sin\phi)d\theta$ and $ds_\phi = R_\phi\,d\phi$. With these substitutions, Eq. 13.3-2 reduces to

$$\text{Equilibrium normal to the midsurface:} \qquad \frac{N_\phi}{R_\phi} + \frac{N_\theta}{R_\theta} = p \qquad (13.3\text{-}3)$$

The second equilibrium equation can be written by cutting the shell in two along a parallel, and considering net axial force on either part. Figure 13.3-1b shows the upper part of the shell in Fig. 13.2-1a. Along an increment ds_θ of the base parallel, N_ϕ exerts force $N_\phi\,ds_\theta$, whose axial component is $N_\phi\,ds_\theta \sin\phi$. Integration around the parallel yields axial force $N_\phi(2\pi r \sin\phi)$. This force must balance the net axial force due to loads, here represented by F. Thus

$$\text{Equilibrium parallel to the axis:} \qquad N_\phi = \frac{F}{2\pi r \sin\phi} \qquad (13.3\text{-}4)$$

Force F accounts for axial components of all loads applied to the shell segment (excluding N_ϕ). Contributions to F may come from internal pressure, liquid contained, snow load, self weight, or line load around a parallel.

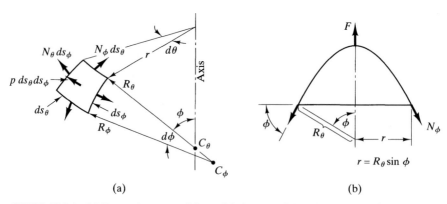

FIGURE 13.3-1 (a) Forces that act on differential element $ABCD$ of Fig. 13.2-1a. (b) Forces N_ϕ and F provide equilibrium of axial forces.

We can easily check that Eqs. 13.3-3 and 13.3-4 yield familiar results for thin-walled cylindrical and spherical pressure vessels of radius R. For a cylinder, and for a sphere if cut around its equator, $F = p\pi R^2$, $r = R$, and $\sin \phi = 1$. Therefore Eq. 13.3-4 yields $N_\phi = pR/2$ (and $\sigma_\phi = N_\phi/t = pR/2t$). For a cylinder, R_ϕ is infinite, so Eq. 13.3-3 yields $N_\theta = pR$. For a sphere, $R_\phi = R_\theta = R$, so Eq. 13.3-3 yields $N_\theta = pR/2$.

Limitations of Membrane Theory. Membrane theory presumes that only membrane forces N_ϕ and N_θ are present in a shell of revolution. It is easy to see that N_ϕ and N_θ alone cannot provide equilibrium in all situations. In Fig. 13.3-2a, N_ϕ exerts no vertical component at A to oppose concentrated load P. At B, N_ϕ exerts no component to oppose the shell-normal component of support reaction. Bending moments and transverse shear forces will arise at and near these locations. Associated flexural stresses are localized, but may be much larger in magnitude than membrane stresses.

Membrane theory would be adequate in Fig. 13.3-2a if load P were distributed over the shell surface near A and the rigid supporting surface near B were made conical so that rollers apply only shell-tangent forces.

One usually encounters bending moments and transverse shear forces in the following situations: near concentrated loads and line loads; near supports; and near sudden changes in curvature, material properties, and thickness. Examples appear in Fig. 13.3-2b. Circumferential line load V produces transverse shear force, and meridional bending moment arises in consequence. The tendency of the vessel to expand is prohibited at fixed support AA, which applies line load and bending moment to the shell, both loads being distributed around parallel AA. At BB there is a sudden change in radius of curvature R_θ and at CC there is a sudden change in radius of curvature R_ϕ. Membrane stresses alone do not provide compatible displacements across BB and CC, so bending stresses arise to maintain continuity.

These remarks do not imply that membrane theory must be discarded, but that membrane theory must be supplemented by bending theory in some regions.

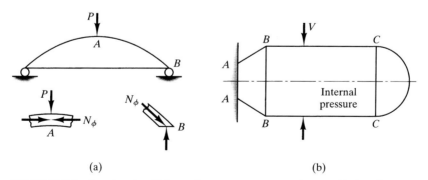

FIGURE 13.3-2 Situations in which membrane stresses are supplemented by bending stresses. (a) Segment of a sphere suppported by rollers around the base parallel. (b) A pressure vessel.

13.4 APPLICATIONS OF MEMBRANE THEORY

Examples illustrate the calculation of membrane forces N_ϕ and N_θ. Supplementary information from bending theory may be needed to complete an analysis.

EXAMPLE 1

The water tank shown in Fig. 13.4-1a is filled to depth H. The tank is conical except for the bottom closure, which is a segment of a sphere. Determine membrane forces in the conical portion, in terms of weight density γ of water, radius r, and other dimensions shown.

This problem is largely an exercise in geometry. Depth z is expressed in terms of r, a, b, and H by means of similar triangles. An expression for R_θ is obtained by trigonometry. Thus

$$z = \frac{a-r}{a-b}H \qquad \phi_o = \arctan\frac{H}{a-b} \qquad R_\theta = \frac{r}{\sin\phi_o} \qquad (13.4\text{-}1)$$

In a cone, radius R_ϕ is infinite and angle ϕ has the constant value ϕ_o. Equation 13.3-3 immediately yields N_θ.

$$p = \gamma z \qquad \text{hence} \qquad N_\theta = \gamma z R_\theta = \frac{\gamma H r(a-r)}{(a-b)\sin\phi_o} \qquad (13.4\text{-}2)$$

In the free body diagram, Fig. 13.4-1b, we see a truncated cone of depth z, whose volume is obtained from a handbook formula. This volume, multiplied by γ, is the weight W of fluid contained. Directed opposite to W is the force provided by pressure p acting on a circle of radius r. Summing axial forces, with positive arbitrarily taken as upward, we write

$$\gamma z(\pi r^2) - \gamma \frac{\pi z}{3}(a^2 + ar + r^2) - 2\pi r N_\phi \sin\phi_o = 0 \qquad (13.4\text{-}3)$$

from which N_ϕ can be expressed with r the only variable by substitution from Eqs. 13.4-1. For $a > b$, N_ϕ will be negative, indicating compression. As a partial check, note that the first two terms in Eq. 13.4-3 cancel if the cone becomes a cylinder ($r = a$). Then $N_\phi = 0$ as long as we ignore the weight of the shell itself.

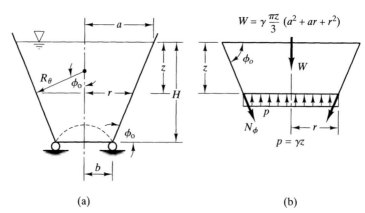

FIGURE 13.4-1 (a) Conical water tank. Symbol ∇ identifies the water surface. (b) Consideration of vertical forces.

Section 13.4 Applications of Membrane Theory

At the base, one should expect that there will be local bending because of circumferential line loads with shell-normal components, applied to the spherical part by the conical part, and vice versa.

EXAMPLE 2

The toroidal shell shown in Fig. 13.4-2a contains gas at pressure p. Determine the membrane forces.

A portion of the torus is isolated by two cutting surfaces, one a cylinder passing through the crown of the torus, the other a plane normal to the axis. Two different forces N_ϕ are thus exposed, but the force at the crown has no axial component. For equilibrium of axial forces in Fig. 13.4-2a,

$$p\pi(r^2 - a^2) - 2\pi r N_\phi \sin \phi = 0 \qquad (13.4\text{-}4)$$

But $r^2 - a^2 = (r-a)(r+a)$, and $r - a = b \sin \phi$. Thus Eq. 13.4-4 yields

$$N_\phi = pb\frac{(r+a)}{2r} \qquad (13.4\text{-}5)$$

Next N_ϕ is substituted into Eq. 13.3-3, with $R_\phi = b$ and $R_\theta = r/\sin \phi$. Thus

$$N_\theta = \frac{pb}{2} \qquad (13.4\text{-}6)$$

For $b \ll a$, the toroidal shell can be regarded as a cylindrical shell of radius b whose axis is slightly curved. For $b \ll a$, Eq. 13.4-5 yields $N_\phi \approx pb$, so that N_ϕ and N_θ then have the values expected for a cylindrical pressure vessel.

A toroidal "knuckle" is sometimes used as part of the end closure on a cylindrical pressure vessel (Fig. 13.4-2b). Force N_θ may be *compressive* in the knuckle, which raises the possibilty of a buckling mode in which ridges and valleys of the buckling pattern lie along meridians. For example, consider the knuckle at parallel BB. Here N_ϕ must balance the force $p\pi R^2$, so Eq. 13.3-4 yields $N_\phi = pR/2$. Then, if N_θ is not to be negative, Eq. 13.3-3 shows that radius b must exceed $R/2$. The possibilty of buckling limits a small-radius knuckle to low pressure use.

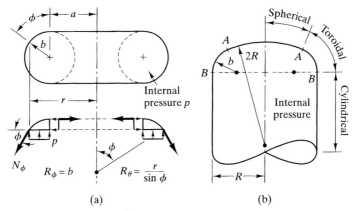

FIGURE 13.4-2 (a) Toroidal shell under internal pressure. (b) Torispherical end cap on a cylindrical pressure vessel.

Although it is not obvious, bending is present in the complete toroidal shell of Fig. 13.4-2a. If N_ϕ and N_θ are used to determine displacements for $0 < \phi < \pi$, and again for $\pi < \phi < 2\pi$, it is found that inner and outer parts of the deformed shell midsurface do not quite fit together. Compatibility is maintained by bending action. Similarly, in Fig. 13.4-2b, compatibility is maintained around parallels AA and BB by local bending action.

EXAMPLE 3

A dome of uniform thickness has the shape of a spherical segment (Fig. 13.4-3a). Simple support at the base is inclined so as to exert only shell-tangent meridional force. Investigate stresses associated with the weight of the dome.

Let the material have weight density γ. A segment of radius r can be isolated. The segment is loaded by its weight and by the vertical component of N_ϕ. Therefore, using a handbook formula for area A of the segment, axial equilibrium requires that

$$-\gamma A t - 2\pi r N_\phi \sin \phi = 0 \quad \text{where} \quad A = 2\pi R^2 (1 - \cos \phi) \tag{13.4-7}$$

Since $r = R \sin \phi$ and $\sin^2 \phi = 1 - \cos^2 \phi$, Eq. 13.4-7 yields

$$N_\phi = -\frac{\gamma R t}{1 + \cos \phi} \tag{13.4-8}$$

The component of weight per unit area directed toward the center of the sphere is $\gamma t \cos \phi$. In Eq. 13.3-3, $p = -\gamma t \cos \phi$, $R_\phi = R_\theta = R$, and N_ϕ is given by Eq. 13.4-8. Therefore

$$N_\theta = \gamma R t \left(\frac{1}{1 + \cos \phi} - \cos \phi \right) \tag{13.4-9}$$

Stresses are $\sigma_\phi = N_\phi/t$ and $\sigma_\theta = N_\theta/t$. Their variation with ϕ is plotted in Fig. 13.4-3b. We see that $\sigma_\phi = \sigma_\theta$ at $\phi = 0$. This conclusion is reasonable: the shell has no cusp at the top, so meridional and tangential directions become indistinguishable there. (For the same reason, $R_\phi = R_\theta$ at the top or bottom of a shell with no cusp.) By trying some numbers, say the γ of concrete and 100 m for R, we find that the stress of greatest magnitude is far below the crushing strength of the material. Circumferential stress at the base becomes tensile if base angle ϕ_o exceeds 51.8°. In masonry we prefer to avoid tensile stress, and so prefer base angles less than 51.8°.

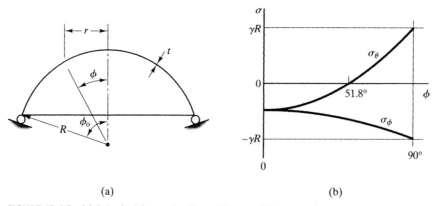

FIGURE 13.4-3 (a) Spherical dome of uniform thickness. (b) Stresses due to self-weight loading.

If the dome were simply supported on a horizontal surface, the support would apply only a vertical line load V_o to the shell. Reaction V_o can be resolved into a meridional component $N_\phi = V_o/\sin\phi_o$ and a radially outward component Q_o in the horizontal plane, $Q_o = V_o \cot\phi_o$. Component Q_o causes bending.

Snow load is similar to self-weight load, except that snow load is defined as force per unit of vertically projected area. Snow of uniform vertical depth h and weight density γ exerts pressure γh per unit of vertically projected area. This pressure is a vertical force $\gamma h \cos\phi$ per unit of shell surface; hence shell-normal pressure p of Eq. 13.3-3 is $p = -\gamma h \cos^2\phi$.

13.5 MEMBRANE SHELLS OF UNIFORM STRENGTH

We define a membrane shell of uniform strength as a shell in which membrane stresses provide the same maximum shear stress at every point. This definition means that where σ_ϕ and σ_θ have opposite sign, their difference must have magnitude $2\tau_{max}$. Or, where they have the same sign, the numerically larger of the two must have magnitude $2\tau_{max}$. In Section 13.4, loading and geometry were prescribed and stresses were required. In the present section, loading and stress τ_{max} are prescribed and shell geometry is required.

Drop-Shaped Tank. Drops of water on a nonwetted horizontal surface assume a shape dictated by gravity and surface tension. In the liquid skin, $N_\phi = N_\theta = S$, where surface tension S is the same everywhere in the skin. If we shape a tank of uniform thickness so that $N_\phi = N_\theta = N_o$, where N_o is a constant, we effectively replace the liquid skin with one of metal, and obtain a drop-shaped tank of uniform strength. Such tanks were first built in 1928.

Let the tank contain liquid of weight density γ. A positive gage pressure is assumed to exist at the top. For convenience, this pressure is represented by γh_o. Thus we imagine that a liquid surface free of gage pressure is a distance h_o above the top of the tank (Fig. 13.5-1a). Equation 13.3-3 becomes

$$\frac{N_o}{R_\phi} + \frac{N_o}{R_\theta} = \gamma h \quad \text{or} \quad \frac{1}{R_\phi} + \frac{1}{R_\theta} = \frac{\gamma h}{N_o} \quad (13.5\text{-}1)$$

where h is depth below the imagined liquid surface.

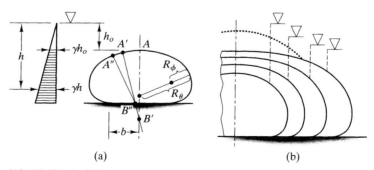

(a)   (b)

FIGURE 13.5-1 (a) Drop-shaped tank, full to the top and with added pressure γh_o at the top. (b) A series of drop-shaped tanks.

From Eq. 13.5-1 and geometrical relations, one can obtain a pair of differential equations, which must be integrated numerically to define the shape of the tank [12.1, 13.1, 13.2]. A graphical interpretation of the procedure is as follows [9.2].

We start at the top, point A in Fig. 13.5-1a. The meridian has no cusp at A, so $R_\phi = R_\theta$ at A. Thus, Eq. 13.5-1 yields $R_\phi = R_\theta = 2N_o/\gamma h_o$ at A. Using radius R_ϕ as a distance, we travel down to B', and use B' as center to draw arc AA'. This arc can be rather large because h does not change much from A to A'. At A' we measure the new h, and use it and the R_θ of arc AA' to calculate a new R_ϕ from Eq. 13.5-1. We travel distance R_ϕ from A' toward B' to locate a new center B'', and use it to draw the next arc $A'A''$. We measure the new h at A'', extend line $A''B''$ to the axis to measure a new R_θ, then calculate a new R_ϕ. The sequence of arc, measurement of h and R_θ, and calculation of R_ϕ continues, using roughly equal increments of h, until we come to a horizontal tangent where the tank meets the ground.

The flat bottom of the tank rests on a horizontal surface such as a concrete slab. The slab applies upward force $\pi b^2 \gamma h_b$, where h_b is the value of h at the bottom. This force must be equal to the weight of fluid contained, γV; therefore $V = \pi b^2 h_b$ is the tank volume. This volume can be compared with the volume dictated by the calculated shape, for a partial check on the correctness of calculations.

The calculated shape is of uniform strength only for the values of γ and h_o used in the calculation. If these are changed, or if the tank is not full, N_ϕ and N_θ are not equal to N_o at all locations, and bending stresses arise near the base.

If h_o is small, the top may be so nearly flat that it is in danger of collapse under its own weight when the tank is emptied. To avoid collapse the top might be replaced by a spherical cap, as suggested by the dotted line in Fig. 13.5-1b. The bottom of the cap can be supported by a circular reinforcing ring, which in turn can be supported by columns that extend up from the horizontal base and are inside the tank [13.1].

Other Shells. A dome loaded by its own weight can be shaped so that σ_ϕ and σ_θ are both equal in magnitude to an allowable compressive stress σ_c [9.2, 12.1, 13.1]. The shape of the meridian resembles a concave-down parabola, and shell thickness t is given by

$$t = t_o \exp\left(\frac{\gamma}{\sigma_c} z\right) \qquad (13.5\text{-}2)$$

where t_o is thickness at the top, z is distance downward from the top, and γ is the weight density of the material. For the present discussion, thickness t_o is arbitrary. (Practical considerations such as ease of construction and possible local loading during maintainance would determine t_o.) Although t grows exponentially it does not become large, since $\gamma z/\sigma_c$ is small for almost any building material, even if the dome is very tall.

A cylindrical pressure vessel can be capped by a membrane shell of uniform strength whose meridian is tangent to the cylinder at the juncture [9.2]. The cap resembles that shown in Fig. 13.4-2b, but radius R_ϕ varies continuously between parallels AA and BB. The uppermost portion of the cap is a segment of a sphere. Membrane forces N_ϕ and N_θ have the same sign (and are equal) above AA, opposite sign between AA and BB, and the same sign below BB. Bending action will arise near parallels AA and BB. Between

AA and BB, N_θ is compressive, so that buckling may be a concern. An alternative shape for which $N_\theta = 0$ can be constructed, but it is not a shell of uniform strength [13.3].

13.6 BENDING THEORY OF CYLINDRICAL SHELLS

Assumptions invoked for the following shell bending analysis are those also invoked for plate bending analysis in Section 12.1, except that a shell midsurface is curved before loading, and the midsurface is not free of strain because N_θ and N_ϕ are in general nonzero. Flexural stresses at shell surfaces are $\sigma'_\phi = \pm 6M_\phi/t^2$ meridionally and $\sigma'_\theta = \pm 6M_\theta/t^2$ circumferentially. (The corresponding equations for plates are Eqs. 12.2-2.) Primes are used in the present chapter to distinguish flexural stresses from membrane stresses.

In the present section we consider a circular cylindrical shell, thin-walled and of uniform thickness. It may be loaded by pressure, temperature gradient, and by shear force and bending moment distributed around a circumference. All loads are considered axisymmetric in what follows.

Figure 13.6-1 shows shell geometry and internal forces and moments per unit length. Force Q_x and moments M_x and M_θ are associated with bending. Force N_θ is produced by displacements associated with bending, and is also produced by membrane action if lateral pressure p is nonzero. Force N_x is membrane force due to axial load (for cylindrical shells, the notation N_x is more common than N_ϕ). For bending analysis we set N_x to zero. Stress and displacement associated with nonzero N_x can be calculated separately by membrane theory and superposed.

Governing Equations. In Fig. 13.6-1b we sum forces normal to the element and moments about a line tangent to the circumference. The only external load is lateral pressure p. With terms that cancel and higher-order terms already discarded, these equilibrium equations are

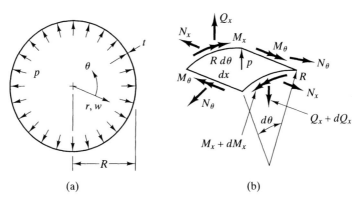

FIGURE 13.6-1 (a) Axial view of a thin-walled cylindrical shell. (b) Differential shell element, loaded by lateral pressure p and by forces and moments per unit length.

$$-dQ_x R d\theta - 2\left(N_\theta dx \sin \frac{d\theta}{2}\right) + pR d\theta dx = 0 \tag{13.6-1a}$$

$$dM_x R d\theta - (Q_x R d\theta) dx = 0 \tag{13.6-1b}$$

where $\sin(d\theta/2) \approx d\theta/2$. With $w = w(x)$ the radial displacement, positive outward, circumferential strain is $\epsilon_\theta = w/R$ (see Fig. 1.4-2b). Also, since $N_x = 0$, we have $\sigma_\theta = E\epsilon_\theta$. Hence $N_\theta = \sigma_\theta t = E\epsilon_\theta t = Etw/R$. This N_θ can be substituted into Eq. 13.6-1a, and both of Eqs. 13.6-1 divided by $R\, d\theta\, dx$. Then dQ_x/dx can be obtained from Eq. 13.6-1b and substituted into Eq. 13.6-1a. The result of these manipulations is

$$\frac{d^2 M_x}{dx^2} + \frac{Et}{R^2} w = p \tag{13.6-2}$$

For a plate, relations between moments and lateral displacement appear in Eq. 12.2-6. For the cylindrical shell, y in Eqs. 12.2-6 can be regarded as the circumferential direction. Thus, since we consider only axisymmetric conditions, w is independent of y. Therefore twist $\partial^2 w/\partial x\, \partial y$ and curvature $\partial^2 w/\partial y^2$ are both zero.* Accordingly, Eqs. 12.2-6 yield, without thermal terms but with a sign change because here w is positive outward,

$$M_\theta = \nu M_x \quad \text{and} \quad M_x = D\frac{d^2 w}{dx^2} \quad \text{where} \quad D = \frac{Et^3}{12(1 - \nu^2)} \tag{13.6-3}$$

If flexural rigidity D is constant, substitution of Eqs. 13.6-3 into Eq. 13.6-2 yields

$$D\frac{d^4 w}{dx^4} + \frac{Et}{R^2} w = p \quad \text{or} \quad \frac{d^4 w}{dx^4} + 4\lambda^4 w = \frac{p}{D} \tag{13.6-4}$$

where, in the latter equation,

$$\lambda^4 = \frac{Et}{4DR^2} \quad \text{or} \quad \lambda = \left[\frac{3(1 - \nu^2)}{R^2 t^2}\right]^{1/4} \tag{13.6-5}$$

Equation 13.6-4 has the same form as the governing equation for a beam on a Winkler elastic foundation, Eq. 5.2-4. Therefore Eq. 13.6-4 has a solution of the same form as Eq. 5.2-6. Specifically, if $p = 0$ on the shell,

$$w = e^{\lambda x}(C_1 \sin \lambda x + C_2 \cos \lambda x) + e^{-\lambda x}(C_3 \sin \lambda x + C_4 \cos \lambda x) \tag{13.6-6}$$

Constants of integration C_1 through C_4 can be evaluated from boundary conditions, which involve prescribed values of w, dw/dx, M_x, and/or Q_x on a circumference. When $w = w(x)$ is known, $M_x = M_x(x)$ can be determined from the second of Eqs. 13.6-3, then $Q_x = Q_x(x)$ determined from Eq. 13.6-1b (which reduces to $Q_x = dM_x/dx$). Formulas for these results, for selected boundary conditions, are stated in Section 13.7. Use of the formulas, in combination with lateral pressure p and thermal load, is illustrated in Section 13.8.

With increasing distance x from an axisymmetric load that causes bending, effects of the load must die out rather than increase, in accord with Saint-Venant's

---

*Uniform radial expansion contributes to circumferential curvature, but this contribution is negligible if the shell is thin (see Problem 13.7-1).

principle. Therefore, if loads are far apart, terms in the solution associated with $e^{\lambda x}$ must be discarded. The remaining terms, associated with $e^{-\lambda x}$, may decay over a surprisingly small distance. For example, let x originate at an end of the shell, where loads are applied. At the end, $x = 0$, so $e^{-\lambda x} = 1$ there. At what x do we find $e^{-\lambda x} = 0.05$? If $R/t = 20$ and $\nu = 0.3$, this x is $0.52R$. That is, at a distance from an end of about half the radius, end effects decay to about 5% of their maximum values. For $R/t = 200$, the 5% decay distance is only $0.16R$. We conclude that a "mathematically long" shell, in which effects of axisymmetric loadings M_x and Q_x at either end are scarcely felt at the middle, may have a length considerably less than its diameter.

Rapid decay occurs only if edge loads are *axisymmetric*. Loads applied over only part of the circumference—such as two diametrally opposed radial forces—create deformations that may still be appreciable many diameters away from the location of loading.

13.7 BENDING FORMULAS FOR EDGE LOADS

The following formulas state displacements, moments, and stresses associated with axisymmetric bending action in circular cylindrical shells and in spherical shell segments. It is assumed that shells are thin-walled and of uniform thickness, and made of material that is homogeneous, isotropic, and linearly elastic. Displacements and rotations are assumed to be small. Use of the formulas is illustrated in Section 13.8.

Notation. The following notation is used in the bending formulas and in applications.

Q_o = radial edge force (force per unit length of circumference directed normal to the axis of revolution)

M_o = meridional edge moment (moment per unit length of circumference)

M_x = meridional bending moment in a cylindrical shell ($M_x = M_o$ at edge $x = 0$)

M_ϕ = meridional bending moment in a spherical shell

M_θ = circumferential bending moment (moment per unit length of meridian)

w = radial displacement (directed normal to the axis of revolution and positive outward)

ψ = meridional rotation ($\psi = dw/dx$ in a cylindrical shell)

σ_θ = circumferential stress (uniform through thickness t, but associated with bending deformation; in a cylindrical shell, $\sigma_\theta = Ew/R$ if membrane stress σ_x is zero)

σ'_x = meridional flexural stress at the shell surface; $\sigma'_x = \pm 6M_x/t^2$ (called σ'_ϕ in a spherical shell; $\sigma'_\phi = \pm 6M_\phi/t^2$)

σ'_θ = circumferential flexural stress at the shell surface (in a cylindrical shell, $\sigma'_\theta = \nu \sigma'_x$)

All the foregoing displacements and stresses are to be superposed on displacements and stresses associated with the membrane theory discussed in preceding sections. For example, in a cylindrical shell, net axial stress at shell surfaces is $(\sigma_x)_{net} = \sigma_x + \sigma'_x$, where, if the shell has closed ends and contains gas at pressure p, axial membrane stress is $\sigma_x = pR/2t$.

424 Chapter 13 Shells of Revolution with Axisymmetric Loads

The following quantities appear in formulas.

$$D = \frac{Et^3}{12(1-\nu^2)} \qquad \lambda = \left[\frac{3(1-\nu^2)}{R^2 t^2}\right]^{1/4} \qquad (13.7\text{-}1)$$

Edge Loads, Cylindrical Shell. The following formulas [1.6, 1.7, 12.1] are obtained from Eq. 13.6-6, with $C_1 = C_2 = 0$. Thus the formulas apply to cylindrical shells for which loaded parallels have sufficient longitudinal separation that bending disturbance, created by external loading or by a support, is almost completely dissipated over the span between affected parallels. In other words, loads are not so close together that they interact, and the shell is "mathematically long." In this case, formulas for radial displacement and meridional bending moment for arbitrary positive x are

$$w = \frac{Q_o}{2D\lambda^3} e^{-\lambda x} \cos \lambda x + \frac{M_o}{2D\lambda^2} e^{-\lambda x}(\cos \lambda x - \sin \lambda x) \qquad (13.7\text{-}2a)$$

$$M_x = \frac{Q_o}{\lambda} e^{-\lambda x} \sin \lambda x + M_o e^{-\lambda x}(\cos \lambda x + \sin \lambda x) \qquad (13.7\text{-}2b)$$

These formulas provide the following results.

Axisymmetric radial edge load Q_o

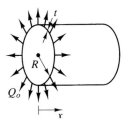

$M_\theta = \nu M_x$

$$\text{Largest } w = \frac{Q_o}{2D\lambda^3} \quad \text{at } x=0 \qquad (13.7\text{-}3a)$$

$$\text{Largest } \psi = -\frac{Q_o}{2D\lambda^2} \quad \text{at } x=0 \qquad (13.7\text{-}3b)$$

$$\text{Largest } M_x = 0.3224 \frac{Q_o}{\lambda} \quad \text{at } x = \frac{\pi}{4\lambda} \qquad (13.7\text{-}3c)$$

$$\text{Largest } \sigma_\theta = \frac{2Q_o \lambda R}{t} \quad \text{at } x = 0 \qquad (13.7\text{-}3d)$$

Axisymmetric edge moment M_o

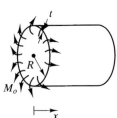

$M_\theta = \nu M_x$

$$\text{Largest } w = \frac{M_o}{2D\lambda^2} \quad \text{at } x=0 \qquad (13.7\text{-}4a)$$

$$\text{Largest } \psi = -\frac{M_o}{D\lambda} \quad \text{at } x=0 \qquad (13.7\text{-}4b)$$

$$\text{Largest } M_x = M_o \quad \text{at } x=0 \qquad (13.7\text{-}4c)$$

$$\text{Largest } \sigma_\theta = \frac{2M_o \lambda^2 R}{t} \quad \text{at } x=0 \qquad (13.7\text{-}4d)$$

Coefficients of Q_o and M_o in Eqs. 13.7-3 and 13.7-4 may be called *influence coefficients*. They are values of w, ψ, and so on for unit values of Q_o and M_o.

Edge Loads, Spherical Shell Segment. With spherical shells, as with cylindrical shells, effects of axisymmetric edge loads decay in exponential fashion with increasing meridional distance from the edge. Formulas for spheres are much more difficult to obtain than formulas for cylinders. Accordingly we do not consider their derivation. The following formulas [1.6, 1.7] provide displacement, rotation, moments, and *midsurface* stresses associated with Q_o and M_o, and apply *at the loaded edge*. The formulas are not exact, but have very little error if ϕ_o is near $\pi/2$. Error increases for larger or smaller values of ϕ_o, but is less than 5% if ϕ_o is in the range $3/\beta < \phi_o < (\pi - 3/\beta)$, where $\beta = \lambda R$. Thus for $R/t = 50$ and $\nu = 0.3$, ϕ_o can be as small as 19° or as large as 161°. The following quantities appear in formulas.

$$C_1 = 1 - \frac{1 - 2\nu}{2\lambda R} \cot \phi_o \qquad C_2 = 1 - \frac{1 + 2\nu}{2\lambda R} \cot \phi_o \qquad (13.7\text{-}5)$$

Axisymmetric radial edge load Q_o, axisymmetric edge moment M_o (formulas apply at $\phi = \phi_o$)

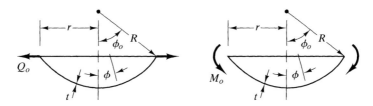

$$w = \frac{Q_o \sin^2 \phi_o}{4D\lambda^3 C_1}(1 + C_1 C_2) + \frac{M_o \sin \phi_o}{2D\lambda^2 C_1} \qquad (13.7\text{-}6a)$$

$$\psi = \frac{Q_o \sin \phi_o}{2D\lambda^2 C_1} + \frac{M_o}{D\lambda C_1} \qquad (13.7\text{-}6b)$$

$$M_\phi = M_o \qquad (13.7\text{-}6c)$$

$$M_\theta = \frac{Q_o t^2 \lambda^2 R \cos \phi_o}{6 C_1} + \frac{M_o}{C_1}\left[\nu + (2 - \nu)\frac{\cot \phi_o}{2\lambda R}\right] \qquad (13.7\text{-}6d)$$

$$\sigma_\phi = \frac{Q_o \cos \phi_o}{t} \qquad (13.7\text{-}6e)$$

$$\sigma_\theta = \frac{Q_o \lambda R \sin \phi_o}{2t}\left(\frac{2}{C_1} + C_1 + C_2\right) + \frac{2M_o \lambda^2 R}{t C_1} \qquad (13.7\text{-}6f)$$

In Eq. 13.7-6a, edge deflection w is directed parallel to radius r. In Eq. 13.7-6b, edge rotation ψ is positive if in the same direction as M_o. Stress σ_ϕ, like stress σ_θ, is uniform through thickness t. Flexural stresses at shell surfaces are $\sigma'_\phi = \pm 6M_\phi/t^2$ and $\sigma'_\theta = \pm 6M_\theta/t^2$. For a hemispherical shell, $\phi_o = \pi/2$, $C_1 = C_2 = 1$, and the foregoing formulas have the same form as corresponding formulas for a cylindrical shell.

13.8 APPLICATIONS OF BENDING THEORY

Problems of axisymmetric shells can often be solved by the "matching deflections and rotations" method used to solve statically indeterminate beam problems. This method has more physical appeal than the approach of solving the differential equations of shell theory. The following examples illustrate the method.

Water Tank, Clamped Edge. The water tank in Fig. 13.8-1a is fixed at its base. We will calculate significant stresses. To do so, we must determine force Q_o and moment M_o, which prevent deformation at the base that would otherwise arise due to membrane stresses. These support reactions are shown in assumed directions in Fig. 13.8-1a.

If self-weight of the tank and the tank support condition are neglected, membrane stresses at the base are $\sigma_x = 0$ and $\sigma_\theta = \gamma h R / t$, where γ is the weight density of water. Due to these membrane stresses, the base of the tank wants to expand an amount w_b and the tank meridian wants to rotate an amount $\psi = -w_b/h$.

$$w_b = R\epsilon_\theta = \frac{R}{E}(\sigma_\theta - \nu\sigma_x) = \frac{\gamma h R^2}{Et} \quad \text{and} \quad \psi = -\frac{\gamma R^2}{Et} \quad (13.8\text{-}1)$$

But in fact the base is fixed, so the net deflection and rotation due to Q_o, M_o, and water contained must be zero. From Eqs. 13.7-3a, 13.7-3b, 13.7-4a, 13.7-4b, and 13.8-1,

$$\text{Zero displacement:} \quad \frac{\gamma h R^2}{Et} + \frac{Q_o}{2D\lambda^3} + \frac{M_o}{2D\lambda^2} = 0 \quad (13.8\text{-}2a)$$

$$\text{Zero rotation:} \quad -\frac{\gamma R^2}{Et} - \frac{Q_o}{2D\lambda^2} - \frac{M_o}{D\lambda} = 0 \quad (13.8\text{-}2b)$$

From Eqs. 13.7-1, $Et/R^2 = 4D\lambda^4$. Hence Eqs. 13.8-2 have the solution

$$M_o = \frac{\gamma}{2\lambda^2}\left(h - \frac{1}{\lambda}\right) \quad \text{and} \quad Q_o = -\frac{\gamma}{\lambda}\left(h - \frac{1}{2\lambda}\right) \quad (13.8\text{-}3)$$

For a numerical example, we take

$$\begin{array}{lll} h = 10 \text{ m} & \gamma = 9800 \text{ N/m}^3 & \nu = 0.3 \\ R = 16 \text{ m} & t = 0.02 \text{ m} & \lambda = 2.272/\text{m} \end{array} \quad (13.8\text{-}4)$$

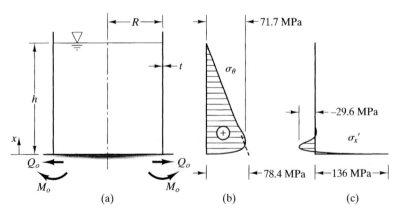

FIGURE 13.8-1 (a) Cylindrical water tank, whose base at $x = 0$ is assumed fixed. (b) Net circumferential membrane stress $\sigma_\theta = E(w/R)$. (c) Meridional flexural stress σ_x' on the inside surface.

Equations 13.8-3 and 13.8-4 yield

$$M_o = 9075 \text{ N} \cdot \text{m/m} \quad \text{and} \quad Q_o = -42{,}185 \text{ N/m} \tag{13.8-5}$$

We see that the correct direction of Q_o is opposite to the direction assumed. The only meridional stress is flexural stress σ'_x. On shell surfaces at the base,

$$(\sigma_x)_{\text{net}} = \sigma_x + \sigma'_x = 0 \pm \frac{6M_o}{t^2} = \pm 136 \text{ MPa} \tag{13.8-6}$$

compressive on the outside and tensile on the inside. On shell surfaces at the base, net circumferential stress is

$$(\sigma_\theta)_{\text{net}} = \sigma_\theta + \sigma'_\theta = \left(\frac{\gamma h R}{t} + \frac{2 Q_o \lambda R}{t} + \frac{2 M_o \lambda^2 R}{t} \right) \pm \nu \frac{6 M_o}{t^2} \tag{13.8-7}$$

Terms in parentheses sum to zero, as might be anticipated, because $\epsilon_\theta = 0$ at the base and $\sigma_x = 0$. From the last term, we obtain $(\sigma_\theta)_{\text{net}} = 40.8$ MPa, compressive on the outside and tensile on the inside.

Let us see how M_x and σ'_x vary with x. Substituting numerical values of M_o and Q_o into Eq. 13.7-2b, we obtain

$$M_x = e^{-\lambda x}(-9492 \sin \lambda x + 9075 \cos \lambda x) \tag{13.8-8}$$

Largest magnitudes of M_x appear at $x = 0$ and where $dM_x/dx = 0$. At $x = 0$, $M_x = 9075$ N·m/m. We find $dM_x/dx = 0$ at $x = 0.682$ m, where $M_x = -1974$ N·m/m, and at $x = 2.064$ m, where $M_x = 85.3$ N·m/m. Meridional flexural stresses at the latter two locations are ± 29.6 MPa and ± 1.3 MPa, respectively. Similar calculations show that radial displacement is maximum at $x = 1.213$ m; here the associated circumferential membrane stress is $\sigma_\theta = E(w/R) = 71.7$ MPa.

Selected stresses are plotted in Fig. 13.8-1. We see that edge-effect flexural stresses are roughly double the largest membrane stress, but decay rapidly with increasing distance from the edge.

The assumption of a completely fixed base is not realistic. An actual base will probably be somewhere between fixed and hinged. Both conditions can be examined in order to bracket actual stresses. (These uncertainties suggest that the small first term in Eq. 13.8-2b might be neglected, in which case we obtain $M_o = 9490$ N·m/m, $Q_o = -43{,}130$ N/m, and $\sigma'_x = \pm 142$ MPa at $x = 0$.)

Thermal Loads, Cylindrical Shell. Let the shell in Fig. 13.8-1a be empty, and stress-free at a reference temperature $T = 0$. Now let the temperature be raised, so that it is constant through thickness t but varies linearly from T_o at $x = 0$ to T_h at $x = h$. If restraint at $x = 0$ were removed, stresses would remain zero for this temperature distribution, but expansion at the base and rotation of the meridian would be

$$w_b = R\alpha T_o \quad \text{and} \quad \psi = R\alpha \frac{T_h - T_o}{h} \tag{13.8-9}$$

where α is the coefficient of thermal expansion. These expressions replace Eqs. 13.8-1, and, in enforcing fixity at the base, replace the two initial terms in Eqs. 13.8-2. Q_o

and M_o can then be determined. In stress calculation, the term $\gamma hR/t$ is absent from Eq. 13.8-7.

Next, let temperature be independent of x but vary linearly through the thickness, from $+T$ on the inner surface to $-T$ on the outer surface (this kind of problem is briefly discussed in connection with Fig. 7.10-2b). Imagine that supports at both ends of the shell prevent ends from rotating. Therefore $\psi = 0$ for all x, and ϵ_x and ϵ_θ are zero throughout the shell. Hence, from Eqs. 7.1-3 we determine stresses at shell surfaces, then the resulting moment M_o applied by a support.

$$\sigma'_x = \sigma'_\theta = \pm \frac{E\alpha T}{1-\nu} \quad \text{hence} \quad M_o = \frac{\sigma'_x t^2}{6} = \frac{E\alpha T t^2}{6(1-\nu)} \quad (13.8\text{-}10)$$

The direction of M_o is opposite to that shown in Fig. 13.8-1a. If ends are in fact free, we make them so in analysis by superposing another moment M_o in the opposite sense. Therefore we direct the added M_o so as to flare an end outward. Stresses due to the added moment are superposed on stresses of Eq. 13.8-10. For example, at a free end, net circumferential stress on the outer surface is, from Eqs. 13.7-4d, 13.8-10, and $\sigma'_\theta = \nu \sigma'_x$,

$$(\sigma_\theta)_{\text{net}} = \frac{E\alpha T}{1-\nu} + \frac{2M_o \lambda^2 R}{t} - \nu \frac{6M_o}{t^2} \quad (13.8\text{-}11)$$

Upon substituting λ from Eq. 13.7-1, we determine that for $\nu = 0.3$, $(\sigma_\theta)_{\text{net}} = 1.25 E\alpha T/(1-\nu)$. Thus, when temperature varies linearly through the thickness, $(\sigma_\theta)_{\text{net}}$ at a free end is 1.25 times the surface stress that prevails away from the end.

Hemispherical End Cap. Let a hemispherical cap be used as an end closure on a cylindrical shell (Fig. 13.8-2). When internal pressure p is applied, the two shells want to expand different amounts. Radial expansion due to membrane stresses is $w = R\epsilon_\theta = R(\sigma_\theta - \nu\sigma_\phi)/E$, where σ_θ is $pR/2t$ in the sphere and pR/t in the cylinder, and σ_ϕ is $pR/2t$ in both. Thus, for spherical and cylindrical shells respectively, due to membrane action alone,

$$w_s = \frac{pR^2}{2E_s t_s}(1-\nu_s) \quad \text{and} \quad w_c = \frac{pR^2}{2E_c t_c}(2-\nu_c) \quad (13.8\text{-}12)$$

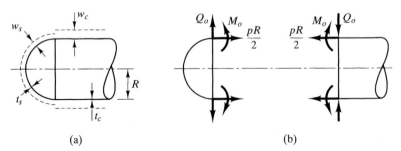

FIGURE 13.8-2 (a) Hemispherical cap on a cylinder, showing deflections associated with internal pressure and membrane stresses only. (b) Loads applied by one shell to another at the juncture.

These deflections would prevail if the cylinder-cap connection could be contrived to resist only axial membrane force. A welded connection causes each shell to apply loads to the other such that continuity of displacement and rotation is preserved across the juncture. Accordingly, we expect loads Q_o and M_o to arise (Fig. 13.8-2b). Clearly Q_o must act in the direction shown. The direction of M_o is assumed. Note that because these loads are action-reaction loads, they act in opposite directions on the two shells.

To match deflections and rotations, a sign convention is needed. For deflection, we will say that positive is outward. For rotation, we arbitrarily decide to look at the top of the juncture and say that counterclockwise is positive. Using w_s and w_c from Eqs. 13.8-12, and formulas from Section 13.7, equations that match deflections and rotations are respectively

$$w_s + \frac{Q_o}{2D_s\lambda_s^3} + \frac{M_o}{2D_s\lambda_s^2} = w_c - \frac{Q_o}{2D_c\lambda_c^3} + \frac{M_o}{2D_c\lambda_c^2} \tag{13.8-13a}$$

$$\frac{Q_o}{2D_s\lambda_s^2} + \frac{M_o}{D_s\lambda_s} = \frac{Q_o}{2D_c\lambda_c^2} - \frac{M_o}{D_c\lambda_c} \tag{13.8-13b}$$

If cylinder and cap have the same elastic properties and the same thickness, Eqs. 13.8-13 have the solution

$$M_o = 0 \quad \text{and} \quad Q_o = D\lambda^3 \frac{pR^2}{2Et} = \frac{p}{8\lambda} \tag{13.8-14}$$

That $M_o = 0$ means Q_o alone serves to restore continuity of displacement and maintain continuity of slope. One could say that this result emerges because edge effects decay in so small a distance that Q_o does not perceive the hemisphere as different from a cylinder.

The largest meridional stress $(\sigma_x)_{\max}$ is a combination of membrane stress $pR/2t$ and flexural stress $6M_x/t^2$, where M_x comes from Eq. 13.7-3c. This stress appears a distance $\pi/4\lambda$ from the juncture. For $\nu = 0.3$, on the outside surface,

$$(\sigma_x)_{\max} = \frac{pR}{2t} + \frac{6}{t^2}\left(0.3224 \frac{p/8\lambda}{\lambda}\right) = 0.65 \frac{pR}{t} \tag{13.8-15}$$

Shallow End Cap. The tank shown in Fig. 13.8-3 is loaded by internal pressure p. This problem is similar to that of Fig. 13.8-2, but an additional radial end force appears because meridional membrane forces in the two shells are not mutually tangent. Membrane force $pR_s/2$ in the spherical cap has axial component $(pR_s/2)\sin\phi_o = pR/2$, which is in equilibrium with axial membrane force $pR/2$ in the cylinder. However, if we imagine that the cylinder exerts force on the cap tangent to the cap meridian, the cap must then apply to the cylinder the inward radial force $Q_p = (pR_s/2)\cos\phi_o$, as shown in Fig. 13.8-3a. Thus, the resultant of forces $pR/2$ and Q_p on the end of the cylinder is $pR_s/2$, directed tangent to the cap. In addition, action-reaction loads Q_o and M_o shown in Fig. 13.8-2b arise in order to preserve continuity of radial displacement w and rotation ψ. (For the cap, $w = w_s \sin\phi_o$, where w_s comes from Eq. 13.8-12.) Equations that describe continuity resemble Eqs. 13.8-13, with Q_o replaced by $Q_o + Q_p$ on right-hand sides. Similarly, in equations for bending-induced stress in the cylinder, Q_o is replaced by $Q_o + Q_p$.

430 Chapter 13 Shells of Revolution with Axisymmetric Loads

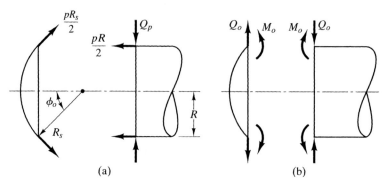

FIGURE 13.8-3 (a) Spherical shell segment used to cap a cylindrical shell, showing loads associated with membrane action in the cap. (b) Additional loads that restore continuity.

As an alternative method, one could regard the cap as applying axial membrane force $pR/2$ to the cylinder, which requires that inward radial force Q_p be applied to the cap rather than to the cylinder. Thus, the resultant of forces $pR_s/2$ and Q_p on the edge of the cap is $pR/2$, directed tangent to the cylinder. Again, Q_o and M_o are also present. This method yields a different Q_o than the foregoing method, but provides the same net radial force on each shell.

In Fig. 13.8-3, if cylinder and cap have the same thickness, and for $R_s = 2R$ and $R = 100t$, the largest net meridional stress (flexural stress included) in the cylindrical shell is about ten times the meridional membrane stress $pR/2t$. That is, peak stresses in Fig. 13.8-3 are much larger than in Fig. 13.8-2. To reduce these stresses, the change in curvature might be made less abrupt (see Fig. 13.4-2b), or the juncture might be provided with a reinforcing ring.

13.9 THE REINFORCING RING

A ring can be used as reinforcement at an opening in a shell or at a juncture between shells. In general, the ring is loaded by radial distributed force, and by distributed moment that tends to "invert" or "roll" the ring. The ring responds by developing circumferential strains and stresses. In what follows, we assume that the ring is homogeneous and has a uniform cross section whose dimensions are small in comparison with the mean radius of the ring. Thus, circumferential strains and stresses have negligible variation with radial coordinate r. We also assume that the ring contains no initial stresses. (If a ring is made by elastically bending a straight bar to a circular shape and welding its ends together, its resistance to rolling is less than that of a stress-free ring. Indeed, if the cross section is circular, or another shape for which principal moments of inertia are equal, the elastically bent and welded ring has *zero* rolling resistance.)

Analysis. Let the ring be loaded by Q_o and M_o as shown in Fig. 13.9-1a. These force and moment loads are assumed to be uniformly distributed around a circle that passes through centroids of cross sections. Figure 13.9-1b shows a free body diagram of half the ring. The statically equivalent load $2Q_oR$ results from integration of $(Q_oR\,d\theta)\sin\theta$

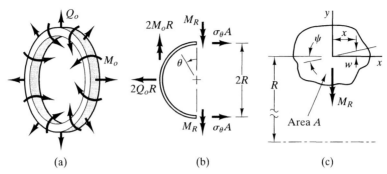

FIGURE 13.9-1 (a) Radial load Q_o and "rolling" moment M_o on a ring. (b) Free body diagram of half the ring. (c) Cross section, showing "rolling" rotation ψ. Axes x and y are centroidal.

from zero to π. Load $2M_oR$ is calculated similarly. Now let us consider the effects of loads Q_o and M_o separately.

Summation of horizontal forces yields $2Q_oR = 2\sigma_\theta A$, where A is the cross-sectional area. It is reasonable to assume that σ_θ is the only significant stress. Accordingly, due to Q_o, circumferential stress σ_θ and radial displacement w are

$$\sigma_\theta = \frac{Q_o R}{A} \quad \text{and} \quad w = \frac{\sigma_\theta}{E} R = \frac{Q_o R^2}{AE} \qquad (13.9\text{-}1)$$

In response to moment M_o, a typical cross section rotates through a small angle ψ, as shown in Fig. 13.9-1c. A typical point in the cross section has radial displacement $w = \psi x$. Circumferential strain at this point is $\epsilon'_\theta = w/R = \psi x/R$. This strain varies linearly with x over the ring cross section, with tensile strain in portions of the cross section that move radially outward and compressive strain in portions that move inward. For linearly elastic conditions, we know from elementary beam theory that linear strain variation implies a neutral axis that passes through the centroid of the cross section if axial force is zero. In Fig. 13.9-1c, the neutral axis coincides with centroidal axis y, and x is distance from it. Circumferential stress σ'_θ due to M_o is

$$\sigma'_\theta = E\epsilon'_\theta = E\frac{\psi x}{R} \qquad (13.9\text{-}2)$$

Equilibrium of moments about the diameter in Fig. 13.9-1b yields $M_R = M_o R$ as the bending moment about the y axis of a cross section. Hence, by applying the flexure formula, we obtain another expression for σ'_θ:

$$\sigma'_\theta = \frac{M_R x}{I_y} = \frac{M_o R x}{I_y} \qquad (13.9\text{-}3)$$

where I_y is the moment of inertia of area A about the centroidal y axis. By eliminating σ'_θ between Eqs. 13.9-2 and 13.9-3, we obtain an expression for rotation ψ.

$$\psi = \frac{M_o R^2}{EI_y} \qquad (13.9\text{-}4)$$

The foregoing equations do not require that the cross section have an axis of symmetry.

EXAMPLE

A ring of rectangular cross section is simply supported (Fig. 13.9-2). The ring is loaded by moment $M_B = 112$ N·mm/mm distributed around the circumference at corner B, and by forces $Q_A = 18$ N/mm and $Q_D = 10$ N/mm distributed around circumferences at corners A and D. We ask for the displacement and rotation of the ring and for stresses at corners A, B, C, and D.

Loads can be transferred to the centroid of cross section $ABCD$ as follows. On a circumferential arc $d\theta$, Q_o at radius R must exert the same radial force as Q_D at radius a. That is,

$$Q_D(a\,d\theta) = Q_o(R\,d\theta) \quad \text{hence} \quad Q_o = \frac{a}{R} Q_D = 1.05\, Q_D$$

$$\text{similarly} \quad Q_1 = \frac{a}{R} Q_A = 1.05\, Q_A \qquad (13.9\text{-}5)$$

Moment M_o is composed of contributions from M_B, Q_A, and Q_D. By the same argument as used for Eq. 13.9-5,

$$M_o(R\,d\theta) = M_B(b\,d\theta) - Q_A(a\,d\theta)s + Q_D(a\,d\theta)\frac{h}{2} \qquad (13.9\text{-}6a)$$

$$M_o = \frac{b}{R} M_B - \left(Q_A \frac{a}{R}\right)s + \left(Q_D \frac{a}{R}\right)\frac{h}{2} \qquad (13.9\text{-}6b)$$

where s and $h/2$ are moment arms of Q_A and Q_D about the centroid of $ABCD$. Using numerical data provided, $Q_o = 10.5$ N/mm and $M_o = 62.1$ N·mm/mm. Surface stresses associated with Q_o and M_o are

$$\sigma_\theta = \frac{Q_o R}{A} = \frac{10.5(78)}{48} = 17 \text{ MPa} \qquad (13.9\text{-}7a)$$

$$\sigma'_\theta = \pm \frac{M_o R(h/2)}{I_y} = \pm \frac{62.1(78)(3)}{144} = \pm 101 \text{ MPa} \qquad (13.9\text{-}7b)$$

where σ_θ is uniform over the cross-sectional area, and σ'_θ is compressive along surface AB and tensile along surface CD. Net stresses due to σ_θ and σ'_θ in combination are -84 MPa along AB and 118 MPa along CD.

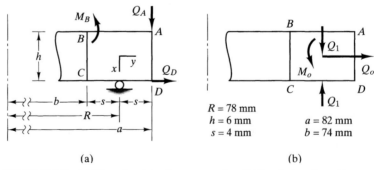

FIGURE 13.9-2 (a) Ring of rectangular cross section. (b) Loads transferred to centroid. Dimensions are those used in the numerical example.

With $E = 200$ GPa, radial displacement w_o at the centroid of $ABCD$ and rotation ψ of the cross section are

$$w_o = \frac{Q_o R^2}{AE} = 0.0067 \text{ mm} \quad \text{and} \quad \psi = \frac{M_o R^2}{EI_y} = 0.0131 \text{ rad} \quad (13.9\text{-}8)$$

Net radial displacement at an arbitrary point in the cross section is $w = w_o + \psi x$. Along surfaces AB and CD, $x = \pm h/2$, and net radial displacements are

$$w_{AB} = -0.0326 \text{ mm} \quad \text{and} \quad w_{CD} = 0.0460 \text{ mm} \quad (13.9\text{-}9)$$

In this example at least, net circumferential stresses and radial displacements at the surfaces are due primarily to rotation ψ.

Remark. If a reinforcing ring is added at a juncture between two shells, there are four unknowns, in contrast to the two unknowns Q_o and M_o that appear in Fig. 13.8-2b or in Fig. 13.8-3b. The four unknowns are force and moment where the first shell joins the ring, and force and moment where the second shell joins the ring. Two pairs of equations are available, each pair analogous to Eqs. 13.8-13. The equations say that at each of the two junctures, shell and ring have the same radial displacement and the same rotation. It is customary to assume that distortion of the cross section of the ring is negligible.

PROBLEMS

13.1-1. How much flexural stress arises because of deformation associated with membrane stress? Answer by considering deformation associated with internal pressure on a cylindrical vessel for which $R/t = 20$ and Poisson's ratio is zero. In the circumferential direction, calculate the ratio of flexural stress to membrane stress. Suggestion: Bending moment has magnitude $M = EI\kappa$, where κ is the change in curvature. The curvature under load is $1/(R + w)$, where w is radial displacement.

13.3-1. Imagine that a sheet of rubberlike material has a network of inextensible fibers embedded in it (see sketch). A closed tube of diameter D is to be made from the sheet by joining edges of length L and capping ends with rigid plates. What should be angle α of the fibers so that the tube will remain cylindrical and of the same length when internal pressure is applied?

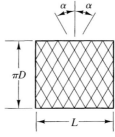

PROBLEM 13.3-1

13.4-1. The toroidal shell shown has an elliptical cross section and is loaded by internal pressure. At top and bottom parallels, show that equilibrium is not possible with membrane stresses alone.

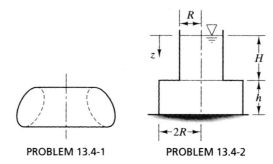

PROBLEM 13.4-1   PROBLEM 13.4-2

13.4-2. The stepped tank shown consists of two coaxial cylindrical shells connected by an annular plate. The tank is filled to depth $H+h$ with water of weight density γ. Determine N_ϕ and N_θ in the two shells in terms of γ, H, R, and z. Indicate where bending stresses are to be expected.

13.4-3. (a) The conical shell shown contains gas at pressure p. Determine N_ϕ and N_θ in terms of p, s, and ϕ_o.

(b) Remove the gas pressure of part (a). Instead let the conical shell be full to the top with liquid of weight density γ. At the top, the liquid has atmospheric pressure. Determine N_ϕ and N_θ in terms of γ, s, and ϕ_o.

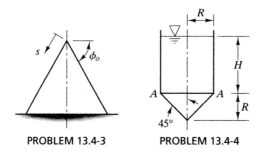

PROBLEM 13.4-3   PROBLEM 13.4-4

13.4-4. The tank shown has a cylindrical top and a conical bottom. Thicknesses of the respective shells are t_t and t_b. The tank is supported by a structure (not shown) that applies only vertical force around parallel AA. The tank is filled to depth $H=2R$ with water of weight density γ.

(a) Obtain expressions for N_ϕ and N_θ just above and just below parallel AA.

(b) For what ratio t_t/t_b do membrane stresses provide the same maximum shear stress just above and just below parallel AA?

(c) If Poisson's ratio is 0.3, for what ratio t_t/t_b is radial (outward) expansion the same in both shells at parallel AA?

13.4-5. In Problem 13.4-4 let $H=0$, so that only the conical shell remains. Fill this shell with water. Let the support at parallel AA apply only meridian-tangent distributed load. At what vertical distances from the bottom do maximum values of N_ϕ and N_θ appear, and what are these maximum values?

13.4-6. The conical shell shown is used as a small rain shelter. It is of uniform thickness and is supported by a vertical center post.
  (a) Determine N_ϕ and N_θ due to the shell's own weight.
  (b) Determine N_ϕ and N_θ due to snow load.

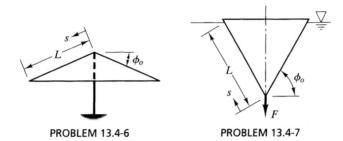

PROBLEM 13.4-6   PROBLEM 13.4-7

13.4-7. The empty conical shell shown is pulled down by force F so that its top is just above the surface of a liquid of weight density γ. Determine membrane forces in terms of γ, L, s, and ϕ_o.

13.4-8. The shell shown is a truncated toroid. It is loaded by vertical line load, F units of force per unit of circumference, around the top opening. Obtain expressions for membrane forces.

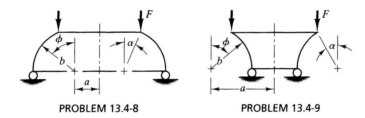

PROBLEM 13.4-8   PROBLEM 13.4-9

13.4-9. Repeat Problem 13.4-8 for the shell shown, which is the *inner* portion of a toroidal shell.

13.4-10. At parallel BB in Fig. 13.4-2b, obtain an expression for the difference in radial (outward) displacement between cylindrical and toroidal shells that tends to appear due to membrane stresses. Use the same thickness and material properties for both shells, and let $b = R/3$.

13.4-11. Two identical spherical shell segments are welded together to form the tank shown. As reinforcement, a flat disk of the same material is welded to the juncture between shells. The disk contains a small central hole to equalize internal pressure. How should disk and shell thicknesses be related if there is to be no bending stress when internal pressure is applied?

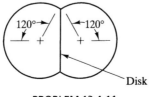

PROBLEM 13.4-11

13.4-12. The shell of revolution shown is supported by rollers on the $\phi = \pi/2$ parallel, and by a central drainpipe. The shell carries snow load.
  (a) Explain why membrane theory demands that the drainpipe carry load. Determine this load.
  (b) Obtain expressions for N_ϕ and N_θ in the range $0 < \phi < \pi/2$.

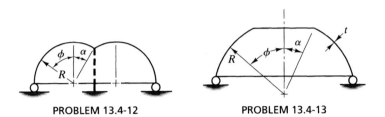

PROBLEM 13.4-12   PROBLEM 13.4-13

13.4-13. The shell shown is a truncated sphere of uniform thickness with an opening cut in the top. The shell is loaded by its own weight.
  (a) Determine membrane forces in terms of weight density γ, R, t, α, and ϕ.
  (b) Plot the variation of N_ϕ and N_θ, in the manner of Fig. 13.4-3b. Let $\alpha = 0.1$ rad.

13.4-14. Repeat Example 3 of Section 13.4, but use snow load rather than self-weight load.

13.4-15. The spherical tank shown contains water of weight density γ. Its surface, at atmospheric pressure, is a distance H from the top. In the region between the water surface and the outlet pipe that supports the tank, obtain expressions for membrane forces in terms of γ, h, b, r, and ϕ. What are h, b, and r in terms of H, R, and ϕ?

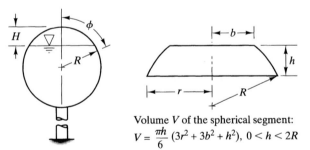

Volume V of the spherical segment:
$$V = \frac{\pi h}{6}(3r^2 + 3b^2 + h^2), \quad 0 < h < 2R$$

PROBLEM 13.4-15

13.4-16. The spherical tank shown is hung from its top and is completely full of liquid of weight density γ. Pressure at $\phi = 0$ is atmospheric. Obtain an expression for membrane force N_ϕ in terms of γ, R, and ϕ. Do so by summing vertical forces on (a) the portion above a parallel, and (b) the portion below a parallel. (c) Show that the two expressions are identical.

13.4-17. The empty spherical tank shown is submerged in water of weight density γ. The tank is held down by a cable. Set up, but do not solve, an equation that will provide an expression for membrane force N_ϕ.

13.4-18. Consider a paraboloidal shell of revolution, $z = cr^2$, where z is the axial coordinate and c is a constant. The base has diameter 320 mm, and the distance from base to vertex is 240 mm (hence constant c can be determined). Shell thickness is 3 mm. Uniform internal pressure $p = 300$ kPa is applied. Express σ_ϕ and σ_θ in terms of angle ϕ from the axis.

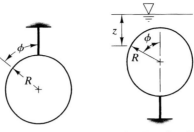

PROBLEM 13.4-16   PROBLEM 13.4-17

13.5-1. Imagine that a drop-shaped tank is roughly half-filled. Show that there must then be a transverse shear force in the shell around the parallel where the shell becomes tangent to the ground.

13.5-2. Using the graphical method, accurately construct meridians of drop-shaped tanks using the following values of radius of curvature at the top in Fig. 13.5-1a: (a) $R_\phi = R_\theta = 10$ m, and (b) $R_\phi = R_\theta = 100$ m. In both cases let $h_o = 30$ m.

13.5-3. A cylindrical tank of radius R contains gas at pressure p. If the tank is capped by a spherical shell segment of radius $2R$ and the same thickness, the construction is of uniform strength because τ_{max} is the same throughout (according to membrane theory). Sketch at least four different ways to configure the spherical shell segment so that the uniform τ_{max} criterion is met.

13.5-4. (a) A shell of revolution contains gas at pressure p. Show that $N_\phi = pR_\theta/2$ and $N_\theta = pR_\theta(1 - 0.5R_\theta/R_\phi)$.

(b) Consider a cap of uniform strength on a cylindrical tank of radius R. Let tank and cap have the same thickness and share a common tangent where they meet (as at parallel BB in Fig. 13.4-2b). Show that, where $R_\theta > 2R_\phi$ in the cap, radius R_ϕ is $R_\phi = R_\theta^2/(R_\theta + 2R)$.

(c) To graphically construct the meridional shape of this cap, we begin at the end of the cylinder, and swing an arc of radius $R_\phi = R/3$. Next we measure the new R_θ (which exceeds R) and calculate a new R_ϕ. Working with arc increments in this way, accurately construct the shape of the cap if $R = 15$ cm. Finish the construction with a spherical arc when $R_\theta = 2R_\phi$.

13.5-5. A cylindrical tank is to be capped in such as way that there is no compressive membrane stress anywhere in the cap when internal pressure is applied [13.3].

(a) If $N_\theta = 0$ for all ϕ, what is the relation between R_ϕ and R_θ?

(b) Using a graphical method, accurately construct the meridian of such a cap if the cylindrical tank has radius $R = 15$ cm. Let cylinder and cap share a common tangent where they meet (as at parallel BB in Fig. 13.4-2b).

13.5-6. Consider a shell of revolution shaped so that it is a dome of uniform strength when loaded by its own weight. Let the material be concrete of density 2400 kg/m³ and compressive strength 62 MPa. For a safety factor of 2, what would be the approximate rise of such a dome, in meters, if it is twice as thick at its base as at its top?

13.7-1. Radial expansion creates circumferential curvature ($\partial^2 w/\partial y^2$ in the notation preceding Eq. 13.6-3). Let a cylindrical shell be loaded by uniformly distributed radial end force Q_o. For this loading, and for $\nu = 0$, determine the ratio of the numerically largest $\partial^2 w/\partial y^2$ to the numerically largest $\partial^2 w/\partial x^2$. Show that this ratio approaches zero as t/R approaches zero. (See also Problem 13.1-1.)

13.7-2. Let $M_o = 0$ in Eq. 13.7-2a. If x_1 is a value of x such that λx_1 defines a half wavelength of a damped cosine function, what is x_1 as a function of R and t, and what is displacement w at $x = x_1$ as a percentage of its value at $x = 0$? Let $\nu = 0.3$, and consider $R/t = 20$ and $R/t = 200$.

13.8-1. Derive an expression for σ_θ as plotted in Fig. 13.8-1b.

13.8-2. At the base of the tank in Fig. 13.8-1, alter the support condition so that radial displacement is prevented but rotation is allowed. Using data of Eqs. 13.8-4, determine significant stresses at the base and at $x = \pi/4\lambda$.

13.8-3. Imagine that the tank of Fig. 13.8-1 is stepped at 5 m from the base, so that the upper half has thickness 0.01 m while the lower half has thickness 0.02 m. Remaining data of the example problem are unchanged. Determine the largest stress at the step. If $(\sigma_x)_{net}$ and $(\sigma_\theta)_{net}$ farther from the step were to be calculated, what effects would contribute to these stresses?

13.8-4. External pressure p is applied to half of a long cylindrical shell, as shown. With $\nu = 0.3$, obtain a numerical value for the ratio of maximum bending stress to maximum magnitude of membrane stress.

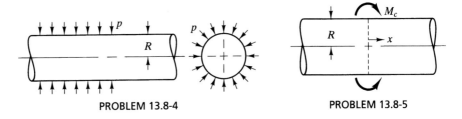

PROBLEM 13.8-4   PROBLEM 13.8-5

13.8-5. A moment load is uniformly distributed around a circumference near the center of a long cylindrical shell, as shown. Obtain expressions for $w, \psi, M_x,$ and σ_θ as functions of x.

13.8-6. Ends of a cylindrical tank are capped with flat disks of the same material and same thickness as the tank, as shown. Let $R = 0.80$ m, $t = 0.03$ m, and $\nu = 0.3$. For internal pressure $p = 0.15$ MPa, determine stresses in cylinder and disk where they meet. (a) Ignore radial expansion in the disks. (b) Account for radial expansion in the disks.

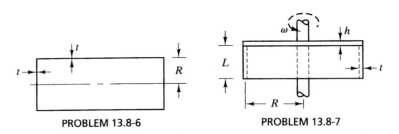

PROBLEM 13.8-6   PROBLEM 13.8-7

13.8-7. A flywheel is constructed by welding a flat steel disc to a short cylindrical steel pipe, as shown. For spinning at angular velocity ω, set up, but do not solve, equations that define loads Q_o and M_o applied to the pipe by the disk. If $R/t = 20$ and $\nu = 0.3$, what is the smallest ratio L/R for which the analysis is reliable?

13.8-8. The upper hemisphere of a spherical shell is uniformly heated to 50°C above the temperature of the lower hemisphere. Investigate stresses at the equator. Also investigate meridional stress close to the equator, using the assumption that edge loads deform a hemisphere like a cylinder. Use $\alpha = 12(10^{-6})/°C$, $E = 200$ GPa, and $\nu = 0.3$.

13.8-9. (a) The spherical shell segment shown is clamped around its base and is loaded by external pressure p. Determine the force and moment loads applied to the base of the shell by the support. Let $\phi_o = \pi/4$, $E = 200$ GPa, $\nu = 0.3$, $R = 1000$ mm, $t = 15$ mm, and $p = 0.80$ MPa.

(b) In part (a), replace the fixed support by rollers that transmit only vertical force. Calculate net meridional and circumferential stresses in the shell at its base.

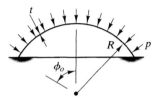

PROBLEM 13.8-9

13.8-10. In Problem 13.4-11, omit the reinforcing disk, so that only the two spherical shell segments remain. Let $R = 600$ mm, $t = 12$ mm, $\nu = 0.3$, and $p = 1.3$ MPa. Determine stresses at the juncture.

13.8-11. Let a cylindrical tank be capped by a hemispherical shell, as in Fig. 13.8-2. If material properties are the same throughout, for what ratio t_c/t_s will there be no bending when internal pressure is applied? Express t_c/t_s in terms of ν.

13.8-12. (a) Let a cylindrical shell be capped by a hemispherical shell, as in Fig. 13.8-2. Let $R = 500$ mm, $t = 10$ mm (for both shells), $E = 200$ GPa, and $\nu = 0.3$. For internal pressure $p = 0.80$ MPa, determine σ_θ at the juncture and σ_x in the cylindrical shell a distance $\pi/4\lambda$ from the juncture. Include both membrane and bending contributions.

(b) Repeat part (a), but let the cap have the shape shown in Fig. 13.4-2b, with $b = R/3$. Assume that the cap acts like a cylinder of radius R in its response to Q_o and M_o.

13.8-13. Use the following data for the cylinder-cap combination of Fig. 13.8-3: $R = 500$ mm, $t = 10$ mm (for both parts), $\phi_o = \pi/4$, $E = 200$ GPa, and $\nu = 0.3$. Solve for M_o and net radial loads on the cylindrical shell and on the cap due to internal pressure $p = 0.80$ MPa.

(a) Use the method described in connection with Fig. 13.8-3.

(b) Use the alternative method described in the text.

(c) Use the results of part (a)—or of part (b); answers should agree—to determine net meridional and circumferential stresses in the cylindrical shell at the juncture.

(d) Similarly, determine stresses in the cap at the juncture.

13.8-14. A spherical shell has a cylindrical outlet pipe, as shown on the next page. Internal pressure p is applied. Use data in the sketch to determine meridional and circumferential stresses of largest magnitude at the juncture. Express results in terms of p.

13.9-1. A homogeneous prismatic bar is bent elastically to a circular shape, then its ends are welded together to form a circular ring of centroidal radius R with initial stresses. Let $I_x = I_y$ (as is the case, for example, with cross sections in Fig. 1.3-2c). Show that such a ring has no resistance to ring-rolling moment M_o.

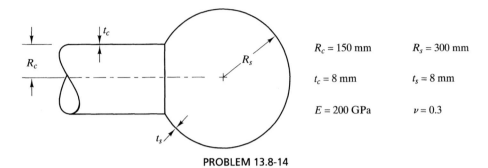

$R_c = 150$ mm $\quad R_s = 300$ mm
$t_c = 8$ mm $\quad t_s = 8$ mm
$E = 200$ GPa $\quad \nu = 0.3$

PROBLEM 13.8-14

13.9-2. In Case 6 of Section 12.7 (Selected Solutions for Axisymmetric Plates), let $a = 160$ mm, $b = 144$ mm, $t = 4$ mm, and $\nu = 0.3$. Use formulas for plates, then formulas for rings, to determine average rotation and circumferential stress on the lower surface at $r = b$ in terms of the load and E. What are percentage errors of the ring formulas?

13.9-3. In Fig. 13.9-2a, change the roller support to a hinge that allows rotation but prevents radial displacement. Let the ring be uniformly heated an amount T relative to its stress-free temperature. Determine the circumferential stress on surface AB and on surface CD.

13.9-4. A roof dome in the shape of a spherical shell segment has base opening angle $\phi_o = 51.8°$ and is loaded by its own weight. Its base is attached to a reinforcing ring of the same material. The ring is supported by rollers on a horizontal surface, in the manner of Fig. 13.9-2a. Only membrane stresses are to be present in the dome. If the ring cross section is a square of side length h, where on the top surface of the ring should the dome-ring intersection appear? And, if $R/t = 500$ for the dome, what should be dimension h in terms of t? Let $\nu = 0.3$.

13.9-5. (a) At its outer edge, a centrally loaded circular plate is fixed to a ring of the same material and of rectangular cross section. The plate is centered in the ring, as shown. The ring is supported by rollers around its inner edge. Determine stresses at the edge of the plate and in the ring in terms of force F.

(b) Repeat part (a) if the plate is attached to the bottom edge of the ring rather than halfway up.

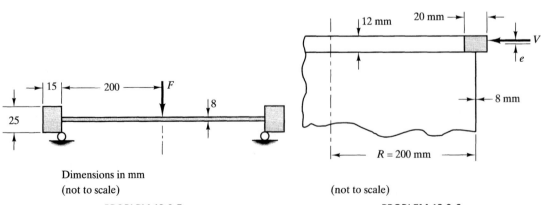

Dimensions in mm
(not to scale)

PROBLEM 13.9-5

(not to scale)

PROBLEM 13.9-6

13.9-6. A ring of rectangular cross section is attached to an end of a cylinder of the same material. The cylinder is radially centered on the ring, as shown. Distributed radial force V is offset a distance e from the center of the ring. For what value of e is there no meridional bending moment at the end of the cylinder? Let $\nu = 0.3$.

13.9-7. The sketch represents one of two pipes that have a flanged connection. Pipe and flange are of the same material. Uniformly distributed force F (N/mm) represents load applied by flange bolts (not shown). A gasket applies force G (also N/mm) that equilibrates force F. In terms of F, determine (a) significant stresses in the flange, and (b) significant stresses in the cylindrical pipe where it meets the flange. Let $\nu = 0.3$.

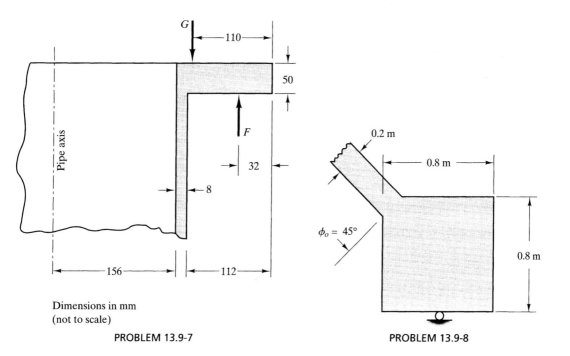

Dimensions in mm
(not to scale)

PROBLEM 13.9-7

PROBLEM 13.9-8

13.9-8. For the shell in Fig. 13.4-3a, let $R = 60$ m, $t = 0.2$ m, $\phi_o = \pi/4$, $\gamma = 23{,}500$ N/m³, and $E = 40$ GPa. Assume that $\nu = 0.3$. At its base, the shell is securely attached to a reinforcing ring of the same material. The ring has a square cross section and is supported by rollers, as shown in the sketch for the present problem. Assume that the shell midsurface passes through a corner of the square cross section of the ring. Determine significant meridional and circumferential stresses.

13.9-9. Ring-rolling moment M_o acts at centroidal radius R of the ring shown on the next page. This ring is *not* slender. Points in plane BB have no radial displacement when a small ring-rolling rotation ψ takes place. Coordinate s is measured from BB.

(a) Obtain an expression that defines the location of plane BB.
(b) Obtain an expression for ψ in terms of M_o, modulus E, and geometric quantities.
(c) Show that the expression for ψ reduces to Eq. 13.9-4 as the ring becomes slender.
(d) For the square cross section shown, determine ψ in terms of M_o and E. For this problem, what is the percentage error of Eq. 13.9-4?

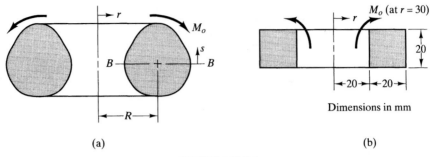

PROBLEM 13.9-9

13.9-10. The slender ring shown is simply supported around its inner edge and carries uniformly distributed line load q around its outer edge. Assume that the material has yield stress σ_Y in uniaxial stress and does not strain harden. In terms of σ_Y and dimensions shown, determine the value of q that makes all material plastic.

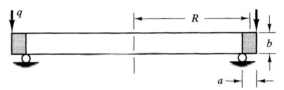

PROBLEM 13.9-10

CHAPTER 14

Buckling and Instability

14.1 INTRODUCTION. ELEMENTARY THEORY

Buckling is associated with loss of the stability of equilibrium. Column buckling is discussed in elementary mechanics of materials. Many additional situations related to buckling and instability arise in practice, such as columns that carry lateral load as well as axial load, plastic action during buckling, and nonlinear buckling and collapse of shells. Some of these problems are difficult to analyze. In this chapter, no difficult analyses of buckling and instability are undertaken. Instead we consider concepts and analysis methods that can be explained with relative ease. Most are not restricted to column problems. Some problems that require more advanced methods are noted. We begin by briefly reviewing a basic analysis procedure, as applied to the elementary problem of column buckling.

Column Buckling. In a typical problem that does *not* involve buckling, the load is prescribed. Unknowns are the resulting stresses and deformations. Stable equilibrium of the structure is presumed, and equilibrium equations are written using the undeformed geometry. In a buckling problem the *load* is unknown. We seek the magnitude of load that will destroy the stability of equilibrium. To solve such a problem, equilibrium equations must be written using the *deformed* geometry.

Consider the column in Fig. 14.1-1a. It is simply supported, uniform, linearly elastic, and initially straight. Axial compressive loads P are collinear and are directed through centroids of cross sections. We seek P_{cr}, the "critical" value of P that produces buckling. Dashed lines indicate the buckled shape $v = v(x)$, whose formula is as yet unknown. Making a cut in the beam (Fig. 14.1-1b) and summing moments, we obtain bending moment $M = -Pv$. The negative sign arises because $M < 0$ when the deflected shape is concave down as shown. (If the deflected shape were concave up, then $M > 0$, but $v < 0$, so again $M = -Pv$.) In buckling and related problems of this chapter, we elect to regard axial load as positive in compression. The small-deflection moment-curvature relation, Eq. 1.7-1, becomes

$$EI\frac{d^2v}{dx^2} = M \quad \text{where} \quad M = -Pv \qquad (14.1\text{-}1)$$

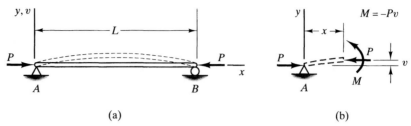

FIGURE 14.1-1 (a) Simply supported column loaded by axial force P. (b) A portion of the deflected column.

in which curvature d^2v/dx^2 is positive if the beam is concave up. Equation 14.1-1 has the solution

$$v = C_1 \sin \lambda x + C_2 \cos \lambda x \quad \text{where} \quad \lambda^2 = \frac{P}{EI} \quad (14.1\text{-}2)$$

and where C_1 and C_2 are constants. The boundary condition $v=0$ at $x=0$ yields $C_2=0$. Hence, the boundary condition $v=0$ at $x=L$ yields $C_1 \sin \lambda L = 0$. Now $C_1 \neq 0$ (because $v \neq 0$), $\lambda \neq 0$ (because $P \neq 0$ and EI is not infinite), and $L \neq 0$. Therefore λL must be π, or 2π, or 3π, and so on. That is, with n a positive integer,

$$\lambda L = n\pi \quad \text{hence, for } n = 1, \quad P_{cr} = \frac{\pi^2 EI}{L^2} \quad (14.1\text{-}3)$$

Buckling load P_{cr} appears for $n=1$ because the column buckles at the smallest P for which buckling is possible: Buckling does not await the larger axial load associated with $n=2$ or greater. For $n=1$, the buckling mode is a half sine wave of amplitude C_1, namely $v = C_1 \sin(\pi x/L)$. The magnitude of C_1 cannot be determined from our analysis because we have used the small-deflection approximation for curvature, namely $1/\rho = d^2v/dx^2$. If instead we use the exact expression for $1/\rho$ we obtain the "elastica" problem [4.5], whose solution relates lateral deflection to P in the post-buckling regime $P > P_{cr}$. The solution of this difficult nonlinear problem was published by Euler in 1744.

The foregoing analysis presumes that the member is perfectly straight, that loads are perfectly axial, and so on. The average stress P_{cr}/A is presumed small enough that there is no yielding before buckling. Practical columns depart from one or more of these ideals. Ideal versus realistic behavior is briefly described as follows (see also Section 14.8).

Consider an ideal column, so slender that its material is always linearly elastic. Load P_{cr} is reached at point B in Fig. 14.1-2a. With yet more axial compression, a perfect column may stay straight (path BC) or it may bend (path BD). Postbuckling path BD is described by the solution of the elastica problem. Postbuckling path BF is described by linearized analysis, Eq. 14.1-1, which is strictly valid only at point B. In reality, path BC is never followed: Even infinitesimal imperfections or disturbances are enough to impose path BD. Larger imperfections result in path OE, for which lateral deflection is noticeable as soon as loading begins. A somewhat slender column behaves similarly, except that even modest lateral deflection v_c may be sufficient to produce yielding of the material, loss of stiffness, and collapse (Fig. 14.1-2b). Nevertheless, the elastic buckling load P_{cr} remains a good estimate of the load-carrying capacity of these columns. The capacity of a less slender column is influenced by the yield strength of the material.

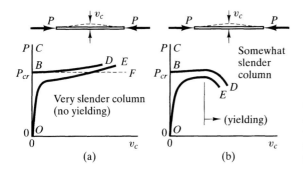

FIGURE 14.1-2 Qualitative plots of axial load P versus central lateral deflection v_c in a column with simply supported ends. (a) Material linearly elastic for all v_c. (b) Material yields at stress slightly greater than P_{cr}/A.

Additional considerations, including the effects of other support conditions and eccentric axial loads, appear in elementary textbooks on mechanics of materials.

A Physical Interpretation. If Eq. 14.1-1 is differentiated twice with respect to x, with EI constant, we obtain

$$EI\frac{d^4v}{dx^4} = q \quad \text{where} \quad q = -P\frac{d^2v}{dx^2} \quad (14.1\text{-}4)$$

In the first equation, q can be regarded as distributed lateral load, positive if in the same direction as v. We know that this is so from Eq. 1.7-2. Equation 14.1-4 has the physical interpretation that, in the buckling mode, a conceptual distributed lateral force q associated with axial force P is equilibrated by elastic bending resistance of the laterally deflected beam. As an example, consider the displacement mode associated with buckling, $v = C_1 \sin(\pi x/L)$. Thus q is a half sine wave, and Eq. 14.1-4 yields $P = \pi^2 EI/L^2$, which is indeed the buckling load.

Critical Speed. A shaft spinning about its axis has critical speed ω_{cr} (also called *whirling* or *whipping* speed). A formula for ω_{cr} can be obtained by an analysis that resembles buckling analysis. Consider the special case of a straight and uniform simply supported shaft of negligible mass, which carries a disk of mass m attached to its center (Fig. 14.1-3a). The mass center of the disk is a distance e from the shaft axis. When the shaft rotates, it develops lateral deflection v, which is elastically resisted by force kv. Stiffness k can be calculated from elementary beam theory as the ratio of a lateral force F to the static displacement produced by F. The present case is that of a simply supported and centrally loaded beam, so that k equals F divided by $FL^3/48EI$, or $k = 48EI/L^3$. Neglecting damping and deflection due to weight, and adapting the physical interpretation associated with Eq. 14.1-4 to the present problem, we write*

$$kv = m(e + v)\omega^2 \quad \text{hence} \quad v = \frac{me\omega^2}{k - m\omega^2} \quad (14.1\text{-}5)$$

---

*The term $m(e + v)\omega^2$ is an "inertia force" or a "D'Alembert force," equal in magnitude to mass times acceleration and directed opposite to acceleration. In the present context it is a "centrifugal force." A D'Alembert force need not be invoked, but many regard it as a helpful conceptual devic

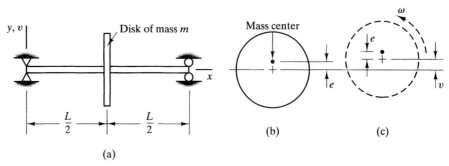

FIGURE 14.1-3 (a) Disk attached to the center of a shaft. (b) End view, disk stationary. (c) End view, showing displacement v due to spinning.

Lateral deflection v is small when $m\omega^2 \ll k$, but becomes unbounded when $m\omega^2 = k$, even if e is very small. Thus we obtain the critical speed

$$\omega_{cr} = \sqrt{\frac{k}{m}} \quad \text{or} \quad \omega_{cr} = \sqrt{\frac{48EI}{mL^3}} \qquad (14.1\text{-}6)$$

Substitution of Eq. 14.1-6 into Eq. 14.1-5 provides

$$v = \frac{e(\omega/\omega_{cr})^2}{1 - (\omega/\omega_{cr})^2} \qquad (14.1\text{-}7)$$

As ω_{cr} is approached, the shaft loses elastic resistance to externally applied lateral load. For $\omega \gg \omega_{cr}$, v approaches $-e$, so that the disk spins stably about its mass center. Therefore speeds greater than ω_{cr} are acceptable, and can be reached if speeds near ω_{cr} are passed through quickly, so that disastrous magnitudes of v do not have time to build up.

Readers familiar with elementary vibration theory will recognize ω_{cr} as the lateral vibration frequency of a lightweight nonrotating beam that carries a mass m at midspan. Similarly, the shaft alone (without the disk) has a critical ω equal to its frequency of lateral vibration when not spinning (see Section 14.2).

14.2 AN ENERGY METHOD

Simple Example. To introduce the energy method simply, we consider the following problem. A rigid tube, Fig. 14.2-1a, contains a linear spring of stiffness k_1. This spring is compressed an amount u by load P. The tube is hinged at A, and supported at B by a second linear spring of stiffness k_2. A plug that slides freely in the tube makes frictionless contact with block C. Gravity is ignored. What magnitude of load P makes the structure buckle in the plane of the paper?

Load P reaches the buckling value P_{cr} before there is any lateral displacement of tube AB. Now let the tube rotate a small amount about A while block C does not move. Axial force remains essentially constant at P_{cr} while the spring of stiffness k_1 lengthens a small amount Δu. Now Δu can be written in terms of lateral displacement

Section 14.2 An Energy Method

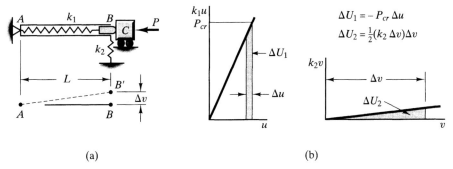

FIGURE 14.2-1 (a) Buckling of a structure built of rigid elements and two springs. (b) Energy relationships, with Δu a decrease in the magnitude of u.

Δv at B by applying first the Pythagorean theorem, then the binomial series, retaining only its first two terms.

$$\Delta u = [L^2 + (\Delta v)^2]^{1/2} - L \approx \left[1 + \frac{1}{2}\left(\frac{\Delta v}{L}\right)^2\right]L - L = \frac{1}{2}\left(\frac{\Delta v}{L}\right)^2 L \quad (14.2\text{-}1)$$

Strain energy U_1 in the spring of stiffness k_1 undergoes a change $\Delta U_1 = -P_{cr}\Delta u$. This energy change is negative because some strain energy in the compressed spring is released by lengthening. The spring of stiffness k_2 gains strain energy in the amount $\Delta U_2 = k_2(\Delta v)^2/2$. External forces do no work during lateral displacement Δv, so the net change in strain energy is zero. That is,

$$\Delta U_1 + \Delta U_2 = 0 \quad \text{hence} \quad P_{cr} = k_2 L \quad (14.2\text{-}2)$$

This value of P_{cr} is easily checked by applying static equilibrium considerations, as is commonly done in elementary textbooks.

Column Buckling. Let the foregoing method be applied to an elastic column that lies along the x axis. For a differential length dx, term $\Delta v/L$ in Eq. 14.2-1 becomes rotation dv/dx, which is a function of x because lateral displacement is $v = v(x)$. Energy change ΔU_2 becomes strain energy of flexure (given by the last term in Eq. 4.9-2). For P positive in compression, the energy conservation relation $\Delta U_1 + \Delta U_2 = 0$ yields

$$\int_0^L \frac{P_{cr}}{2}\left(\frac{dv}{dx}\right)^2 dx = \int_0^L \frac{EI}{2}\left(\frac{d^2v}{dx^2}\right)^2 dx \quad (14.2\text{-}3)$$

In the first integral, P_{cr} need not be constant. For example, if there is distributed axial load, P depends on x. Then P_{cr} is replaced by P times a critical multiplier that controls the magnitude of the load but not its distribution. The first integral incorporates the same small-deflection approximation as used in Eq. 14.2-1.

One can easily check that Eq. 14.2-3 yields the correct buckling load when the correct buckling mode is substituted. For the column of Fig. 14.1-1a, using Eq. 14.2-3,

$$v = \bar{v}\sin\frac{\pi x}{L} \quad \text{yields} \quad P_{cr} = \frac{\pi^2 EI}{L^2} \quad (14.2\text{-}4)$$

where $\bar{v}$ is the amplitude of the half sine wave. Like the analysis of Section 14.1, the present analysis says nothing about amplitude. We can only say that it is "small," since the analysis is based on small-deflection theory.

Remarks. A physical interpretation of the energy method is as follows. Prior to sudden buckling, strain energy in columns, plates, and some shells is mainly associated with membrane strains. These strains are produced by stresses directed along the length of a bar or tangent to the surface of a plate or a shell, and are constant through the thickness, in contrast to bending strains that vary through the thickness. Membrane stiffness greatly exceeds bending stiffness, and a large amount of strain energy can be stored in the membrane mode with deformations so small that they are not noticed. Buckling occurs when membrane strain energy is large enough that an alternative deformation mode associated with lateral displacement, such as $v = \bar{v} \sin(\pi x/L)$ in Eq. 14.2-4, allows some of the membrane energy to be converted into an equal amount of bending energy. Deformations associated with buckling can be large, and failure can be catastrophic because it is sudden and unexpected.

An important feature of energy methods is that one need not know the correct buckling mode in order to calculate a critical load. An approximate displacement mode yields an approximate critical load. For example, the buckling mode of the column in Fig. 14.1-1a might be approximated as a parabola. Thus Eq. 14.2-3 provides

$$v = \frac{4\bar{v}x(L-x)}{L^2} \quad \text{yields} \quad P_{cr} = \frac{12EI}{L^2} \tag{14.2-5}$$

This result is not a good approximation because the assumed mode is poor: it says that curvature d^2v/dx^2 is constant along the length. An improved approximation can be obtained by assuming that d^2v/dx^2 has the form $x(L-x)$, then integrating to obtain dv/dx. Another alternative is to substitute $d^2v/dx^2 = M/EI$, so that the second integral in Eq. 14.2-3 has the integrand $M^2/2EI$, where $M = -P_{cr}v$ for a simply supported column with $P = P_{cr}$. Thus, when M rather than d^2v/dx^2 appears in the strain energy integral, the form $x(L-x)$ for v provides a better approximation. However, for other column problems, this approach may not be as simple. For example, if one end of the simply supported column is now made fixed, the expression for M must include end moment and lateral force terms, but the end moment and lateral force are not initially known in terms of P_{cr}.

Equation 14.2-3 provides an upper bound on P_{cr} if $v = v(x)$ is not exact. This upper bound happens because an approximate displacement mode constrains the mathematical model to displace into a shape it does not prefer. Constraint stiffens the model and therefore raises P_{cr}. Accuracy improves as the approximating mode is improved.

An energy-balance equation can be written in forms applicable to any structural form and any manner of buckling. The resulting equations can be discretized, by finite elements or other numerical methods. In doing so, approximate displacement modes are automatically introduced. Computed buckling loads are approximate, but can be indefinitely improved by refining the discretization.

Critical Speed of a Shaft. Let the disk be removed from the shaft of Fig. 14.1-3, and let the shaft be uniform and have mass ρ per unit length. The critical rotational speed can be determined by an energy argument in which work done by centrifugal forces is equated to energy stored in bending.

When a length increment dx has lateral displacement v, the associated force increment produced by spinning is $dF = (\rho\, dx)v\, \omega^2$. Since dF varies linearly with v, the associated work increment is $dW = dF(v/2)$. Accordingly

$$dW = \frac{1}{2}dF\, v = \frac{1}{2}(\rho\, dx\, v\, \omega^2)v \quad \text{hence} \quad W = \frac{\rho\omega^2}{2}\int_0^L v^2\, dx \quad (14.2\text{-}6)$$

This W must be equal to the strain energy integral on the right-hand side of Eq. 14.2-3. Hence, substitution of the correct displacement mode yields the correct critical speed:

$$v = \bar{v}\sin\frac{\pi x}{L} \quad \text{yields} \quad \omega_{cr} = \frac{\pi^2}{L^2}\sqrt{\frac{EI}{\rho}} \quad (14.2\text{-}7)$$

which is also the frequency of lateral vibration for a simply supported and nonrotating uniform beam. An approximate lateral displacement mode $v = v(x)$ would provide an approximate ω_{cr}.

More generally, L in Eq. 14.2-7 is the length of a half-wave in the deflected shape. The shaft has another (higher) critical frequency for two half-waves, another for three, and so on.

14.3 BEAM-COLUMNS: CLASSICAL THEORY

In practice, a column is rarely loaded in pure axial compression. A column that carries lateral load or moment load in addition to axial compression is known as a *beam-column*. In a beam-column, deflection and bending moment produced by lateral loading are amplified by axial compressive load.

Consider the problem of Fig. 14.3-1a. Let $M_{\text{lat}} = M_{\text{lat}}(x)$ be bending moment due to distributed lateral load q, calculated in the usual way, as if $P = 0$. (In general, M_{lat} accounts for all lateral loads and for moment loads.) We presume that P is positive in compression and is less than the buckling load P_{cr}. The differential equation that describes the problem is Eq. 14.1-1, where now $M = M_{\text{lat}} - Pv$. This equation, and its solution for constant EI, are

$$\frac{d^2v}{dx^2} + \lambda^2 v = \frac{M_{\text{lat}}}{EI} \quad \text{where} \quad \lambda^2 = \frac{P}{EI} \quad (14.3\text{-}1)$$

$$v = C_1 \sin \lambda x + C_2 \cos \lambda x + v_p \quad (14.3\text{-}2)$$

where C_1 and C_2 are constants of integration and v_p is a particular solution of Eq. 14.3-1 associated with M_{lat}.

For the problem of Fig. 14.3-1b, $M_{\text{lat}} = -M_B x/L$ and $v_p = -M_B x/(EI\lambda^2 L)$. Constants C_1 and C_2 are evaluated from the conditions $v = 0$ at $x = 0$ and $v = 0$ at $x = L$. Thus

$$v = \frac{M_B}{EI\lambda^2}\left[\frac{\sin \lambda x}{\sin \lambda L} - \frac{x}{L}\right] \quad \text{and} \quad M = EI\frac{d^2v}{dx^2} = -M_B \frac{\sin \lambda x}{\sin \lambda L} \quad (14.3\text{-}3)$$

That M is negative means there is tension on the $+y$ beam surface when M_B is directed as shown. As a numerical example, we examine lateral deflection and bending moment

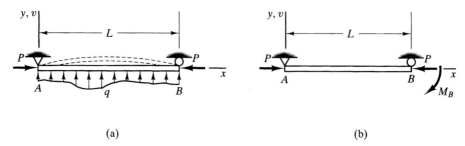

FIGURE 14.3-1 (a) Simply supported beam loaded by axial force P and lateral load q. (b) Simply supported beam loaded by axial force P and end moment M_B.

at midspan, $x = L/2$. For $P = 0$, from elementary beam theory, these quantities are $v_{L/2} = M_B L^2/16EI$ and $M_{L/2} = -M_B/2$. For $P \neq 0$, each result can be stated as an "amplification factor," which is the ratio of v (or M) for the given P to v (or M) for $P = 0$. Also, P can be stated as a fraction of P_{cr} where, in the present example, $P_{cr} = \pi^2 EI/L^2$. Various values of P/P_{cr} provide the following midspan amplification factors for the problem of Fig. 14.3-1b.

P/P_{cr}:	0.00	0.30	0.60	0.80	0.90	0.95
$v_{L/2}$ amplification:	1.00	1.44	2.55	5.13	10.29	20.61
$M_{L/2}$ amplification:	1.00	1.53	2.88	6.06	12.42	25.15

The foregoing solution is based on the usual approximation for curvature, $1/\rho = d^2v/dx^2$. This expression is satisfactory if dv/dx is small in comparison with unity. A numerical result should be regarded with suspicion if the magnitude of dv/dx exceeds roughly 0.2 anywhere along the beam.

Formulas for various beam-columns appear in [1.6, 1.9, 4.5]. For the special case of zero lateral load, but with bending present because axial forces P are not directed through centroids of cross sections, "secant formulas" may be found in the columns chapter of textbooks on elementary mechanics of materials.

Tensile Axial Load. If load P is tensile, the algebraic sign in Eq. 14.3-1 becomes negative, and the solution contains hyperbolic functions rather than trigonometric functions. Results are tabulated for many cases [1.6]. The physical effect of tensile axial force is to reduce deflection and bending moment produced by lateral loading. Tensile axial force in a beam can be produced by lateral loading if beam ends are restrained against axial displacement. The mechanism is as described for the analogous plate problem in the latter part of Section 12.6. However, in a beam the resulting tensile axial force and its consequences are usually small. It is more common for compressive axial force to be of concern, as when a compression member must also resist significant lateral load.

Superposition. In Eq. 14.3-3, we see that v and M are directly proportional to M_B. Solutions of other beam-column problems [1.6, 1.9, 14.1] also show results propor-

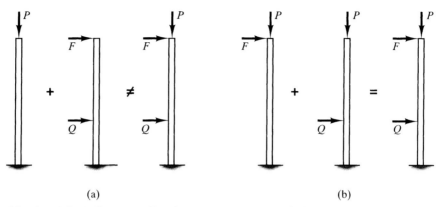

FIGURE 14.3-2 Deflections and bending moments *cannot* be obtained by superposition in the manner shown in part (a), but *can* be obtained in the manner shown in part (b).

tional to transverse load. So the question arises: Can the principle of superposition be applied to beam-column problems?

In most other practical problems of stress analysis, the principle of superposition is applicable. Thus, if L_1 and L_2 are two sets of loads, deformation due to loads L_1 alone, plus deformation due to loads L_2 alone, is equal to deformation due to loads L_1 and L_2 acting simultaneously. Superposition is valid if deflections are small, the material is linearly elastic, and conditions of support and loading are unchanged by deformation. In a beam-column problem, the latter restriction is violated because bending moment due to axial load P is a function of deformation. This effect of P does not mean that superposition cannot be applied to a beam-column problem. It means that P must be excluded from loads superposed.

As an example, consider the problem of Fig. 14.3-2. In Fig. 14.3-2a, there is no lateral deflection when $P < P_{cr}$ is applied alone. If loads F and Q are then added, superposition predicts that P does not influence the resulting deformation. This incorrect conclusion is the result of superposition misapplied. Superposition is correctly applied in Fig. 14.3-2b, provided that P is not changed. The mathematical justification is that F and Q affect only the particular solution of the differential equation, v_p in Eq. 14.3-2. Whether v_p is constructed from F, then Q, and results added, or from F and Q acting simultaneously, makes no difference.

14.4 BEAM-COLUMNS: APPROXIMATIONS

For many loadings of practical interest, beam-column formulas are complicated and tedious to derive. Approximate methods may be useful even though they are not exact. A simple approximate method is now described.

The differential equation of the beam-column problem is Eq. 14.3-1. Rearranged, this equation is

$$EI\frac{d^2v}{dx^2} = M_{\text{lat}} - Pv \qquad (14.4\text{-}1)$$

where M_{lat} is bending moment due to lateral load and/or moment load alone, when $P = 0$. Accordingly, for $P = 0$,

$$EI\frac{d^2v_{lat}}{dx^2} = M_{lat} \qquad (14.4\text{-}2)$$

where v_{lat} is lateral displacement associated with M_{lat}. Elimination of M_{lat} between Eqs. 14.4-1 and 14.4-2 yields

$$EI\frac{d^2v}{dx^2} = EI\frac{d^2v_{lat}}{dx^2} - Pv \qquad (14.4\text{-}3)$$

We now make the crucial assumption on which the accuracy of results depends: that the deflection curve has practically the same shape whether or not axial load is applied. Axial load P is assumed to change only the amplitude of lateral deflection. Thus, for a simply supported beam of length L, one might approximate the actual shape by a half sine wave:

$$v = \bar{v}\sin\frac{\pi x}{L} \quad \text{and} \quad v_{lat} = \bar{v}_{lat}\sin\frac{\pi x}{L} \qquad (14.4\text{-}4)$$

where $\bar{v}$ and $\bar{v}_{lat}$ are amplitudes of lateral deflection, respectively for $P > 0$ and $P = 0$. Substitution of Eqs. 14.4-4 into Eq. 14.4-3 yields

$$\frac{\pi^2 EI}{L^2}\bar{v} = \frac{\pi^2 EI}{L^2}\bar{v}_{lat} + P\bar{v} \qquad (14.4\text{-}5)$$

We recognize $\pi^2 EI/L^2$ as the column buckling load P_{cr}. Therefore lateral deflection amplitude $\bar{v}$ of the beam-column is

$$\text{Approximate:} \quad \bar{v} = \frac{\bar{v}_{lat}}{1 - (P/P_{cr})} \qquad (14.4\text{-}6)$$

Assuming that bending moment is directly proportional to lateral deflection, we write the expression for bending moment in the beam-column as

$$\text{Approximate:} \quad M = \frac{M_{lat}}{1 - (P/P_{cr})} \qquad (14.4\text{-}7)$$

Bending moments M and M_{lat} pertain to the same location on the beam, as do $\bar{v}$ and $\bar{v}_{lat}$.

Equation 14.4-6 was derived for simply supported ends. Equations 14.4-6 and 14.4-7 can also be used with other support conditions; however, accuracy may decrease, as can be shown by comparison with exact solutions [1.6].

The reciprocal of $1 - (P/P_{cr})$ is an approximate amplification factor that "converts" a beam problem to a beam-column problem. Accuracy is problem-dependent. For a simply supported beam loaded by moment at one end (Fig. 14.3-1b), and the respective P/P_{cr} values tabulated below Fig. 14.3-1, values of the approximate amplification factor are 1.00, 1.43, 2.50, 5.00, 10.00, and 20.00. We see that (for this problem) the approximation is quite good for lateral displacement and fairly good for bending moment. However, if instead the beam were loaded by almost-equal moments at each end, both directed clockwise, $\bar{v}_{lat}$ (for $P = 0$) would display two loops of almost equal span, while v for $P \approx P_{cr}$ would display two loops of quite unequal span. For this case the approximation would be poor.

For situations in which support reactions are statically determinate, an improved bending moment formula is easily obtained by using statics to calculate moment due to P. For example, in a simply supported beam, at the location where $\bar{v}$ appears,

$$\text{Improved approximation:} \quad M = M_{\text{lat}} - P\bar{v} \tag{14.4-8}$$

where $\bar{v}$ is obtained from Eq. 14.4-6. The negative sign arises because a negative (downward) $\bar{v}$ causes positive (compressive) P to contribute a positive bending moment. For $P = 0.95 P_{cr}$ in the problem of Fig. 14.3-1b, $M_{\text{lat}} = -M_B/2$, and Eq. 14.4-8 yields 24.44 as the amplification factor for M at midspan.

14.5 INELASTIC BUCKLING

When a column buckles, a cross section of area A carries average compressive stress $\sigma_{cr} = P_{cr}/A$. If an elastic column formula provides $\sigma_{cr} > \sigma_Y$, where σ_Y is the yield strength in compression, the calculated σ_{cr} is incorrect; a revised formula is needed that accounts for plastic action.

Conceptually, in obtaining a formula for the elastic critical load of a perfectly straight and axially loaded column, we assume that a small lateral displacement takes place only after P has been increased to P_{cr}. When there is plastic action, this conceptual sequence of events leads to difficulties as follows.

Let the simply supported column of Fig. 14.5-1a have a material whose compressive stress-strain relation is shown in Fig. 14.5-1b. Here tangent modulus $E_t = d\sigma/d\epsilon$ is the slope of the curve beyond the proportional limit. Increasing axial load levels 1, 2, 3, and so on produce increasing levels of axial strain. Assume that load level 4 is associated with plastic action. If we imagine that load level 4 is reached, after which a small lateral displacement takes place at the same load, then the magnitude of axial strain decreases at B and increases at C (Fig. 14.5-1c). Material that unloads does so

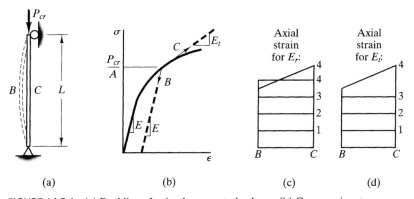

FIGURE 14.5-1 (a) Buckling of a simply supported column. (b) Compressive stress-strain relation of the material. (c, d) Distributions of compressive axial strain, according to double-modulus theory and Shanley theory, respectively.

elastically. Thus, near the middle of the column, buckling occurs with elastic modulus E on the convex side and tangent modulus E_t on the concave side (Fig. 14.5-1b). This "double-modulus" concept, which dates from 1895, leads to a "reduced modulus" E_r, where $E_t < E_r < E$, and to the buckling load

$$\text{Reduced modulus load:} \quad P_{cr} = \frac{\pi^2 E_r I}{L^2} \qquad (14.5\text{-}1)$$

On the other hand, when the straight column is about to buckle, all material has the same stress, and therefore has effective modulus E_t. Thus we would expect the buckling load to be given by the tangent modulus formula, which dates from 1889:

$$\text{Tangent modulus load:} \quad P_{cr} = \frac{\pi^2 E_t I}{L^2} \qquad (14.5\text{-}2)$$

Now difficulties appear. Since $E_t < E_r$, the column should buckle before the double-modulus buckling load is reached. Also, if modulus E_r is to exist there must be elastic unloading on the convex side, yet the column is presumed straight until it buckles, which means that to reach the double-modulus load the column must "buckle before it buckles." These difficulties were noted and resolved by Shanley in 1946.

Shanley realized that one may imagine that the last increment of axial load is applied *simultaneously* with the appearance of a small lateral displacement. Thus we obtain axial strain distributions shown in Fig. 14.5-1d. There need be no unloading in going from load level 3 to load level 4: Strains at B and C both increase in magnitude, with modulus E_t. With yet more displacement, there is some unloading at B and the effective modulus becomes greater than E_t, although it does not reach E_r. In summary, the axial load at which buckling begins is given by the tangent-modulus formula. The sustainable load may then increase slightly, but will not reach the double-modulus load, even when tests are carefully done. References for plastic buckling and for Shanley's work include [11.3, 11.9, 14.2, 14.3] and others cited in [14.1].

Tangent-modulus theory is entirely satisfactory for plastic analysis of columns that buckle by overall bending (rather than by local buckling). Trial calculation can be used to determine P_{cr} for a given column as follows. Assume a value of average axial stress σ, obtain the associated E_t from a stress-strain plot, then calculate P_{cr} from Eq. 14.5-2. If P_{cr}/A agrees with the σ assumed, the calculated P_{cr} is considered satisfactory. If substantial disagreement is encountered another trial is needed.

In elastic buckling, end conditions may be taken into account by applying a multiplier to the critical load of a simply supported column of equal length. This practice does not work if imposition of constraints on a simply supported column changes buckling from the elastic range to the plastic range, because E changes to E_t. For example, if we multiply the elastic P_{cr} of a simply supported column by 4 because ends are in fact fixed, we may considerably overestimate the actual buckling load.

Residual Stress. Implicit in the foregoing plastic buckling analysis is the assumption that the same stress-strain relation prevails throughout the cross section. Such is not the case if residual stresses are present, due perhaps to uneven cooling after hot rolling of structural steel or welding of built-up sections. For example, in a rolled wide-flange

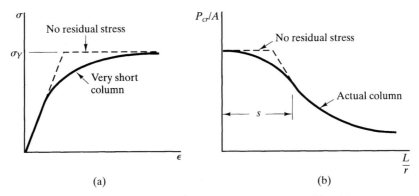

FIGURE 14.5-2 (a) Compressive stress-strain relation of a material. (b) Critical stress versus slenderness ratio of a column. Dashed lines pertain to a material with a definite yield stress and no strain hardening.

section, axial residual stresses are compressive at flange tips and the middle of the web, and tensile near web-flange intersections (which cool more slowly). In magnitude, residual stresses may approach half the yield strength in a rolled section, and may be even higher in welded sections. Material already in compression due to residual stress does not have as far to go to enter the plastic regime when axial compressive force is applied to the column. Thus, the effect of residual stress is to reduce the buckling load, to perhaps 70% of the value that would be expected from calculations based on material properties obtained from testing a coupon of material cut from the member [14.4]. Strength reduction is most significant for columns of intermediate slenderness ratio: Long columns still buckle elastically, and very short columns tend to yield in compression rather than buckle.

The tangent-modulus formula can be used for buckling analysis if the stress-strain relation is obtained by using a very short column with residual stresses intact as the compression test specimen. The resulting stress-strain relation is an average that includes the effect of residual stresses. Thus we can use the tangent-modulus formula to predict the actual buckling load in span s of Fig. 14.5-2b. Dashed lines in Fig. 14.5-2 pertain to columns whose residual stresses have been removed by annealing, and whose material displays a well-defined yield stress and no strain hardening. Such columns buckle when axial stress P/A equals the elastic buckling stress σ_{cr} or the yield stress σ_Y, whichever is reached first.

14.6 PLATES: BUCKLING, EFFECTIVE WIDTH, AND FAILURE

Buckling. An elastic flat plate under in-plane loading remains flat until a critical stress is reached, whereupon the lateral displacements of buckling appear. Thus, plate buckling is similar to column buckling, although plate buckling is more complicated because of biaxiality of the stress field and the variety of possible plate geometries, loadings, and support conditions. A simply supported rectangular plate of uniform thickness in uniform uniaxial compression buckles as shown in Fig. 14.6-1b.

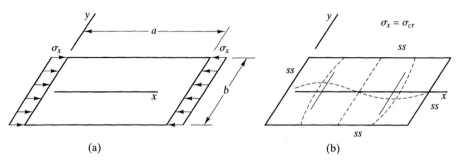

FIGURE 14.6-1 (a) A rectangular plate in uniaxial compression. (b) Dashed lines show the buckling mode for simply supported edges.

There is one half-wave in the y direction. In the x direction there is one half-wave for $a/b < \sqrt{2}$, two half-waves for $\sqrt{2} < a/b < \sqrt{6}$, and so on [4.5]. The (compressive) buckling stress σ_{cr} may be taken as

$$\text{For } \frac{a}{b} > 1 \text{ and } \nu = 0.3: \quad \sigma_{cr} = 3.62E\left(\frac{t}{b}\right)^2 \quad (14.6\text{-}1)$$

where t is the plate thickness. This value is exact if a/b is 1, 2, 3, and so on. If a/b is not an integer, the correct σ_{cr} is only slightly greater, so Eq. 14.6-1 is a good approximation for any a/b greater than 1 [4.5].

In Fig. 14.6-1, σ_x is uniformly distributed. Next, let σ_x be applied via a rigid block (Fig. 14.6-2a). Then axial stress σ_x is uniform across dimension b until buckling at critical load $P_{cr} = bt\,\sigma_{cr}$. Load *greater* than P_{cr} can be applied: Although the central portion of the plate can carry little additional stress because it is buckled, outer portions can

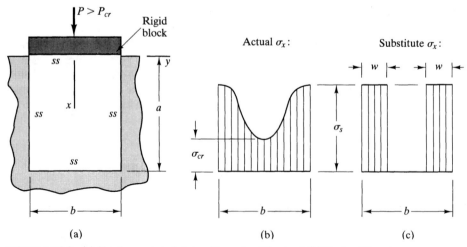

FIGURE 14.6-2 (a) Simply supported plate with postbuckling load. (b) Postbuckling stress distribution. (c) Postbuckling stress distribution in the effective width idealization.

carry increasing stress because they are constrained against failure by nearby supports. Stress in the central portion remains approximately σ_{cr}, while stress near vertical edges increases as load P increases. The postbuckling stress distribution is approximately as shown in Fig. 14.6-2b.

In practice, such behavior may appear in sheet metal attached to two or more parallel bars (or beams). Where sheet metal is attached to bars, it is restrained from buckling; midway between bars, buckling may occur. Analysis of such a problem is greatly simplified by empirical formulas now described.

Effective Width. For convenience in calculation, von Kármán suggested in 1932 that the stress distribution of Fig. 14.6-2b might be replaced by that of Fig. 14.6-2c, where edge stress σ_s is presumed uniform over an "effective width" w that depends on buckling stress σ_{cr}, edge stress σ_s, and overall width b. Load P can be expressed in terms of either σ_x or σ_s.

$$P = \int_{-b/2}^{b/2} \sigma_x t\, dy \quad \text{or} \quad P = \sigma_s(2wt) \tag{14.6-2}$$

If σ_x is known or approximated, these equations provide an expression for w. For example, a cosine variation might be assumed for σ_x:

$$\sigma_x = \frac{\sigma_s + \sigma_{cr}}{2} - \frac{\sigma_s - \sigma_{cr}}{2} \cos \frac{2\pi y}{b} \quad \text{yields} \quad w = \frac{b}{4}\left(1 + \frac{\sigma_{cr}}{\sigma_s}\right) \tag{14.6-3}$$

Extensive testing, using various b/t ratios, support conditions, and load levels, has given rise to many other formulas for w [1.9, 12.5, 14.5]. A simple formula is

$$\text{For } \sigma_s > \sigma_{cr}: \quad w = 0.447b \sqrt{\frac{\sigma_{cr}}{\sigma_s}} \quad \text{or} \quad w = 0.850t \sqrt{\frac{E}{\sigma_s}} \tag{14.6-4}$$

where the latter form is obtained by use of Eq. 14.6-1. Equations 14.6-2 and 14.6-4 provide an expression for load carried by the buckled panel:

$$\text{For } \sigma_s > \sigma_{cr}: \quad P = 1.70t^2 \sqrt{E\sigma_s} \tag{14.6-5}$$

For curved panels, other formulas for buckling and effective width must be used. Such formulas find application in aircraft structures, where low weight is paramount. In aircraft, panels are attached to beams or other stiffeners and serve as stressed skin.

EXAMPLE

The box beam in Fig. 14.6-3 is loaded in bending by axial forces F. The beam consists of thin flat panels of thickness t attached to bars of cross-sectional areas A_1, A_2, and A_3. Determine stresses and effective widths in the upper (compression) flange for $F = 1500$ N and for $F = 6000$ N. Assume that bars do not buckle in compression.

To keep things simple, we assume that vertical webs serve to maintain flange separation but do not contribute to bending stiffness, that individual bars have no bending stiffness, and that

458   Chapter 14   Buckling and Instability

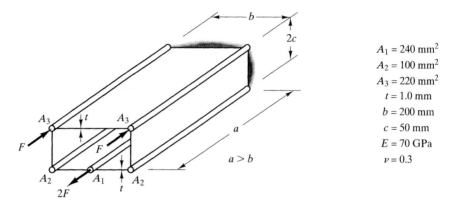

FIGURE 14.6-3 Cantilever box beam loaded by bending moment $M = 4Fc$.

preceding equations of the present section are applicable. For $F = 1500$ N, bending moment is $M = 300$ N·m. We tentatively assume that there is no buckling at this load. Then, for the dimensions given, the neutral axis is at mid-depth, and elementary beam theory is applicable. Thus

$$I = 2[(2A_3 + 200)(50^2)] = 3.20(10^6) \text{ mm}^4 \tag{14.6-6a}$$

$$\sigma_s = \frac{Mc}{I} = \frac{300{,}000(50)}{3.20(10^6)} = 4.69 \text{ MPa} \tag{14.6-6b}$$

where σ_s is the compressive stress all across the upper flange and in the two compression bars. From Eq. 14.6-1, $\sigma_{cr} = 6.34$ MPa (in compression). Since $\sigma_s < \sigma_{cr}$, the compression flange has not yet buckled, and $\sigma_s = 4.69$ MPa may be accepted as correct.

The second load, $F = 6000$ N, is four times as large as the first load. Clearly it will cause buckling. Accordingly, from Eq. 14.6-5,

$$2F = 2A_3\sigma_s + 1.70t^2\sqrt{E\sigma_s} \quad \text{hence} \quad \sigma_s = 22.43 \text{ MPa} \tag{14.6-7}$$

Stress σ_s exists in compression bars and in immediately adjacent strips of the compression panel. Note that quadrupling F more than quadruples the compressive stresses.

The total effective width in the buckled panel is $2w$. From Eq. 14.6-4,

$$2w = 2(0.850)(1)\sqrt{\frac{70{,}000}{22.43}} = 95.0 \text{ mm} \tag{14.6-8}$$

The tension flange is fully effective, not having buckled. Tensile force $2F$ acts on area $A_1 + 2A_2$ plus cross-sectional area of the tension panels. Thus, $\sigma = 18.75$ MPa in the tension flange (except near the end where load $2F$ is applied; see the following subsection entitled *Shear Lag*).

In a more realistic structure one might see several compression panels, probably curved rather than flat, and not all at the same distance from centroidal axes of the beam. Then statics alone is not sufficient to determine panel loads. An iterative analysis is required. In the first calculation cycle all material can be considered fully effective and stresses determined by beam theory. From these stresses one can obtain effective widths and therefore a new cross section, whose size and outline are like the original but which lacks panel material between effective widths. Centroidal axes of the new cross section can be located and beam theory reapplied. Calculation cycles repeat until the cross section ceases to change.

Tension-Field Beam. Another structure in which buckling appears is the *tension-field beam*, Fig. 14.6-4a (also called a *Wagner beam*, after its primary investigator, in 1929). Web panels are thin and are loaded in shear. For the direction of load P shown, and before buckling, pure shear in the web demands web compressive stress in directions AF, BG, and CH. For sufficiently high load, webs buckle, with crests and troughs of the buckling mode in directions EB, FC, and GD. Thus the buckled web transfers load much like a set of parallel wires, as suggested by Fig. 14.6-4a. From Fig. 14.6-4b we see that individual segments of a flange (AB, BC, CD) resemble fixed-fixed beams under distributed load. They also carry axial force. Vertical stiffeners are loaded in compression.

Buckling may not indicate failure. This important concept is illustrated by the box beam of Fig. 14.6-3 and by the tension field beam. In these structures buckling is localized and produces redistribution of stresses rather than failure. Load can continue to increase until overall buckling or some other mode of failure results in collapse of the structure as a whole. This knowledge may be a comfort to passengers in light aircraft when they see buckle patterns appear on the skin of a wing in turbulent conditions.

Shear Lag. Figure 14.6-5a is a plan view of the lower (tensile) flange of the box beam in Fig. 14.6-3. At $x = 0$, outer bars are unstressed. Load is transferred from middle to outer bars by shear stress in the panels. If flange length a is great enough, all bars come to have practically the same stress, as shown at $x = a$. The behavior near $x = 0$, where load is transferred by shear and stress in one part "lags" stress in another, is known as *shear lag*. Shear lag can also appear in a beam of I-shaped cross section with wide flanges. For example, if the I-section beam is a cantilever loaded by transverse tip force, each flange is loaded along its midline by shear stress from the web. At a given cross section, material at flange tips is not as highly stressed as predicted by the flexure formula $\sigma = Mc/I$. Material near web-flange intersections must then be *more* highly stressed, because the stress distribution must still supply the bending moment M demanded by statics. These higher stresses can be calculated from the flexure formula if I is based on a reduced width of cross section. The reduction factor depends on proportions of the cross section and how the beam is loaded and supported [1.6].

For the structure in Fig. 14.6-5a, simplified analysis [10.3] credits panels with shear stiffness only and bars with axial stiffness only. Results show that stresses σ in bars and τ in panels have rates of change proportional to e^{-kx}, where k depends on the

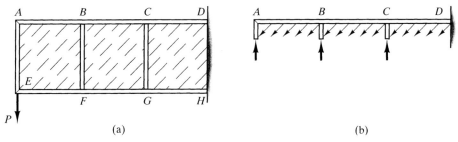

FIGURE 14.6-4 (a) Web buckling in a member built of flanges, vertical stiffeners, and thin webs. (b) Forces that act on the upper flange and the vertical stiffeners.

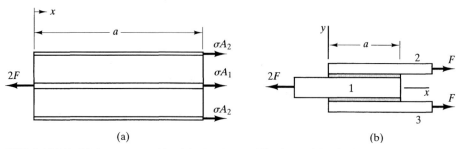

FIGURE 14.6-5 (a) Bottom assembly of the box beam of Fig. 14.6-3. (b) A lap joint. Members 2 and 3 are glued to member 1.

geometry and elastic moduli of bars and panels. For data given in Fig. 14.6-3, this analysis shows that stresses in inner and outer bars differ by less than 10% when $x = 357$ mm, which is a distance less than twice width b. The simplified analysis requires shear stress in panels to be maximum at $x = 0$. However, edge $x = 0$ is in fact free, so shear stress must vanish there. This contradiction clearly shows the analysis to be approximate.

A related problem is shown in Fig. 14.6-5b. Bar 1 is attached to bars 2 and 3 by thin layers of glue, which transfer load by shear stress τ_{xy}. Near $x = a/2$, τ_{xy} approaches zero if a is large enough (see the argument associated with Fig. 7.2.-2). But near $x = 0$ and $x = a$, analysis [14.6] shows that $\partial \tau_{xy}/\partial x$ is large in the glue. Therefore, from the second differential equation of equilibrium, Eq. 7.3-2b, we see that therefore $\partial \sigma_y/\partial y$ is also large. This result in turn implies that a large σ_y develops. A large "peel stress" σ_y may cause cracking and failure of the glue. It is of no help that large σ_y stresses are localized near ends of the glue layer, since a crack that may develop there only moves the effective end of joint toward its center.

14.7 ADDITIONAL BUCKLING PROBLEMS

Instability may occur in a variety of structures, under many conditions of loading and support, and may display various buckling modes that appear singly or in combination. What actually happens depends on geometry, relative thickness of components, and so on. An appreciation of the possibilities helps ensure that they are not overlooked as failure modes that must be considered. A sampling of buckling problems appears in Fig. 14.7-1. References include [1.6, 1.7, 1.9, 4.5, 14.7, 14.8, 14.9].

Although all buckling involves displacements lateral to an axis, plane, or surface of a structure, the kind of beam buckling shown in Fig. 14.7-1a is known as *lateral buckling*. Prior to buckling, load P bends the beam only in the usual way, about the horizontal (stronger) axis of its cross section, and with only vertical displacement. Lateral buckling involves bending about the vertical axis as well, and also twisting about the longitudinal axis. Analysis of lateral buckling problems makes use of formulations for torsion discussed in Chapter 9.

A buckling mode often does not span an entire member or structure. Channel sections in Fig. 14.7-1b are shown buckled by flange wrinkling rather than by bowing of the entire member. The second (plastic) case involves yield lines, as discussed in

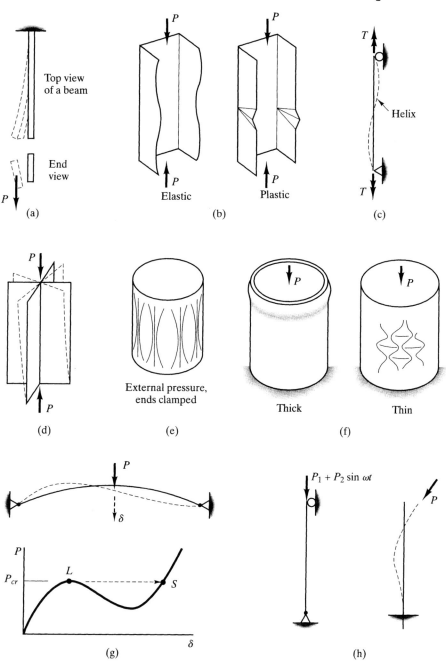

FIGURE 14.7-1 Various buckling problems. (a) Lateral. (b) Local. (c, d) Torsional. (e) Cylindrical shell, external pressure. (f) Cylindrical shells, axial load. (g) Snap-through. (h) Dynamic.

Section 12.9. Local buckling may also appear in a thin web of a beam (Section 14.6) and in the end closure of a cylindrical tank under internal pressure (Section 13.4). Local buckling may or may not be associated with collapse, depending on the nature of the structure.

"Torsional buckling" may occur under torque loading (Fig. 14.7-1c) or under axial loading (Fig. 14.7-1d). In these examples, buckling modes involve bending under torsional load and twisting under axial load. One should expect that in most cases bending will arise when there is torsional buckling.

Figure 14.7-1e depicts buckling of a thin cylindrical shell under external pressure. Ends are clamped so they remain circular. The number of circumferential waves into which the shell buckles increases as the length and thickness of the shell decrease. Pressure always acts normal to a shell surface, so analysis procedures must acknowledge that a pressure load changes direction as a shell assumes its buckled shape.

Axial compression of a rather thick-walled cylindrical shell produces a symmetric buckle pattern that begins at an end (Fig. 14.7-1f). If the shell is thin-walled, the buckle pattern resembles a set of triangular facets that have rotated with respect to one another, as shown. The latter pattern also appears on the compression side of a thin-walled cylindrical shell that buckles locally under bending load.

Snap-through buckling is illustrated by a slender arch whose rise is much less than its span (Fig. 14.7-1g). If deflections are not small, an almost-central load P produces a decidedly asymmetric deflected shape, as shown by dashed lines. The "tangent" stiffness seen by load P is $k = dP/d\delta$, which is the slope of the load P versus deflection δ curve. We see that k becomes zero at point L, then negative. If an object whose weight is equal to P_{cr} is placed on the center of the arch, the arch is on the verge of suddenly "snapping" downward. On the P versus δ curve, this behavior is represented by a jump from L to S.

Dynamic buckling is said to occur when time-dependent displacements become unacceptably large. For the first column in Fig. 14.7-1h, certain combinations of P_1, P_2, frequency ω, and static buckling capacity P_{cr} cause dynamic buckling. This kind of loading may appear if a column supports unbalanced rotating machinery. Other time-dependent loads can result from blast or earthquake. The second column in Fig. 14.7-1h is loaded by a "follower force" P of constant magnitude that is always directed tangent to the column at its tip. The column may "flutter" if P is large enough. Dynamic load greater than the static P_{cr} may not cause buckling if its duration is short. For example, a slender nail can be hammered into wood, but pushing the nail in slowly would make it buckle. In dynamic buckling, mass must be considered; in static buckling, mass plays no role.

14.8 REMARKS ABOUT STABILITY, BUCKLING, AND COLLAPSE

As the term is commonly used, "buckling" means general bowing or local wrinkling that occurs without fracture of the material, or at least prior to fracture, and is somehow related to compression or compressive stress. Thus, a load that causes buckling by local wrinkling may be much less than the load that ultimately causes collapse of the member or structure. In turn, the load that causes collapse may have little relation to the "critical load" calculated by the kind of analysis that provides the buckling load of a straight column or a flat plate. These matters are discussed with the aid of Fig. 14.8-1,

Section 14.8 Remarks About Stability, Buckling, and Collapse

which shows some possible shapes of P versus δ (P-δ) curves for structures whose material remains linearly elastic. Here P is either the load or is representative of its magnitude, and δ represents some displacement of interest, either tangent or lateral to a surface of the member or structure.

Figure 14.8-1a shows behavior representative of straight columns under axial load and flat plates under in-plane loading. Here δ represents axial or in-plane displacement. At the *bifurcation point*, point B, two equilibrium configurations are possible under load P_B: the straight or flat configuration, and an infinitesimally close alternative configuration that involves lateral displacement. Load P_B is called the *bifurcation load*. For $P > P_B$, the "primary" path BC is possible only in theory or when the loading is of short duration. Actual columns and plates under gradually increasing load, no matter how carefully made and tested, always follow the "secondary" path BD. The slope of path BD is zero at B, which indicates neutral equilibrium at B. Path BD gradually rises with increasing δ, which means that loads greater than P_B can be supported (provided the material does not yield), but with greatly increased deflection.

The *critical load* of a structure may be defined as the smallest load at which the structure as a whole ceases to be stable as load is gradually increased from zero [14.7]. With this definition, most practical structures are on the verge of collapse when the critical load is applied. In the foregoing example the critical load is $P_{cr} = P_B$. A critical load can also appear when no bifurcation point exists, as in the following example.

The P-δ curve of Fig. 14.8-1b is possible when local buckling appears at a load less than the load that causes the entire structure to collapse [14.7, 14.8]. Here δ is the magnitude of a displacement normal to a plate or shell midsurface. Critical load P_{cr} corresponds to point L, which is called a *limit point*. A limit point is defined as a relative maximum on the P-δ curve for which there is no infinitesimally close alternative configuration. (The arch of Fig. 14.7-1g displays a limit point.) The P-δ curve of Fig. 14.8-1b, which approximates the behavior of some shell structures, displays deformation-dependent stiffness and no bifurcation point. A linear analysis, based on stiffness parameters of the undeformed structure, predicts bifurcation point B and critical load P_B, which for some shell structures may be only 10% of the actual collapse load P_{cr} [14.9].

In some other situation, collapse may occur at a load *less* than P_B, as in Fig. 14.8-1c. The P-δ relations shown in Fig. 14.8-1c are approximately representative of a thin,

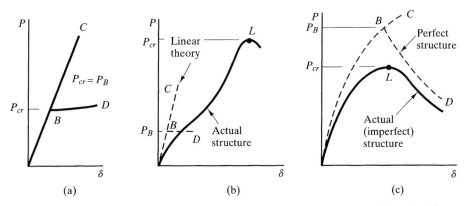

FIGURE 14.8-1 Possible load versus displacement relations for structures susceptible to buckling.

shallow dome loaded by external pressure. Unlike a column, which remains straight until it buckles, this structure displays prebuckling rotations of its midsurface. Nonlinearities associated with these rotations are taken into account in the calculation of P_B, but calculations assume a "perfect structure." An actual structure has imperfections, not accounted for in classical analysis, that cause it to display a P_{cr} less than P_B. For some shell structures, load P_{cr} may be less than 25% of P_B [14.9]. It is common for a thin-walled structure to be *imperfection-sensitive*, which means that the maximum load it can support is strongly affected by small imperfections of geometry, loading, or support conditions. A structure whose secondary path BD trends steeply downward is imperfection-sensitive.

The complexity of buckling problems often dictates a computational approach. Imperfections, present in any real structure, can be deliberately inserted into a numerical analysis. By computing the load versus deflection relation, rather than seeking only a bifurcation load P_B, actual behavior of a structure with nonlinearities can be more reliably anticipated. Pitfalls of numerical analysis include using few degrees of freedom because structure geometry is simple, when many degrees of freedom are needed to represent important short-wavelength buckles [14.8]. Also, even when using the most appropriate software, nonlinear analysis often presents many more difficulties than linear analysis [4.6].

In any case, analysis tools are available. It is up to the engineer to understand the physical situation well enough to realize what kind of analysis should be undertaken. Such is the case for buckling, for other stress analysis problems, and for design in general.

PROBLEMS

14.1-1. Instead of choosing a circle for the cross section of a uniform column, one might choose a square or an equilateral triangle. For a given cross-sectional area, would the square or the triangle provide a larger buckling load than the circle? How much larger? Assume that all cross sections are solid rather than hollow.

14.1-2. A horizontal beam BD is supported by columns AB and CD, of flexural stiffnesses EI and $4EI$ as shown. End A is fixed. There are hinge connections at B, C, and D. For buckling in the plane of the paper, determine the critical value of load P if it is placed at (a) end B of the beam, and (b) end D of the beam.

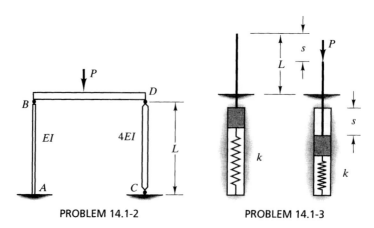

PROBLEM 14.1-2   PROBLEM 14.1-3

14.1-3. A slender and uniform elastic bar is embedded in a cylindrical plug, as shown. The plug is supported by a spring of stiffness k. Bar and plug ride in close-fitting but frictionless guides. The spring carries no force when $s = 0$.
  (a) What should k be in terms of E, I, and L if the upper portion of the bar is not to buckle as load P is gradually increased until $s = L$?
  (b) For this value of k, will the lower portion of the bar buckle?

14.1-4. A uniform column of length L is fixed at top and bottom ends. It is fabricated in such a way that its flexural stiffness is EI_1 in portions bent concave left and EI_2 in portions bent concave right, where $I_1 > I_2$. What is axial load P_{cr} for buckling in the plane of the paper, in terms of E, L, I_2, and the ratio I_1/I_2?

14.1-5. Use Eq. 14.1-4 to obtain buckling loads for the following uniform columns.
  (a) One end fixed, the other free.
  (b) Both ends fixed.

14.1-6. For the uniform column in Fig. 14.1-1a, obtain an approximate P_{cr} from Eq. 14.1-4. Use the assumption $d^2v/dx^2 = Cx(L-x)$, where C is a constant. With this assumption, Eq. 14.1-4 cannot be satisfied at all values of x, so treat the curvature as follows.
  (a) Use the maximum magnitude of curvature.
  (b) Use an average curvature, equal to $1/L$ times the integral of d^2v/dx^2 over length L.

14.1-7. A straight vertical shaft, 500 mm long and 12 mm in diameter, is simply supported at each end. A 15 kg disk is attached at midspan. The mass center of the disk is 0.50 mm from the axis of the shaft. Determine the range of steady rotational speed for which flexural stress in the shaft will exceed 120 MPa. Let $E = 200$ GPa. Neglect the mass of the shaft and stress concentrations.

14.1-8. Uniform and linearly elastic bars are welded together and to a central shaft, as shown. Bars AB and CD are parallel to the shaft; the other two bars are collinear and extend radially from the shaft. Four particle masses m are attached, at A, B, C, and D. The shaft spins at angular velocity ω. If $L \gg a$ and bars have negligible mass, for what ω is the shaft-parallel orientation of AB and CD no longer stable?

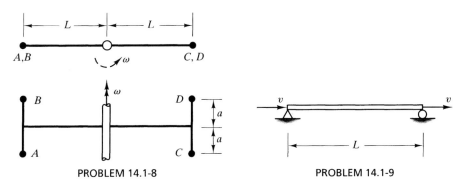

PROBLEM 14.1-8   PROBLEM 14.1-9

14.1-9. A long pipe is simply supported, as shown. A frictionless fluid of mass density ρ flows through the pipe at velocity v. Show that the pipe may buckle into a half sine wave when $v = (\pi/L)\sqrt{EI/\rho A}$, where A is the area of the fluid cross section.

14.2-1. A wooden column has a central axial hole that fits closely but without friction around a steel rod, as shown. The rod is tightened by turning nuts at either end. Will the column buckle?

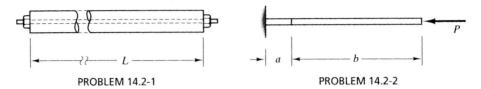

PROBLEM 14.2-1   PROBLEM 14.2-2

14.2-2. The column shown is fixed at one end and free at the other. In length a, the bending stiffness is EI. Length b may be considered infinitely stiff. For $a \ll b$, obtain an expression for P_{cr} by use of Eq. 14.2-3. Check the result by equilibrium considerations.

14.2-3. Obtain P_{cr} for a uniform simply supported column by pursuing the suggestion made in text following Eq. 14.2-5. That is, replace the second integrand in Eq. 14.2-3 by a moment expression, where $M = -P_{cr}v$.
   (a) Use the correct buckling mode, Eq. 14.2-4.
   (b) Use an approximate buckling mode, Eq. 14.2-5.

14.2-4. (a) A uniform vertical column of length L is fixed at the base and free at the top. It has weight γ per unit length. Use Eq. 14.2-3 to determine the critical length L_{cr} for which buckling impends. Assume that the buckled shape is one quarter of a complete sine or cosine wave. [The exact result is $L_{cr} = 1.986(EI/\gamma)^{1/3}$].
   (b) A uniform vertical wooden pole is immersed in water, free at the bottom and held fixed at the top. How can the solution of part (a) be applied to this problem?

14.2-5. The uniform column shown is fixed at one end and simply supported at the other. Assume that lateral displacement has the form $v = a_1 x^2 + a_2 x^3$, where a_1 and a_2 are constants. Use end conditions to express one constant in terms of the other, then use Eq. 14.2-3 to approximate P_{cr}. Why is the approximation not very good?

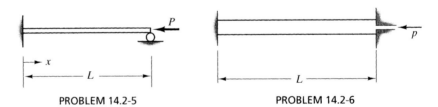

PROBLEM 14.2-5   PROBLEM 14.2-6

14.2-6. A uniform thin-walled tube is clamped between rigid walls, as shown. Explain why the tube may buckle like a column when internal pressure p is applied to fluid contained by the pipe. Determine p_{cr} in terms of E, ν, L, mean diameter D, and wall thickness t.

14.2-7. (a) A uniform beam rests on a Winkler elastic foundation of modulus k N/mm/mm. The beam is very long, and buckles into a large number of half sine waves in a plane normal to the foundation. Assuming that the beam does not detach from the foundation, determine the wavelength and the buckling load P_{cr} in terms of E, I, and k.
   (b) Consider a steel railroad rail of cross-sectional area $A = 8400$ mm^2 and moment of inertia $I = 37(10^6)$ mm^4 for bending in the vertical plane. The foundation modulus is $k = 1.7$ N/mm/mm. Let $E = 204$ GPa. Approximately what temperature increase relative to the stress-free condition will cause buckling in the vertical plane?

14.2-8. Approximate the ω_{cr} of Eq. 14.2-7 by using the lateral displacement assumption $v = Cx(L - x)$, where C is a constant.

14.3-1. The uniform bar shown has a square cross section and is axially loaded by collinear forces P. Each force P acts at the center of an edge of the cross section. If $P = 0.7P_{cr}$, what is the stress of largest magnitude? Let $E = 200$ GPa.

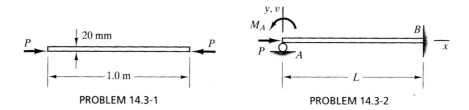

PROBLEM 14.3-1   PROBLEM 14.3-2

14.3-2. Obtain an expression for $v = v(x)$ of the uniform beam-column shown. Suggestion: Use $v = v(x)$ from Eq. 14.3-3, superposition, and the support condition at $x = L$.

14.3-3. Let a tensile axial load P be applied to a simply supported beam loaded by moment at one end (as in Fig. 14.3-1b, except for the direction of P). Obtain expressions for $v = v(x)$ and $M = M(x)$, analogous to Eqs. 14.3-3. (Note that a tensile P is negative in the sign convention of the present chapter.)

14.3-4. **(a)** Let a uniform beam of length L be simply supported at its ends. Load the beam by compressive axial force P and uniformly distributed lateral load q. Obtain the deflection equation $v = v(x)$ (analogous to the first of Eqs. 14.3-3).

(b) Evaluate lateral deflection and bending moment at midspan. Use this information to construct a table of amplification factors like that in Section 14.3.

14.3-5. A uniform shaft of length L is simply supported at its ends. The shaft has mass per unit length ρ, spins about its axis at angular velocity ω, and is axially compressed by end forces P. Obtain an expression for the instability frequency ω_{cr} in terms of E, I, ρ, L, P, and the static buckling load P_{cr}. Suggestion: Use Eq. 14.1-4 and the assumed mode $v = \bar{v} \sin(\pi x/L) \sin \omega t$.

14.4-1. **(a)** Repeat Problem 14.3-1, now using Eq. 14.4-7.

(b) Repeat Problem 14.3-1, now using Eqs. 14.4-6 and 14.4-8.

14.4-2. In Problem 14.3-3, evaluate v and M at midspan $(x = L/2)$. For $|P| = P_{cr}/2$, compare these exact results with approximate results at midspan from Eqs. 14.4-6 and 14.4-7.

14.4-3. Pipes AB and BC are welded together at B, as shown on the next page. Structure ABC is hinged at A and kept upright by cable BD. Assume that the structure deforms only in the plane of the paper. In terms of $E, I,$ and L, approximately what load W produces the same magnitude of bending moment at B as midway between A and B?

14.4-4. The column shown on the next page has a hollow circular cross section of wall thickness 5.0 mm. Let outer diameter $d = 100$ mm, $E = 200$ GPa, and $\sigma_Y = 500$ MPa (the material yield stress). For a safety factor of 2.0 based on the onset of yielding, approximately what axial load P can be allowed if its eccentricity is (a) $e = 20$ mm, and (b) $e = 40$ mm?

14.4-5. Let the column of Problem 14.4-4 have a solid circular cross section of diameter d. Also let $E = 200$ GPa, $e = 40$ mm, and $P = 70$ kN. Approximately what diameter d is required if the allowable compressive stress is 180 MPa?

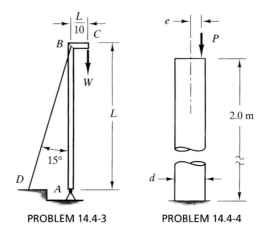

PROBLEM 14.4-3   PROBLEM 14.4-4

14.4-6. The uniform simply supported beam shown has a T-shaped cross section whose centroid is 78 mm from the lower edge and 34 mm from the upper edge, and for which $A = 2400$ mm^2 and $I = 2.896(10^6)$ mm^4 about the centroidal axis. The 120 kN axial forces are centroidal. The allowable stress magnitude is 130 MPa in tension or in compression, and $E = 200$ GPa. Approximate the allowable transverse force F if the beam is (a) oriented as shown, with stem of the T down, and (b) inverted, with stem of the T up.

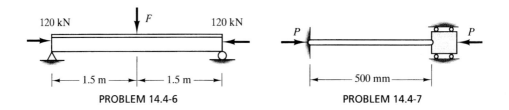

PROBLEM 14.4-6   PROBLEM 14.4-7

14.4-7. The column shown has a solid circular cross section 12 mm in diameter. Ends are rounded and rest in spherical seats. It is estimated that for each newton of compressive axial force P, a frictional couple of 0.25 N·mm must be overcome before an end can rotate in its seat. If axial force $P = 12$ kN is applied, what lateral force applied at midspan will produce collapse? Let $E = 200$ GPa.

14.4-8. An energy method can be applied to beam-column problems: In a lateral displacement, work is done by lateral load and by axial load, and strain energy is stored in bending. Consider a uniform simply supported beam of length L, loaded by uniformly distributed lateral load q and axial compressive load P. Use the energy method to obtain an expression for midspan lateral deflection in terms of q, L, P, E, and I.

(a) Assume that lateral deflection is $v = \bar{v} \sin(\pi x/L)$.

(b) Assume that lateral deflection is $v = 4\bar{v} x (L - x)/L^2$.

14.5-1. Imagine that compressive stress-strain data have been used to obtain the plot of stress versus tangent modulus shown. Consider a uniform simply supported column of length L, for which r is the least radius of gyration of its cross-sectional area A. Use the given plot to construct a graph of $\sigma_{cr} = P_{cr}/A$ versus L/r. Do so by evaluating and plotting points in both elastic and plastic regimes.

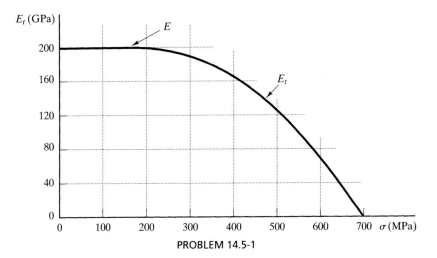

PROBLEM 14.5-1

14.5-2. A column has length 550 mm and a solid circular cross section 20 mm in diameter. Both ends are fixed. What is the buckling load? Use the σ versus E_t plot provided in Problem 14.5-1.

(a) Use the trial method suggested in Section 14.5.

(b) Determine the required slope E_t/σ_{cr} and read the corresponding σ_{cr} from the plot.

14.5-3. A column is 400 mm long and 24 mm in diameter. The material has the σ versus E_t plot depicted in Problem 14.5-1. The column is fixed at the base and free at the top. It is proposed that the buckling load be increased by a factor of 16 by fixing the top as well. Why is the goal not achieved? What is the actual factor of increase?

14.5-4. A uniform column 300 mm long is fixed at both ends and is to have a solid circular cross section. If the column must carry an axial load of 40 kN before it buckles, what should be the radius of the cross section? Use the σ versus E_t plot depicted in Problem 14.5-1.

14.6-1. If $b \gg a$ in Fig. 14.6-1, edges of length a provide little restraint. For such a geometry, what equation for σ_{cr} in terms of E, t, and a replaces Eq. 14.6-1?

14.6-2. In Fig. 14.6-1, let an x-direction force P create the compressive stress $\sigma_x = P/bt$. Imagine that a panel for which $a = 520$ mm and $b = 260$ mm must carry force $P = 10$ kN without buckling. What should be thickness t if the material is (a) aluminum, or (b) wood? What would each of these panels weigh? Use typical material properties, but assume that $\nu = 0.3$ for both, and ignore the anisotropy of wood.

14.6-3. (a) What is the maximum bending moment that can be applied to the beam whose cross section is shown in Fig. 14.6-3 if $\sigma = 400$ MPa is the allowable magnitude of stress? What then is the effective width of the compression panel? Again assume that vertical webs do not contribute to bending stiffness.

(b) Repeat part (a) with the compression panel doubled in thickness to 2.0 mm.

(c) In part (a), revise cross-sectional areas A_1 and A_2 so as to achieve a more efficient design. Maintain the existing ratio $A_1/A_2 = 2.40$.

14.6-4. The sketch shows a uniform bar of cross-sectional area A and elastic modulus E, attached to two thin panels. Each panel has width c, uniform thickness t, and shear modulus G. Assume that panels carry only shear flow $q = \tau t$, which is shown in an assumed positive sense. From the second part of the sketch one can relate q to dF/dx, where $F = F(x)$ is axial force in the bar. From the third part of the sketch (region B enlarged and shown deformed), one can obtain an expression for dy/dx and hence an expression for dq/dx.

(a) Show that the foregoing expressions can be combined to produce the differential equation $d^2F/dx^2 - k^2F = 0$, where $k^2 = 2tG/cAE$.

(b) Solve the differential equation. Then express F in terms of P, k, and x. Assume that $L \gg c$.

(c) Obtain expressions for q and the displacement of load P.

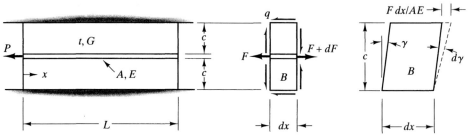

PROBLEM 14.6-4

14.6-5. Analyze the assembly in Fig. 14.6-5a by adapting the approximate method described in Problem 14.6-4 to the present problem. Let bars and panels be made of the same material. Obtain an expression for axial force $F_2 = F_2(x)$ in the outer bars. Then, using data in Fig. 14.6-3, determine the distance x for which F_2 reaches 95% of its value at large x. Assume that $a \gg b$. (Note that dimension c in Problem 14.6-4 corresponds to dimension $b/2$ in Fig. 14.6-3.)

14.7-1. A uniform simply supported column 3.0 m long must carry a 400 kN axial load before buckling. The cross section is thin-walled and circular, of mean radius R and wall thickness t. Assume that local buckling is possible when axial stress P/A reaches $0.2Et/R$ in compression. Choose R and t so as to make the column as light as possible. Let $E = 200$ GPa.

14.8-1. The structure shown consists of rigid weightless bar AB that moves without friction on a vertical surface at A and in a horizontal guide at B. The spring of stiffness k is unstressed in the unloaded configuration shown. If an upward load P is applied at A, point A displaces v units upward and point B displaces u units rightward, compressing the spring.

(a) Determine, and sketch, the relation between P and v, in terms of c, k, and L. Assume that $L \gg c$.

(b) Determine the limit point value of v in terms of c.

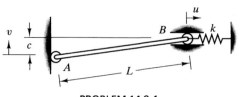

PROBLEM 14.8-1

References

CHAPTER 1

1.1 R. D. Adams and W. C. Wake, *Structural Adhesive Joints in Engineering,* Elsevier Applied Science Publishers, London, 1984.

1.2 A. Blake, *Design of Mechanical Joints,* Marcel Dekker, New York, 1985.

1.3 G. L. Kulak, J. W. Fisher, and J. H. A. Struik, *Guide to Design Criteria for Bolted and Riveted Joints,* 2nd ed., John Wiley & Sons, New York, 1987.

1.4 R. Narayanan, ed., *Structural Connections: Stability and Strength,* Elsevier Science Publishers, Barking, England, 1989.

1.5 J. H. Bickford, *An Introduction to the Design and Behavior of Bolted Joints,* 3rd ed., Marcel Dekker, New York, 1995.

1.6 W. C. Young, *Roark's Formulas for Stress and Strain,* 6th ed., McGraw-Hill Book Co., New York, 1989.

1.7 W. D. Pilkey, *Formulas for Stress, Strain, and Structural Matrices,* John Wiley & Sons, New York, 1994.

1.8 E. F. Megyesy, *Pressure Vessel Handbook,* 9th ed., Pressure Vessel Handbook Publishing Co., Tulsa, OK, 1992.

1.9 Column Research Committee of Japan, eds., *Handbook of Structural Stability,* Corona Publishing Co., Tokyo, 1971.

1.10 R. D. Blevins, *Formulas for Natural Frequency and Mode Shape,* Van Nostrand Reinhold Co., New York, 1979.

1.11 J. Feld and K. L. Carper, *Construction Failure,* 2nd ed., John Wiley & Sons, New York, 1997.

1.12 *ASCE Journal of Performance of Constructed Facilities* [papers examine the causes and costs of failures and other performance problems].

1.13 A. M. Greene, Jr., *History of the ASME Boiler Code,* American Society of Mechanical Engineers, New York, 1956.

CHAPTER 2

2.1 C. V. G. Vallabhan, "A New Formula for Principal Axes," *Mechanics Research Communications,* vol. 24, no. 4, 1997, pp. 443–445.

- **2.2** A. P. Boresi and O. M. Sidebottom, *Advanced Mechanics of Materials*, 4th ed., John Wiley & Sons, New York, 1985.
- **2.3** A. C. Ugural and S. K. Fenster, *Advanced Strength and Applied Elasticity*, 3rd ed., American Elsevier Publishing Co., New York, 1995.
- **2.4** F. Osweiller, "Evolution and Synthesis of the Effective Elastic Constants Concept for the Design of Tubesheets," *ASME Journal of Pressure Vessel Technology*, vol. 111, no. 3, 1989, pp. 209–217.
- **2.5** W. D. Pilkey, *Peterson's Stress Concentration Factors*, 2nd ed., John Wiley & Sons, New York, 1997.
- **2.6** D. A. Hills, D. Nowell, and A. Sackfield, *Mechanics of Elastic Contacts*, Butterworth Heinemann, Oxford, 1993.
- **2.7** T. E. Tallian, *Failure Atlas for Hertz Contact Machine Elements*, ASME Press, New York, 1992.
- **2.8** R. C. Juvinall and K. M. Marshek, *Fundamentals of Machine Component Design*, 2nd ed., John Wiley & Sons, New York, 1991.

CHAPTER 3

- **3.1** A. Nadai, *Theory of Flow and Fracture of Solids*, vol. I, 2nd ed., McGraw-Hill, New York, 1963.
- **3.2** J. H. Faupel and F. E. Fisher, *Engineering Design*, 2nd ed., John Wiley & Sons, New York, 1981.
- **3.3** A. A. Ilyushin and V. S. Lensky, *Strength of Materials*, Pergamon Press, Oxford, 1967.
- **3.4** R. S. Sandhu, *A Survey of Failure Theories of Isotropic and Anisotropic Materials*, U.S. Air Force, Wright-Patterson Air Force Base, Ohio, Report AFFDL-TR-72-71, January 1972.
- **3.5** R. M. Jones, *Mechanics of Composite Materials*, McGraw-Hill Book Co., New York, 1975.
- **3.6** D. P. Rooke and D. J. Cartwright, *Compendium of Stress Intensity Factors*, Her Majesty's Stationery Office, London, 1976.
- **3.7** D. Broek, *Elementary Engineering Fracture Mechanics*, Martinus Nijhoff Publishers, Dordrecht, 1986.
- **3.8** D. Broek, *The Practical Use of Fracture Mechanics*, Kluwer Academic Publishers, Dordrecht, 1988.
- **3.9** N. E. Dowling, *Mechanical Behavior of Materials*, Prentice Hall, Englewood Cliffs, NJ, 1993.
- **3.10** P. Polak, *Designing for Strength*, Macmillan Press, Oxford, 1982.
- **3.11** A. Blake, *Practical Fracture Mechanics in Design*, Marcel Dekker, New York, 1996.

CHAPTER 4

- **4.1** H. L. Langhaar, *Energy Methods in Applied Mechanics*, John Wiley & Sons, New York, 1962.
- **4.2** J. T. Oden and E. A. Ripperger, *Mechanics of Elastic Structures*, 2nd ed., McGraw-Hill Book Co., New York, 1981.
- **4.3** U. Schram, et al., "On the Shear Deformation Coefficient in Beam Theory," *Finite Elements in Analysis and Design*, vol. 16, no. 2, 1994, pp. 141–162.
- **4.4** S. Timoshenko, *Strength of Materials*, Part II, 2nd ed., D. Van Nostrand Co., New York, 1941.

4.5 S. Timoshenko and J. M. Gere, *Theory of Elastic Stability,* 2nd ed., McGraw-Hill Book Co., New York, 1961.

4.6 R. D. Cook, *Finite Element Modeling for Stress Analysis,* John Wiley & Sons, New York, 1995.

CHAPTER 5

5.1 C. J. Burgoyne and R. Dilmaghanian, "Bicycle Wheel as Prestressed Structure," *ASCE Journal of Engineering Mechanics,* vol. 119, no. 3, 1993, pp. 439–455.

5.2 G. W. Housner and T. Vreeland, *The Analysis of Stress and Deformation,* Macmillan Publishing Co., New York, 1966.

5.3 A. D. Kerr and H. W. Shenton III, "Railroad Track Analyses and Determination of Parameters," *ASCE Journal of Engineering Mechanics,* vol. 112, no. 11, 1986, pp. 1117–1134.

5.4 M. Hetenyi, *Beams on Elastic Foundation,* University of Michigan Press, Ann Arbor, 1946.

CHAPTER 6

6.1 S. P. Timoshenko and J. N. Goodier, *Theory of Elasticity,* 3rd ed., McGraw-Hill Book Co., New York, 1970.

6.2 A. M. Wahl, *Mechanical Springs,* 2nd ed., McGraw-Hill Book Co., New York, 1963.

6.3 R. D. Cook, "Circumferential Stresses in Curved Beams," *ASME Journal of Applied Mechanics,* vol. 59, no. 1, 1992, pp. 224–225.

6.4 V. Feodosyev, *Strength of Materials,* Izdatelstvo Mir, Moscow, 1968.

6.5 R. D. Cook, "Axisymmetric Finite Element Analysis for Pure Moment Loading of Curved Beams and Pipe Bends," *Computers & Structures,* vol. 33, no. 2, 1989, pp. 483–487.

6.6 T. S. Wang, "Shear Stress in Curved Beams," *Machine Design,* vol. 39, no. 28, 1967, pp. 175–178.

6.7 W. C. Young and R. D. Cook, "Radial Stress Formula for Curved Beams," *ASME Journal of Vibration, Acoustics, Stress, and Reliability in Design,* vol. 111, no. 4, 1989, pp. 491–492.

6.8 H. Bleich, "Die Spannungsverteilung in den Gurtungen gekrümmter Stabe mit T- und I-förmigem Querschnitt," *Der Stahlbau* (appendix to *Die Bautechnik*), vol. 6, no. 1, 1933, pp. 3–6.

6.9 *Penstock Analysis and Stiffener Design,* U.S. Dept. of the Interior, Bureau of Reclamation, Boulder Canyon Project Final Reports, Part V, Bulletin 5, 1940.

6.10 L. Sobel, "In-Plane Bending of Elbows," *Computers & Structures,* vol. 7, no. 6, 1977, pp. 701–715.

6.11 W. G. Dodge and S. E. Moore, "Stress Indices and Flexibility Factors for Moment Loadings on Elbows and Curved Pipes," *Welding Research Council Bulletin 179,* December 1972.

6.12 R. D. Cook, "Pure Bending of Curved Beams of Thin-Walled Rectangular Box Section," *ASME Journal of Applied Mechanics,* vol. 58, no. 1, 1991, pp. 154–156.

CHAPTER 7

7.1 W. J. O'Donnell, "The Additional Deflection of a Cantilever Due to the Elasticity of the Support," *ASME Journal of Applied Mechanics,* vol. 27, no. 3, 1960, pp. 461–464.

7.2 B. A. Boley and J. H. Weiner, *Theory of Thermal Stresses,* John Wiley & Sons, New York, 1960.

7.3 R. R. Archer, *Growth Stresses and Strains in Trees,* Springer-Verlag, Berlin, 1987.

CHAPTER 8

8.1 R. V. Southwell, *An Introduction to the Theory of Elasticity for Engineers and Physicists*, Clarendon Press, Oxford, 1936.

8.2 A. Stacy and G. A. Webster, "Determination of Residual Stress Distributions in Autofrettaged Tubing," *International Journal of Pressure Vessels & Piping*, vol. 31, no. 3, 1988, pp. 205–220.

8.3 J. H. Faupel, "Yield and Bursting of Heavy-Wall Cylinders," *Transactions of the ASME*, vol. 78, no. 5, 1956, pp. 1031–1064.

8.4 G. F. Lake, "A Simplified Method of Determining Hoop Stresses in Fan Rotors," *ASME Journal of Applied Mechanics*, vol. 12, no. 2, 1945, pp. A65–A68.

CHAPTER 9

9.1 G. W. Trayer and H. W. March, "The Torsion of Members Having Sections Common in Aircraft Construction," *NACA Report No. 334*, 1929.

9.2 J. P. Den Hartog, *Advanced Strength of Materials*, McGraw-Hill, New York, 1952.

9.3 J. H. Huth, "Torsional Stress Concentration in Angle and Square Tube Fillets," *ASME Journal of Applied Mechanics*, vol. 17, no. 4, 1950, pp. 388–390.

9.4 S. Fairman and C. S. Cutshall, *Mechanics of Materials*, John Wiley & Sons, New York, 1953.

9.5 V. Z. Vlasov, *Thin-Walled Elastic Beams*, 2nd ed., Israel Program for Scientific Translations, Jerusalem, 1961.

9.6 W. D. Pilkey and P. Y. Chang, *Modern Formulas for Statics and Dynamics*, McGraw-Hill, New York, 1978.

9.7 N. G. Stephen and X. L. Shi, "The Centre of Twist or the Centre of Flexure?" *Zeitschrift für Angewandte Mathematik und Mechanik*, vol. 67, no. 6, 1987, pp. 271–273.

9.8 K. Zbirohowski-Koscia, *Thin-Walled Beams*, Crosby Lockwood & Son, London, 1967.

9.9 S. Krenk, "A Linear Theory for Pretwisted Elastic Beams," *ASME Journal of Applied Mechanics*, vol. 50, no. 1, 1983, pp. 137–142.

9.10 R. T. Shield, "Extension and Torsion of Elastic Bars with Initial Twist," *ASME Journal of Applied Mechanics*, vol. 49, no. 4, 1982, pp. 779–786.

9.11 W. Carnegie, "Static Bending of Pre-Twisted Cantilever Blading," *Proceedings of the Institution of Mechanical Engineers*, vol. 171, 1957, pp. 873–894.

9.12 Chen Chu, "The Effect of Initial Twist on the Torsional Rigidity of Thin Prismatical Bars and Tubular Members," *Proceedings of the First U.S. National Congress of Applied Mechanics*, 1951, pp. 265–269.

9.13 Y.-L. Pi and N. S. Trahair, "Plastic-Collapse Analysis of Torsion," *ASCE Journal of Structural Engineering*, vol. 121, no. 10, 1995, pp. 1389–1395.

CHAPTER 10

10.1 W. E. Mason and L. R. Herrmann, "Elastic Shear Analysis of General Prismatic Beams," *ASCE Journal of the Engineering Mechanics Division*, vol. 94, no. EM4, 1968, pp. 965–983.

10.2 M. J. French, "A Simple Introduction to Shear Centers," *International Journal of Mechanical Engineering Education*, vol. 2, no. 2, 1974, pp. 55–57.

10.3 P. Kuhn, *Stresses in Aircraft and Shell Structures*, McGraw-Hill Book Co., New York, 1956.

10.4 M. Paz, C. P. Strehl, and P. Schrader, "Computer Determination of the Shear Center of Open and Closed Sections," *Computers & Structures*, vol. 6, no. 2, 1976, pp. 117–125.

CHAPTER 11

11.1 B. G. Neal, *The Plastic Methods of Structural Analysis*, 3rd ed., Chapman and Hall, London, 1977.

11.2 W. Johnson and P. B. Mellor, *Engineering Plasticity*, Van Nostrand Reinhold Co., London, 1973.

11.3 M. R. Horne, *Plastic Theory of Structures*, 2nd ed., Pergamon Press, Oxford, 1979.

11.4 W. F. Chen and D. J. Han, *Plasticity for Structural Engineers*, Springer-Verlag, New York, 1988.

11.5 S. Kaliszky, *Plasticity: Theory and Engineering Applications*, Elsevier, Amsterdam, 1989.

11.6 J. Heyman, *Plastic Design of Frames, Vol. 2: Applications*, Cambridge University Press, Cambridge, 1971.

11.7 T. X. Yu and L. C. Zhang, *Plastic Bending: Theory and Applications*, World Scientific Publishing Co., Singapore, 1996.

11.8 R. M. Haythornthwaite, "Beams with Full End Fixity," *Engineering*, vol. 183, no. 4742, 1957, pp. 110–112.

11.9 F. R. Shanley, *Strength of Materials*, McGraw-Hill Book Co., New York, 1957.

11.10 J. Y. R. Liew, D. W. White, and W. F. Chen, "Second-Order Refined Plastic-Hinge Analysis for Frame Design. Part II," *ASCE Journal of Structural Engineering*, vol. 119, no. 11, 1993, pp. 3217–3237.

CHAPTER 12

12.1 S. Timoshenko and S. Woinowsky-Krieger, *Theory of Plates and Shells*, 2nd ed., McGraw-Hill Book Co., New York, 1959.

12.2 D. G. Bellow, G. Ford, and J. S. Kennedy, "Anticlastic Behavior of Flat Plates," *Experimental Mechanics*, vol. 5, no. 7, 1965, pp. 227–232.

12.3 C. R. Calladine, *Engineering Plasticity*, Pergamon Press, Oxford, 1969.

12.4 M. A. Save and C. E. Massonnet, *Plastic Analysis and Design of Plates, Shells, and Disks*, North-Holland Publishing Co., Amsterdam, 1972.

12.5 R. Szilard, *Theory and Analysis of Plates: Classical and Numerical Methods*, Prentice Hall, Englewood Cliffs, NJ, 1974.

CHAPTER 13

13.1 W. Flügge, *Stresses in Shells*, Springer-Verlag, Berlin, 1960.

13.2 R. Royles and A. B. Sofoluwe, "Membrane Approximation of the Behavior of the Drop-Shaped Tank Under Symmetrical Loading," *Computers & Structures*, vol. 14, nos. 5–6, 1981, pp. 423–425.

13.3 E. H. Mansfield, "An Optimum Surface of Revolution for Pressurized Shells," *International Journal of Mechanical Sciences*, vol. 23, no. 1, 1981, pp. 57–62.

CHAPTER 14

14.1 J. M. Gere and S. P. Timoshenko, *Mechanics of Materials,* 4th ed., PWS Publishing Co., Boston, 1997.

14.2 B. Johnston, "Column Buckling Theory: Historic Highlights," *ASCE Journal of Structural Engineering,* vol. 109, no. 9, 1983, pp. 2086–2096.

14.3 F. R. Shanley, *Weight-Strength Analysis of Aircraft Structures,* Dover Publications, New York, 1960.

14.4 L. S. Beedle, "Basic Column Strength," *ASCE Journal of the Structural Division,* vol. 85, no. ST7, 1960, pp. 139–173.

14.5 E. E. Sechler and G. Dunn, *Airplane Structural Analysis and Design,* Dover Publications, New York, 1963.

14.6 G. A. O. Davies, *Virtual Work in Structural Analysis,* John Wiley & Sons, Chichester, 1982.

14.7 D. O. Brush and B. O. Almroth, *Buckling of Bars, Plates, and Shells,* McGraw-Hill Book Co., New York, 1975.

14.8 D. Bushnell, *Computerized Buckling Analysis of Shells,* Martinus Nijhoff Publishers, Dordrecht, 1985.

14.9 B. O. Almroth and F. A. Brogan, "Bifurcation Buckling as an Approximation of the Collapse Load for General Shells," *AIAA Journal,* vol. 10, no. 4, 1972, pp. 463–467.

Index

Admissible configuration, 79, 110
Airy stress function, 199, 207
Anisotropy, 38, 58
Anticlastic surface, 384
Arches, 95–97, 100–101, 363, 412
   see also Beams, curved
Autofrettage, 247, 251

Beam columns, 449–453
Beams, curved:
   circumferential stress, 164–165
   deflections of, 94–97, 99–100, 103–105, 180–183
   deformation and strains, 162–163
   distortion of cross section, 161, 169–170, 174–180
   equilibrium equations, 94, 171
   radial stress, 171–174
   transverse shear stress, 172–173
   statically indeterminate, 99–101
   strain energy, 106
   strain energy, complementary, 181
   thin walled, 161, 174–180
Beams, straight:
   axial load added, 449–453
   buckling, 458–459, 460
   collapse, plastic, 355–361
   deflections, elastic, 14–15, 17, 90–92, 98, 105, 111, 115
   deflections, plastic, 354–355, 364
   elastically supported:
      concentrated loading, 138–147, 150–152
      discrete springs, 134, 143–144
      distributed loading, 147–150
      infinite and semi-infinite, 138–150
      intermediate length, 151–152
      short length, 150–152
      Winkler foundation, 133–134

   elementary theory, 12–17
   plastic capacity, 355–361
   point load, stresses near, 212
   shear center, 327–337
   shear lag, 459
   statically indeterminate, 98–99
   strain energy, 106
   strain energy, complementary, 84–87
   tension field, 459
   transverse shear deformation, 86–87
   transverse shear stress, 13, 323–327
   twisting of, 278–283, 293–295, 305, 336
   unsymmetric bending:
      defined, 315
      deflections, 319–323
      flexural stress, 316–319
      transverse shear stress, 323–327
   Wagner beam, 459
Bearings, 46
Bending, *see* Beams
Bifurcation point, 463
Biharmonic operator, 199, 207
Bimoment, 295–298
Body forces, 192, 196, 198, 205–207
Boundary conditions:
   beam, in elasticity theory, 197–198, 205
   defined, 3
   general, in elasticity theory, 190, 196–197
   natural, 113
   plate bending, 385–386, 391
   surface tractions, 197
   torsion problem, 218, 219, 280, 297
Bredt's formulas, 273
Brittle failure, 52–55
   see also Fracture mechanics

Buckling:
  beam columns, 449–453
  bifurcation point, 463
  columns, 443–445
  critical load, 443, 463
  defined, 51, 462
  dynamic, 462
  effective width, 457–458
  energy method for, 446–448
  imperfections, 444, 464
  inelastic, 453–455
  lateral, 460
  limit point, 463
  local, 459, 460–462
  plates, 455–458
  pressure vessel head, 417, 421
  residual stress, effect of, 454–455
  shells, 463–464
  snap-through, 462
  torsional, 462
  versus collapse or failure, 459, 462–464
Burst pressure, 249, 253
Burst speed, 253

Castigliano's theorem, 89–90, 102–103, 181–183
Center of twist, 271, 336
Codes, 21
Cold working, defined, 3
Collapse analysis, plastic:
  arches, 363–364
  beams, straight, 355–361
  bound theorems, 365–366
  deflections, 354–355, 364
  interaction of loads, 368–371
  number of hinges, 360–361
  parallel bars, 348–349
  plane frames, 361–363
  plastic hinge, 355
  plate bending, 399–404
  virtual work, 355–356, 401
Collapse, versus buckling, 459, 462–464
Compatibility equation, 194, 200–201, 207
Computational methods, dangers of, 2, 118
Connections, 20, 194–195, 361
Conservation of energy, 81–82, 110
Conservative system, 107–108
Constitutive relations, see Stress-strain relations
Contact stress, 44–46
Crack arrester, 59
Cracks, see Fracture mechanics
Creep, 52, 222
Critical load, 443, 463
Critical speed, 445–446, 448–449
Cumulative damage, fatigue, 66
Curved beams, see Beams, curved

d.o.f., 77
Definitions, selected, 3–5
Degrees of freedom, 77
Derivations, why study, 2
Design load, defined, 4, 346
Deviatoric stresses, 40
Dimensional homogeneity, 19
Disks, spinning:
  governing equations, 231–233, 245
  plastic deformation in, 251–253
  shrink fits, 244
  stability of, 445–446
  uniform strength, 246–247
  uniform thickness, 243–244
  varying thickness, 245–247
Drop-shaped tank, 419–420
Dummy load method, 89

Effective stress, 41
Effective width, 457–458
Elastic axis, 327–328
Elastic foundation, see Beams, straight, elastically supported
Elastic modulus, 3
Elastic range of loading, 18–19, 350
Elastica problem, 444
Endurance limit, 65
Energy methods:
  buckling analysis, 446–448
  Castigliano's theorem, 89–90, 102–103, 183
  compared, 109–110
  complementary strain energy, 79
  conservation of energy, 81–82, 110
  dummy load, 89
  finite element, 116–118
  Maxwell-Mohr, 89
  potential energy, stationary, 107–110
  Rayleigh-Ritz, 110–113
  reciprocal theorems, 82–84
  unit load, 87–90, 109
  virtual work, 79–81, 349, 355–356
  work and energy, 78
Equivalent spring, 39, 82, 134

Factor of safety, see Safety factor
Failure criteria:
  anisotropic materials, 58–59
  brittle behavior:
    maximum normal stress, 53
    Mohr, 53–55
  cyclic loading, 65–66
  ductile behavior:
    maximum shear stress, 56–58
    octahedral shear stress, 56
    strain energy of distortion, 56

Index

Tresca, 56
von Mises, 56–58
in general, 52
Failure, defined, 51
Failure, versus buckling, 459, 462–464
Fatigue:
    contact stress, 45–46
    cumulative damage, 66
    damage reduction, 66–67
    defined, 51, 63
    life, 65
    limit, 65
    mean stress, 65–66
    mechanism of, 64
    Palmgren-Miner rule, 66
    size effect, 64, 67, 68
    strength, 65
    stress concentration and, 42, 44, 66, 68
Finite element analysis:
    beam, curved, 170
    beam, elastically supported, 134, 144–145
    beam, shear center, 337
    dangers of, 2, 118
    theory, beam elements, 116–118
Flexural rigidity, 383, 442
Flywheel, 242–243
Foundation, elastic, 133–134
    see also Beams, straight, elastically supported
Fourier inequality, 79–80
Fracture, defined, 51
Fracture mechanics, 59–63
Fracture toughness, 60

Generalized coordinates, 78

Handbooks, 21
Harmonic operator, 199, 207
Hertz contact stress, 44–46
Hinge, plastic, 355

Imperfection sensitivity, 464
Influence coefficients, 424
Instability, rotating shaft, 445–446, 448–449
    see also Buckling
Instantaneous center, 362
Interaction formulas, 367–371
Invariants, of stress, 33

Jourawski's formula, 325

Kinematically admissible, 110

Lagrange multiplier, 103
Lamé formulas, 234
Large deflections, *see* Nonlinear elastic problems

Limit design, 346
Limit point, 463

Maxwell-Mohr method, 89
Mechanics of materials method, 1
Membrane analogy, 263–269
Midsurface, 379
Moment-area method, 105
Moment-curvature relations:
    beams, curved, 165, 181
    beams, straight, 14, 180, 321
    column buckling, 443
    cylindrical shell, 422
    plate bending, 383, 391
Moments of inertia, 5

Nonlinear elastic problems:
    buckling, 444, 463–464
    defined, 4
    plate bending, 393–395
    torsion, 298–301
    see also Collapse analysis, plastic; Plastic deformation
Notch sensitivity, fatigue, 68

Orthotropic material, 4, 38
Ovalization, 179

Parallel axis theorem, 6
Peel stress, 460
Plane stress, described, 37, 192
Plastic collapse, 346
Plastic deformation:
    axial loading, 347–349
    beam bending, 350–355
    buckling, 453–455
    deflection calculation, 354, 364
    elastic unloading, 18, 249, 253, 349, 352
    fatigue, damage reduction, 67
    interaction of loads, 367–371
    plastic hinge, 355
    plate bending, 399–400
    pressurized cylinder, 247–251
    spinning disks, 251–253
    stress concentration and, 42
    theory of, 221–222
    torsion, 18, 302–305
    yielded zone, beam bending, 353–354
    yield lines, plate bending, 400
    see also Collapse analysis, plastic; Residual stress
Plate bending:
    assumptions for theory, 380
    boundary conditions, 385–386, 391
    circular, solutions, 392–399
    cylindrical bending, 384

480 Index

Plate bending (*continued*)
  governing equations, 381–383, 389–391
  Kirchhoff theory, 380–381
  large deflections, 393–395
  nonlinear elastic problems, 393–395
  plastic collapse, 399–404
  pure twist, 384–385
  rectangular, solutions, 388–389
  series solution, 388
  spherical bending, 392
  thermal loading, 386–387
Plates, buckling of, 455–459
Potential energy, stationary principle, 107–111
Poynting effect, 298
Pressure vessels:
  cylindrical, thick walled:
    autofrettage, 247
    bursting, 249
    compound, 237–242
    dimensional changes, 237, 239
    governing equations, 231–233
    plastic deformation in, 247–251
    shrink-fit theory, 237–240
    stresses in, 233–241
  thin walled, 10, 236
  see also Shells of revolution
Principal axes of area, 7
Principal axes of stress, 34–35
Prismatic, defined, 4
Product of inertia, 5

Rayleigh-Ritz method, 110–113
Reaction, defined, 78
Reciprocal theorems:
  applied, 83–84, 88, 302, 336
  explained, 82–83
Redundant, defined, 97
Relaxation, 52, 222
Residual stress:
  axial loading and, 349–350
  bending and, 352
  buckling and, 454–455
  cold working and, 3, 67
  disks, spinning and, 253
  fatigue damage and, 67
  shrink fits and, 237–241, 247, 249
  torsional loading and, 18
Rings, as reinforcement, 430–433

Safety factor, defined, 4
Saint-Venant's principle:
  applied, 197, 200, 202, 205, 213, 234, 386, 422–423
  discussed, 4, 282, 297
Sand hill analogy, 303–304
Section modulus, 350

Sectorial area:
  discussed, 283–285
  principal, 286–287
  sectorial properties, 286–289
Semi-inverse method, 200, 207–209, 217–218
Series:
  polynomial, 111, 112
  trigonometric, 114–116, 388
Shape factor, 350
Shear center:
  center of symmetry and, 332
  center of twist and, 271, 336
  closed sections, 336
  discussed, 286, 327–329, 336–337
  symmetric open sections, 330–332
  unsymmetric open sections, 333–336
Shear flow:
  torsional load and, 272, 275, 292
  transverse load and, 13, 323–327
Shear lag, 459–460
Shear modulus, 5, 38
Shear stress:
  maximum, 36
  octahedral, 35–36
  transverse, in beams, 13, 323–327
  transverse, in plates, 381–382, 390
Shells of revolution:
  bending theory, 421–423
  clamped edge, 426–427
  drop-shaped tank, 419–420
  edge loading, 423–425
  geometry, 412–413
  joining of, 428–430, 433
  limitations and restrictions, 412, 415, 423, 424
  membrane solutions, 416–420
  membrane theory, 413–415
  pressure vessel head, 417, 420, 428–430
  ring, as reinforcement, 430–433
  terminology, 411–413
  thermal loads, 427–428
  uniform strength, 419
  see also Pressure vessels
Shrink fits, 237–241, 244, 247
Sidesway, 346, 364–365
Size effect, 55, 64, 67, 68
Slabs, 379, 399, 403
Snow load, 419
Soap-film analogy, 263
Spalling, 46, 64
Springback, 352–353
Springs, massively coiled, 163
Stability, *see* Instability; Buckling
Strain energy:
  basic formulas, 39–41
  beams, curved, 106–107, 181

beams, straight, 84–87, 105–106
complementary, 79, 84–87, 181
distortional, 40–41
in buckling, 446–448
in fracture mechanics, 60
see also Energy methods
Strain-displacement relations:
plane, 193–194, 207
plate bending, 383, 391
Stress concentration factor:
fatigue and, 42, 44, 66, 68
hole in plate, 42, 207–210
reduction of, 43
spinning disk, 243
torsion, 269, 273
Stresses, principal:
defined, 4, 29
direction, 31, 34–35
three dimensions, 31–35
two dimensions, 30–31
Stress intensity factor, 60–61
Stress transformation, 19, 29
Stress-strain relations, 37–38, 192, 205
see also Moment-curvature relations
Superposition, defined, 5
Support conditions, *see* Boundary conditions
Surface tractions, 197
Symmetry, discussed, 15–17

Theory of elasticity:
Airy stress function, 199, 207
body forces, 192, 196, 198, 205–207
boundary conditions, 190, 196–197
Cartesian coordinates, 193–205
compatibility equations, 194, 200–201, 207
cylinder, pressurized, 231–233
described, 2, 190–191
disk, spinning, 231–233, 245
displacements, calculation of, 203–205
equilibrium equations, 196, 206
force on edge or wedge, 210–212
inverse method, 200
polar coordinates, 205–212, 214–215, 231–233
Prandtl stress function, 219–221
semi-inverse method, 200, 207–209, 217–218
solution methods, 200–201
strain-displacement relations, 193–194, 207
surface tractions, 197
thermal loading, 213–216
torsion, 216–221, 261–262
Theory of failure, *see* Failure criteria

Thermal loading:
cylindrical shell, 427–428
disk, 214–215
hot spot, 215
numerical calculation procedure, 109
one-dimensional, 213
plates in bending, 386–387
shrink fits, 237–241, 244
stress-free states, 215–216
Torsion:
axial load and, 300–302
bimoment loading and, 296–298
center of twist, 271, 336
elasticity formulation, 216–221
elementary theory, 11–12
equations, summary of, 261–262
large rotation, 298–301
membrane analogy, 263–269
nonlinear elastic problems, 298–301
open sections, unrestrained, 265–270, 274
plastic, 18, 302–305
Prandtl stress function, 219–221
pretwisted member, 301–302
rate of twist, defined, 217, 261
rate of twist, formulas, 266, 270, 273, 276, 281, 297
stiffness in twisting, 270, 281, 302
stress concentration, 269, 273
tubes, multicell, 275–277, 336
tubes, single cell, 271–274, 290–291
warping displacements, 216–217, 289–291
warping function, 217–218
warping, restraint of, 278–283, 292–295, 305
Truss:
deflections of, 92–93, 108
statically indeterminate, 101–102, 108

Unit load method, 87–89
Units, 20

Virtual work, 79–81, 349, 355–356, 401
von Mises stress, *see* Effective stress

Warping constant, 286, 289
Warping, in torsion, 216–217, 289–291
Winkler foundation, 133–134
Working load, defined, 4

Yield criterion, 56
Yield lines, 400
Yield strength, yield stress, 5
Yielding, *see* Plastic deformation

Properties of plane areas   *Point g is the centroid*

Rectangle	Right triangle
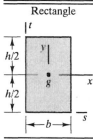 $A = bh$ $I_x = \dfrac{bh^3}{12}$ $I_s = \dfrac{bh^3}{3}$ $I_{st} = \dfrac{b^2h^2}{4}$	$A = \dfrac{bh}{2}$ $I_x = \dfrac{bh^3}{36}$ $I_s = \dfrac{bh^3}{12}$ $I_{xy} = -\dfrac{b^2h^2}{72}$
Complete circle	Quarter circle
$A = \pi R^2$ $I_x = \dfrac{\pi R^4}{4}$ $I_s = \dfrac{5\pi R^4}{4}$ $J_g = \dfrac{\pi R^4}{2}$	$A = \dfrac{\pi R^2}{4}$ $I_x = 0.0549\, R^4$ $I_s = \dfrac{\pi R^4}{16}$ $I_{st} = \dfrac{R^4}{8}$
Ellipse	Thin quarter-ring ($R \gg t$)
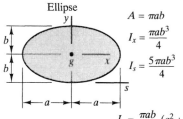 $A = \pi ab$ $I_x = \dfrac{\pi ab^3}{4}$ $I_s = \dfrac{5\pi ab^3}{4}$ $J_g = \dfrac{\pi ab}{4}(a^2 + b^2)$	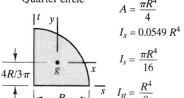  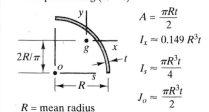 $A = \dfrac{\pi R t}{2}$ $I_x \approx 0.149\, R^3 t$ $I_s \approx \dfrac{\pi R^3 t}{4}$ $J_o \approx \dfrac{\pi R^3 t}{2}$ $R = $ mean radius
Parabola $t = h - \dfrac{hs^2}{b^2}$	Parabola $t = \dfrac{hs^2}{b^2}$
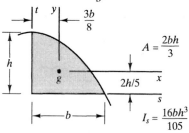 $A = \dfrac{2bh}{3}$ $I_s = \dfrac{16 bh^3}{105}$	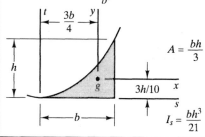 $A = \dfrac{bh}{3}$ $I_s = \dfrac{bh^3}{21}$

Narrow rectangle ($L \gg t$)

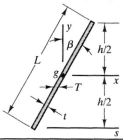 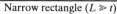

$T = \dfrac{t}{\cos \beta}$

$A = Lt$ or $A = hT$

$I_x \approx \dfrac{tL^3}{12} \cos^2 \beta$ or $I_x \approx \dfrac{Th^3}{12}$

$I_s \approx \dfrac{tL^3}{3} \cos^2 \beta$ or $I_s \approx \dfrac{Th^3}{3}$